Trends in Mathematics is a series devoted to the publication of volumes arising from conferences and lecture series focusing on a particular topic from any area of mathematics. Its aim is to make current developments available to the community as rapidly as possible without compromise to quality and to archive these for reference.

Proposals for volumes can be sent to the Mathematics Editor at either

Birkhäuser Verlag
P.O. Box 133
CH-4010 Basel
Switzerland

or

Birkhäuser Boston Inc.
675 Massachusetts Avenue
Cambridge, MA 02139
USA

Material submitted for publication must be screened and prepared as follows:

All contributions should undergo a reviewing process similar to that carried out by journals and be checked for correct use of language which, as a rule, is English. Articles without proofs, or which do not contain any significantly new results, should be rejected. High quality survey papers, however, are welcome.

We expect the organizers to deliver manuscripts in a form that is essentially ready for direct reproduction. Any version of TeX is acceptable, but the entire collection of files must be in one particular dialect of TeX and unified according to simple instructions available from Birkhäuser.

Furthermore, in order to guarantee the timely appearance of the proceedings it is essential that the final version of the entire material be submitted no later than one year after the conference. The total number of pages should not exceed 350. The first-mentioned author of each article will receive 25 free offprints. To the participants of the congress the book will be offered at a special rate.

Real and Complex Singularities

São Carlos Workshop 2004

Jean-Paul Brasselet
Maria Aparecida Soares Ruas
Editors

Birkhäuser Verlag
Basel · Boston · Berlin

Editors:

Jean-Paul Brasselet
Institut de Mathématiques de Luminy – CNRS
Campus de Luminy, Case 907
F-13288 Marseille Cedex
France
e-mail: jpb@iml.univ-mrs.fr

Maria Aparecida Soares Ruas
Instituto de Ciências Matemáticas e de
Computação – USP
Av. Trabalhador São-carlense, 400
Caixa Postal 668
13560-970 São Carlos, SP
Brasil
e-mail: maasruas@icmc.usp.br

2000 Mathematical Subject Classification 14D, 14H, 14J, 20F, 32

Library of Congress Control Number: 2006936540

Bibliographic information published by Die Deutsche Bibliothek
Die Deutsche Bibliothek lists this publication in the Deutsche Nationalbibliografie; detailed bibliographic data is available in the Internet at http://dnb.ddb.de

ISBN 978-3-7643-7775-5 Birkhäuser Verlag, Basel – Boston – Berlin

Part of Springer Science+Business Media
Printed on acid-free paper produced from chlorine-free pulp. TCF ∞
Printed in Germany
ISBN-10: 3-7643-7775-5
ISBN-13: 976-3-7643-7775-5
e-ISBN-10: 3-7643-7776-3
e-ISBN-13: 978-3-7643-7776-2

9 8 7 6 5 4 3 2 1
www.birkhauser.ch

Contents

Preface

The *Workshop on Real and Complex Singularities* is a series of biennial workshops organized by the singularity group at the Instituto de Ciências Matemáticas e de Computação, of the Universidade de São Paulo, São Carlos (ICMC-USP, São Carlos), Brazil. Its main purpose is to bring together specialists in the vanguard of singularities and its applications. Initiated in 1990, it became a key international event for people working in the field.

For the first time not in São Carlos, "The 8th workshop on real and complex singularities" took place at the Centre International de Rencontres Mathématiques, Luminy, France, from 19 to 23 of July, 2004. A total of 94 mathematicians from 19 countries (Brazil, France, Belgium, Canada, Denmark, England, Germany, Hungary, Israel, Italy, Japan, Mexico, Netherlands, New Zealand, Poland, Portugal, Russia, Spain, U.S.) participated in this very successful event, among these, 24 from Brazil.

The workshop program consisted of 14 plenary sessions, 44 ordinary sessions, and 2 mini-courses given by Maxim Kazarian on "Calculations on Thom polynomials", and Victor Goryunov on "Lagrangian and Legendrian Singularities ".

We could not have organized the workshop without the help of many people and institutions. We start by thanking the members of the Scientific Committee: Jose Manuel Aroca, Jim Damon, Gert Martin Greuel, Abramo Hefez, Heisuke Hironaka, Alcides Lins Neto, Marcio Soares, Bernard Teissier, Terry Wall for their support for the event. The workshop was funded by Institutions from Brasil: FAPESP, CNPq, CAPES, USP, and from France: Ministère de l'Enseignement Supérieur et de la Recherche, Université de la Méditerranée, CNRS (Délégation Régionale Provence, FRUMAM, IML, LATP), Conseil Général des Bouches-du-Rhône, Ville de Marseille. Their support we gratefully acknowledge. We are also very grateful to the staff of the IML and the CIRM for their help to organize the event.

It is a pleasure to thank the speakers and the participants whose presence was the real success of the 8*th Workshop*.

Marseille, São Carlos

Jean-Paul Brasselet, Maria Ruas

São Carlos-sur-mer, VIIIth Workshop on Real and Complex Singularities, Marseille, Luminy, July 19–23, 2004

Introduction

"Singularities are all over the place. Without singularities, you cannot talk about shapes. When you write a signature, if there is no crossing, no sharp point, it's just a squiggle. It doesn't make a signature. Many phenomena are interesting, or sometimes disastrous, because they have singularities. A singularity might be a crossing or something suddenly changing direction. There are many things like that in the world, and that's why the world is interesting. Otherwise, it would be completely flat. If everything were smooth, then there would be no novels or movies. The world is interesting because of the singularities. Sometimes people say resolving the singularities is a useless thing to do – it makes the world uninteresting! But, technically it is quite useful, because when you have singularities, computation of change becomes very complicated. If I can make some model that has no singularities but that can be used as a computation for the singularity itself, then that's very useful. It's like a magnifying glass. For smooth things, you can look from a distance and recognize the shape. But when there is a singularity, you must come closer and closer. If you have a magnifying glass, you can see better. Resolution of singularities is like a magnifying glass. Actually, it's better than a magnifying glass.

A very simple example is a roller coaster. A roller coaster does not have singularities – if it did, you would have a problem! But if you look at the shadow that the roller coaster makes on the ground, you might see cusps and crossings. If you can explain a singularity as being the projection of a smooth object, then computations become easier. Namely, when you have a problem with singularities in evaluation or differentiation or whatever, you can pull back to the smooth thing, and there the calculation is much easier. So you pull back to the smooth object, you do the computation or analysis, and then pull back to the original object to see what it means in the original geometry."

Heisuke Hironaka

Notices of the AMS, v. 52, *no.* 9.

The papers presented here are a selection of those submitted to the editors. They are grouped into three categories:

The first set, dedicated to local singularity theory, starts with the notes of the mini-course on Lagrangian and Legendrian Singularities, by V. Goryunov and V. Zakalyukin, followed by the papers by F. Aroca, *Valuations compatible with a projection*, D. Barlet, *Quelques résultats sur certaines fonctions à lieu singulier de dimension 1*, T. Gaffney, *The multiplicity of pairs of modules and hypersurface singularities*, I. Gregorio and D. Mond, *F-manifolds from composed functions*, K. Houston, *On Equisingularity of Images of Corank* 1 *Maps*, V.H. Jorge-Perez, E.C. Rizziolli and M.J. Saia, *Whitney equisingularity, Euler obstruction and invariants of map germs from* C^n *to* C^3, A. Libgober and A. Dimca, *Local topology of reducible divisors*, B. Martin, *Modular spaces of singularities of the* T*-series*, A. Pratoussevitch, *On the Link Space of a* Q*-Gorenstein Quasi-Homogeneous Surface Singularity*, and D. Siersma and M. Tibar, *Singularity exchange at the frontier of the space.*

The second group, dedicated to singular varieties both in affine and projective spaces, includes the papers *Celestian integration, stringy invariants, and Chern-Schwartz-MacPherson classes*, by P. Aluffi, *Versality properties of projective hypersurfaces* and *Minimal intransigent hypersurfaces*, by A. du Plessis, *Classification of rational unicuspidal projective curves whose singularities have one Puiseux pair*, by J. Bobadilla, A. Nemethi, I. Luengo and A. Melle, *Bounding from below the degree of an algebraic differential system having a prescribed algebraic solution*, by D. Lehmann and V. Cavalier, and *Mackey functors on provarieties*, by S. Yokura.

The third category includes applications of singularity theory to differential geometry, robotics, symmetric functions and bifurcation problems, and Goursat distributions. It contains the papers by A. Diatta and P. Giblin, *Vertices and inflexions of plane sections of surfaces in* R^3, by P. Donelan, *Trajectory singularities for a class of parallel motions* , by D. Dreibelbis, *The Geometry of Flecnodal Pairs*, by I. Labouriau and E. Pinho, by A. Sitta and J.C. Ferreira Costa, *Path formulation for* $Z_2 + Z_2$*-equivariant bifurcation problems*, and by P. Mormul, *Do moduli of Goursat distributions appear on the level of nilpotent approximations?*.

We thank the staff members of Birkhäuser, involved with the preparation of this book, and all those who have contributed in whatever way to these proceedings. All the papers here have been refereed.

J.P. Brasselet
M. Ruas
Editors

Real and Complex Singularities
Trends in Mathematics, 1–13

Celestial Integration, Stringy Invariants, and Chern-Schwartz-MacPherson Classes

Paolo Aluffi

Abstract. We introduce a formal integral on the system of varieties mapping properly and birationally to a given one, with value in an associated Chow group. Applications include comparisons of Chern numbers of birational varieties, new birational invariants, 'stringy' Chern classes, and a 'celestial' zeta function specializing to the topological zeta function.

In its simplest manifestation, the integral gives a new expression for Chern-Schwartz-MacPherson classes of possibly singular varieties, placing them into a context in which a 'change of variable' formula holds.

The formalism has points of contact with motivic integration.

1. Introduction

1.1. In this note I review the notion of *celestial integration,* and sketch a few applications; for proofs and further details, the reader is addressed to [Alu05]. I am very grateful to Jean-Paul and Cidinha for organizing the *VIIIème Rencontre Internationale de* São Carlos *sur les singularités réelles et complexes au CIRM,* relocating for the occasion the idyllic surroundings of *São Carlos, Brasil* to the idyllic surroundings of *Luminy, France.* Perfect weather, exceptionally interesting talks, and spirited conversations made the conference a complete success. What follows was the subject of my lecture at the São Carlos/Luminy meeting, and preserves (for better or worse) the informal nature of a seminar talk.

I thank the Max-Planck-Institut für Mathematik in Bonn, where much of this material was conceived and where this note was written.

1.2. Summary: to a variety X I will associate a large group $A_*\mathcal{C}_X$ (containing the Chow group A_*X of X); for certain data $\mathcal{D}$, $\mathcal{S}$ (arising, for example, from a divisor D and a constructible subset S of X) I will define a distinguished element

$$\int_{\mathcal{S}} \mathbb{1}(\mathcal{D})\, d\mathfrak{c}_X \in A_*\mathcal{C}_X \quad .$$

These are the *celestial integrals* in the title; they are not defined as integrals, but satisfy formal properties justifying the terminology. The *celestial* qualifier is meant to evoke the fact that the *modification systems* on which this operation is defined are close relatives of Hironaka's *voûte étoilée.*

Applications of this construction:

- Comparison of Chern classes of birational varieties;
- Birational invariants;
- 'Celestial' zeta functions;
- Invariants of singular varieties (*'Stringy Chern classes'*);
- Relations with the theory of Chern-Schwartz-MacPherson classes.

1.3. Digression: **motivic integration**. This is technically not necessary for the rest of the talk, but useful nonetheless as 'inspiration' for the main construction.

The *Grothendieck group of varieties* is the free abelian group on symbols $[X]$, where X is a complex algebraic variety up to isomorphism, modulo relations

$$[X] = [Y] + [X \setminus Y]$$

for each closed subvariety $Y \subset X$. This group may be made into a ring by setting $[X] \cdot [Y] = [X \times Y]$.

Example 1.1. The class of a point is the identity for this multiplication. The class $[\mathbb{A}^1]$ is denoted $\mathbb{L}$; thus

$$[\mathbb{P}^n] = [\mathbb{A}^0] + [\mathbb{A}^1] + \cdots + [\mathbb{A}^n] = \frac{\mathbb{L}^{n+1} - 1}{\mathbb{L} - 1} \quad .$$

As Eduard Looijenga writes ([Loo02], p. 269), this ring – or rather its localization at $\mathbb{L}$ – is *interesting, big, and hard to grasp.* In practice, it is necessary to further tweak this notion, by a suitable completion with respect to a dimension filtration; I will glibly ignore such important 'details'.

Mapping X to its class in this ring gives a *universal Euler characteristic:* anything satisfying the basic relations – e.g., topological Euler characteristic, Hodge polynomials and structures,..., must factor through this map. This is motivation to 'compute' $[X]$ for given X.

Through *motivic integration,* one can determine an element

$$\int_{\mathcal{S}} \mathbb{L}^{-\operatorname{ord} D} d\mu$$

in the completed Grothendieck ring of varieties, from the information of a divisor D and a constructible subset $\mathcal{S}$ of the arc space $\mathcal{X}$ of X. If X is nonsingular, then choosing $D = 0$, $\mathcal{S} = \mathcal{X}$ gives $\int_{\mathcal{X}} \mathbb{L}^0 d\mu = [X]$.

Motivic integration was defined and developed by Maxim Kontsevich, Jan Denef, and François Loeser, and it is of course much deeper than this brief summary can begin to suggest. There are many good surveys of this material, for example: [Cra04], [DL01], [Loo02], [Veya]. For an explanation of what makes motivic integration *motivic,* see the appendix in [Cra04].

The definition of the integral $\int_{\mathcal{S}} \mathbb{L}^{-\operatorname{ord} D} d\mu$ relies on the study of the *arc spaces* of a variety; $d\mu$ is a measure on this space, with value in the completed Grothendieck ring; the integral is an honest integral with respect to this measure, and as such it satisfies a *change of variable* formula: for example, if $\pi : Y \to X$ is proper and birational, then

$$\int_{\mathcal{X}} \mathbb{L}^{-\operatorname{ord} D} d\mu = \int_{\mathcal{Y}} \mathbb{L}^{-\operatorname{ord}(\pi^{-1}D+K_\pi)} d\mu \quad ,$$

where K_π denotes the relative canonical sheaf.

This formula is at the root of spectacular applications of motivic integration. For example, suppose X, Y are *birational* nonsingular, complete Calabi-Yau varieties; resolve a birational morphism between them:

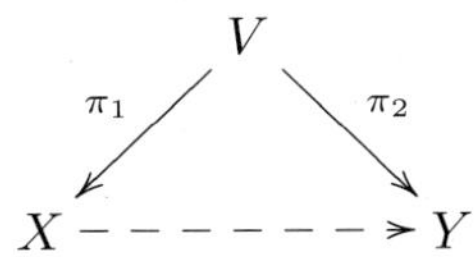

with V nonsingular and π_1, π_2 proper and birational. Then the Calabi-Yau condition implies $K_{\pi_1} = K_{\pi_2}$; denoting this by K gives

$$[X] = \int_{\mathcal{X}} \mathbb{L}^0 \, d\mu = \int_{\mathcal{V}} \mathbb{L}^{-\operatorname{ord} K} \, d\mu = \int_{\mathcal{Y}} \mathbb{L}^0 \, d\mu = [Y] \quad .$$

Hence: such varieties must have the same topological Euler characteristic, Betti numbers, Hodge polynomials, etc.

1.4. Motivic integration only serves as *motivation* for the rest of this lecture, or maybe more correctly as a motivating analogy. The basic relation in the Grothendieck ring holds (in a suitable sense) for *Chern-Schwartz-MacPherson* classes; 'hence' there should be a 'motivic' theory of such classes: it should be possible to deal with the classes within the framework of an integration theory, satisfying a suitable change-of-variable formula; one should be able to play tricks such as the application sketched above at the level of Chern classes.

This is the guiding theme in what follows.

2. Modification systems

2.1. The task is to define an 'integral' carrying information about Chern classes. Taking at heart the lesson learned in motivic integration, we should start by defining an appropriate context in which this integral may take its value.

Let X be a variety over an algebraically closed field of characteristic zero (the precise requirement is that embedded resolution à la Hironaka should work).

Definition 2.1. I will denote by $\mathcal{C}_X$ the category of proper birational maps

$$\begin{array}{c} V_\pi \\ \downarrow{\scriptstyle \pi} \\ X \end{array}$$

with morphisms given in the obvious way by commutative triangles

$$\begin{array}{ccc} V_{\pi_1} & \xrightarrow{\alpha} & V_{\pi_2} \\ & {\scriptstyle \pi_1}\searrow \quad \swarrow{\scriptstyle \pi_2} & \\ & X & \end{array}$$

with α proper and birational.

Proper birational maps are often called *modifications,* and the natural way to think of $\mathcal{C}_X$ is as an inverse system, so it seems appropriate to call this category the *modification system* of X. Also, it is useful to take this notion up to the following equivalence relation: say that $\mathcal{C}_X$ and $\mathcal{C}_Y$ are equivalent if there exist objects in $\mathcal{C}_X$ and $\mathcal{C}_Y$ with a common source. For example, if X and Y are birational and complete then their modification systems are equivalent in this sense.

2.2. I will (usually) denote by V_π the source of the object π of $\mathcal{C}_X$. It is hard to resist the temptation to think of the object π really in terms of its corresponding V_π, and of $\mathcal{C}_X$ as a system of varieties birational to X.

Morally I would like to take the inverse limit of this system, and define ordinary data such as divisors, Chow group, etc. for the resulting *pro*variety. In practice, it is more straightforward to simply define these data as appropriate limits of the corresponding data on the individual V_π's. For example, denote by A_*V_π the Chow group of V_π, with rational coefficients; then

$$\mathcal{A}_X := \{A_*V_\pi \,|\, \pi \in \mathrm{Ob}(\mathcal{C}_X)\}$$

is an inverse system of abelian groups under proper push-forward.

Definition 2.2. The *Chow group* of $\mathcal{C}_X$ is the inverse limit of this system:

$$A_*\mathcal{C}_X := \varprojlim \mathcal{A}_X \quad .$$

Thus, an element $a \in A_*\mathcal{C}_X$ consists of the data of a class $(a)_{\mathrm{id}}$ in the Chow group of X and of compatible lifts $(a)_\pi$ for all $\pi \in \mathrm{Ob}(\mathcal{C}_X)$. I call $(a)_\pi$ the π-*manifestation* of a.

Note that any class $\alpha \in A_*X$ determines a 'silly' class $a \in A_*\mathcal{C}_X$: just set $(a)_\pi := \pi^*\alpha$. One intriguing (to me, at least) consequence of the construction given in this paper is that certain classes on X have *other,* more interesting, lifts to $A_*\mathcal{C}_X$. These lifts call for rational coefficients, hence the need for rational coefficients in the definition of $A_*\mathcal{C}_X$.

Equivalent modification systems have isomorphic Chow groups.

2.3. Other standard notions may be defined similarly. Divisors and constructible sets of sources V_π are organized by *direct* systems, under pull-backs; the corresponding notions for a modification system are defined as direct limits of these systems.

For example, a divisor $\mathcal{D}$ of $\mathcal{C}_X$ is represented by a pair (π, D_π) with D_π a divisor of V_π, and where pairs (π, D_π), $(\pi \circ \alpha, D_{\pi\circ\alpha})$ are identified whenever $\alpha : V_{\pi\circ\alpha} \to V_\pi$ is a proper birational map and $D_{\pi\circ\alpha} = \alpha^{-1}(D_\pi)$:

$$V_{\pi\circ\alpha} \xrightarrow{\alpha} V_\pi \xrightarrow{\pi} X \quad .$$

An obvious way to get a divisor $\mathcal{D}$ is by pulling back a divisor of X through the whole system; but note that there are *many other* divisors: for example, every subscheme S of X determines a divisor of $\mathcal{C}_X$ (represented by the exceptional divisor in the blow-up of X along S). As a bonus, equivalent modification systems have the same divisors, while birational varieties don't.

2.4. The story is entirely analogous for *constructible subsets* of a modification system. The 'obvious' such object is determined by a constructible (for example, closed) subset of X, by taking inverse images through the system. While this is our main example, one should keep in mind that the notion is considerably more general.

Of course, equivalent systems have the same constructible subsets. For example, if V maps properly and birationally to both X and Y:

$$\begin{array}{ccccc} & & V & & \\ & \swarrow^{\pi_X} & & \searrow^{\pi_Y} & \\ X & & \dashrightarrow & & Y \end{array}$$

then (π_X, V) determines the same subset of $\mathcal{C}_X$ as (id, X) and the same subset of $\mathcal{C}_Y$ as (id, Y); abusing language I may denote this object by $\mathcal{C}_X$ or $\mathcal{C}_Y$ according to the context, but the reader should keep in mind that these constructible subsets of different systems may be identified.

Details about all these notions, and natural definitions (such as sums of divisors, or unions of constructible subsets) are left to the interested reader, and may be found in [Alu05].

3. Celestial integrals

3.1. The main result of this note is that for a variety X, a divisor $\mathcal{D}$ of $\mathcal{C}_X$, and a constructible subset $\mathcal{S}$, there is an element (the 'celestial integral' of $\mathcal{D}$ over $\mathcal{S}$, in $\mathcal{C}_X$)

$$\int_{\mathcal{S}} \mathbb{1}(\mathcal{D})\, d\mathfrak{c}_X$$

of the Chow group $A_*\mathcal{C}_X$, satisfying interesting properties.

The actual definition of this element is uninspiring; I'll give it at the end of the paper for the sake of completeness. The properties satisfied by this 'integral' are more important, so they get the honor of prime time.

Of course the notion is additive with respect to disjoint unions of constructible subsets, as should be expected from an integral. What makes the notion interesting is that it computes interesting objects for suitable choices of the input data, and that it satisfies a change-of-variable formula (again, as should be expected from an integral!).

More explicitly:

Theorem 3.1.

1. (Normalization) *Assume X is nonsingular. If $\mathcal{S}$ is represented by (id, S) with $S \subset X$ a nonsingular subvariety, then*

$$\left(\int_{\mathcal{S}} \mathbb{1}(0)\, d\mathfrak{c}_X\right)_{id} = c(TS) \cap [S] \quad ,$$

the total homology Chern class *of S, viewed as an element of A_*X.*

2. (Change-of-variables) *If $\rho : Y \to X$ is proper and birational, then*

$$\int_{\mathcal{S}} \mathbb{1}(\mathcal{D})\, d\mathfrak{c}_X = \int_{\mathcal{S}} \mathbb{1}(\mathcal{D} + K_\rho)\, d\mathfrak{c}_Y \quad ,$$

where K_ρ denotes the relative canonical divisor of ρ.

The normalization property (1) is self-explanatory; it will serve as a point of depart for extensions to possibly *singular* subsets S of a nonsingular variety, in §§5 and 6. By contrast, (2) deserves an immediate clarification.

3.2. First of all, note that under the given hypotheses we have that $\mathcal{C}_X$ and $\mathcal{C}_Y$ are equivalent systems; thus we may indeed treat $\mathcal{S}$ and $\mathcal{D}$ as data belonging to either.

Secondly, I have to clarify what I mean by *the relative canonical divisor* K_ρ of $\rho : Y \to X$. It in fact turns out that there are more than one sensible such notions, according to the context. In the simplest case, when X and Y are nonsingular, K_ρ is the divisor corresponding to the determinant of the differential $d\rho : TY \to \rho^*TX$. In the general case there is a choice as to what is the correct generalization of TX; I'll come back to this point in §5.2.

Further, in general the scheme corresponding to the vanishing of the determinant may not be locally principal. This is not a problem in our context, however, since every subscheme of every variety in the modification system $\mathcal{C}_X$ determines a divisor of the system, as observed in §2.3.

4. Sketch of applications

4.1. Invariance of Chern classes. Exactly as in the case of motivic integration, the change-of-variable formula yields an invariance statement for celestial integration across birational morphisms preserving the canonical class.

Two varieties X, Y (nonsingular, for simplicity) are *K-equivalent* if their modification systems are equivalent, and the canonical divisors K_X, K_Y agree in the system(s): that is, if the pull-backs of K_X, K_Y to a common source agree:

$$\begin{array}{ccc} & V & \\ {}^{\pi_X}\swarrow & & \searrow^{\pi_Y} \\ X & \dashrightarrow & Y \end{array}$$

$$\pi_X^* K_X = \pi_Y^* K_Y \quad .$$

In this situation, letting $K = K_{\pi_X} = K_{\pi_Y}$ and applying the change-of-variable formula (2) gives:

$$\int_{\mathcal{S}} \mathbb{1}(\mathcal{D})\, d\mathfrak{c}_X = \int_{\mathcal{S}} \mathbb{1}(\mathcal{D}+K)\, d\mathfrak{c}_V = \int_{\mathcal{S}} \mathbb{1}(\mathcal{D})\, d\mathfrak{c}_Y \quad .$$

Therefore:

Theorem 4.1. *Celestial integrals on K-equivalent varieties agree as elements of the* (*common*) *Chow group of the corresponding modification systems.*

4.2. For example, applying this observation with $\mathcal{D} = 0$, $\mathcal{S} = \mathcal{C}_X$ ($= \mathcal{C}_Y$ on Y) and using (1) from §3 shows that $c(TX) \cap [X]$ and $c(TY) \cap [Y]$ are manifestations (in A_*X, A_*Y, respectively) of *the same* class in the Chow group of the modification system.

This recovers the fact (known, for example, through motivic integration) that the Euler characteristics of K-equivalent varieties must agree. More generally, it shows (via a simple application of the projection formula) that all numbers

$$c_1^i \cdot c_{n-i}$$

with $n = \dim X = \dim Y$ must agree for K-equivalent varieties.

Incidentally, these numbers must therefore be invariant through classical flops, and hence (as shown by Burt Totaro, [Tot00]) they must factor through the complex elliptic genus. It is a pleasant exercise to verify this fact directly: these numbers can be assembled into a genus (which I would like to call the *cuspidal* genus, for reasons which will likely be apparent to many readers), corresponding to the *characteristic* $e^{xT}(1 + xU)$; it is straightforward to check directly that the cuspidal genus factors through the complex elliptic genus.

The invariance of Chern numbers mentioned above is of course only a very particular case of similar results accessible through celestial integration. Every choice of a divisor and a constructible subset yields an analogous (but, unfortunately, usually much less transparent) invariance statement.

4.3. Birational invariants. As another application, we can extract new birational invariants from the integral. For example, let

$$\mathrm{dCan}(X) := \{\deg \int_{\mathcal{C}_X} \mathbb{1}(\mathcal{K})\, d\mathfrak{c}_X \mid \mathcal{K} \text{ effective canonical divisor of } X\} \subset \mathbb{Z} \quad .$$

Theorem 4.2. *If X, Y are birational complete varieties, then* $\mathrm{dCan}(X) = \mathrm{dCan}(Y)$.

For example: if X is birational to a Calabi-Yau variety Y, then $\mathrm{dCan}(X) = \{\chi(Y)\}$.

An interesting question is whether an analogous (and nontrivial) invariant can be defined for varieties without effective canonical divisors. A few natural candidates for such invariants, involving negative representatives, must be ruled out at the moment because of a sticky technical obstacle to the definition of celestial integrals for noneffective divisors (see §7).

4.4. Zeta functions. In analogy with motivic integration, zeta functions can be concocted from motivic integrals. For example, for a divisor D of X (say defined by $f = 0$) set

$$Z(D, m) := \int_{\mathcal{C}_X} \mathbb{1}(m\mathcal{D})\, d\mathfrak{c}_X \quad ,$$

a series in the variable m, with coefficients in $\mathcal{A}_*\mathcal{C}_X$; here $\mathcal{D}$ is the divisor in $\mathcal{C}_X$ determined by D. Then

Theorem 4.3. *The degree of $Z(D, m)$ equals the topological zeta function of f.*

This connection makes it possible to formulate analogs of the *monodromy conjecture* (see for example [Veya], §6.8) for celestial zeta functions. I hope that the celestial viewpoint will add something to the circle of ideas surrounding zeta functions. For example, conceivably the relationship between celestial integration and the theory of Chern-Schwartz-MacPherson classes (§6) may give a tool to compute local contributions to the zeta function of a hypersurface in terms of the Segre class of its singularity subscheme.

5. Stringy invariants

5.1. If X has sufficiently mild singularities, there is a notion of *stringy Euler characteristic* of X, introduced by Batyrev. For example, in the particular case in which X admits a *crepant* resolution V, the stringy Euler characteristic of X may be defined to be the ordinary Euler characteristic $\chi(V)$ of V: remarkably, this turns out to be independent of the chosen crepant resolution.

Celestial integration extends this notion to a whole class in A_*X. By the normalization property (Theorem 3.1 (1)),

$$\left(\int_{\mathcal{C}_X} \mathbb{1}(0)\, d\mathfrak{c}_X\right)_{\mathrm{id}} = c(TX) \cap [X]$$

if X is nonsingular; but the expression on the left-hand-side defines an element of A_*X *even if* X *is singular* (in fact, the celestial integral defines this expression together with distinguished lifts to all varieties mapping to X). If X admits a crepant resolution $\pi : V \to X$, it is easy to check that this definition produces the push-forward $\pi_* c(TV) \cap [V]$. By the Poincaré-Hopf theorem, therefore, the degree of this class recovers the stringy Euler characteristic of X in this case.

This in fact holds for any X for which the stringy Euler characteristic is defined, justifying the following:

Definition 5.1. The *stringy Chern class* of X is the identity manifestation

$$\left(\int_{\mathcal{C}_X} \mathbb{1}(0)\, d\mathfrak{c}_X \right)_{\mathrm{id}} .$$

Coincidentally, a notion of stringy Chern class was produced simultaneously as the one presented above, by Tommaso de Fernex, Ernesto Lupercio, Thomas Nevins, and Bernardo Uribe (in fact, the preprint [dFLNU] appeared on the arXiv during the São Carlos/Luminy conference!). While the approaches to the two notions differ somewhat, the two stringy classes agree.

5.2. There is a subtlety here, which I can only touch upon in this note. For singular X, the notion of celestial integral depends on the choice of a good notion of *relative canonical divisor*. The 'usual' notion is constructed starting from the *double dual* ω_X of the Kähler differentials $\Omega_X^{\dim X}$ of X; this 'ω flavor' of the celestial integral is what leads to the stringy Chern class recovering the usual stringy Euler characteristic, as explained above, and agreeing with the class introduced by deFernex et al.

The ω flavor leads to a technical difficulty, which may make the celestial integral (and hence stringy Chern classes) *undefined* if the singularities of X are not mild enough – the technical condition is that they should be *log terminal.* Whether stringy classes (or more generally celestial integrals) may be defined for varieties with more general singularities is an open question, see §7.

One way out of this bind is to choose a different notion of relative canonical divisor in the main set-up. For example, one can *avoid* taking the double-dual, leading to the ω flavor as mentioned above; this leads to a different notion (which I call the Ω flavor of the integral), which is defined for arbitrarily singular varieties. While this yields a stringy Chern class for arbitrary varieties, the meaning of this class (for example vis-a-vis the stringy Euler characteristic) has not been explored.

6. Chern-Schwartz-MacPherson classes from celestial integrals

6.1. The stringy notion presented in §5.1 amounts to taking the identity manifestation of the integral of 0 over the whole modification system $\mathcal{C}_X$ of the variety X. By the normalization property ((1) in Theorem 3.1), this yields the usual Chern class of the tangent bundle of X when X is nonsingular.

There is a different natural way to use the same tool and define a class generalizing $c(TX) \cap [X]$: embed X into an ambient nonsingular variety M, then compute the identity manifestation of the celestial integral of 0 *over the constructible subset* $\mathcal{X}$ determined by X:

$$\left(\int_{\mathcal{X}} \mathbb{1}(0)\, d\mathfrak{c}_M \right)_{\mathrm{id}} \quad .$$

With due care, this class can be defined in A_*X (our definition of the celestial integral would only place it in A_*M); remarkably, as such it does *not* depend on the ambient variety M. In fact:

$$\left(\int_{\mathcal{X}} \mathbb{1}(0)\, d\mathfrak{c}_M \right)_{\mathrm{id}} \stackrel{!}{=} c_{\mathrm{SM}}(X) \quad ,$$

the *Chern-Schwartz-MacPherson* class of X.

This is a famous notion, going back to Marie-Hélène Schwartz ([Sch65]) and Robert MacPherson ([Mac74]). In MacPherson's construction (as recalled, for example, in [Ful84], §19.1.7), one obtains in fact a natural transformation c_* from the functor of constructible functions (with proper push-forward defined by Euler characteristic of fibers) to the Chow group functor; applying c_* to the constant function $\mathbb{1}_X$ defines the class $c_{\mathrm{SM}}(X)$.

6.2. The connection between celestial integrals and Chern-Schwartz-MacPherson classes mentioned above goes in fact much deeper. Given any divisor $\mathcal{D}$ and any constructible subset $\mathcal{S}$ of a modification system $\mathcal{C}_X$, one may define a constructible function by

$$p \mapsto I_X(\mathcal{D}, \mathcal{S}) := \deg \left(\int_{\mathcal{S} \cap p} \mathbb{1}(\mathcal{D})\, d\mathfrak{c}_X \right) \quad ;$$

here, $\mathcal{S} \cap p$ is the constructible subset of $\mathcal{C}_X$ obtained by intersecting $\mathcal{S}$ with inverse images of p through the system.

Theorem 6.1.

$$\left(\int_{\mathcal{S}} \mathbb{1}(\mathcal{D})\, d\mathfrak{c}_X \right)_{id} = c_*(I_X(\mathcal{D}, \mathcal{S})) \quad .$$

Thus, celestial integrals and Chern-Schwartz-MacPherson classes are, in a sense, equivalent information: each can be obtained from the other.

Classes such as the stringy Chern class considered in §5.1 correspond, via MacPherson's natural transformation, to specific constructible functions. These 'stringy' characteristic functions deserve much further study.

The apparatus of Chern-Schwartz-MacPherson classes is an important ingredient in the construction in [dFLNU].

6.3. It should be noted that the definition of celestial integration (which I will finally summarize in §7) does *not* rely on Chern-Schwartz-MacPherson classes; the latter are an honest subproduct of the former. Thus, one could try to recover the main defining features of Chern-Schwartz-MacPherson classes from celestial properties.

For example, I would like to venture the guess that *the covariance property of Chern-Schwartz-MacPherson classes is a facet of the change-of-variable formula for celestial integrals.* This should mean that the change-of-variable formula is a Riemann-Roch theorem in disguise. As things stand now I don't even have a precise version of this 'guess' to offer, and I will have to leave it at the stage of half-baked speculations.

7. The definition

A summary of celestial integration *without* a definition of this notion would be incomplete, even though I have tried to defend the idea that the definition itself is less important than the fact alone that such a notion exists – in practice, the normalization and change-of-variable properties suffice for interesting applications and do not require the actual definition of the integral to be appreciated.

In any case, here is the definition. Given a divisor $\mathcal{D}$ and a constructible (say closed and proper, for simplicity) subset $\mathcal{S}$ of the modification system $\mathcal{C}_X$, embedded resolution of singularities ensures that there is an object $\pi : V_\pi \to X$ in $\mathcal{C}_X$, a normal crossing divisors E with nonsingular components E_j, $j \in J$, and divisors D_π, $S_\pi = \cup_{j\in J_S} E_j$ of V_π such that:

- $\mathcal{D}$ is represented by (π, D_π);
- $\mathcal{S}$ is represented by (π, S_π);
- $D_\pi + K_\pi = \sum_{j\in J} m_j E_j$, with $m_j \in \mathbb{Q}$.

This set of data depends on the chosen notion of relative canonical divisor K_π. *Assume that all coefficients m_j are > -1.*

Definition 7.1.
$$\left(\int_{\mathcal{S}} \mathbb{1}(\mathcal{D})\, d\mathfrak{c}_X\right)_\pi := c(\Omega_{V_\pi}(\log E)^\vee) \cap \sum_{I\subset J,\, I\cap J_S\neq\emptyset} \frac{[\cap_{i\in I} E_i]}{\prod_{i\in I}(1+m_i)}$$

This expression defines the manifestation of the integral on all varieties such as V_π, in which the data $\mathcal{D}$, $\mathcal{S}$ are 'resolved' by a normal-crossing divisor. The manifestation on any other variety is obtained by push-forward, compatibly with the requirement that the celestial integral is an element of the inverse limit $A_*\mathcal{C}_X$.

The obvious difficulty with this definition is that it is not at all clear that it should not depend on the chosen π used to resolve the given data. In motivic integration, similar expressions are obtained *a posteriori,* and compute intrinsically defined objects, hence it is clear that they do not depend on the choices. In celestial integration I have to prove the necessary independence explicitly, directly from Definition 7.1.

Theorem 7.2. *If all m_j are > -1, then the given expression does define an element of the inverse limit $A_*\mathcal{C}_X$.*

This is proved by applying the factorization theorem of [AKMW02], which reduces this claim to a computation across blow-ups along nonsingular centers.

Manipulating the expressions is a somewhat messy, but manageable, exercise in standard intersection theory.

The independence requires that all $m_j > -1$ (even though the expression in Definition 7.1 makes sense as soon as no m_j is $= -1$); this is where the singularities of X may play a rôle for the particular case $\mathcal{D} = 0$, as I discussed in §5.2: the restriction $m_j > -1$ in this case amounts to the requirement that X be log terminal.

The difficulty arising if some $m_j \leq -1$ is that in a chain of varieties connecting two varieties where the data is resolved, one may appear for which the expression in Definition 7.1 does not make sense, for the mundane reason that one of the denominators in the expression may vanish.

This problem arises in many different contexts, of which celestial integration is but one instance (see for example [Veya], §8, Question I). While there is a feeling that the obstacle is technical rather than conceptual, it has opposed stubborn resistance to the attempts made so far to overcome it, and examples such as the one presented in §3.4 in [Veyb] suggest that the issue may be more fundamental than initially expected.

The question of exactly which celestial integrals are well defined outside the range specified in Theorem 7.2 is subtle and difficult. Answering this question is a worthwhile challenge: the present state of affairs limits the scope of the definition of certain key celestial integrals and hence, as pointed out in §4 and §5, of some potentially interesting applications.

References

[AKMW02] Dan Abramovich, Kalle Karu, Kenji Matsuki, and Jarosław Włodarczyk. Torification and factorization of birational maps. *J. Amer. Math. Soc.*, 15(3):531–572, 2002.

[Alu05] Paolo Aluffi. Modification systems and integration in their Chow groups. *Selecta Math. (N.S.)*, 11(2):155–202, 2005.

[Cra04] Alastair Craw. An introduction to motivic integration. In *Strings and geometry*, volume 3 of *Clay Math. Proc.*, pages 203–225. Amer. Math. Soc., Providence, RI, 2004.

[dFLNU] Tommaso de Fernex, Ernesto Lupercio, Thomas Nevins, and Bernardo Uribe. Stringy Chern classes of singular varieties, arXiv:math.AG/0407314. To appear in Adv. Math.

[DL01] Jan Denef and François Loeser. Geometry on arc spaces of algebraic varieties. In *European Congress of Mathematics, Vol. I (Barcelona,* 2000), volume 201 of *Progr. Math.*, pages 327–348. Birkhäuser, Basel, 2001.

[Ful84] William Fulton. *Intersection theory.* Springer-Verlag, Berlin, 1984.

[Loo02] Eduard Looijenga. Motivic measures. *Astérisque*, (276):267–297, 2002. Séminaire Bourbaki, Vol. 1999/2000.

[Mac74] Robert D. MacPherson. Chern classes for singular algebraic varieties. *Ann. of Math.* (2), 100:423–432, 1974.

[Sch65] Marie-Hélène Schwartz. Classes caractéristiques définies par une stratification d'une variété analytique complexe. I. *C. R. Acad. Sci. Paris*, 260:3262–3264, 1965.

[Tot00] Burt Totaro. Chern numbers for singular varieties and elliptic homology. *Ann. of Math.* (2), 151(2):757–791, 2000.

[Veya] Willem Veys. Arc spaces, motivic integration and stringy invariants. In *Singularity Theory and its Applications*, Sapporo (Japan), September 16–25, 2003. Adv. Stud. Pure Math. Kinokuniya, Tokyo, 2005.

[Veyb] Willem Veys. On motivic principal value integrals, arXiv:math.AG/0604167.

Paolo Aluffi
Mathematics Department
Florida State University
Tallahassee
Florida, 32306, USA
e-mail: `aluffi@math.fsu.edu`

Real and Complex Singularities
Trends in Mathematics, 15–22

Valuations Compatible with a Projection

Fuensanta Aroca

Abstract. Given an N-dimensional germ of analytic hypersurface $\mathcal{H}$, a finite projection $\pi : \mathcal{H} \longrightarrow \mathbb{C}^N$ and a valuation ν on the ring of convergent series in N variables, we study the valuations on the ring $\mathcal{O}_\mathcal{H}$ that extend $\pi^*\nu$. All these valuations are described when ν is a monomial valuation whose weight vector is not orthogonal to any of the faces of the Newton Polyhedron of the discriminant of the projection π. This description is done in terms of the Puiseux parameterizations of $\mathcal{H}$ with exponents in a cone.

Mathematics Subject Classification (2000). Primary 32B10; Secondary 13B02.

Keywords. Newton Polyhedron, valuation, discriminant of a projection, Puiseux parametrizations.

1. Introduction

Let $\mathcal{H}$ be an irreducible germ of analytic hypersurface at the origin in $\mathbb{C}^{N+1}$, let $\pi : \mathcal{H} \longrightarrow (\mathbb{C}^N, 0)$ be a finite projection and denote by $\mathcal{R}$ the ring of convergent series in N variables. Given a valuation $\nu_\sim : \mathcal{R} \longrightarrow \mathbb{R}_{\geq 0} \cup \{\infty\}$, we want to describe all the valuations $\nu : \mathcal{O}_\mathcal{H} \longrightarrow \mathbb{R}_{\geq 0} \cup \{\infty\}$ that *extend* $\nu_\sim$. That is, that make the diagram

$$\begin{array}{ccc} \mathcal{O}_\mathcal{H} & \xrightarrow{\nu} & \mathbb{R}_{\geq 0} \cup \{\infty\} \\ & {}^{\pi^*}\nwarrow \quad \nearrow^{\nu_\sim} & \\ & \mathcal{R} & \end{array} \tag{1.1}$$

commute.

In this note all these valuations are described when $\nu_\sim$ is a monomial valuation whose weight vector is not orthogonal to any of the faces of the Newton Polyhedron of the discriminant of the projection π. This description is done in terms of the Puiseux parameterizations of $\mathcal{H}$.

Supported by grant IN105806 from PAPIIT, UNAM. The research was carried out while the author was in Sao Carlos profiting from a DTI fellowship given by IM-AGIMB processo 381461/2004-1 (NV).

The question was posed to me by Bernard Teissier at the congress "Singularity theory and Applications" held at Sapporo in 2003. I thank Daniel Levcovitz for fruitful discussions during the preparation of this note.

In all what follows $\mathcal{H}$ will be a germ of analytic hypersurface embedded in $(\mathbb{C}^{N+1}, 0)$ defined by $F(x_1, \ldots, x_N, y) = 0$, where $F = y^d + a_{d-1}y^{d-1} + \cdots + a_0$ is an irreducible polynomial of degree d in y and $a_i \in \mathcal{R}$. The projection

$$\begin{array}{cccc} \pi : & \mathcal{H} & \longrightarrow & \mathbb{C}^N \\ & (x_1, \ldots, x_N, y) & \mapsto & (x_1, \ldots, x_N) \end{array} \tag{1.2}$$

will be supposed to be finite. The discriminant of the projection π will be denoted by δ. That is

$$\delta(x_1, \ldots, x_N) = \text{Resultant}_y(F, \frac{\partial}{\partial y}F).$$

The ideal of $\mathcal{R}[y]$ generated by F will be denoted by $\mathcal{I}$. So that $\mathcal{O}_{\mathcal{H}} = \frac{\mathcal{R}[y]}{\mathcal{I}}$ and π^* is the natural inclusion $\mathcal{R} \hookrightarrow \frac{\mathcal{R}[y]}{\mathcal{I}}$.

2. Monomial valuations

Given a Laurent series $\varphi = \sum_{\alpha \in \mathbb{Z}^N} a_\alpha x^\alpha$, the *set of exponents of* φ is the set

$$\mathcal{E}(\varphi) := \{\alpha \in \mathbb{Z}^N \mid a_\alpha \neq 0\}.$$

When $\mathcal{E}(\varphi)$ is finite, φ is a Laurent polynomial. When $\mathcal{E}(\varphi)$ is contained in the first orthant, φ is a series with non-negative exponents.

A subset of $\mathbb{R}^N$ of the form $\sigma = \{\lambda_1 v_1 + \cdots + \lambda_r v_r \mid \lambda_i \in \mathbb{R}_{\geq 0}\}$ for some $v_1, \ldots, v_r \in \mathbb{Q}^N$ is called a (rational convex polyhedral) *cone*. A cone is said to be *strongly convex* when it contains no non-trivial linear subspace.

Let $\sigma \subset \mathbb{R}^N$ be a strongly convex cone. The set of *formal Laurent series with exponents in* σ

$$\mathbb{C}[[\sigma]] = \{\varphi \mid \mathcal{E}(\varphi) \subset \sigma\}$$

is a ring with the natural sum and product.

Definition 2.1. The *Newton polyhedron* of a series $\phi \in \mathcal{R}$ is the convex hull in $\mathbb{R}^N$ of the set $\mathcal{E}(\phi) + {\mathbb{R}_{\geq 0}}^N$ and is denoted by NPϕ. Let V be a vertex of NPϕ. The *cone of NPϕ associated to* V is the cone

$$\sigma_V = \{v \in \mathbb{R}^N \mid (V + \lambda v) \in \text{NP}\phi \quad \text{for some positive real number} \quad \lambda\}.$$

Remark 2.2. The cone σ_V is always strongly convex and contains the positive orthant.

Now for $k \in \mathbb{N}$ consider the ring of series

$$\mathbb{C}[[\sigma]]_{\frac{1}{k}} := \left\{ \sum_{\alpha \in \left(\frac{1}{k}\mathbb{Z}\right)^N \cap \sigma} a_\alpha x^\alpha \mid a_\alpha \in \mathbb{C} \right\}.$$

When k divides k' there is a natural inclusion $\mathbb{C}[[\sigma]]_{\frac{1}{k}} \hookrightarrow \mathbb{C}[[\sigma]]_{\frac{1}{k'}}$. So, it makes sense to consider the *ring of formal Puiseux series with exponents in* σ

$$\mathbb{C}[[\sigma]]^{\wp} := \bigcup_{k\in\mathbb{N}} \mathbb{C}[[\sigma]]_{\frac{1}{k}}.$$

Given a formal series with rational exponents $\phi = \sum_{\alpha\in\mathbb{Q}^N} a_\alpha x^\alpha$ the set of exponents of ϕ is the set $\mathcal{E}(\phi) := \{\alpha \in \mathbb{Q}^N \mid a_\alpha \neq 0\}$. The set of exponents of a formal Puiseux series with exponents in a cone is contained in a lattice.

We will work only with cones σ containing the first orthant, for such σ we have the following diagram of inclusions:

$$\begin{array}{ccccc} \mathcal{R} & \longrightarrow & \mathbb{C}[[\sigma]] & \longrightarrow & \mathrm{Fr}(\mathbb{C}[[\sigma]]) \\ & & \downarrow & & \downarrow \\ & & \mathbb{C}[[\sigma]]^{\wp} & \longrightarrow & \mathrm{Fr}(\mathbb{C}[[\sigma]]^{\wp}) \end{array}$$

where $\mathrm{Fr}(\mathcal{A})$ stands for the field of fractions of $\mathcal{A}$.

The *dual* of a cone σ is the cone $\sigma^\vee := \{v \in \mathbb{R}^N \mid u \cdot v \geq 0, \forall u \in \sigma\}$. A cone $\sigma \subset \mathbb{R}^N$ is strongly convex if and only if the interior of $\sigma^\vee$ (as a subset of $\mathbb{R}^N$) is not empty.

Definition 2.3. Let σ be a strongly convex cone. Given $w \in \sigma^\vee$ the map

$$\nu_w : \mathbb{C}[[\sigma]]^{\wp} \longrightarrow \mathbb{R} \cup \{\infty\}$$

defined by

$$\nu_w \phi = \min_{\alpha\in\mathcal{E}(\phi)} w \cdot \alpha \tag{2.1}$$

is a valuation called the *monomial* valuation with weight w.

By restriction, ν_w induces a valuation in all subrings of $\mathbb{C}[[\sigma]]^{\wp}$. The extension of $\nu_w|_{\mathcal{R}}$ from $\mathcal{R}$ to $\mathbb{C}[[\sigma]]^{\wp}$ is not unique. Anyhow the following holds:

Lemma 2.4. *There exists a unique way to extend $\nu_w|_{\mathbb{C}[[\sigma]]}$ from the ring of series with exponents in σ to the ring of Puiseux series with exponents in σ.*

Proof.

$$\mathbb{C}[[\sigma]]_{\frac{1}{k}} = \mathbb{C}[[\sigma]][x_1{}^{\frac{1}{k}}, \ldots, x_N{}^{\frac{1}{k}}] = \frac{\mathbb{C}[[\sigma]][t_1, \ldots, t_N]}{(\{t_i{}^k - x_i\}_{i=1,\ldots,N})}.$$

The field extension

$$\mathrm{Fr}\left(\mathbb{C}[[\sigma]]\right) \hookrightarrow \mathrm{Fr}\left(\mathbb{C}[[\sigma]]\right)(x_1{}^{\frac{1}{k}}, \ldots, x_N{}^{\frac{1}{k}})$$

is a normal algebraic extension. Its Galois group is formed by the automorphisms

$$g_{(j_1,\ldots,j_N)} : x_i{}^{\frac{1}{k}} \mapsto \xi^{j_i} x_i{}^{\frac{1}{k}}; \qquad \xi^k = 1, \quad (j_1, \ldots, j_N) \in \mathbb{N}^N$$

by a theorem of Ostrowski and Krull [5, F,Theorem 1] two extensions of a given valuation are conjugate. Since $\nu_w(g_{(j_1,\ldots,j_N)}(\phi)) = \nu_w\phi$ we have the result. □

Remark 2.5. Let $\mathbb{C}[\sigma]$ denote the ring of Laurent polynomials with exponents in σ. That is $\mathbb{C}[\sigma] = \{\varphi \in \mathbb{C}[[\sigma]] \mid \#\mathcal{E}(\varphi) < \infty\}$. Then $\mathbb{C}[[\sigma]]$ is the completion of $\mathbb{C}[\sigma]$ with respect to the valuation ν_w for any w in the interior of $\sigma^\vee$. The field of fractions of $\mathbb{C}[\sigma]$ is the same as the field of fractions of $\mathbb{C}[x_1, \dots, x_N]$. Then, $\mathbb{C}[[\sigma]]$ is a ring of dimension N.

3. Parametrizations with exponents in a cone

J. McDonald showed in [4] that given a polynomial P in $\mathbb{C}[x_1, \dots, x_N][y]$, for any $w \in \mathbb{R}^N$ of rationally independent coordinates, there exists a strongly convex cone σ, with $w \in \sigma^\vee$ such that P has a root in $\mathbb{C}[[\sigma]]^\wp$. Moreover he gives an algorithm to compute such series. Then, P. González Pérez showed in [3] that σ may be chosen to be a cone of the Newton polyhedron of the discriminant of P with respect to y.

Definition 3.1. Let $\sigma \subset \mathbb{R}^N$ be a cone. For $\varrho \in (\mathbb{R}_{>0})^N$, the σ-*wedge* of polyradius ϱ is the set

$$\mathrm{W}(\sigma, \varrho) := \left\{ z \in (\mathbb{C}^*)^N ;\ \tau(z)^u \leq \varrho^u,\ \forall\, u \in \sigma \cap \mathbb{Z}^N \right\}$$

where $\tau(z_1, \dots, z_N) = (|z_1|, \dots, |z_N|)$.

Denote by δ the discriminant of F with respect to y. For each vertex V of the Newton Polyhedron of δ let σ_V be the cone of NPδ associated to V (Definition 2.1). By [1, Proposition 5.1], there exists $\varrho_V \in (\mathbb{R}_{>0})^N$ such that the σ_V-wedge of polyradius ϱ_V does not intersect the zero locus of δ.

Definition 3.2. A connected component of $\pi^{-1}(\mathrm{W}(\sigma_V, \varrho_V)) \cap \mathcal{H}$ will be called a σ_V-*branch* of $\mathcal{H}$.

Given a σ_V-branch C of $\mathcal{H}$ the degree of the covering $\pi : \mathrm{C} \longrightarrow \mathrm{W}(\sigma_V, \varrho_V)$ will be denoted by d_{C}.

Remark 3.3. Let $\mathcal{B}_V$ be the set of σ_V-branches of $\mathcal{H}$, $d = \sum_{\mathrm{C} \in \mathcal{B}_V} d_{\mathrm{C}}$, d being the degree of F in y.

The following proposition is proved in [1]:

Proposition 3.4. *Let V be a vertex of NPδ and let σ_V be the cone of NPδ associated to V. Given a σ_V-branch C of $\mathcal{H}$, there exist a σ_V-wedge W and a series $\varphi_{\mathrm{C}} \in \mathbb{C}[[\sigma_V]]$, convergent on W such that*

$$\begin{array}{cccc} \Phi_{\mathrm{C}} : & W & \longrightarrow & \mathbb{C}^{N+1} \\ & (x_1, \dots, x_N) & \mapsto & ({x_1}^{d_{\mathrm{C}}}, \dots, {x_N}^{d_{\mathrm{C}}}, \varphi_{\mathrm{C}}(x)) \end{array}$$

parameterizes C. *That is:*

$$\left\{ ({x_1}^{d_{\mathrm{C}}}, \dots, {x_N}^{d_{\mathrm{C}}}, \varphi_{\mathrm{C}}(x)) \mid (x_1, \dots, x_N) \in W \right\} = \mathrm{C}. \tag{3.1}$$

Remark 3.5. Let ζ be a primitive d_{C}-root of unity, the set of functions with property (3.1) is

$$\{\varphi_{\mathrm{C}}(\zeta^{i_1}x_1,\ldots,\zeta^{i_N}x_N) \mid i_j \in \{1,\ldots,d_{\mathrm{C}}\}\} = \left\{\varphi_{\mathrm{C}}^{(1)},\ldots,\varphi_{\mathrm{C}}^{(d_{\mathrm{C}})}\right\}$$

and it has exactly d_{C} elements.

Remark 3.6. The irreducibility of $\mathcal{H}$ implies that a polynomial $h \in \mathcal{R}$ is an element of $\mathcal{I}$ if and only if it vanishes on C. That is

$$h({x_1}^{d_{\mathrm{C}}},\ldots,{x_N}^{d_{\mathrm{C}}},\varphi_{\mathrm{C}}(x_1,\ldots,x_N)) = 0$$

which is equivalent to $h(x_1,\ldots,x_N,\varphi_{\mathrm{C}}({x_1}^{\frac{1}{d_{\mathrm{C}}}},\ldots,{x_N}^{\frac{1}{d_{\mathrm{C}}}})) = 0$.

4. σ-branches and primary decomposition

Let V be a vertex of NPδ and let C be a σ_V-branch of $\mathcal{H}$. We will denote by $\mathcal{J}_{\mathrm{C}}$ the kernel of the morphism

$$\begin{array}{ccc} \mathbb{C}[[\sigma_V]][y] & \longrightarrow & \mathbb{C}[[\sigma_V]] \\ h(x_1,\ldots,x_N,y) & \mapsto & h({x_1}^{d_{\mathrm{C}}},\ldots,{x_N}^{d_{\mathrm{C}}},\varphi_{\mathrm{C}}(x)) \end{array}$$

where φ_{C} is as in Proposition 3.4.

Since $\mathbb{C}[[\sigma_V]]$ is an integral domain, $\mathcal{J}_{\mathrm{C}}$ is a prime ideal of $\mathbb{C}[[\sigma_V]][y]$. By Remark 3.6 we have $\mathcal{I} = \mathcal{J}_{\mathrm{C}} \cap \mathcal{R}[y]$ for any $\mathrm{C} \in \mathcal{B}_V$.

Proposition 4.1. *Let $\mathcal{I}^{\sigma_V}$ be the extension of $\mathcal{I}$ to the ring $\mathbb{C}[[\sigma_V]][y]$ via the natural inclusion. Then*

$$\mathcal{I}^{\sigma_V} = \bigcap_{\mathrm{C}\in\mathcal{B}_V} \mathcal{J}_{\mathrm{C}}.$$

Proof. Let ${\mathcal{I}^{\sigma_V}}^{\wp}$ be the extension of $\mathcal{I}$ to the ring $\mathbb{C}[[\sigma_V]]^{\wp}[y]$ via the natural inclusion. Since $\mathbb{C}[[\sigma_V]][y] \hookrightarrow \mathbb{C}[[\sigma_V]]^{\wp}[y]$ is an integral extension, by the "Going-up theorem" [2, Theorem 5.10] there exists an ideal $\mathcal{L} \subset \mathbb{C}[[\sigma_V]]^{\wp}[y]$ such that $\mathcal{I}^{\sigma_V} = \mathcal{L} \cap \mathbb{C}[[\sigma_V]][y]$. We have that

$${\mathcal{I}^{\sigma_V}}^{\wp} \cap \mathbb{C}[[\sigma_V]][y] = \mathcal{L}^{\mathrm{cec}} = \mathcal{L}^{\mathrm{c}} = \mathcal{I}^{\sigma_V} \tag{4.1}$$

where ^{c} stands for contraction and ^{e} for extension.

Set $\phi_{\mathrm{C}}^{(i)} := \varphi_{\mathrm{C}}^{(i)}({x_1}^{\frac{1}{d_{\mathrm{C}}}},\ldots,{x_N}^{\frac{1}{d_{\mathrm{C}}}})$, $i = 1,\ldots d_{\mathrm{C}}$, where the $\varphi_{\mathrm{C}}^{(i)}$ are as in Remark 3.5. Each $\varphi_{\mathrm{C}}^{(i)}$ is a root of F as a polynomial in y. Then $\prod_{i=1}^{d_{\mathrm{C}}}(y-\phi_{\mathrm{C}}^{(i)})$ divides F as an element of $\mathbb{C}[[\sigma_V]]^{\wp}[y]$. This, together with Remark 3.3, implies

$$F = \prod_{\mathrm{C}\in\mathcal{B}_V}\prod_{i=1}^{d_{\mathrm{C}}}(y-\phi_{\mathrm{C}}^{(i)}). \tag{4.2}$$

Let $\mathcal{K}_{\mathrm{C}}^{(i)}$ be the kernel of the morphism

$$\begin{array}{ccc} \mathbb{C}[[\sigma_V]]^{\wp}[y] & \longrightarrow & \mathbb{C}[[\sigma_V]]^{\wp} \\ h(x,y) & \mapsto & h(x,\phi_{\mathrm{C}}^{(i)}(x)). \end{array}$$

By Remark 3.6 we have

$$\mathcal{J}_{\mathrm{C}} = \mathcal{K}_{\mathrm{C}}^{(i)} \cap \mathbb{C}[[\sigma_V]][y] \quad \text{for any} \quad i \in \{1,\dots,d_{\mathrm{C}}\}. \tag{4.3}$$

Equation (4.2) implies $\mathcal{I}^{\sigma_V{}^{\wp}} = \bigcap_{\mathrm{C}\in\mathcal{B}_V} \bigcap_{i=1}^{d_{\mathrm{C}}} \mathcal{K}_{\mathrm{C}}^{(i)}$, and the conclusion follows from (4.1) and (4.3). □

Proposition 4.2. *The only prime ideals $\mathcal{P}$ of $\mathbb{C}[[\sigma_V]][y]$ with the property $\mathcal{P}\cap\mathcal{R}[y] = \mathcal{I}$ are of the form $\mathcal{J}_{\mathrm{C}}$ with $\mathrm{C} \in \mathcal{B}_V$.*

Proof. It follows from Proposition 4.1 and Remark 2.5. □

5. The theorem

Given $w \in \mathbb{R}_{>0}{}^N$, take a vertex V of the Newton polyhedron of δ such that $w \in \sigma^\vee$. Each σ_V-branch C of $\mathcal{H}$ induces a valuation

$$\begin{array}{cccc} \nu_{\mathrm{C}}: & \mathcal{O}_{\mathcal{H}} & \longrightarrow & \mathbb{R}\cup\infty \\ & h(x_1,\dots,x_N,y) & \mapsto & \frac{1}{d_{\mathrm{C}}}\nu_w h(x_1{}^{d_{\mathrm{C}}},\dots,x_N{}^{d_{\mathrm{C}}},\varphi_{\mathrm{C}}) \end{array}$$

that extends ν_w.

Theorem 5.1. *Let $w \in \mathbb{R}_{>0}{}^N$ be a vector non-orthogonal to any of the faces of the Newton Polyhedron of δ, and let V be the only vertex of $NP\delta$ such that w belongs to the dual of σ_V. Then all the valuations that extend ν_w are the ones induced by the σ_V-branches of $\mathcal{H}$.*

We start by proving a lemma:

Lemma 5.2. *Let w be a vector in the interior of $\sigma_V{}^\vee$, and let $\nu : \mathcal{O}_{\mathcal{H}} \longrightarrow \mathbb{R}_{\geq 0}\cup\{\infty\}$ be a valuation that extends ν_w. There exists a σ_V-branch C of $\mathcal{H}$ and a valuation $\bar{\nu} : \frac{\mathbb{C}[[\sigma_V]][y]}{\mathcal{J}_{\mathrm{C}}} \longrightarrow \mathbb{R}\cup\{\infty\}$ that makes the diagram*

$$\begin{array}{ccccc} \mathbb{C}[[\sigma_V]] & \hookrightarrow & \frac{\mathbb{C}[[\sigma_V]][y]}{\mathcal{J}_{\mathrm{C}}} & \hookleftarrow & \mathcal{O}_{\mathcal{H}} \\ & \searrow \nu_w & \downarrow \bar{\nu} & \nu \swarrow & \\ & & \mathbb{R}\cup\{\infty\} & & \end{array}$$

commutative.

Proof. An element $h \in \mathbb{C}[[\sigma]][y]$ is written as

$$h = \sum_{i=0}^{\deg h} \psi_i y^i, \quad \text{where} \quad \psi_i = \sum_{\alpha \in \mathbb{Z}^N \cap \sigma} a_\alpha^{(i)} x^\alpha.$$

For each $i \in \{0, \dots, \deg h\}$ and $j \in \mathbb{Z}$ set

$$\psi_i^{(j)} := \sum_{\substack{\alpha \in \mathbb{Z}^N \cap \sigma \\ j \leq w \cdot \alpha < j+1}} a_\alpha^{(i)} x^\alpha.$$

Since w is in the interior of $\sigma^\vee$, for all j, the set $\{\alpha \in \mathbb{Z}^N \cap \sigma \mid j \leq w \cdot \alpha < j+1\}$ is finite and then the $\psi_i^{(j)}$'s are Laurent polynomials. We have

$$\psi_i = \sum_{j=0}^{\infty} \psi_i^{(j)} \quad \text{for all} \quad i \in \{1, \dots, \deg h\}.$$

Let $\mathcal{R}[y]_{\mathcal{I}}$ be the localization of $\mathcal{R}[y]$ with respect to $\mathcal{I}$. Since $\mathcal{I} \cap \mathcal{R} = \{0\}$, we have

$$\psi_i^{(j)} \in \mathcal{R}[y]_{\mathcal{I}}, \qquad \forall j \in \mathbb{Z} \,\text{and}\, \forall i \in \{1, \dots, \deg h\}$$

Let $\hat{\nu} : \mathcal{R}_{\mathcal{I}} \longrightarrow \mathbb{R}$ be the extension of the morphism given by the composition

$$\mathcal{R}[y] \longrightarrow \mathcal{O}_{\mathcal{H}} \stackrel{\nu}{\longrightarrow} \mathbb{R} \cup \infty.$$

Since ν extends ν_w and $\psi_i^{(j)} \in \mathrm{Fr}(\mathcal{R})$, we have $\hat{\nu}\left(\psi_i^{(j)}\right) = \nu_w\left(\psi_i^{(j)}\right) \geq j$. Then

$$\hat{\nu}\left(\sum_{i \in \{1, \dots, \deg h\}} \psi_i^{(K)} y^i\right) \geq K + \min\{0, \nu(y) \deg h\}, \qquad \forall K. \tag{5.1}$$

Set

$$\tau_K := \hat{\nu}\left(\sum_{j=0}^{K} \sum_{i \in \{1, \dots, \deg h\}} \psi_i^{(j)} y^i\right).$$

Inequality (5.1) implies that either $\tau_K \geq K + \min\{0, \nu(y) \deg h\}$ for all K or there exists K such that $\tau_l = \tau_K$ for all $l > K$. Then it makes sense to define:

$$\tilde{\nu} h := \lim_{K \longrightarrow \infty} \tau_K.$$

By construction, $\tilde{\nu}(hh') = \tilde{\nu}(h) + \tilde{\nu}(h')$ and $\tilde{\nu}(h + h') \geq \min\{\tilde{\nu}(h), \tilde{\nu}(h')\}$, then $\tilde{\nu}$ induces a valuation $\bar{\nu} : \dfrac{\mathbb{C}[[\sigma]][y]}{\tilde{\nu}^{-1}(\infty)} \longrightarrow \mathbb{R} \cup \infty$.

The ideal $\tilde{\nu}^{-1}(\infty)$ is prime and $\tilde{\nu}^{-1}(\infty) \cap \mathcal{R}[y] = \mathcal{I}$. Then, by Proposition 4.2 there exists $\mathrm{C} \in \mathcal{B}_V$ such that $\tilde{\nu}^{-1}(\infty) = \mathcal{J}_{\mathrm{C}}$. □

Proof of the theorem. Let C and $\bar{\nu} : \frac{\mathbb{C}[[\sigma]][y]}{\mathcal{J}_{\mathrm{C}}} \longrightarrow \mathbb{R} \cup \{\infty\}$ be as in the lemma. Let $\mathcal{K}_{\mathrm{C}}^{(1)}$ be as defined in the proof of Proposition 4.1, and let $\bar{\bar{\nu}}$ be an extension of $\bar{\nu}$ to $\dfrac{\mathbb{C}[[\sigma_V]]^{\wp}[y]}{\mathcal{K}_{\mathrm{C}}^{(1)}}$.

Given $h \in \mathcal{R}[y]$,

$$h(x,y) = h(x, \phi_{\mathrm{C}}^{(1)}) + (y - \phi_{\mathrm{C}}^{(1)})g(x,y), \quad \text{where} \quad g(x,y) \in \mathbb{C}[[\sigma]]^{\wp}.$$

So,

$$\nu(h) = \bar{\bar{\nu}}(h) = \bar{\bar{\nu}}(h(x, \phi_{\mathrm{C}}^{(1)})) \overset{\text{lemma 2.4}}{=} \nu_w h(x, \phi_{\mathrm{C}}^{(1)})) = \nu_{\mathrm{C}} h. \qquad \square$$

References

[1] Fuensanta Aroca. Puiseux parametric equations of analytic sets. *Proc. Amer. Math.*, 132:3035–3045, 2004.

[2] M.F. Atiyah and I.G. MacDonald. *Introduction to Commutative Algebra.* Addison-Wesley series in mathematics. Perseus books, Cambridge, Massachusetts, 1969.

[3] P.D. González Pérez. Singularités quasi-ordinaires toriques et polyèdre de Newton du discriminant. *Canad. J. Math.*, 52(2):348–368, 2000.

[4] John McDonald. Fiber polytopes and fractional power series. *J. Pure Appl. Algebra*, 104(2):213–233, 1995.

[5] Paulo Ribenboim. *Théorie des valuations.* Séminaire de Mathématiques Supérieures de l'Université de Montréal, (Été 1964). Les Presses de l'Université de Montréal, Montreal, Que., 1965.

Fuensanta Aroca
Instituto de Matemáticas (Unidad Cuernavaca)
Universidad Nacional Autónoma de México.
A.P. 273, Admon. de correos no. 3
C.P. 62251
Cuernavaca, Morelos, Mexico
e-mail: `fuen@matcuer.unam.mx`

Real and Complex Singularities
Trends in Mathematics, 23–30

Quelques Résultats sur Certaines Fonctions à Lieu Singulier de Dimension 1

Daniel Barlet

Abstract. This text is a survey on my recent work [B.04] on some holomorphic germs having a one dimensional singular locus. An analogous of the Brieskorn module of an isolated singularity is defined and a finiteness theorem is proved using Kashiwara's constructibility theorem. A bound for the (finite dimensional) torsion is also obtained. Non existence of torsion is proved for curves (reduced or not) an this property is stable by "Thom-Sebastiani" adjunction of an isolated singularity. This provides a lot of examples in any dimension where our formula $r = \mu(f) + \nu(f)$ generalizing the Milnor number formula, is valid.

1. Le complexe $(\text{Ker } df^\bullet, d^\bullet)$

Nous introduisons, de façon générale pour une fonction holomorphe f sur une variété complexe X, le complexe $(\text{Ker } df^\bullet, d^\bullet)$ de faisceaux sur $Y := f^{-1}(0)$ en posant:

$$\text{Ker}\, df^p := \text{Ker}\, \big[\wedge\, df : \Omega^p_X|_Y \to \Omega^{p+1}_X|_Y\big]$$

et où la différentielle $d^\bullet$ est induite par la différentielle de de Rham.
Les faisceaux de cohomologie de ce complexe seront notés par $\mathcal{H}^\bullet$.
De façon générale[1] ces faisceaux de cohomologie sont naturellement munis de deux opérations a et b. La première est simplement donnée par la multiplication par f. La seconde est donnée par l'inverse de la connexion de Gauss-Manin, c'est à dire par $b = df \wedge d^{-1}$. On a le résultat général suivant

Proposition 1.1. *Soit* $f : (\mathbb{C}^{n+1}, 0) \to (\mathbb{C}, 0)$ *un germe non nul de fonction holomorphe. Considérons pour* $p \in [1, n+1]$ *le* $\mathbb{C}$*-espace vectoriel*

$$\mathcal{H}^p_0 := \big((\text{Ker } df)^p \cap \text{Ker } d\big)_0 / d((\text{Ker } df)^{p-1})_0$$

---

[1] c'est à dire sans hypothèse sur f.

muni des $\mathbb{C}$-endomorphismes a et b définis respectivement par la multiplication par f et $df \wedge d^{-1}$.
Alors $\mathcal{H}_0^p$ vérifie les conditions 0), 1), 2), *et* 3) *de la définition d'un pré-(a,b)-module donnée ci-dessous.*

Pour les démonstrations, le lecteur se reportera de façon générale à mon preprint [B.04] qui est accessible sur matharxiv.

Définition 1.2. *Soit E un espace vectoriel complexe muni d'endomorphismes a et b.*
On pose:

$$B(E) = \bigcup_{m\in\mathbb{N}} \mathrm{Ker}\ b^m \quad \text{et} \quad A(E) = \{x \in E \ / \ \mathbb{C}[b].x \subset \bigcup_{m\in\mathbb{N}} \mathrm{Ker}\ a^m\ \}$$

On dira que E est un **pré-(a,b)-module** *lorsque les conditions suivantes sont vérifiées*

0) $a.b - b.a = b^2$.
1) *Pour tout* $\lambda \in \mathbb{C}^*, b - \lambda$ *est bijectif dans* E.
2) $\exists N \in \mathbb{N}\ /a^N.A(E) = 0$ *et on a* $B(E) \subset A(E)$.
3) $\bigcap_{m\in\mathbb{N}} b^m(E) \subset A(E)$.
4) *Le noyau et le conoyau de b sont de dimensions finies sur $\mathbb{C}$.* □

On dira que E est **sans torsion**[2] si on a de plus $\mathrm{Ker}\ b = 0$, ce qui équivaut à $B(E) = 0$ et donc à l'absence de b-torsion. Mais ceci implique aussi $A(E) = 0$ (voir [B.04] Lemme (1.3)). Dans le cas considéré dans la proposition ci-dessus le théorème de positivité de B. Malgrange [M. 74] donne alors l'absence de a-torsion[3] (voir aussi [B.S.04]).

A tout pré-(a,b)-module E on associe un (a,b)-module, c'est à dire un module libre de type fini sur l'anneau $\mathbb{C}[[b]]$ muni d'un endomorphisme $\mathbb{C}$-linéaire a qui est continu pour la topologie b-adique et vérifie

$$ab - ba = b^2.$$

Nous le noterons par $\mathcal{L}(E)$ et c'est le complété b-adique du quotient $E/B(E)$. On remarquera que grace à la relation 3) (et à l'égalité entre $B(E)$ et $A(E)$) on a b-séparation de ce quotient. Donc $E/B(E)$ s'injecte toujours dans $\mathcal{L}(E)$. Quand E est sans torsion, on en déduit que la dimension complexe de E/bE est le rang du (a,b)-module $\mathcal{L}(E)$. En général ce rang est donné par $\dim E/bE - \delta$ où $\delta := \dim \mathrm{Ker}\ b$.
Les (a,b)-modules que l'on associe à des singularités sont loin d'être les plus "généraux"; en particulier ils sont réguliers, propriété qui reflète la regularité de la connexion de Gauss-Manin. Pour plus de détails la-dessus consulter par exemple [B. 93], [B.04] ou [B.S.04]).

[2] même en l'absence de la condition 4).
[3] ce qui est plus fort, en général, que la nullité de $A(E)$.

2. L'hypothèse (H)

Soit $\tilde{f} : (\mathbb{C}^{n+1}, 0) \longrightarrow (\mathbb{C}, 0)$ un germe non constant de fonction holomorphe et soit $f : X \longrightarrow D$ un représentant de Milnor de $\tilde{f}$. Nous ferons les hypothèses suivantes (auxquelles nous nous référons sous le nom de "l'hypothèse (H)")

H a) Le lieu singulier $S := \{x \in X / df_x = 0\}$ est une courbe contenue dans $Y := f^{-1}(0)$, dont chaque composante irréductible contient l'origine et est non singulière en dehors de 0.

H b) En chaque point x de $S - \{0\}$ il existe un germe en x de champ de vecteur holomorphe V_x , non nul en x, tel que $V_x \cdot f \equiv 0$.

L'hypothèse H b) est assez restritive puisqu'elle implique que le long de chaque composante connexe de $S - \{0\}$ la singularité $\{f = 0\}$ est une déformation localement analytiquement triviale de la singularité hyperplane transverse (qui est une singularité isolée de $\mathbb{C}^n$).

Cependant il est facile de voir que cette hypothèse est toujours vérifiée pour $n = 1$ (courbes planes réduites ou non) et qu'il y a beaucoup d'exemples en dimensions supérieures, comme le montreront les théorèmes 5 et 7.

Proposition 2.1. *Sous l'hypothèse* (H) *on a les propriétés suivantes:*

1) *Les faisceaux de cohomologie* $\mathcal{H}^p$ *du complexe* $(\mathrm{Ker}\, df^\bullet, d^\bullet)$ *sont nuls pour* $p \neq 1, n, n+1$.
2) *Pour* $n \geq 2$ *le faisceau* $\mathcal{H}^1 \simeq \mathrm{Ker}\, df^1 \cap \mathrm{Ker}\, d$ *est constant sur* Y *de fibre isomorphe à* $E_1 = \Omega^1_{\mathbb{C},0}$ *muni des opérations* a *et* b *"naturelles" de multiplication par* z *et de "primitive sans constante" .*
3) *Pour* $n \geq 2$ *le faisceau* $\mathcal{H}^n$ *a son support contenu dans* S*. Il est localement constant sur* $S^* := S \setminus \{0\}$*. Sa fibre en un point d'une composante connexe* S_j^* *de* S^* *est le module de Brieskorn de la singularité obtenue par section hyperplane transverse. De plus le faisceau* $\mathcal{H}^n$ *n'a pas de torsion (même à l'origine).*
4) *Le faisceau* $\mathcal{H}^{n+1}$ *est supporté par l'origine.*
5) *Pour* $n = 1$ *on a une suite exacte de faisceaux sur* Y*, compatible aux opérations* a *et* b*:*

$$0 \to E_1 \otimes \underline{\mathbb{C}}_Y \to \mathcal{H}^1 \to \tilde{\mathcal{H}}^1 \to 0$$

où le faisceau $\tilde{\mathcal{H}}^1$ *a son support dans* S *et induit sur* S^* *un système local évident à décrire*[4].

Introduisons maintenant deux ingrédients importants dans cette situation.

2.1. L'idéal $\widehat{J}(f)$

Notons par $J(f)$ l'idéal jacobien de f et soit $i : X \setminus \{0\} \hookrightarrow X$ l'inclusion, où X désigne ici une boule de Milnor pour f à l'origine. Définissons alors l'idéal (cohérent) $\widehat{J}(f) := i_* i^*(J(f))$. Donc un germe g de fonction holomorphe à

[4]la singularité transverse se réduit à la multiplicité transverse dans ce cas.

l'origine sera dans $\widehat{J}(f)_0$ si et seulement s' il induit une section de $J(f)$ en dehors de 0[5]. Le faisceau cohérent $\widehat{J}(f)/J(f)$ est à support l'origine. C'est donc un espace vectoriel de dimension finie. Nous poserons

$$\mu(f) := \dim \widehat{J}(f)/J(f)$$

ce qui définit l'analogue du nombre de Milnor pour le cas d'une fonction à singularité isolée[6].

2.2. Le $\mathcal{D}_X$-module $\mathcal{M}$

Définissons maintenant l'idéal à gauche $\mathcal{I}$ de $\mathcal{D}_X$ comme l'déal engendré par $\widehat{J}(f)$ et par l'annulateur de f dans le faisceau Θ_X des champs de vecteurs holomorphes sur X:

$$\mathrm{Ann}(f) = \{V \in \Theta_X /\ V.f \equiv 0\}.$$

Posons alors $\mathcal{M} := \mathcal{D}_X/\mathcal{I}$.
Il est clair que l'hypothèse (H) implique immédiatement que $\mathcal{M}$ est un $\mathcal{D}_X$-module holonome de support contenu dans S. D'après le théorème de constructibilité de M. Kashiwara [K. 75] son complexe de de Rham est à cohomologie constructible. En particulier, le faisceau $DR^{n+1}(\mathcal{M})$ est concentré à l'origine et se réduit à un espace vectoriel de dimension finie. Nous poserons

$$\nu(f) := \dim DR^{n+1}(\mathcal{M}).$$

On remarquera que pour f à singularité isolée à l'origine on a $\mathcal{M} = 0$ puisque $\widehat{J}(f) = \mathcal{O}_X$. On a donc $\nu(f) = 0$ dans ce cas.

Théorème 2.2. *Sous l'hypothèse* (H) *l'espace vectoriel*

$$E' := \mathcal{H}_0^{n+1} \simeq H^0_{\{0\}}(Y, \mathcal{H}^{n+1})$$

est un pré-(a,b)-module de rang r *vérifiant*

$$\dim E'/b.E' = \mu(f) + \nu(f) - \gamma + \delta \quad \text{et} \quad r = \mu(f) + \nu(f) - \gamma$$

où γ *et* δ *sont les dimension de* $\mathrm{Ker}\ j$[7] *et de* $\mathrm{Ker}\ b$. *De plus on a les majorations:*

$$\delta \ \leq \gamma \leq \dim H^1_{\{0\}}(S, \mathcal{H}^n/b.\mathcal{H}^n).$$

On a $\gamma = 0$ *et donc l'égalité* $\dim E'/b.E' = \mu(f) + \nu(f) = r$ *sous la condition* (P) *décrite ci-dessous.*

On remarquera que comme $\mathcal{H}^n/b.\mathcal{H}^n$ est un système local d'espaces vectoriels de dimensions finies sur S^*, on a bien une majoration finie de γ. Par ailleurs ce majorant est en pratique assez facile à estimer.

---

[5] nous supposerons toujours $n \geq 1$ ce qui permet d'utiliser Hartogs.
[6] dans ce cas $\widehat{J}(f) = \mathcal{O}_X$.
[7] Voir plus loin la définition de j.

3. Etude de la torsion

3.1. La condition (P)

Comme le fait E. Brieskorn dans le cas d'une fonction à singularité isolée à l'origine, il est intéressant d'introduire également le faisceau

$$\mathcal{E}'' := \Omega_X^{n+1}/df \wedge d\Omega_X^{n-1}$$

en plus du faisceau $\mathcal{E}' \equiv \mathcal{H}^{n+1}$. Il est également muni d'opérations a et b naturelles vérifiant la relation de commutation $ab - ba = b^2$. On a alors une surjection naturelle compatible à a et b

$$j : \mathcal{E}'' \to \mathcal{E}'$$

de noyau $d(\text{Ker } df)^n/df \wedge d\Omega_X^{n-1}$. Mais le faisceau $\mathcal{E}''$ n'est pas, en général, supporté par l'origine sous l'hypothèse (H)[8].
Posons:

$$E' := H^0_{\{0\}}(X, \mathcal{E}') \quad \text{et} \quad E'' := H^0_{\{0\}}(X, \mathcal{E}'').$$

La suite exacte de cohomologie à support l'origine déduite du morphisme j donne une suite exacte (a,b)-linéaire

$$0 \to H^0_{\{0\}}(X, \text{Ker } j) \to E'' \xrightarrow{\hat{j}} E' \to \cdots$$

On a également une suite exacte de faisceaux

$$0 \to \mathcal{E}' \xrightarrow{\tilde{b}} \mathcal{E}'' \to \Omega_X^{n+1}/df \wedge \Omega_X^n \to 0 \tag{μ}$$

où le morphisme $\tilde{b}$ est donné par $df \wedge d^{-1}$. Comme le faisceau $\mathcal{E}'$ est à support l'origine, on en déduit la suite exacte

$$0 \to E' \xrightarrow{\tilde{b}} E'' \to H^0_{\{0\}}(X, \Omega_X^{n+1}/df \wedge \Omega_X^n) \to 0$$

et l'application $b : E' \to E'$ est la composée $\hat{j} \circ \tilde{b}$. Comme $\tilde{b}$ est injective, l'injectivité de $\hat{j}$ implique celle de b dans E'. La condition suivante est necessaire et suffisante pour que l'application $\hat{j}$ soit injective:

$$d(\text{Ker } df^n) \cap \widehat{J}(f).\Omega_X^{n+1} \subset df \wedge d\Omega_X^{n-1} \tag{P}$$

alors qu'une condition nécessaire et suffisante pour avoir l'injectivité de $b : E' \to E'$ est donnée par l'inclusion:

$$d(\text{Ker } df^n) \cap (df \wedge \Omega_X^n) \subset df \wedge d\Omega_X^{n-1}. \tag{P'}$$

Bien sur, la condition (P) implique la condition (P'). Mais il est bon de préciser, qu'à ce jour, je ne connais pas d'exemple de germe vérifiant l'hypothèse (H) et ne vérifiant pas la condition (P). Malheureusement, je n'ai pas non plus de preuve que l'hypothèse (H) implique toujours la condition (P). C'est cependant vrai (et non trivial) pour $n = 1$:

[8]comme le montre la suite exacte (μ) ci-dessous.

Théorème 3.1. *Pour* $n = 1$, *c'est à dire pour les germes à l'origine de* $\mathbb{C}^2$, *réduits ou non, la condition* (P) *est toujours vérifiée.*

La preuve de ce résultat utilise la récurrence tordue suivante qui permet de tenir compte des multiplicités d'annulation d'une fonction sur les différentes composantes irréductibles de Y[9].

Lemme 3.2 (récurrence tordue.). *Soient* $p_1, \dots, p_k$ *des entiers* ≥ 2 *et soient* $\phi_1 \cdots \phi_k$ *des fonctions strictement croissantes*

$$\phi_j : [0, p_j - 1] \cap \mathbb{N} \longrightarrow [0, 1] \qquad \textit{vérifiant} \qquad \phi_j(p_j - 1) = 1, \forall j \in [1, k].$$

Considérons des propositions $A(\sigma_1, \dots, \sigma_k)$ *indéxées par les entiers*

$$\sigma_j \in [0, p_j - 1] \cap \mathbb{N}.$$

On suppose

1) $A(0, \dots, 0)$ *est vraie .*
2) *l'implication* $A(\sigma_1, \dots, \sigma_k) \Rightarrow A(\sigma_1, \dots, \sigma_j + 1, \dots, \sigma_k)$ *est vraie si les deux conditions suivantes sont satisfaites*
   a) $\sigma_j \leq p_j - 2$
   b) $\phi_j(\sigma_j) = \min_{l \in [1,k]}\{\phi_l(\sigma_l)\}$

Alors la proposition $A(p_1 - 1, \dots, p_k - 1)$ *est vraie.*

On prendra garde que nous n'affirmons pas ici que $A(\sigma_1, \dots, \sigma_k)$ **est vraie pour toutes les valeurs de** $(\sigma_1, \dots, \sigma_k)$.

Le résultat suivant fournit, à partir de ce théorème, une bonne quantité d'exemples de fonctions vérifiant l'hypothèse (H) pour lesquelles la condition (P) est vraie:

Théorème 3.3. *Soit* f *un germe de fonction à singularité isolée à l'origine de* $\mathbb{C}^{n+1}$ *et soit* g *un germe de fonction holomorphe à l'origine de* $\mathbb{C}^{p+1}$ *vérifiant l'hypothèse* (H). *Alors le germe de fonction holomorphe* F *à l'origine de* $\mathbb{C}^{n+p+2}$ *défini par*

$$F(x, y) := f(x) + g(y) \quad \text{pour} \quad (x, y) \in \mathbb{C}^{n+1} \times \mathbb{C}^{p+1}$$

vérifie (H). *De plus, le (a,b)-module associé à* F *à l'origine est le produit tensoriel*[10] *des (a,b)-modules associés à l'origine à* f *et* g *respectivement.*
Si, de plus, le germe g *vérifie la condition* (P), *il en est de même pour* F.

Dans la situation du théorème ci-dessus, on peut également décrire le faisceau $\mathcal{H}^{n+p+1}$ associé à F[11] à partir du module de Brieskorn de f et du faisceau $\mathcal{H}^n$ associé à g.

---

[9] rappelons que l'on est ici dans $\mathbb{C}^2$ et donc que Y est une courbe et que S est la réunion des composantes irréductibles de Y de multiplicité ≥ 2.

[10] au sens des (a,b)-modules ; voir ci-dessous.

[11] ou plus exactement le faisceau de (a,b)-module qui lui est associé après complétion formelle.

3.2. Produits tensoriels de (a,b)-modules

Nous allons nous contenter ici de définir le produit tensoriel de deux (a,b)-modules, ce qui permet de comprendre le résultat à la “Thom-Sébastiani” énoncé ci-dessus.

Définition 3.4. *Etant donné deux (a,b)-modules* E *et* F *nous poserons*

$$E \otimes_{a,b} F := E \otimes_{\mathbb{C}[[b]]} F$$

comme $\mathbb{C}[[b]]$*-module, et nous définirons l'endomorphisme* $\mathbb{C}$*-linéaire* a *en posant*

$$a := a_E \otimes 1_F + 1_E \otimes a_F.$$

On vérifie alors immédiatement la relation de commutation $ab - ba = b^2$ sur $E \otimes_{a,b} F$. Comme un (a,b)-module est complètement déterminé par la donnée d'une $\mathbb{C}[[b]]$-base ainsi que l'action de a sur cette base (grâce à la relation de commutation et à la continuité de a pour la topologie b-adique) il est facile de décrire en ces termes le produit tensoriel de deux (a,b)-modules donnés de cette façon.

En fait dans [B.04] on définit également la notion de produit tensoriel de deux pré-(a,b)-modules et on montre, sous une condition très large[12], que l'on obtient à nouveau un pré-(a,b)-module. On montre alors que le (a,b)-module associé au produit tensoriel est le produit tensoriel des (a,b)-modules associés.

3.3. Un exemple

Considérons le polynôme homogène à deux variables[13]

$$f(X,Y) = X^3(X^3 + Y^3).$$

Il est facile de voir que $\mathrm{Ker}\, df^1/\mathcal{O}.df$ est engendré par la 1-forme associé au champ de vecteur

$$V = XY^2\frac{\partial}{\partial X} - (2X^3 + Y^3)\frac{\partial}{\partial Y}$$

annulant f et de divergence $\mathrm{div}\, V = -2Y^2$. Posons $\tilde{V} = V - div(V)$.
On a $J(f) = X^2.(2X^3 + Y^3, XY^2)$, $\widehat{J(f)} = (X^2)$ et
$\dim_{\mathbb{C}} \widehat{J(f)}/J(f) = \mu(f) = 9$.
Par ailleurs, en explicitant $DR^2(\mathcal{M})$ on trouve:

$$\nu(f) = \dim \mathcal{O}_X/(\widehat{J}(f)) + \tilde{V}(\mathcal{O}_X).$$

Pour calculer $\dim_{\mathbb{C}} \Big(\mathbb{C}\{X,Y\}/(X^2) + \widetilde{V}\big(\mathbb{C}\{X,Y\}\big)\Big)$ on regarde l'identité

$$\widetilde{V}(X^pY^q) = (p - q - 2)X^pY^{q+2} - 2qX^{p+3}Y^{q-1} \ ;$$

on peut donc réduire X^aY^b à $X^{a+3}Y^{b-3}$ pourvu que $a \neq b$ et $b \geq 2$.

[12] l'hypothèse “locale”, à savoir l'existence d'un entier l tel que $a^l.E \subset b.E$; elle est toujours satisfaite dans les cas que nous considérons.

[13] pour $n = 1$ l'hypothèse (H) est toujours vérifiée, ainsi que la condition (P) comme nous l'avons vu plus haut.

Si $a = b \geq 2$ on est dans (X^2). Il reste donc $1, X, Y$ et X^pY pour $p < 2$. Donc $1, X, Y, XY$ donne une base et $\nu(f) = 4$.
Alors $\mathcal{L}(E')$[14] est un (a,b)-module de rang 13. Une base de $E'/_{bE'}$ est donnée par

$$1, X, Y, XY, X^2, X^3, X^4, X^5, X^2Y, X^3Y, X^4Y, X^5Y, X^2Y^2.$$

Comme on a

$$b(X^pY^q)dX \wedge dY = df \wedge \frac{X^{p+1}Y^q}{p+1}dY = -df \wedge \frac{X^pY^{q+1}}{q+1}dX,$$

on en déduit que

$$a[X^pY^q] = \frac{p+q+2}{6}.b[X^pY^q].$$

Pour calculer le (a,b)-module associé à l'origine au polynôme à trois variables $F(X, Y, Z) = X^3(X^3 + Y^3) + Z^2$ il nous suffit donc de calculer le (a,b)-module de Brieskorn associé à la fonction $Z \to Z^2$ et de faire le produit tensoriel avec celui associé à $X^3(X^3 + Y^3)$ à l'origine. Le (a,b)-module de Brieskorn associé à $Z \to Z^2$ est de rang 1 et engendré par un générateur e vérifiant $a.e = \frac{1}{2}.b.e$. Le produit tensoriel est donc de rang 13 et engendré par les $X^pY^q \otimes e$ mais avec

$$a[X^pY^q \otimes e] = \frac{p+q+5}{6}b[X^pY^q \otimes e].$$

Références

[B.93] Barlet, D. *Théorie des (a, b)-modules I,* in Complex Analysis and Geometry, Plenum Press, (1993), p. 1–43.

[B.04] Barlet, D. *Sur certaines singularités non isolées d'hypersurfaces I*, preprint Institut E. Cartan 2004/ n^0 03, 47 pages.

[Br.70] Brieskorn, E. *Die Monodromie der isolierten Singularitäten von Hyperflächen,* Manuscripta Math. 2 (1970), p. 103–161.

[B.S.04] Barlet, D. and Saito, M. *Brieskorn modules and Gauss-Manin systems for non isolated hypersurface singularities*, preprint Institut E. Cartan 2004/ n^0 54, 16 pages.

[K.75] Kashiwara, M. *On the maximally over determined systems of differential equations,* Publ. R.I.M.S. vol.10 (1975), p. 563–579.

[M.74] Malgrange,B. *Intégrale asymptotique et monodromie,* Ann. Sc. Ec. Norm. Sup. 7 (1974), p. 405–430.

Daniel Barlet
Universitée H. Poincaré et Institut Universitaire de France
Institut E. Cartan, UMR 7502 CNRS/UHP/INRIA
B.P. 239
F-54506 Vandoeuvre-les-Nancy, France
e-mail: `barlet@iecn.u-nancy.fr`

[14]Le complété b-adique de E', puisque $\mathrm{Ker}\, b = 0$ et donc $B(E) = A(E) = 0$.

Real and Complex Singularities
Trends in Mathematics, 31–45

Classification of Rational Unicuspidal Projective Curves whose Singularities Have one Puiseux Pair

Javier Fernández de Bobadilla, Ignacio Luengo,
Alejandro Melle Hernández and Andras Némethi

Abstract. It is a very old and interesting open problem to characterize those collections of embedded topological types of local plane curve singularities which may appear as singularities of a projective plane curve C of degree d. The goal of the present article is to give a complete (topological) classification of those cases when C is rational and it has a unique singularity which is locally irreducible (i.e., C is unicuspidal) with one Puiseux pair.

Mathematics Subject Classification (2000). Primary. 14B05,14J17,32S25 Secondary: 14E15,32S4599Z99.

Keywords. Cuspidal rational plane curves, logarithmic Kodaira dimension.

1. Introduction

It is a very old and interesting open problem to characterize those collections of embedded topological types of local plane curve singularities which may appear as singularities of a projective plane curve C of degree d. (We invite the reader to consult the articles of Fenske, Flenner, Miyanishi, Orevkov, Sugie, Tono, Zaidenberg, Yoshihara or [3] and the references therein, for recent developments.) The goal of the present article is to give a complete (topological) classification of those cases when C is rational and it has a unique singularity which is locally irreducible (i.e., C is unicuspidal) with one Puiseux pair.

In fact, as a second goal, we also wish to present some of the techniques which are/might be helpful in such a classification, and we invite the reader to join us

The first author is supported by the Netherlands Organization for Scientific Research (NWO). The first three authors are partially supported by BFM2001-1488-C02-01. The forth author is partially supported by NSF grant DMS-0304759.

in our effort to produce a classification for all the cuspidal rational plane curves. In fact, this effort also motivates that decision, that in some cases (in order to have a better understanding of the present situation), we produce more different arguments for some of the steps.

In the next paragraph we formulate the main result. We will write d for the degree of C and (a,b) for the Puiseux pair of its cusp, where $1 < a < b$. We denote by $\{\varphi_j\}_{j\geq 0}$ the Fibonacci numbers $\varphi_0 = 0$, $\varphi_1 = 1$, $\varphi_{j+2} = \varphi_{j+1} + \varphi_j$.

Theorem 1.1 (Main Theorem). *The Puiseux pair (a,b) can be realized by a unicuspidal rational plane curve of degree d if and only if (d,a,b) appears in the following list.*

(a) $(a,b) = (d-1,d)$;
(b) $(a,b) = (d/2, 2d-1)$, *where d is even;*
(c) $(a,b) = (\varphi_{j-2}^2, \varphi_j^2)$ *and* $d = \varphi_{j-1}^2 + 1 = \varphi_{j-2}\varphi_j$, *where j is odd and* ≥ 5;
(d) $(a,b) = (\varphi_{j-2}, \varphi_{j+2})$ *and* $d = \varphi_j$, *where j is odd and* ≥ 5;
(e) $(a,b) = (\varphi_4, \varphi_8 + 1) = (3,22)$ *and* $d = \varphi_6 = 8$;
(f) $(a,b) = (2\varphi_4, 2\varphi_8 + 1) = (6,43)$ *and* $d = 2\varphi_6 = 16$.

All these cases are realizable: (a) e.g., by $\{zy^{d-1} = x^d\}$, (b) by $\{(zy-x^2)^{d/2} = xy^{d-1}\}$, or by the parametrization $[z(t) : x(t) : y(t)] = [1 + t^{d-1} : t^{d/2} : t^d]$. The existence of (c) and (d) is guaranteed by the results by Miyanishi-Sugie in [7] by Miyanishi and or by Kashiwara classification [5], Corollary 11.4. These two cases can be realized by a rational pencil of type $(0,1)$: the generic member of the pencil is (c), while the special member of the pencil is of type (d) (cf. also with the last paragraphs of the present article). Orevkov in [9] provides a different construction for curves which realize the case (d) (denoted by him by C_j). Similarly, the cases (e) and (f) are realized by the sporadic cases C_4 and C_4^* of Orevkov [9].

1.1. Remarks

(1) Since C is rational and its singular locus p has Milnor number $\mu = (a-1)(b-1)$, the genus formula says that

$$(a-1)(b-1) = (d-1)(d-2). \tag{1}$$

On the other hand, not any triple (d,a,b) with $(a-1)(b-1) = (d-1)(d-2)$ can be geometrically realized. E.g., $(5,3,7)$ or $(17,6,49)$ cannot.

(2) There are two integers which coordinate the above classification. The first one is defined as follows. Let $\pi : X \to \mathbb{P}^2$ be the minimal good embedded resolution of $C \subset \mathbb{P}^2$, and let $\bar{C}$ be the strict transform of C. Clearly, $(\pi^*C, \bar{C}) = C^2 = d^2$, and $\pi^*C = \bar{C} + abE_{-1} + \cdots$ (where E_{-1} is the unique -1 exceptional curve of π), hence $d^2 = \bar{C}^2 + ab$. Using (1), we get:

$$\begin{cases} a+b = 3d - 1 - \bar{C}^2 \\ ab = d^2 - \bar{C}^2. \end{cases} \tag{2}$$

Then $\bar{C}^2$ in the above cases is as follows: it is positive for (a) and (b), it is zero for (c), equals -1 for (d), and $= -2$ for (e) and (f).

(3) The second guiding integer is the logarithmic Kodaira dimensions $\bar{\kappa} := \bar{\kappa}(\mathbb{P}^2 \backslash C)$ (cf. [4]). Its values are the following (cf. [9]): $-\infty$ for (a)-(d), and 2 for the last two sporadic cases. (In particular, $\bar{\kappa}$ depends only on the integers (d, a, b), and it is independent on the analytic type of C which realizes these integers.)

In particular, the above classification shows that $\bar{\kappa} = -\infty$ if and only if $\bar{C}^2 > -2$.

In fact, after we finished the manuscript, we learned from the introduction of [10] that in [16] (written in Japanese) it is proved that for any unicuspidal rational curve C, $\bar{\kappa} = -\infty$ if and only if $\bar{C}^2 > -2$. Using [16] (i.e., this equivalence), a possible 'quick' classification for $\bar{C}^2 > -2$ would run as follows: Since for all these cases $\bar{\kappa} = -\infty$, we just have to separate in Kashiwara's classification [5] those unicuspidal curves with exactly one Puiseux pair. Their numerical invariants (d, a, b) are exactly those listed in (a)–(d).

On the other hand, this argument probably does not show what is really behind the classification of this case. Therefore, we decided to keep the structure of our manuscript, and provide an independent classification.

Note also that in [2] we list the complete topological classification of the cuspidal rational curves with $\bar{\kappa} < 2$. In fact $\bar{\kappa} = 0$ cannot occur because of a result of Tsunoda's [11], see also Orevkov's paper [9]. Moreover, Tono in [10] provides all the possible curves C with $\bar{\kappa} = 1$: there is no one with one Puiseux pair.

Hence, in our case, the remaining part of the classification corresponds to $\bar{C}^2 \leq -2$, or equivalently, to $\bar{\kappa} = 2$. In general, the classification of this ('general') case is the most difficult; and in our case it is not clear at all at the beginning (and, in fact, it is rather surprising) that there are only two (sporadic) cases satisfying these data.

(4) Let $\alpha = (3 + \sqrt{5})/2$ be the root of $\alpha + \frac{1}{\alpha} = 3$. Notice that in family (d) d/a and b/d asymptotically equals α. In fact, for j odd, $\{\varphi_j / \varphi_{j-2}\}_j$ are the increasing convergents of the continued fraction of α. Using this, another remarkable property of the family (d) can be described as follows (cf. [9], page 658). The convex hull of all the pairs $(m, d) \in \mathbb{Z}^2$ satisfying $m + 1 \leq d < \alpha m$ (cf. with the sharp Orevkov inequality [9], or 2.4) coincides with the convex hull of all pairs (m, d) realizable by rational unicuspidal curves C (where $d = \deg(C)$ and $m = mult(C, p)$) with $\bar{\kappa}(\mathbb{P}^2 \setminus C) = -\infty$; moreover, this convex hull is generated by curves with numerical data (a) and (d).

(5) It is clear that the families (a)–(d) are organized in nice series of curves. It is less clear from the statement of the theorem, but rather clear from the proof, that also (e)–(f) form a 'series': they are the only curves with $3d = 8a$ (cf. also with the next remark).

(6) A hidden message of the classification (and some of the steps of the proof) is that there is an intimate relationship between the semigroup of $\mathbb{N}$ generated by the elements a and b, and the intervals of type $(\,(l-1)d\,, ld\,]$. The endpoints d and $3d$ play crucial roles in some of the arguments. (E.g., $\bar{C}^2 \leq -2$ if and only if

$a + b > 3d$; see also 2.5.) This fact is deeply exploited in [2]. In fact, that paper strongly motivated the present manuscript.

(7) The (part of the) proof in section 4 clearly shows the deficiencies of the known restrictions, bounds which connect the local data (a, b) with the degree d – although we list and try to use a large number of them. On the other hand, the above classification fits perfectly with the conjectured restriction proposed by the authors in [2] (valid in a more general situation), which, in fact, alone would provide the classification.

2. Restrictions and bounds

In the present section we list some general results which impose some restrictions for the integers (d, a, b). We start with a trivial one: (1) and (2) clearly imply:

Lemma 2.1 (The 'trivial' bound.). *In any situation $b \geq d$. Moreover, if $b = d$ then $(a, b) = (d - 1, d)$.*

If $b > d$ then $a < d - 1$. The next result proves a 'gap' for a: if $a < d - 1$ then $a \leq d/2$ too.

Lemma 2.2 (The 'dual curve bound'). *If $b > d$ then $d \geq 2a$ (hence $b > 2a$ too).*

Proof. Let (C, p) be the germ of the singular point p of C, and let $\{m_i\}_i$ be the multiplicity sequence of (C, p). We will use the symbol $^\vee$ for the corresponding invariants of the dual curve $C^\vee$ of C. By a result of C.T.C. Wall [15] Proposition 7.4.5, the blow ups of the singularities (C, p) and $(C^\vee, p^\vee)$ (where $p^\vee$ corresponds to the tangent cone of (C, p)) are equisingular. Assume that $b < 2a$. Then $m_2 = b - a$, hence $m_2^\vee = b - a \leq m_1^\vee$. But, according [15], the intersection multiplicity of the tangent cone of (C, p) with C at p is $i = m_1 + m_1^\vee$, hence $d \geq i = m_1 + m_1^\vee \geq a + b - a = b$, a contradiction. In particular, $b \geq 2a$. In this case $m_2 = a$, hence $m_2^\vee = a$ as well. The above argument gives: $d \geq i \geq m_1 + m_1^\vee \geq 2a$. □

2.1. The semicontinuity of the spectrum

The very existence of the curve C shows that the local plane curve singularity (C, p) is in the deformation of the local plane curve singularity $(U, 0) := (x^d + y^d, 0)$ (see, e.g., [1] (3.24)). In particular, we can use the semicontinuity of the spectrum for this pair [13, 14]. More precisely, this assures that in any interval $(c, c + 1)$, the number of spectral numbers of (C, p) is not larger than the number of spectral numbers of $(U, 0)$. E.g., for the intervals $(-1 + l/d, l/d)$ $(l = 2, 3, \ldots, d)$ one has the following inequality:

$$\#\{\frac{i}{a} + \frac{j}{b} < \frac{l}{d};\ i \geq 1,\ j \geq 1\} \leq 1 + 2 + \cdots + l - 2 = \frac{(l-2)(l-1)}{2}. \qquad (SS_l)$$

Notice that the inequality (SS_d) is automatically satisfied (with equality), since for both singularities the number of spectral numbers strict smaller than 1 is $(d - 1)(d - 2)/2$.

2.2. Example. The inequality (SS_{d-1})

We denote by $\#_{d-1}$ the number of lattice points at the left-hand side of (SS_{d-1}). Since $i/a < (d-1)/d$ and $a < d$, one gets that $1 \leq i \leq a-1$. Therefore,

$$\#_{d-1} = \sum_{i=1}^{a-1} \#\{j \ : \ 1 \leq j < b(\frac{d-1}{d} - \frac{i}{a})\} = \sum_{i=1}^{a-1} \left\lceil b - \frac{b}{d} - \frac{ib}{a} \right\rceil - 1,$$

hence

$$\#_{d-1} = (a-1)(b-1) - \sum_{i=1}^{a-1} \left\lfloor \frac{b}{d} + \frac{ib}{a} \right\rfloor.$$

This expression can be computed explicitly. Indeed, since (a, b) is a lattice point and $\gcd(a, b) = 1$, one has:

$$\sum_{i=1}^{a} \left\lfloor \frac{ib}{a} \right\rfloor = \frac{(a+1)(b+1)}{2} - a,$$

hence

$$\sum_{i=1}^{a} \left\lfloor \frac{ib}{a} + \frac{b}{d} \right\rfloor = \frac{(a+1)(b+1)}{2} - a + a\left\lfloor \frac{b}{d} \right\rfloor + \sum_{i=1}^{a} \left\lfloor \left\{\frac{ib}{a}\right\} + \left\{\frac{b}{d}\right\} \right\rfloor.$$

Notice that the set $\{ib/a\}$ for $i = 1, \ldots, a$ is the same as the set r/a for $r = 0, \ldots, a-1$. Moreover, $r/a + \{b/d\} \geq 1$ if and only if $a - 1 \geq r \geq \lceil a(1 - \{b/d\}) \rceil$, hence the number of possible r's is $\lfloor a\{b/d\} \rfloor$. Therefore,

$$\sum_{i=1}^{a} \left\lfloor \frac{ib}{a} + \frac{b}{d} \right\rfloor = \frac{(a+1)(b+1)}{2} - a + \left\lfloor \frac{ab}{d} \right\rfloor.$$

Hence

$$\sum_{i=1}^{a-1} \left\lfloor \frac{ib}{a} + \frac{b}{d} \right\rfloor = \frac{(a+1)(b+1)}{2} - a - b - \left\lfloor \frac{b}{d} \right\rfloor + \left\lfloor \frac{ab}{d} \right\rfloor,$$

or

$$\#_{d-1} = \frac{(a-1)(b-1)}{2} + \left\lfloor \frac{b}{d} \right\rfloor - \left\lfloor \frac{ab}{d} \right\rfloor,$$

Then, using (1) and (2), (SS_{d-1}) becomes:

$$\left\lfloor \frac{b}{d} \right\rfloor + \left\lceil \frac{\bar{C}^2}{d} \right\rceil \leq 2. \tag{3}$$

2.2.1. Other examples of (SS_l)**.** (SS_2) is equivalent with $1/a + 1/b \geq 2/d$. This is true automatically, since $1/a + 1/b \geq 1/d + 1/(d-1) > 2/d$. The next inequality (SS_3) is equivalent with $2/b + 1/a \geq 3/d$, which also is satisfied automatically.

If $b > d$ then $a + 2b > 3a + b$ (cf. 2.2), hence (SS_4) is equivalent with the pair of inequalities: $a + 2b \geq 4ab/d$ and $4a + b \geq 4ab/d$. Or, via (2):

$$\min\{3a, b\} \geq d + 1 + \frac{d-4}{d}\bar{C}^2.$$

This with an (absolute) lower bound for $\bar{C}^2$ already is interesting: $3a > d+\text{const}$, which has the flavor of the Matsuoka-Sakai inequality $3a > d$ (see 2.3) proved by different methods.

By a similar method as above, one can verify that (SS_{d-2}) is equivalent with:

$$\left\lfloor \frac{2b}{d} \right\rfloor + \left\lceil \frac{2\bar{C}^2}{d} \right\rceil \leq 5.$$

and (SS_{d-3}) is equivalent with:

$$\left\lfloor \frac{3b}{d} \right\rfloor + \left\lfloor \frac{3b}{d} - \frac{b}{a} \right\rfloor + \left\lceil \frac{3\bar{C}^2}{d} \right\rceil \leq 8.$$

In general, one expects that the set of all inequalities (SS_l) is really strong.

2.3. Matsuoka-Sakai inequality

The next set of restrictions are provided by *Bogomolov-Miyaoka-Yau type* inequalities in [6] which in our case reads as $d < 3a$ (valid for any $\bar{\kappa}$).

2.4. Remark. Orevkov's inequality

Orevkov in [9] obtained different improved versions of 2.3. Below $\alpha = (3+\sqrt{5})/2 \approx 2.618$ and $\beta = 1/\sqrt{5}$.

(a) [9] Theorem B(a): If $\bar{\kappa} = -\infty$, then $d < \alpha a$.
(b) [9] Theorem B(b): If $\bar{\kappa} = 2$, then $d < \alpha(a+1) - \beta$.
(c) [9] (2.2)(4): If $\bar{\kappa} = 2$, then

$$-\bar{C}^2 \leq -2 + \frac{a}{b} + \frac{b}{a}. \tag{4}$$

Finally, we end with the following:

2.5. The 'semigroup density property' [2]

Let Γ be the semigroup of (C, p), i.e., the semigroup (with 0) of $\mathbb{N}$ generated by the integers a and b. Then for any $0 \leq l < d$ the following inequality holds:

$$\#\Gamma \cap [0, ld] \geq (l+1)(l+2)/2.$$

Proof. It is instructive to sketch the proof for $l = 3$ case: we wish to prove $\#\Gamma \cap [0, 3d] \geq 10$. Recall that a cubic is determined by 9 parameters. Therefore, $\#\Gamma \cap [0, 3d] \leq 9$ would imply the existence of a cubic with intersection multiplicity with C at p strict greater than $3d$, which contradicts Bézout's theorem. □

In the classical theory, many 'candidates' (d, a, b) were eliminated by different geometric constrictions using ingenious Cremona transformations. We will exemplify this in 4.3.

3. The classification in the case $\bar{C}^2 > 0$

Theorem 3.1. *If $\bar{C}^2 > 0$ then either $(a, b) = (d-1, d)$ or $(a, b) = (d/2, 2d-1)$.*

Proof. Since $b \geq d$ (cf. Lemma 2.1), by (3) we get that $\bar{C}^2 \leq d$. Clearly, equality holds if and only if $(a, b) = (d-1, d)$. Next, assume that $0 < \bar{C}^2 < d$. Then again by (3) one has $\lfloor b/d \rfloor \leq 1$, or $b < 2d$. But notice that $b < 2d-1$ would imply (by (1)) that $a > d/2$ which contradicts Lemma 2.2. Hence $b = 2d-1$. □

4. Classification in the case $\bar{C}^2 \leq -2$

4.0.1. Our first goal is to prove that $3d \geq 8a$. For this we apply 2.5 for $l = 3$. Since $a+b > 3d$ (cf. (2)) and $9a > 3d$ (cf. 2.3), the needed 10 elements of $\Gamma \cap [0, 3d]$ must be $b, 0, a, \ldots, 8a$, hence $8a \leq 3d$.

Corollary 4.1. $\bar{\kappa}(\mathbb{P}^2 \setminus C) = 2$.

Proof. $\bar{\kappa}$ cannot be $-\infty$ because of 2.4(a); cannot be 0 because of [9], Theorem B(c) (see also [11]). Unicuspidal rational curves with $\bar{\kappa} = 1$ are classified by K. Tono [10], the corresponding splice diagrams are listed in [2]: there is no example with one Puiseux pair. □

Now, the classification for $\bar{C}^2 \leq -2$ can be finished in two different ways.

4.1. First proof. Using the computer

The first version is based on the inequality 2.4(b). Notice that in the case of a geometric realization one must have

$$3\alpha(a+1) - 3\beta > 3d \geq 8a,$$

which is true only if $a \leq 44$, (or, by using again $d < \alpha(a+1) - \beta$), only if $d \leq 117$. Hence, we have only to analyze the *finite* family determined by, say, $d \leq 117$. Then, one can search with the computer for 3-uples (d, a, b) verifying all the restrictions considered above. E.g., we used the conditions $d \leq 117$, $\gcd(a, b) = 1$, $a < d < b$, $d < 3a$, $3d \geq 8a$, $2 \leq -\bar{C}^2 \leq -2 + \frac{a}{b} + \frac{b}{a}$, $b < \alpha(d-1)(d-2)/(d-2\alpha+\beta)) + 1$, $(d - \alpha + \beta)/\alpha < a$, and $(SS_{d-1}), (SS_{d-2}), (SS_{d-3}), (SS_{d-4}), (SS_4)$. Using the inequality $3d \geq 8a$ and a similar computation as in the case of (SS_{d-1}), we obtain that (SS_{d-4}) is equivalent with

$$\left\lfloor \frac{4b}{d} \right\rfloor + \left\lfloor \frac{4b}{d} - \frac{b}{a} \right\rfloor + \left\lceil \frac{4\bar{C}^2}{d} \right\rceil \leq 13. \qquad (5)$$

Then the only triplets satisfying all these are listed below (in the list appears $(d, a, b; \bar{C}^2)$):

$$\begin{aligned} C_1 &:= (\ 8, 3, 22; -2), \\ C_2 &:= (11, 4, 31; -3), \\ C_3 &:= (16, 6, 43; -2), \\ C_4 &:= (17, 6, 49; -5), \\ C_5 &:= (19, 7, 52; -3), \\ C_6 &:= (20, 7, 58; -6). \end{aligned}$$

Next, notice that the curves C_1 and C_3 exist, they are listed in our classification theorem. The others do not exist: C_2 is eliminated by Orevkov in [9], page 2 (see also 4.4 (b)); C_4 and C_6 can be excluded by the semicontinuity property of the spectrum (applied for all the intervals of type $(l/d, l/d+1)$, $-d < l < d$), finally C_5 can be eliminated by the 'nodal cubic trick', see Example 4.4 (a). (Notice also that C_2 and C_5 cannot be eliminated by the semicontinuity property.)

4.2. Second proof. Resolving diophantic equations

Next we show how one can analyze the case $3d \geq 8a$ (cf. 4.0.1) by a diophantic equation (for the convenience of the reader, later we will make more precise the geometry behind this equation, cf. 4.2 and 4.3). Our goal is to eliminate everything excepting C_1 and C_3, and to emphasize that C exists if and only if $3d = 8a$, and C_1 and C_3 are the only solutions with $3d = 8a$.

Let us write $x := 3d - 8a \geq 0$. Then clearly $3|a - x$. Moreover,

$$\begin{aligned}
-\bar{C}^2(a-1) &= -(3d-1-a-b)(a-1) \\
&= -(x+7a-1-b)(a-1) \\
&= (b-1)(a-1) - (x+7a-2)(a-1) \\
&= (d-1)(d-2) - (x+7a-2)(a-1).
\end{aligned}$$

Using again $d = (x+8a)/3$ one gets

$$-9\bar{C}^2(a-1) = x^2 + 7ax + a^2 + 9a. \tag{6}$$

4.2.1. The case $x = 0$. (6) implies the divisibility $a - 1|10$. Since one also has $3|a$, the only solutions are $a = 3$ and $a = 6$, corresponding to C_1 and C_3 above.

4.2.2. Facts. *$-\bar{C}^2 \leq 7$ and $x \leq 5$.*

Proof. First we verify $-\bar{C}^2 \leq 7$. It is easy to verify (using (1), (2) and $d/3 < a \leq d/2$, cf. 2.3 and 2.2) that for $6 \leq d \leq 10$ this is true. Hence assume that $d \geq 11$. Notice that if for some (positive) k one has $kd \leq -\bar{C}^2 < (k+1)d$, then (3) gives $b/d \leq 3 + k$. But $d/a < 3$ by 2.3, hence $b/a < 3(3+k)$. Using 2.4(c) one gets $-\bar{C}^2 \leq 3k + 7$. Since for $k > 0$ and $d \geq 11$ one has $3k + 7 < dk$, one should have $k = 0$.

Using this and $x \geq 6$, from (6) we get $63(a-1) \geq 36 + 42a + a^2 + 9a$, which has no solution. □

Now, we consider the above equation (6) for $x \geq 1$. By 4.2.2 we only have to analyze the cases $1 \leq x \leq 5$, and eliminate all the solutions.

The case $x = 1$. In this case one has $-9(\bar{C}^2 + 2)(a-1) = (a-1)^2 + 18$, hence $3|a-1|18$ but $9 \not| a - 1$. In particular, $a = 4$ or 7 corresponding to C_2 and C_5 above.

The case $x = 2$. Similarly as above, $a - 1|28$ and $3|a - 2$, hence $a - 1 = 4, 7$ or 28. In fact, if $a = 5$ then $d = 12$ and $b \notin \mathbb{Z}$. The next case $(d, a, b; \bar{C}^2) = (22, 8, 61; -4)$ can be eliminated by (5); the last $(78, 29, 210; -6)$ by 2.4(c).

The case $x = 3$. Now $a-1|40$ and $3|a$. The possible a's are $a = 3$ which gives $d = 9$ contradicting 2.3; $a = 6$ providing C_4; $a = 9$ providing $(25, 9, 70; -5)$ which can be eliminated by (5), and $a = 21$ providing $(57, 21, 155; -6)$ which is eliminated by 2.4(c).

The case $x = 4$. (6) has two solutions: C_6: $(20, 7, 58; -6)$ and $(28, 10, 79; -6)$, the second one can be eliminated by (5).

The case $x = 5$ provides two integral solutions: $(23, 8, 67; -7)$ and $(31, 11, 88; -7)$. Both can be eliminated by 2.4(c).

We end this section by the description of the promised geometric construction (used also in [9] and by E. Artal-Bartolo as well).

Lemma 4.2 (The existence of a specific nodal cubic.). *There exists a (unique) irreducible cubic $N \subset \mathbb{P}^2$ with a node singularity at p such that N and C share the first seven infinitely near points at p.*

Proof. A cubic is determined by nine parameters. The multiplicity sequence of N at p should be $[2, 1_6]$. Passing through p and having multiplicity 2 provides 3 conditions. The remaining six conditions are imposed by the remaining six infinitely near points. The condition which would imply that the singularity (N, p) is a cusp would involve another equation (the vanishing of the determinant of the quadratic part at p), and the corresponding system of equations would not have any solution. Similar arguments eliminates other type of singularities (two smooth branches with contact two, or (N, p) with multiplicity 3). Hence (N, p) is a node.

Next we prove that N cannot be a product of three linear forms. Indeed, the tangent line L_0 of C at p goes just through the first two infinitely near points because $d < 3a$ and $d = L_0 \cdot C$. Any other line has less tangency than L_0. This also shows that N cannot be $L_0 \cdot Q$ for some Q (transversal to L_0 at p).

The remaining possibility is $N = LQ$ where Q is a smooth conic and L and Q meets transversally at p. Since Q is determined by five conditions (five infinitely near points) then Q and C must be tangent and share the seven infinitely near points at p. In particular by Bezout $2d = Q \cdot C \geq 6a$ which is in contradiction with $d < 3a$, cf. 2.3. □

4.3. The Cremona transformation associated with the nodal cubic N

Consider the nodal cubic N given in Lemma 4.2. First we verify that C and N share exactly the first seven infinitely near points. Indeed, assume that this is not the case. If $b \leq 8a$ then the multiplicity sequence of (C, P) is $[a_7, b-7a, \dots]$, hence $3d \geq 2a+6a+b-7a = a+b = 3d-1-\bar{C}^2 > 3d$, a contradiction. If $b > 8a$ then the multiplicity sequence of (C, P) is $[a_8, \dots]$, hence $3d \geq 9a$ which contradicts 2.3.

In particular, the intersection multiplicity of C and N at P is $8a$. Assume that $C \cap N = \{P, P_1, \dots, P_r\}$. Notice that at P_i $(1 \leq i \leq r)$ both curves C and N are smooth, let k_i be their intersection multiplicity at P_i. By Bezout's theorem one has $3d = 8a + \sum_i k_i$. We prefer to write $x := \sum_i k_i$, hence $3d = 8a + x$ (and the notation is compatible with above).

Blow up the common seven infinitely near points. We get seven irreducible exceptional divisors $\{E_i\}_{i=1}^7$. Let $\tilde{C}$ and $\tilde{N}$ be the strict transforms of C and N. One has the following intersections: $E_1^2 = \cdots = E_6^2 = -2$, $E_7^2 = -1$, $\tilde{N}^2 = -1$, $E_1 \cdot E_2 = E_2 \cdot E_3 = \cdots = E_6 \cdot E_7 = 1$, $E_1 \cdot \tilde{N} = E_7 \cdot \tilde{N} = 1$. Also, $\tilde{C}$ intersects E_7 (but not the other irreducible exceptional divisors) at a point P', and the singularity $(\tilde{C}, P')$ has exactly one Puiseux pairs of type $(b-7a, a)$. The intersection of $\tilde{N}$ with E_7 is not P'.

Consider now the curve $\tilde{N} \cup \cup_{i=1}^6 E_i$. Clearly, this can be blown down, and after this modification π we get another copy of $\mathbb{P}^2$. Let the image of $\tilde{C}$ via this projection be C'. By standard (intersection) argument one gets that the degree d' of C' is

$$d' = 8d - 21a \quad \text{(which also satisfies } 3d' = 8x + a\text{)}.$$

The curve C' has at most two singular points. One candidate is the (isomorphic) image of the germ at P' with one Puiseux pair $(b-7a, a)$. The other is the common image of the points $\{P_i\}$ $(1 \leq i \leq r)$. Clearly, if $x = 0$ then this point does not exist, if $x = 1$ then this is a smooth point, but otherwise it is singular. One can find its embedded resolution graph by blowing up (for each i) k_i times the point P_i. Hence, by A'Campo's formula one can determine its Milnor number, which is $\mu = 7x^2 - 7x - r + 1$ (provided that $x \geq 1$). Since it has r local irreducible components, the delta-invariant is $(7x^2 - 7x)/2$. Then one can verify that (6) corresponds to the genus formula of C'.

4.4. Example

(a) Let us start with $(d, a, b) = (19, 7, 52)$. Then $x = 1$, hence C' is again rational and unicuspidal with $(d', a', b') = (5, 3, 7)$. But such a curve does not exist because of 3.1 (one can also check the classification of rational curves of degree five, e.g., in [8]).

(b) Let us consider now the curve C_2 above with data $(d, a, b) = (11, 4, 31)$. Then $x = 1$, hence C' is rational unicuspidal, say at Q_1, with $(d', a', b') = (4, 3, 4)$. Notice that a curve with this triplet may exists – although C_2 does not. The image $\bar{N}$ under the modification π of the exceptional curve E_7 is a (rational) nodal cubic with a node, say at $Q_2 (\neq Q_1)$. Moreover, $\bar{N} \cdot C' = 4Q_1 + 8Q_2$. At Q_1, $\bar{N}$ is non-singular and with the same tangent as C', and at Q_2 the quartic C' has intersection multiplicity 7 with one of the branches of the node of $\bar{N}$ and 1 with the other. To show that C_2 does not exist we will prove that such configuration of the rational curves C' and $\bar{N}$ in $\mathbb{P}^2$ does not exist.

Choosing affine coordinates we may assume that C' is given by the zero locus of $ay^3 + a_1y^3x + a_2y^2x^2 + a_3yx^3 + x^4 + a_0y^4$; with $a \neq 0$. In such a case $Q_1 = (0, 0)$ and its tangent line $L_1 = \{y = 0\}$ verifies $L_1 \cdot C' = 4Q_1$. The curve C' has a parametrization given by $[z(\lambda, t) : x(\lambda, t) : y(\lambda, t)] = [\lambda^4 + a_3t\lambda^3 + a_2t^2\lambda^2 + a_1t^3\lambda + a_0t^4 : -at^3\lambda : -at^4]$.

To have $I_{Q_1}(\bar{N}, C') = 4$ then $\bar{N}$ must be the zero locus of a polynomial $y + f_2(x,y) + f_3(x,y)$ (see the parametrization of C'), where $f_2(x,y) = m_{1,1}xy + m_{2,0}x^2 + m_{0,2}y^2$ and $f_3(x,y) = n_{1,2}xy^2 + n_{2,1}x^2y + n_{3,0}x^3 + n_{0,3}y^3$.

Next one imposes that, in the affine plane $\mathbb{P}^2 \setminus L_1 = \{y \neq 0\}$, the curves C' and $\bar{N}$ must meet at only one point Q_2 (with intersection multiplicity 8). The parametrization of C' in this affine chart is $(z,x) = (s^4+a_3s^3+a_2s^2+a_1s+a_0, -as)$ and the equation of $\bar{N}$ is given by $z^2 + f_2(x,1)z + f_3(x,1) = 0$. Imposing to have a solution of the form $(As+B)^8$ gives $B = a_3A/4$ which means $s = -a_3/4$. We have two possibilities: firstly, if $a_3 = 0$ then $s = 0$ and $Q_2 = (z,x) = (a_0, 0)$. The solutions are given by $m_{1,1} = 2a_1/a$; $m_{2,0} = -2a_2/a^2$; $m_{0,2} = -(2a_0 - a_2^2)$; $n_{1,2} = (-2a_0+a_2^2)a_1/a$; $n_{2,1} = (a_1^2+2a_0a_2-a_2^3)/a^2$; $n_{3,0} = -2a_1a_2/a^3$; $n_{0,3} = (a_0-a_2^2)a_0$. To have $\bar{N}$ a node at Q_2 implies a_2 vanishes and therefore $\bar{N}$ must be a conic which is a contradiction.

In the other case, i.e., $a_3 \neq 0$ then $s = -a_3/4$ and $Q_2 = (z,x) = (z_0, aa_3/4)$. The solutions are given by:

$$m_{1,1} = (16a_1 - a_3^3)/(8a);$$

$$m_{2,0} = (3/4a_3^2 - 2a_2)/a^2;$$

$$m_{0,2} = -2a_0 + a_2^2 + 19a_3^4/128 - 3a_2a_3^2/4;$$

$$n_{1,2} = -\frac{4096a_0a_1 - 2048a_1a_2^2 - 304a_1a_3^4 + 1536a_1a_2a_3^2 - 256a_0a_3^3 + a_3^7}{2048a};$$

$$n_{2,1} = \frac{2^{11}a_0a_2 + (2^5a_1)^2 - 2^{10}a_2^3 - 152a_2a_3^4 + 768a_2^2a_3^2 - 128a_1a_3^3 - 768a_0a_3^2 + 7a_3^6}{(2^5a)^2};$$

$$n_{3,0} = (-64a_1a_2 + 32a_3a_2^2 + 3a_3^5 - 20a_2a_3^3 + 24a_1a_3^2)/(32a^3);$$

$$n_{0,3} = a_0^2 - a_0a_2^2 - (19/128)a_0a_3^4 + (3/4)a_0a_2a_3^2 + (1/65536)a_3^8.$$

In order $\bar{N}$ to have multiplicity two at Q_2 one needs $a_2 = 3a_3^2/8$ but this condition also impose that the tangent cone of $\bar{N}$ at Q_2 is a double line and therefore Q_2 cannot be a node. Hence this configuration also does not exist.

5. The case $\bar{C}^2 = 0, -1$

In this section we find all the integer solution (d,a,b) of (2) with $\bar{C}^2 = 0, -1$ and we show that all of them can be realized by some unicuspidal rational plane curve of degree d and Puiseux pair (a,b). Let φ_j be the ith Fibonacci number, that is $\varphi_0 = 0$, $\varphi_1 = 1$ and $\varphi_{j+2} := \varphi_{j+1} + \varphi_j$. They share many interesting properties, see, e.g., [12]. We will use here the following:

$$3\varphi_j = \varphi_{j-2} + \varphi_{j+2}, \quad \text{and} \quad \varphi_j^2 = (-1)^{j+1} + \varphi_{j-1}\varphi_{j+1}. \tag{7}$$

Let $\Phi = \frac{1+\sqrt{5}}{2}$ be the positive solution of the equation $\Phi^2 - \Phi - 1 = 0$. For every integer $j > 0$ one has:

$$\Phi^j = \frac{\varphi_{j+1} + \varphi_{j-1} + \varphi_j\sqrt{5}}{2}. \tag{8}$$

5.1. The Pell equation

The system of equations (2) for $\bar{C}^2 = 0, -1$ can be transformed (see below) into the *Pell equation*:

$$x^2 - 5y^2 = -4,\ x, y \in \mathbb{Z}. \tag{9}$$

Consider the number field $K = \mathbb{Q}[\sqrt{5}]$ and its ring of integers $R = \mathbb{Z}[\sqrt{5}]$, which is a UFD. If $\gamma = x + y\sqrt{5}$ is a solution of (9) then its norm is $N_K(\gamma) = \gamma\bar{\gamma} = -4$. Consider $\eta = 1 + \sqrt{5}$, then $N_K(\eta) = -4$ and -4 has a prime decomposition $-4 = \eta\bar{\eta}$. Since the fundamental unit of K turns out to be $u = 2 + \sqrt{5}$ and γ is associated either to η or $\bar{\eta}$ then γ is either $\pm u^r\eta$ or $\pm\bar{u}^r\bar{\eta}$ (since $\bar{u} = -1/u$) for $r \in \mathbb{Z}$. Moreover $N_K(u) = -1$ which implies that r must be even, that is $r = 2j$ for $j \in \mathbb{Z}$. Then $\eta = 2\Phi$ and from the identity $\Phi^2 = \Phi + 1$ one gets $\Phi^3 = u$.

Thus solutions of (9) are either $\gamma = \pm u^{2j}\eta = \pm 2\Phi^{6j+1}$ or $\gamma = \pm\bar{u}^{2j}\bar{\eta} = \pm 2\bar{\Phi}^{6j+1}$ with $j \in \mathbb{Z}$. Using $\Phi\bar{\Phi} = -1$, γ is either $\pm 2\Phi^{6j+1}$, $\pm 2\Phi^{6j-1}$, or their conjugates $\pm 2\bar{\Phi}^{6j+1}, \pm 2\bar{\Phi}^{6j-1}$ with $j \geq 0$.

Using (7) and (8) the set of solutions of (9) is given by

(A) $\pm\left(\varphi_{6j+2} + \varphi_{6j} + \varphi_{6j+1}\sqrt{5}\right)$, with $j \geq 0$,

(B) $\pm\left(\varphi_{6j} + \varphi_{6j-2} + \varphi_{6j-1}\sqrt{5}\right)$, with $j \geq 0$,

(C) $\pm\left(\varphi_{6j+2} + \varphi_{6j} - \varphi_{6j+1}\sqrt{5}\right)$, with $j \geq 0$,

(D) $\pm\left(\varphi_{6j} + \varphi_{6j-2} - \varphi_{6j-1}\sqrt{5}\right)$, with $j \geq 0$.

5.2. The case $\bar{C}^2 = 0$

Since $\gcd(a, b) = 1$ and $ab = d^2$ then $a = m^2, b = n^2$ and $d = mn$ for some positive integers m, n with $\gcd(m, n) = 1$. Thus $a + b = 3d - 1$ transforms into

$$m^2 + n^2 = 3mn - 1. \tag{10}$$

5.3. The case $\bar{C}^2 = -1$

The system (2) provides the equation

$$a^2 + d^2 = 3ad - 1. \tag{11}$$

Thus, any solution (ω, v) of $\omega^2 + v^2 = 3\omega v - 1$ is a solution of $(2\omega - 3v)^2 - 5v^2 = -4$. Hence, with the transformation $x = 2\omega - 3v$, $y = v$, one gets the solutions of (9).

Case A. If $\gamma = \pm\left(\varphi_{6j+2} + \varphi_{6j} + \varphi_{6j+1}\sqrt{5}\right)$, $j \geq 0$, is a solution of (9) then $v = \pm\varphi_{6j+1}$ and $\omega = \pm(\varphi_{6j+2} + \varphi_{6j} + 3\varphi_{6j+1})/2 = \pm\varphi_{6j+3}$ is a solution of (10) and (11) (for the last equality use (7)). Since $1 < a < d$, if $\bar{C}^2 = -1$, then $a = \varphi_{6j+1}$, $d = \varphi_{6j+3}$ and $b = 3d - a = 3\varphi_{6j+3} - \varphi_{6j+1} = \varphi_{6j+5}$ for some $j > 0$, by property (7) of Fibonacci numbers. Similarly, if $\bar{C}^2 = 0$, then ω and v are both either positive or negative which implies $a = \varphi_{6j+1}^2$, $b = \varphi_{6j+3}^2$ and $d = \omega v = \varphi_{6j+1}\varphi_{6j+3} = \varphi_{6j+2}^2 + 1$.

Case B. If $\gamma = \pm\left(\varphi_{6j} + \varphi_{6j-2} - \varphi_{6j-1}\sqrt{5}\right)$, $j \geq 0$, is a solution of (9) then $v = \pm(-\varphi_{6j-1})$ and $\omega = \pm(\varphi_{6j} + \varphi_{6j-2} - 3\varphi_{6j-1})/2 = \pm(-\varphi_{6j-3})$ is a solution of (10) and (11). In the case $\bar{C}^2 = -1$, one gets $a = \varphi_{6j-3}$, $d = \varphi_{6j-1}$ and $b =$

$3d - a = 3\varphi_{6j-1} - \varphi_{6j-3} = \varphi_{6j+1}$ with $j > 0$. If $\bar{C}^2 = 0$, then ω and v are both either positive or negative which implies $a = \varphi_{6j-3}^2$, $b = \varphi_{6j-1}^2$ and $d = \omega v = \varphi_{6j-1}\varphi_{6j-3} = \varphi_{6j-2}^2 + 1$ with $j > 0$.

Case C. If $\gamma = \pm\left(\varphi_{6j+2} + \varphi_{6j} - \varphi_{6j+1}\sqrt{5}\right)$, $j \geq 0$, is a solution of (9) then $v = \pm(-\varphi_{6j+1})$ and $\omega = \pm(\varphi_{6j+2} + \varphi_{6j} - 3\varphi_{6j+1})/2 = \pm(-\varphi_{6j-1})$ is a solution of (10) and (11). If $\bar{C}^2 = -1$, then $a = \varphi_{6j-1}$, $d = \varphi_{6j+1}$ and $b = \varphi_{6j+3}$ with $j > 0$. If $\bar{C}^2 = 0$, then ω and v are both either positive or negative which implies $a = \varphi_{6j-1}^2$, $b = \varphi_{6j+1}^2$ and $d = \varphi_{6j-1}\varphi_{6j+1} = \varphi_{6j}^2 + 1$ with $j > 0$.

Case D. Any solution in this case is included in the previous cases.

Hence, we determined all the possible integer solutions.

Theorem 5.1 (Classification for $\bar{C}^2 = -1$). *If $\bar{C}^2 = -1$ then $(a, b) = (\varphi_{j-2}, \varphi_{j+2})$ and $d = \varphi_j$, with j odd ≥ 5. For every such j there exists a unicuspidal rational plane curve of degree with such invariants.*

Theorem 5.2 (Classification for $\bar{C}^2 = 0$). *If $\bar{C}^2 = 0$ then $(a, b) = (\varphi_{j-2}^2, \varphi_j^2)$ and $d = \varphi_{j-1}^2 + 1$, with j odd ≥ 5. For every such j there exists a unicuspidal rational plane curve with such invariants.*

Proof. We only need to give equations for such curves. We will rely on [5], Corollary 11.4. Let (x, y) be a system of affine coordinates in $\mathbb{P}^2$ and consider

$$P_{-1} = y - x^2,\ Q_{-1} = y,\ P_0 = (y - x^2)^2 - 2xy^2(y - x^2) + y^5,$$

$$Q_0 = y - x^2,\ G = xy - x^3 - y^3,\ Q_s = P_{s-1},\ P_s = \left(G^{\varphi_{2s+1}} + Q_s^3\right)/Q_{s-1}.$$

Then P_s is a polynomial in x and y of degree φ_{2s+3} and defines a rational unicuspidal curve whose unique singularity p has exactly one characteristic pair of type $(a, b) = (\varphi_{2s+1}, \varphi_{2s+5})$. The curves $P_s = 0$ and $Q_s = 0$ only meet at p. The rational pencil with only one base point determined by the rational function $R_s = (P_s)^{\varphi_{2s+1}}/(Q_s)^{\varphi_{2s+3}}$ has only two special fibres $P_s = 0$ and $Q_s = 0$, and the other fibres are rational unicuspidal plane curves of degree $\varphi_{2s+3}\varphi_{2s+1} = \varphi_{2s+2}^2 + 1$. The singularity of a generic fiber has one characteristic pair $(a, b) = (\varphi_{2s+1}^2, \varphi_{2s+3}^2)$. □

Acknowledgment

The authors thanks Maria Aparecida Ruas and Jean-Paul Brasselet for the wonderfull atmosphere in the "VIIIéme Rencontre Internationale de São Carlos sur les singularités reélles et complexes au CIRM". We also thank J. Stevens for pointing us out the full strength of the semicontinuity of the spectrum to eliminate some cases in the classification. Finally we thank David Lehavi who wrote for the fourth author some computer programs to help in the classification.

References

[1] Dimca, A.: *Singularities and Topology of Hypersurfaces*, Universitext, Springer-Verlag, New York, 1992.

[2] Fernández de Bobadilla J., Luengo I., Melle-Hernández A., Némethi A.: On rational cuspidal projective plane curves, *Proc. London Math. Soc.* (3), **92**, (2006), 99–138.

[3] Fernández de Bobadilla J., Luengo I., Melle-Hernández A., Némethi A.: On rational cuspidal curves, open surfaces and local singularities, arXiv.math.AG/0604421.

[4] Fujita, T.: On the topology of non-complete algebraic surfaces, *J. Fac. Sci. Univ. Tokyo (Ser1A)*, **29** (1982), 503–566.

[5] Kashiwara, H.: Fonctions rationelles de type (0,1) sur le plan projectif complexe, *Osaka J. Math.*, **24** (1987), 521–577.

[6] Matsuoka, T. and Sakai, F.: The degree of of rational cuspidal curves, *Math. Ann.*, **285** (1989), 233–247.

[7] Miyanishi, M. and Sugie, T.: On a projective plane curve whose complement has logarithmic Kodaira dimension $-\infty$, *Osaka J. Math.* , **18** (1981), 1–11.

[8] Namba, M.: *Geometry of projective algebraic curves.* Monographs and Textbooks in Pure and Applied Mathematics, **88** Marcel Dekker, Inc., New York, 1984.

[9] Orevkov, S. Yu.: On rational cuspidal curves, I. Sharp estimate for degree via multiplicities, *Math. Ann.* **324** (2002), 657–673.

[10] Tono, K.: On rational unicuspidal plane curves with logarithmic Kodaira dimension one, preprint.

[11] Tsunoda, Sh.: The complements of projective plane curves, *RIMS-Kôkyûroku*, **446** (1981), 48–56.

[12] Vajda, S.: *Fibonacci and Lucas numbers, and the Golden Section: Theory and Applications*, Ellis Horwood Series: Mathematics and its Applications. Ellis Horwood Ltd., Chichester; Halsted Press New York, 1989.

[13] Varchenko, A.N.: On the change of discrete characteristics of critical points of functions under deformations, *Uspekhi Mat. Nauk,* **38:5** (1985), 126–127.

[14] Varchenko, A.N.: Asymptotics of integrals and Hodge structures. Science reviews: current problems in mathematics 1983. **22**, 130–166; *J. Sov. Math.* **27** (1984).

[15] Wall, C.T.C.: *Singular Points of Plane Curves*, London Math. Soc. Student Texts **63**, Cambridge University Press, 2004.

[16] Yoshihara, Y.: Rational curves with one cusp (in Japanese), *Sugaku*, **40** (1988), 269–271.

J. Fernández de Bobadilla
Department of Mathematics
University of Utrecht,
Postbus 80010,
NL-3508TA Utrecht, The Netherlands
e-mail: `bobadilla@math.uu.nl`

I. Luengo and A. Melle Hernández
Facultad de Matemáticas
Universidad Complutense
Plaza de Ciencias
E-28040, Madrid, Spain
e-mail: iluengo@mat.ucm.es
e-mail: amelle@mat.ucm.es

A. Némethi
Department of Mathematics
Ohio State University
Columbus, OH 43210, USA

and

Rényi Institute of Mathematics
Budapest, Hungary
e-mail: nemethi@math.ohio-state.edu; nemethi@renyi.hu

Real and Complex Singularities
Trends in Mathematics, 47–51

Bounding from below the Degree of an Algebraic One-dimensional Foliation Having a Prescribed Algebraic Solution

Vincent Cavalier and Daniel Lehmann

In this paper, we summarize [CaLe2] without proof.

Let $\mathcal{F}$ be a one-dimensional algebraic foliation of degree d over the complex projective space $\mathbb{P}_n$: this means (see for instance [LiSo]), that $\mathcal{F}$ is given by a morphism $\ell : \mathcal{O}(1-d) \to T\mathbb{P}_n$ $(d \geq 0)$ of holomorphic vector-bundles, where $\mathcal{O}(k)$ denotes as usually the $|k|^{th}$ tensorial power of the tautological holomorphic line-bundle over $\mathbb{P}_n$ when k is negative (resp. of its dual when k is positive), and $T\mathbb{P}_n$ the complex tangent space to $\mathbb{P}_n$. [Notice that, for $d < 0$, there is no morphism but 0 from $\mathcal{O}(1-d)$ into $T\mathbb{P}_n$.] The singularities of $\mathcal{F}$ are the points $m \in M$ where ℓ_m vanishes. Let Γ be a (compact connected) algebraic curve of degree δ in $\mathbb{P}_n$, invariant by $\mathcal{F}$ ("invariant" means that, over the regular part Γ_0 of Γ, the restriction of ℓ factorizes through $T\Gamma_0$). It is not possible in general, as is well known, to bound from above δ in function of d, without further assumptions on F or on Γ.

However, for $n = 2$, the inequality $d + 2 - \delta \geq 0$ has been proved

- by Cerveau and Lins-Neto ([CeLi] when Γ has only nodal singularities (see also [So3]),
- by Carnicer ([C]), when the foliation has only non-dicritical singularity (in dimension 2, a singularity is said to be "non-dicritical" if there is only a finite number of separatrices through it).

Moreover, Brunella ([B]) recovered Carnicer's result by observing that the negativity of the GSV-indices (see [GSV]) is an obstruction to the above inequality, and proving that these indices are always non-negative in the non-dicritical case. Carnicer and Campillo ([CC]) proved also that there exists some non-negative integer a, depending on conditions imposed to $\mathcal{F}$ or to Γ, such that $d + 2 - \delta \geq -a$.

In higher dimension n, the inequality $\left(d+n-\sum_{\lambda=1}^{n-1} \delta_\lambda\right) \geq 1$ has been proved by Soares ([So2]), when Γ is the complete intersection $\bigcap_{\lambda=1}^{n-1} S_\lambda$ of $n-1$ algebraic hypersurfaces S_λ of degree δ_λ, under the further conditions that Γ be smooth, and the restriction of the foliation to Γ be non-degenerate. (He gave more generally in [So1][So2] a lower bound for the degree of the algebraic foliations leaving invariant a smooth submanifold of $\mathbb{P}_n$, under conditions of non-degeneracy of the foliation). Also, the inequality $(d-1)(\delta-1)-2g \geq 1-r(\Gamma)$, has been proved by Esteves and Kleiman ([EK]), g denoting the geometrical genus of Γ and $r(\Gamma)$ the number of its globally irreducible components.

We consider in this paper the case of curves with any kind of singularity, in any dimension. The normal bundle N_{Γ_0} to the non-singular part Γ_0 of Γ in M has a stable class which always admits a natural extension $[N_\Gamma]$ in the Grothendieck group $K^0(\Gamma)$. If Γ is moreover a locally complete intersection (LCI) in M, $[N_\Gamma]$ may even be realized as the stable class of a natural bundle N_Γ which is a natural extension of N_{Γ_0} to all of Γ. (See for instance [LS1][LSS] for the LCI case, and [CaLeSo1][CaLeSo2] in general). Denote by $\Sigma = \mathrm{Sing}\Gamma \cup (\mathrm{Sing}\mathcal{F} \cap \Gamma)$ the union (made of isolated points) of the singular part of Γ with the set of singular points of $\mathcal{F}$ which are in Γ. To each point m_α in Σ, we can associate an integer $GSV_{m_\alpha}(\mathcal{F},\Gamma)$ (defined in [CaLe2], but already in [LS1][LSS] for LCI's and in [CaLeSo2] in the general case, under the notation $BB(c_n,\mathcal{F}_\Gamma)$), generalizing the index of GomezMont-Seade-Verjovski ([GSV]), such that we get:

Theorem 1.

(i) *The following formula holds:*

$$(d+n)\delta - \Big(c_1([N_\Gamma]) \frown [\Gamma]\Big) = \sum_\alpha GSV_{m_\alpha}(\mathcal{F},\Gamma).$$

(ii) *The following inequality holds:*

$$\sum_\alpha GSV_{m_\alpha}(\mathcal{F},\Gamma) \geq \mathcal{B}(\Gamma) - \mathcal{E}(\Gamma),$$

- *where $\mathcal{B}(\Gamma)$ denotes the total number of locally irreducible branches through singular points of Γ when Γ has singularities, and $\mathcal{B}(\Gamma)=1$ (instead of 0) when Γ is smooth,*
- *and $\mathcal{E}(\Gamma) = 2-2g+\Big(c_1([N_\Gamma]) \frown [\Gamma]\Big)-(n+1)\delta$ denotes the correction term in the genus formula (g being the geometrical genus of Γ).*

Equivalently, we get $(d-1)\delta+2-2g \geq \mathcal{B}(\Gamma)$.

When $\Gamma = \bigcap_{\lambda=1}^{n-1} S_\lambda$ is the complete intersection (no more necessarily smooth) of $n-1$ algebraic hyper-surfaces S_λ in $\mathbb{P}_n$ of respective degree δ_λ $(1 \leq \lambda \leq n-1)$, we get the

Corollary. *The following inequality holds:*

$$\Big(d+n-\sum_{\lambda=1}^{n-1}\delta_\lambda\Big)\delta \geq \mathcal{B}(\Gamma)-\mathcal{E}(\Gamma),$$

and in particular: $(d+2-\delta)\delta \geq \mathcal{B}(\Gamma)-\mathcal{E}(\Gamma)$ for $n=2$.

Let $\pi : \Gamma' \to \Gamma$ be the normalisation of Γ: this means that Γ' is a non-singular complex curve, that the composition $\hat{\pi} : \Gamma' \to M$ of π with the the natural inclusion $\Gamma \hookrightarrow M$ is holomorphic, and that the restriction of π to $\pi^{-1}(\Gamma_0)$ is a biholomorphism $\pi^{-1}(\Gamma_0) \to \Gamma_0$. When Γ has singularities, the number $\mathcal{B}(\Gamma)$ above is still equal to the number of points in $\pi^{-1}\big(\mathrm{Sing}(\Gamma)\big)$.

When Γ is reducible, we can refine Theorem 1 as follows. For any irreducible component C of Γ, denote more generally:

- by δ_C its degree,
- by g_C its geometrical genus,
- by $\mathcal{B}_C(\Gamma)$ the number of points in $\pi^{-1}\big(\mathrm{Sing}(\Gamma)\big) \cap C'$, where $C' = \pi^{-1}(C)$,
- and set: $\mathcal{E}_C(\Gamma) = 2 - 2g_C + \Big(c_1([N_\Gamma]) \frown [C]\Big) - (n+1)\delta_C$.

Theorem 2. *For any irreducible component C of Γ, the following inequality holds:*

$$(d+n)\delta_C - \Big(c_1([N_\Gamma]) \frown [C]\Big) \geq \mathcal{B}_C(\Gamma) - \mathcal{E}_C(\Gamma),$$

or equivalently: $(d-1)\delta_C + 2 - 2g_C \geq \mathcal{B}_C(\Gamma)$. In particular, for complete intersections, we get, with the same notations as above:

$$\Big(d + n - \sum_{\lambda=1}^{n-1} \delta_\lambda\Big)\delta_C \geq \mathcal{B}_C(\Gamma) - \mathcal{E}_C(\Gamma).$$

Examples of applications

A: Case n=2

1) If Γ has an irreducible component C such that all singularities of Γ which are in C are non-dicritical, then $d + 2 - \delta \geq 0$. This uses a slight improvement (given in [CaLe1]) of the positivity of GSV proved in [B], and refines the result of [C] when Γ is reducible.

2) We shall say that some r-multiple point of Γ ($r \geq 2$) is "elementary", when the r local branches through this point are all smooth and have distinct tangents. Assume that Γ has only elementary singularities, and let n_r be the number of r-multiple points. Then, the following formulae hold:

$$(d-1)\delta + 2 - 2g \geq \sum_{r\geq 2} n_r r, \quad \textit{and} \quad (d+2-\delta)\delta \geq -\sum_{r\geq 2} n_r r(r-2).$$

More generally, if Γ is reducible and if there is an irreducible component C such that all singularities of Γ which are in C are elementary singularities of Γ, denote by $n_r(C)$ the total number of local branches included into C through singular r-multiple points of Γ. Then, the following formula holds:

$$(d-1)\delta_C + 2 - 2g_C \geq \sum_{r\geq 2} n_r(C), \quad \textit{and} \quad (d+2-\delta)\delta_C \geq -\sum_{r\geq 2} n_r(C)(r-2).$$

If $n_r = 0$ when $r \neq 2$, we recover the result of [CeLi][So3], and refine it in the reducible case when there exists some irreducible component C such that $n_r(C) = 0$ for $r \neq 2$.

3) Let Γ_1 be a non-degenerate conic in $\mathbb{P}_2$, m_0 be a point of Γ_1, and $(\Delta_i)_i$ a family of $\delta - 2$ projective straight lines Δ_i through m_0 in the plane ($\delta \geq 3$), none of them being tangent to Γ_1. Taking for Γ the union of this conic and of these straight lines, all singularities of Γ are nodal (double point), except m_0 which is a $\delta - 1$-uple point, and the degree of the curve is δ. We get:

$d \geq 1$ when applying Theorem 2 to Δ_i,

$d \geq \frac{\delta-1}{2}$ when applying Theorem 2 to the conic,

and $d \geq 2 - \frac{3}{\delta}$ when applying Theorem 1.

The strongest of these inequalities is of course the second one.

B: Higher dimension

1) Assume again that that $\Gamma = \bigcap_{\lambda=1}^{n-1} S_\lambda$ be a complete intersection, with the same notations as above. Assume moreover that there is an irreducible component C of Γ, which is smooth. We then get:

$$\Big(d + n - \sum_{\lambda=1}^{n-1} \delta_\lambda\Big)\delta_C \geq \mathcal{B}_C(\Gamma).$$

Since $\mathcal{B}_C(\Gamma)$ is always greater or equal to 1, we recover the result of [So2], and refine it in the reducible case, or in the case of degenerate singularities.

2) Let Γ be the (non-LCI) rational quintic parametrized by the map

$$[u, v] \mapsto [X(u, v) = u^3v^2, Y(u, v) = u^4v, Z(u, v) = u^5, T(u, v) = v^5]$$

from $\mathbb{P}_1$ (with homogeneous coordinates $[u, v]$) into $\mathbb{P}_3$ (with homogeneous coordinates $[X, Y, Z, T]$). It has only the origin for singular point with one local branch at this point, hence $\mathcal{B}(\Gamma) = 1$. According to [CaLeSo1], $c_1(N_\Gamma) \frown [\Gamma] = 21$, hence $\mathcal{E}(\Gamma) = 3$. The lower-bound 1 of d is in fact reached, since Γ is invariant by the foliation of degree 1 defined by the vector field $3x\frac{\partial}{\partial x} + 4y\frac{\partial}{\partial y} + 5z\frac{\partial}{\partial z}$. Notice that this foliation has dicritical singularities. We can prove that the minimal degree of the foliations without dicritical singularities leaving this quintic invariant is 2.

References

[B] M. Brunella, *Some remarks on indices of holomorphic vector fields*, Publ. Mat. 41-2, 1997, 527–544.

[CC] A. Campillo, and M. Carnicer *Proximity inequalities and bounds for the degree of invariant curves by foliations of* $\mathbb{P}^2_{\mathbb{C}}$, Transactions of the American Mathematical Society, 349-6, 1997, 2211–2228.

[C] M. Carnicer, *The Poincaré problem in the nondicritical case*, Annals of Mathematics, 140, 1994, 289–294.

[CaLe1] V. Cavalier et D. Lehmann, *Localisation des résidus de Baum-Bott, Courbes généralisées et K-théorie. Première partie: Feuilletages dans* $\mathbf{C}^2$, Commentarii Math. Helv. 76, 2001, 1–19.

[CaLe2] V. Cavalier et D. Lehmann, *On the Poincaré inequality for one-dimensional foliations*, Compositio Math. 142 (2006) 529–540.

[CaLeSo1] V. Cavalier, D. Lehmann et M.G. Soares, *Classes de Chern des ensembles analytiques*, C.R. Acad. Sci. Paris, Ser. I, 338 (2004), 879–884.

[CaLeSo2] V. Cavalier, D. Lehmann et M.G. Soares, *Chern classes of analytical sets and applications*, K-theory 34 (2005) 35–69.

[CeLi] D. Cerveau and A. Lins Neto, *Holomorphic foliations in CP(2) having an invariant algebraic curve*, Ann. Inst. Fourier, Grenoble, 41-4, 1991, 883–903.

[EK] E. Esteves and S. Kleiman, *Bounds on leaves of one-dimensional foliations*, Bull. Braz. Mat. Soc. (NS) 34/1, (2003), 145–169.

[GSV] X. Gomez-Mont, J. Seade, and A. Verjovski, *The index of a holomorphic flow with an isolated singularity*, Math. Ann. 291, (1991), 737–751.

[LS1] D. Lehmann and T. Suwa, *Residues of holomorphic vector fields relative to singular invariant subvarieties*, J. Differential Geometry, 42-1, 1995, 367–384.

[LS2] D. Lehmann and T. Suwa, *Generalization of variations and Baum-Bott residues for holomorphic foliations on singular varieties*, International J. of Mathematics, 10-3, 1999, 367–384.

[LSS] D. Lehmann, M.G. Soares and T. Suwa, *On the index of a holomorphic vector field tangent to a singular variety*, Bol. Soc. Bras. Mat., vol 26, 1995, 183–199.

[So1] M.G. Soares, *The Poincaré's problem for hypersurfaces invariant by one-dimensional foliations*, Inventiones, 1997.

[So2] M.G. Soares, *Projective varieties invariant by one-dimensional foliations*, Annals of Mathematics, 152, 2000, 369–382.

[So3] M.G. Soares, *On the Geometry of Poincaré's problem for one-dimensional projective foliations*, An. Acad. Brasil Ciênc., 73-4, 2001, 475–482.

[Su] T. Suwa, *Indices of holomorphic vector fields relative to invariant curves on surfaces*, Proc. of the Am. Math. Soc. 123, 1995, 2989–2997.

Vincent Cavalier and Daniel Lehmann
Département des Sciences Mathématiques
CP 051, Université de Montpellier II
Place Eugène Bataillon
F-34095 Montpellier Cedex 5, France
e-mail: cavalier@math.univ-montp2.fr
e-mail: lehmann@math.univ-montp2.fr

Real and Complex Singularities
Trends in Mathematics, 53–70

Trajectory Singularities for a Class of Parallel Motions

Matthew W. Cocke, Peter Donelan and Christopher G. Gibson

Abstract. A rigid body, three of whose points are constrained to move on the coordinate planes, has three degrees of freedom. Bottema and Roth [2] showed that there is a point whose trajectory is a solid tetrahedron, the vertices representing corank 3 singularities. A theorem of Gibson and Hobbs [9] implies that, for general 3-parameter motions, such singularities cannot occur generically. However motions subject to this kind of constraint arise as interesting examples of parallel motions in robotics and we show that, within this class, such singularities can occur stably.

Mathematics Subject Classification (2000). Primary 70B15; Secondary 57R45.

Keywords. Screw system, trajectory singularity, parallel motion.

1. Introduction

The presence of singularities in the motion of a robot manipulator presents both difficulties in terms of control and potential advantages in task performance. Understanding the singularity types that arise is valuable therefore for optimal manipulator design. While it is true that engineers frequently use special design geometry to bring about specific outcomes, nevertheless stable phenomena are still desirable if one is to allow some tolerance in the manufacture and operation of a manipulator. In this paper, the authors explore the occurrence of certain highly degenerate singularities in a class of parallel manipulators from the perspective of stability.

The International Organisation for Standardisation (ISO) [13] defines a *manipulator* to be:

> a machine, the mechanism of which usually consists of a series of segments, jointed or sliding relative to one another, for the purpose of grasping and/or moving objects (pieces or tools) usually in several degrees of freedom. It may be controlled by an operator, a programmable electronic controller, or any logic system (for example cam device, wired, etc.)

Of special interest is the motion of the *end-effector*, the grasping component or component to which the operative tool is attached, which may be represented as a function of the joint variables.

From the point of view of kinematics we are interested in the geometry of the manipulator, rather than its control. The class of *parallel* manipulators is distinguished from *serial* manipulators in that they include closed chains of linked or jointed components. Characteristic advantages of parallel manipulators are that they generally can handle heavier payloads and have greater accuracy. Their inverse kinematics – determining joint variables from knowledge of the configuration of the end-effector – is generally easier than their forward kinematics.

In [6], the authors showed that trajectory singularities of tracing points in the end-effector of a manipulator form *instantaneous singular sets* whose geometry is determined by the screw system of the manipulator at that configuration. The Hunt–Gibson class of screw system [10] also provides first-order information, i.e., the corank, for the singular trajectories.

In this paper, the implications are explored for a particular class of parallel manipulators, namely those for which a given set of k points is each constrained to lie on a surface, in particular, the case $k = 3$. We will call such a device a *k-point mechanism*. The contact of one point of a rigid body with the surface of another constitutes the simplest kind of joint. Although, in the terminology of mechanisms, it is a higher pair, in many cases it is possible to synthesize the resulting motion using lower pairs (joints with contacting surfaces). Each joint of this kind imposes, in general, one constraint, or the loss of one degree of freedom, on the moving body.

The motivating example is due initially to Darboux. He considered the 3 degree-of-freedom rigid-body motion, generated by three points in the body being constrained to lie in three planes in general position, and whether a fourth point could move in a plane, which he answered in the negative. Bottema and Roth [2] analyzed this motion more carefully in the case that the planes are mutually orthogonal. They showed that if the triangle formed by the constrained points is acute, then there is a point in the body whose trajectory is a solid tetrahedron. Such a situation arises when the normals to the constraining planes at the points of contact mutually intersect. The point of intersection in the moving body lies at the vertex of its tetrahedral trajectory and is at a corank 3 singularity.

Two practical manipulators that incorporate 3-point mechanisms are:

Remote Centre Compliance Device: a device attached to a robot-arm end-effector to facilitate peg-in-hole insertion tasks. The device was invented by Watson, Nevins and Whitney [14, 16, 17]. It is not actively controlled but passively responsive to forces and torques at the end-effector tip, where the peg is held. In a simplified form, three rigid rods of equal length connect the vertices of an equilateral triangle in the base to a similar smaller triangle in the the end-effector, by means of ball joints. Hence, the three joints in the end-effector are effectively constrained to move on the surfaces of spheres. In its relaxed configuration the axes of the rods intersect at the end-effector tip. This component handles rotation of the peg in response to torques.

HVRam mechanism: designed for control of telescope mirror focussing. The device is analyzed in detail by Carretero *et al.* [3, 4]. Three hydraulically extensible (P joint) arms in the plane of the base are each connected by revolute (R) joints, with axes perpendicular to the arms in the same plane, to legs of fixed and equal length. These in turn are connected to the mirror by ball (S) joints, thus forming a 3-PRS architecture. The optimal positioning of the component arms is the subject of the second paper, but in its simplest form, they are symmetrically placed at angles of $2\pi/3$. The three joints in the mirror (end-effector) are constrained to move on three planes which intersect in the focal axis of the mirror when in its home configuration (mirror parallel to the base plane). The associated singularities mean that the tracking motion of the mirror is an order of magnitude smaller than the input through extension of the hydraulic legs, leading to superior control.

Section 2 of this paper describes the mathematical formulation for analyzing rigid-body motions, together with basic results on screw systems and instantaneous singular sets, including the Genericity Theorem. Section 3 establishes the relationship between the configuration of contact normals for a 3-point mechanism and the associated screw system. The key result in section 4 is to establish conditions under which high-corank singularities appear stably. Although the theorem disproves a natural genericity hypothesis for this class of parallel motions, the phenomenon explains why such motions provide a valuable class from the point of view of mechanical advantage and control. Application of the result to the examples above appears in Section 5.

2. Motions, screw systems and ISSs

2.1. Motions and trajectories

We shall restrict our attention to those spatial motions for which the underlying joint space, encoding all the feasible combinations of joint variables for a given manipulator, is a smooth manifold. In practice, for most design geometries, the joint space is in fact an algebraic variety, and is smooth for almost all choices of design parameters (e.g., component lengths). Let $SE(3)$ denote the Euclidean isometry group $SE(3)$, combining rotations and translations in the semi-direct product $SO(3) \ltimes \mathbb{R}^3$. It is a 6-dimensional Lie group. By assigning orthonormal coordinate frames to the rigid body (moving coordinates) and the ambient space (fixed coordinates), configurations of the body can be represented by elements of the group.

Definition 2.1. A **spatial rigid-body motion** is a smooth function $\lambda : M \to SE(3)$, where M, the joint space of the motion, is a manifold. The rank d of (the derivative of) λ at a given configuration $x \in M$ is the **infinitesimal degree of freedom at** x. The maximum value of the infinitesimal degrees of freedom over M is the **degree of freedom** of the motion.

The rigid-body motion of the end-effector of a mechanical devices is frequently called a *kinematic mapping*, but we stick to the terminology of motions as it confers greater generality. A rigid-body motion $\lambda : M \to SE(3)$ may be represented by $\lambda(x) = (A(x), \mathbf{a}(x))$, where $A(x) \in SO(3)$ and $\mathbf{a}(x) \in \mathbb{R}^3$. Given a point $\mathbf{w} \in \mathbb{R}^3$ of the rigid body (in moving coordinates), the *trajectory* of $\mathbf{w}$ is determined by the action of the $SE(3)$ on $\mathbb{R}^3$, that is by the function

$$\tau_w : M \to \mathbb{R}^3, \qquad \tau_w(x) = \lambda(x).\mathbf{w} = A(x)\mathbf{w} + \mathbf{a}(x). \tag{2.1}$$

Note that τ_w can be thought of as a composition of the action with the motion λ itself. That is, if $ev_w : SE(3) \to \mathbb{R}^3$ is the map

$$ev_w(A, \mathbf{a}) = A\mathbf{w} + \mathbf{a}, \tag{2.2}$$

then $\tau_w = ev_w \circ \lambda$. It is valuable to regard the trajectories as forming a family parametrised by $\mathbf{w} \in \mathbb{R}^3$:

$$\tau : M \times \mathbb{R}^3 \to \mathbb{R}^3; \qquad \tau(x, \mathbf{w}) = \tau_w(x). \tag{2.3}$$

Given positive integers r and k, there is an induced multijet extension ${}_rj^k\tau_w : M^{(r)} \to {}_rJ^k(M, \mathbb{R}^3)$, where $M^{(r)}$ is the manifold of r-tuples of distinct points in M. Since τ depends smoothly on $\mathbf{w}$, there is also a three-parameter family of multijets:

$${}_rj_1^k\tau : M^{(r)} \times \mathbb{R}^3 \longrightarrow {}_rJ^k(M, \mathbb{R}^3),$$

where the subscript 1 indicates we are taking jets with respect to the first component only. The following Genericity Theorem [9] concerns general kinematics of rigid-body motions.

Theorem 2.2. *Let $\mathcal{S}$ be a finite stratification of ${}_rJ^k(M, \mathbb{R}^3)$. The set of rigid-body motions $\lambda : M \to SE(3)$ with ${}_rj_1^k\tau$ transverse to $\mathcal{S}$ is residual in $C^\infty(M, SE(3))$, endowed with the Whitney C^∞ topology.*

A relevant example is to apply the theorem to 3-dof spatial motions ($\dim M = 3$) and to consider monogerms of 1-jets ($r = k = 1$) with stratification by corank of the derivative. The corank 1 stratum has codimension 1so for a generic motion, given any tracing point $\mathbf{w} \in \mathbb{R}^3$, there would be a surface in M (or possibly no points) where the trajectory τ_w has a corank 1 singularity. There would be a surface of tracing points $w \in \mathbb{R}^3$ whose trajectories possess isolated corank 2 singularities (codimension 4 stratum), and there would be no tracing points with corank 3 singularities (codimension 9 stratum).

A working hypothesis is that such a theorem holds true for classes of motion arising from specific mechanism geometries, such as the k-point mechanisms under consideration here. The difficulty is that such classes typically depend only on finitely many parameters. So the validity of the hypothesis depends on how that finite-dimensional family sits within the infinite-dimensional space of all motions.

2.2. Screw systems

The Lie algebra $se(3)$ of the Euclidean group inherits its semi-direct product structure: $so(3) \ltimes t(3)$. The Lie algebra $so(3)$, corresponding to the rotations, consists of the skew-symmetric 3×3 matrices. That means we can write $B \in so(3)$ in the form

$$B = \begin{pmatrix} 0 & -u_3 & u_2 \\ u_3 & 0 & -u_1 \\ -u_2 & u_1 & 0 \end{pmatrix},$$

which we identify with the vector $\mathbf{u} = (u_1, u_2, u_3)$. Note that $\mathbf{u}$ spans the kernel of B, so long as $B \neq 0$. The instantaneous translations may be represented by a 3-vector $\mathbf{v} = (v_1, v_2, v_3)$. In the engineering literature, elements of the Lie algebra are referred to as *motors* and elements of the corresponding 5-dimensional projective space $Pse(3)$ are called *screws*. Thus $(\mathbf{u}, \mathbf{v})$ are referred to as *motor coordinates* and the homogeneous version $(\mathbf{u}; \mathbf{v})$ as *screw coordinates.*

Given a rigid-body motion $\lambda : M \to SE(3)$, the instantaneous motion at a configuration $x \in M$ is given by the image of the derivative of λ at x, a subspace of $T_{\lambda(x)}SE(3)$. By suitable choice of moving and fixed coordinates, we may assume this to be a subspace of the Lie algebra $se(3)$. The corresponding projective subspace in $Pse(3)$ is called a *screw system.* If λ has k-dof at x then it is a *k-system.*

The classification of screw systems was originally proposed by Hunt [12], and given a firm mathematical basis by Gibson and Hunt [10]. They introduced the pencil of pitch quadrics $Q_h = 0$, where $h \in \mathbb{R} \cup \{\infty\}$ is the *pitch.* Explicitly, Q_h is a quadratic form, given in motor coordinates by:

$$Q_h(\mathbf{u}, \mathbf{v}) = \mathbf{u}.\mathbf{v} - h(\mathbf{u}.\mathbf{u}), \quad h \neq \infty; \quad Q_\infty(\mathbf{u}, \mathbf{v}) = \mathbf{u}.\mathbf{u} \tag{2.4}$$

Since the forms are homogeneous, the corresponding varieties (quadrics) are well defined in screw coordinates. The special case $Q_0 = 0$ corresponds to the classical Klein quadric of lines in projective 3-space, where we identify a pure rotation with its axis; in this case, the coordinates correspond to Plücker line coordinates of the axis (see, for example, [15]). We note for later reference that by viewing a line as an intersection of a pencil of planes, it may also be represented by its *dual* Plücker coordinates $(\mathbf{u}^*; \mathbf{v}^*) = (\mathbf{v}, \mathbf{u})$. Application of duality enables geometric assertions about points and lines to be transformed into statements about planes and lines, and vice versa.

The quadric $Q_\infty = 0$, corresponding to infinitesimal translations, degenerates to a (projective) plane. Thus the family S_h of sets of screws of each pitch h, in the screw system S, is just its pencil of intersections with the hypersurfaces $Q_h = 0$ – in the case of 3-systems, this gives a pencil of conics, as had been recognized by Ball [1]. Each pitch quadric $Q_h = 0$, $h \neq \infty$, has a pair of rulings: by the α-planes corresponding to all screws of pitch h whose axis passes through a given point, and by the β-planes corresponding to all screws whose axis lies in a given plane.

The quadratic forms (2.4) have associated bilinear forms; in particular, for a pair of motors $\mu_i = (\mathbf{u}_i, \mathbf{v}_i)$, $i = 1, 2$, we have $Q_0(\mu_1, \mu_2) = \frac{1}{2}(\mathbf{u}_1.\mathbf{v}_2 + \mathbf{u}_2.\mathbf{v}_1)$.

Given a k-system S, define the *reciprocal* $(6-k)$-system

$$S^{\perp} = \{\$ \mid Q_0(\$, \$') = 0, \forall \$' \in S\}.$$

The broad classification of 3-systems is on the following basis. Type I systems do not lie wholly in a single pitch quadric and therefore they intersect the pitch quadrics in a pencil of conics as indicated above. Type II systems are those contained within a single pitch quadric. The subtypes A, B, C, D distinguish the projective dimension of intersection with Q_∞: subtype A denoting empty intersection, up to subtype D denoting a 2-dimensional intersection. Within the type I systems, further distinction is made on the basis of the projective type of the pencil of conics.

Further refinement of the classification was provided in [7, 8] where screw systems were placed in the context of the Lie group approach to rigid body motions. The Hunt–Gibson classification can be derived from equivalence under the action induced on the Grassmannian of screw systems of a given dimension by the adjoint action of the Lie group on its Lie algebra. The pitch arises as the fundamental invariant of the projectivised adjoint action. The benefit of this approach is that the classes of screw systems arise as submanifolds of a Grassmannian manifold. The codimensions and adjacencies were determined in [8] and the stratification was shown to be Whitney regular.

The intersection of the screw system with Q_0 is also of significance both mathematically and from the engineering point of view. In the fine classification of 3-systems in Table 1 from [8], superscripts refer to the signs of the principal pitches and other moduli.

The Thom Transversality Theorem ensures that motions in a residual set have 1-jet transverse to these stratifications. Therefore, generically, one would expect to encounter, for example, a smooth surface (codimension 1 manifold) in the 3-dimensional jointspace M where the screw system is of type IA_1^{+0-}, since this stratum has codimension 1, but not one of type IIA^0, as it has codimension 6.

2.3. Instantaneous singular sets

Given a motion $\lambda : M \to SE(3)$ and a configuration $x \in M$, define the *instantaneous singular set at x* to be

$$I(\lambda, x) = \{\mathbf{w} \in \mathbb{R}^3 \mid x \text{ is a singular point of } \tau_w\} \tag{2.5}$$

where the trajectory function τ_w associated to λ was defined in (2.1). Given a tracing point $\mathbf{w} \in \mathbb{R}^3$, let A_w denote the α-plane in Q_0 consisting of lines (or screws of pitch zero) through $\mathbf{w}$. We have the following central theorem, stated here for 3-dof motions. The general result appears in [6].

Theorem 2.3. *Let $\lambda : M \to SE(3)$ be a 3-dof motion and let S be the screw system at $x \in M$. A tracing point $\mathbf{w}$ belongs to $I(\lambda, x)$ if and only if $S \cap A_w$ is non-empty.*

Broad class	Intermediate class	Codim	Fine classes
IA_1	IA_1	0	IA_1^{+++}, IA_1^{++-}, IA_1^{+--}, IA_1^{---}
	IA_1^0	1	IA_1^{++0}, IA_1^{+0-}, IA_1^{0--}
IA_2	IA_2	2	$IA_2^{(++)+}$, $IA_2^{+(++)}$, $IA_2^{(++)-}$, $IA_2^{+(--)}$, $IA_2^{(--)-}$, $IA_2^{-(--)}$
	IA_2^0	3	$IA_2^{(++)0}$, $IA_2^{+(00)}$, $IA_2^{(00)-}$, $IA_2^{0(--)}$
IB_0	IB_1	1	IB_0^+, IB_0^-
	IB_1^0	2	IB_0^0
	IB_2	2	$IB_0^{0,+}$, $IB_0^{0,-}$
	IB_2^0	3	$IB_0^{0,0}$
IB_3	IB_3	3	IB_3^{++}, IB_3^{+-}, IB_3^{--}
	IB_3^0	4	IB_3^{+0}, IB_3^{0-}
IC	IC	4	IC^+
	IC^0	5	IC^0
IIA	IIA	5	IIA^+, IIA^-
	IIA^0	6	IIA^0
IIB	IIB	5	IIB^+, IIB^-
	IIB^0	6	IIB^0
IIC	IIC	6	IIC^+, IIC^-
	IIC^0	7	IIC^0
IID		9	no finer subtypes

TABLE 1. Classification of 3-systems.

Elementary corollaries of Theorem 2.3 are:

1. $I(\lambda, x)$ depends only on the associated screw system S – it is a first-order invariant of the motion. We may therefore write $I(S)$ for the ISS associated in this way to the screw system S.
2. The projective dimension of $S \cap A_w$ plus one is the corank of the singularity of τ_w at x.
3. A point $\mathbf{w}$ is in $I(S)$ if and only if S contains a screw of pitch zero, and $\mathbf{w}$ lies on its axis. Hence, $I(S)$ is *ruled*, in the sense that for any point $\mathbf{w} \in I(S)$, there is a line through $\mathbf{w}$ contained in $I(S)$.

type of 3-system	intersection	ISS	max corank
IA_1^{+++}, IA_1^{---}	empty	empty	0
IA_1^{++-}, IA_1^{+--}	conic	elliptic 1-sheet hyperboloid	1
IA_1^{++0}, IA_1^{0--}	point	line	1
IA_1^{+0-}	line pair	plane pair	2
$IA_2^{(++)+}$, $IA_2^{+(++)}$	empty	empty	0
$IA_2^{(--)-}$, $IA_2^{-(--)}$	empty	empty	0
$IA_2^{(++)-}$, $IA_2^{+(--)}$	conic	circular 1-sheet hyperboloid	1
$IA_2^{(++)0}$, $IA_2^{0(--)}$	point	line	1
$IA_2^{+(00)}$, $IA_2^{(00)-}$	repeated line	plane	2
IB_0^{+}, IB_0^{-}, $IB_0^{0,+}$, $IB_0^{0,-}$	conic	hyperbolic paraboloid	1
IB_0^{0}, $IB_0^{0,0}$	line pair	plane pair	2
IB_3^{++}, IB_3^{--}	point in Q_∞	empty	0
IB_3^{+-}	line pair	parallel planes	1
IB_3^{+0}, IB_3^{0-}	line	plane	1
IC, IC^0	line	plane	1
IIA^+, IIA^-	empty	empty	0
IIA^0	α-plane	whole space	3
IIB^+, IIB^-	point in Q_∞	empty	0
IIB^0	β-plane	plane	2
IIC^+, IIC^-	line in Q_∞	empty	0
IIC^0	α-plane	whole space	1
IID	Q_∞	empty	0

TABLE 2. Instantaneous singular sets for 3-systems.

4. The geometric form of this ISS is determined by the intersection of the screw system with Q_0 which, in turn, can be determined from the fine classification of 3-systems. These are given in Table 2, together with the maximum corank of any singular trajectory.

The following theorem in [6] provides a fundamental connection between a screw system and its reciprocal and is important for the analysis of 3-point motions.

Theorem 2.4. *Let S be a 3-system, and let $S^\perp$ be the reciprocal 3-system. Then $I(S) = I(S^\perp)$.*

3. k-point motions

3.1. Configuration space

We formalize the notion of a k-point motion.

Definition 3.1. A **k-point motion** ($1 \leq k \leq 6$) is a rigid-body motion in which a set of points $W_1, \ldots, W_k$ of the rigid body, satisfying the condition that any subset of four or less points is affinely independent, is constrained to lie, respectively, on a set of k smooth surfaces, $N_1, \ldots, N_k$. The points W_i are the **contact points** and the surfaces N_i the **contact surfaces**. The points W_1, W_2, W_3 define the **coupler triangle** where, if $k < 3$, introduce $3 - k$ additional points W_j, $j = k+1, \ldots, 3$ so that W_1, W_2, W_3 are affinely independent.

Not all choices of contact points and surfaces result in a proper rigid-body motion, that is one for which the configuration space is a manifold. Sufficient conditions for this are established below. For clarity we shall call such a motion *regular* and otherwise *singular* (though, strictly, it is not a motion at all by our definition).

By a smooth surface, in the definition, is meant an embedded, orientable, 2-dimensional submanifold in $\mathbb{R}^3$. Globally, therefore, $N_i = \phi_i(M_i)$, where M_i is an orientable 2-dimensional manifold and $\phi_i : M_i \to N_i \subset \mathbb{R}^3$ an embedding, for each i. If one is only interested in what happens locally, then each contact surface N_i can be parametrised by a smooth function $\phi_i : U_i \to N_i$, where U_i is some open subset of $\mathbb{R}^2$. Sometimes it may be more convenient to represent the contact surfaces implicitly by $N_i = f_i^{-1}(0)$, where $f_i : \mathbb{R}^3 \to \mathbb{R}$ is a smooth function and $0 \in \mathbb{R}$ a regular value, for each $i = 1, \ldots, k$.

Let the moving coordinates of the contact points W_i, $i = 1, \ldots, l$ (where $l = \min\{3, k\}$) be $\mathbf{w}_i = (w_{i1}, w_{i2}, w_{i3})$. Then the rigidity of the moving body gives rise to an equation for each pair i, j such that $1 \leq i < j \leq k$:

$$||\phi_i(x_i) - \phi_j(x_j)||^2 - ||\mathbf{w}_i - \mathbf{w}_j||^2 = 0. \tag{3.1}$$

This gives $\frac{1}{2}k(k-1)$ equations on points $(x_1, \ldots, x_k) \in M_1 \times \cdots \times M_k$, a $2k$-dimensional manifold. For $k < 3$, we add further variables, $\mathbf{z} = (z_{i1}, z_{i2}, z_{i3})$ for $i = (k+1), \ldots, 3$, denoting the fixed coordinates of the unconstrained vertices of the coupler triangle, and further equations of the form:

$$\begin{aligned} ||\phi_i(x_i) - \mathbf{z}_j||^2 - ||\mathbf{w}_i - \mathbf{w}_j||^2 &= 0, \quad 1 \leq i \leq k,\ k+1 \leq j \leq 3 \\ ||\mathbf{z}_i - \mathbf{z}_j||^2 - ||\mathbf{w}_i - \mathbf{w}_j||^2 &= 0 \quad k+1 \leq i < j \leq 3. \end{aligned} \tag{3.2}$$

In the case $k \geq 5$ there is some dependence between the equations and it suffices to restrict to three equations respecting the distance between the vertices of the coupler triangle and, for each W_j, $j > 3$, the distances from W_j to each vertex of the coupler triangle. Further care is needed in the cases $k \geq 4$ as here the equations do not distinguish between the possible orientations or combinations of orientations of contact tetrahedra. Thus, we must choose the component of the solution set of equations (3.1) for which the orientation of each contact tetrahedron over the coupler triangle corresponds to that of the moving contact points.

In summary, the configuration space has the form $F_k^{-1}(0)$, with $F_k : M_1 \times \cdots \times M_k \times \mathbb{R}^{p_k} \to \mathbb{R}^{q_k}$, where p_k is the number of additional variables for non-constrained contact points, and q_k is the number of equations, taking the values given in Table 3. For a regular motion, 0 is required to be a regular value of this map. In that case dimension of the configuration space is $2k + p_k - q_k = 6 - k$ for each k.

Contacts	k	1	2	3	4	5	6
Surface variables	$2k$	2	4	6	8	10	12
Non-constraint variables	p_k	6	3	0	0	0	0
Total variables	$2k + p_k$	8	7	6	8	10	12
Equations	q_k	3	3	3	6	9	12

TABLE 3. Variables and equations for parametric k-point motions.

3.2. Regularity conditions

The following theorem determines sufficient conditions on contact surfaces and points for the configuration space to be a manifold and hence for the motion to be regular. A direct proof using local parametrisations is reasonably straightforward too, but here a simpler implicit surface approach is given.

Theorem 3.2. *The configuration space for a k-point motion (with contact surfaces defined implicitly) is a smooth manifold unless, in some realizable configuration, the surface normals at the contact points, thought of as screws of pitch zero, fail to span a k-system.*

Proof. Suppose the contact surfaces are defined implicitly by $N_i = f_i^{-1}(0)$, $i = 1, \ldots, k$. Let the moving coordinates of the contact points W_i, $i = 1, \ldots, k$, be $\mathbf{w}_i = (w_{i1}, w_{i2}, w_{i3})$. Then the configuration space M is defined as a subset of $SE(3)$ by the equation $G(\mu) = 0$, where the components of $G : SE(3) \to \mathbb{R}^k$ in terms of $\mu = (A, \mathbf{a}) \in SE(3)$ are:

$$G_i(A, \mathbf{a}) = f_i(A\mathbf{w}_i + \mathbf{a}), \qquad i = 1, \ldots, k. \tag{3.3}$$

M is a manifold unless, for some $\alpha \in M$, the rank of G at α is less than k. Suppose we are at such an α and the fixed contact points are $\alpha(\mathbf{w}_i) = \mathbf{x}_i$, $i = 1, \ldots, k$. By the rank formula of linear algebra, the dimension of the kernel S of the derivative $DG(\alpha)$ must be $> (6 - k)$. Note that S is just a screw system. In fact, since $M = \bigcap_{i=1}^k G_i^{-1}(0)$, we have $S = \bigcap_{i=1}^k S_i$, where S_i is the kernel of $DG_i(\alpha)$.

Now $G_i = f_i \circ ev_{w_i}$, where ev_{w_i} is defined in equation (2.2). Since ev_{w_i} has no singular points and, by assumption, 0 is a regular value of f_i, it follows that

0 is a regular value of G_i and hence S_i is a 5-system. Its elements are just those screws for which the instantaneous direction of motion of $\mathbf{w}_i$ lies in the tangent space to N_i at $\mathbf{x}_i$.

It is clear that S_i must contain the α-plane of pitch-zero screws whose axes pass through $\mathbf{x}_i$, since these fix $\mathbf{x}_i$, and also the pitch infinity screws parallel to the tangent plane. This gives us five independent screws, which therefore span S_i. It is now an easy exercise to show that the pitch-zero screw with axis normal to the tangent plane is reciprocal to S_i, and hence to $S \subseteq S_i$. Thus the contact surface normals all lie in the reciprocal screw system $S^\perp$ and in fact span it. Since the dimension of S is greater than $6 - k$ then $S^\perp$, spanned by the surface normals, must have dimension less than k. □

We spell out the import of Theorem 3.2 for 3-point motions.

Corollary 3.3. *The configuration space of a 3-point motion is a manifold unless, for some configuration, either*

1. *there is a common normal to two of the contact surfaces at the points of contact or*
2. *the three surface normals are coplanar and coincident or*
3. *the three surface normals are coplanar and parallel.*

Proof. By Theorem 3.2 we require the three surface normals in each configuration to span a 3-system. This can only fail if they span a 2-system; a 1-system is not possible as this would require the three normals to coincide, but we have assumed the contact points are affinely independent, so cannot be collinear. We have a 2-system if either two normals coincide, and hence we are in case (a), or they are (projectively) collinear as points in Q_0. In the latter case, three distinct points of a 2-system lie on Q_0 so the 2-system must be of type IIA^0 or IIB^0 [6] and projectively span a line in an α-plane (or β-plane) corresponding to a planar pencil of lines in $\mathbb{R}^3$, with either finite or infinite vertex, giving rise to cases (b) and (c) respectively. □

Theorem 3.4. *Given k contact surfaces in $\mathbb{R}^3$, $1 \le k \le 6$, for a residual set of k contact points in $\mathbb{R}^{3k}$ the configuration space for the resulting k-point motion is a manifold of dimension $6 - k$ or empty.*

Proof. Treat the coordinates of the contact points as variables in the equations (3.1). The resulting function can readily be shown to be a submersion, so 0 is a regular value. The result now follows by a standard transversality theorem (e.g., Golubitsky and Guillemin [11], Chapter 2, §4). □

3.3. Singularities of 3-point motions

The key to determining the instantaneous singular sets and singular trajectory types lies in the observation, in the proof of Theorem 3.2, that the normals to the contact surfaces at the points of contact lie in the reciprocal screw system at a given configuration. We have the following result.

Theorem 3.5. *For a 3-point motion, the contact surface normals lie in the ISS at each configuration.*

Proof. Let S be the 3-system of the motion in a given configuration. Then the reciprocal system $S^{\perp}$ is spanned by the surface normals. As screws these have pitch zero and so, by Theorem 2.3, the normals belong to $I(S^{\perp})$. The invariance of the ISS under reciprocity (Theorem 2.4) establishes the result. □

A similar statement is true for k-point motions, $k = 1, 2$, but more care is needed for $k \geq 4$ as a screw belonging to Q_0 does not ensure that its axis is in the ISS; rather we require the k-system to intersect an α-plane in a projective line at least.

The knowledge that any ISS of a regular 3-point motion contains three distinct lines, together with the information in Table 2, enables us to exclude immediately a number of possible screw types, namely all those for which the ISS is empty or a line. It is also possible to eliminate types $\mathrm{IA}_2^{+(00)}$, $\mathrm{IA}_2^{(00)-}$, IB_3^{+0}, IB_3^{0-} and IC. For in those cases, the ISS is a plane which would require the surface normals to be coplanar. Then, either the normals form a planar pencil in which case the configuration space is singular, or the entire 3-system lies in Q_0 and hence is of type II. Details can be found in [5].

The following lemma establishes a simple relationship between the configuration of the surface normals and screw system types.

Lemma 3.6. *Given a 3-point motion and some configuration, if the direction vectors of the surface normals in that configuration:*

1. *span $\mathbb{R}^3$, then the associated 3-system has type A;*
2. *span a plane, then the associated 3-system has type B;*
3. *span a line, then the associated 3-system has type C.*

A type D system is not possible.

Proof. The screw system contains a screw of infinite pitch if and only if it corresponds to an infinitesimal translation perpendicular to all the surface normal directions. The result follows. □

Further consideration of the ISSs for each type enables us to establish a precise correspondence between the screw system type and the configuration of the surface normals. This is summarized in Table 4.

It can be noted immediately that the special configurations, described in the Introduction, for the classical Darboux motion and for the RCC device, are ones in which the surface normals are coincident, so the instantaneous screw type is IIA^0. From Table 2, the trajectory of the point in question has corank 3 and every point in the moving body is instantaneously singular. This screw type has codimension 6 amongst 3-systems, so it and the corresponding corank 3 singularities should not occur generically.

For the HVRam device, the home configuration is one in which the contact normals are coplanar but not coincident, and hence the screw system type is IIB^0.

type of 3-system	configuration of surface normals
IA_1^{++-}, IA_1^{+--}	3 mutually skew lines
IA_1^{+0-}	2 intersect in finite point, 3rd skew to others
$\mathrm{IA}_2^{(++)-}$, $\mathrm{IA}_2^{+(--)}$	3 mutually skew lines [1]
IIA^0	3 lines intersect in finite point
IB_0^+, IB_0^-	3 mutually skew lines
IB_0^0,	2 intersect in finite point, 3rd in parallel plane
$\mathrm{IB}_0^{0,+}$, $\mathrm{IB}_0^{0,-}$	3 mutually skew lines with common perpendicular
$\mathrm{IB}_0^{0,0}$	2 intersect in finite point, 3rd in parallel plane with common perpendicular through intersection
IB_3^{+-}	2 parallel, 3rd in parallel plane
IIB^0	3 coplanar not meeting in a point
IIC^0	3 parallel but not coplanar

[1] The distances between each pair of lines in the direction parallel to the third are equal [5].

TABLE 4. Screw system types for 3-point motions.

This type has codimension 6, the ISS is the plane of the mirror joints and every point in the plane has a trajectory with a corank 2 singularity. Again this is not generic in the space of all motions.

4. Stability of type II screw systems

There are several ways of perturbing a k-point motion:

- by altering the dimensions of the coupler triangle;
- by altering design parameters within a given family of contact surfaces;
- by altering the contact surfaces in a general way.

From an engineering perspective, the first two are of most interest. Mathematically, the last, of which the second is a special case, provides greatest leeway (one is perturbing in an infinite-dimensional space) and would be most likely to perturb away degenerate behavior.

Within the class of 3-point motions, we show that certain high-codimension screw types (IIA^0 and IIB^0) may occur stably – so the class fails to satisfy the genericity hypothesis referred to in Section 2.3. Essentially, this is because the condition in Table 4, concerning the surface normal arrangements giving rise to these screw types, has only codimension 3 among triples of lines in 3-space.

Let $N_i = \phi_i(M_i)$, $i = 1, 2, 3$, be the three contact surfaces, as in Section 3.1. We are interested in the space of triples of embeddings $\Gamma_3 = \mathrm{Emb}(M_1, \mathbb{R}^3) \times$

$\mathrm{Emb}(M_2, \mathbb{R}^3) \times \mathrm{Emb}(M_3, \mathbb{R}^3)$ endowed with the (product) Whitney C^∞ topology. Let $\hat{\Gamma}_3$ denote the open subset for which the associated 3-point motion is regular.

If $W_1W_2W_3$ is a coupler triangle, then the resulting motion is characterized by the lengths of its sides, $d_{ij} = ||\mathbf{w}_i - \mathbf{w}_j|| > 0$, $(i, j) = (1, 2), (2, 3), (3, 1)$. We denote this set of design parameters by $\delta = (d_{23}, d_{31}, d_{12}) \in \mathbb{R}^3_+ = \{(x, y, z) \in \mathbb{R}^3 \mid x, y, z > 0\}$. Given $\phi = (\phi_1, \phi_2, \phi_3) \in \hat{\Gamma}_3$, and $\delta \in \mathbb{R}^3_+$, define $F^{\phi,\delta} : M_1 \times M_2 \times M_3 \to \mathbb{R}^3$ by

$$F_k^{\phi,\delta}(x_1, x_2, x_3) = ||\phi_i(x_i) - \phi_j(x_j)||^2 - d_{ij}^2, \quad k = 1, 2, 3$$

where (i, j, k) is a cyclic permutation of $(1, 2, 3)$. Then $M = (F^{\phi,\delta})^{-1}(0)$ is the configuration space of the corresponding 3-point motion. Treating ϕ and δ as variables gives rise to a continuous map

$$F : \hat{\Gamma}_3 \times \mathbb{R}^3_+ \to C^\infty(M_1 \times M_2 \times M_3, \mathbb{R}^3); \quad (\phi, \delta) \mapsto F^{\phi,\delta}.$$

We now characterize the configurations corresponding to the screw systems of interest. Given any $X = (x_1, x_2, x_3) \in M_1 \times M_2 \times M_3$, let $L_i(X)$ denote the normal line to $N_i = \phi_i(M_i)$ at the point $\mathbf{y}_i = \phi_i(x_i)$, $i = 1, 2, 3$. If $\mathbf{n}_i(x_i)$ denotes a smooth choice of normal vector to N_i at $\mathbf{y}_i$, then we may represent $L_i(X)$, in motor coordinates, by $(\mathbf{n}_i(x_i), \mathbf{y}_i(x_i) \times \mathbf{n}_i(x_i)) = (\mathbf{n}_i(x_i), \mathbf{v}_i(x_i))$, say.

In terms of the bilinear form Q_0, define $G_k^\phi : M_1 \times M_2 \times M_3 \to \mathbb{R}$, $k = 1, 2, 3$, by

$$G_k^\phi(X) = Q_0(L_i(X), L_j(X))$$

where (i, j, k) is a cyclic permutation of $(1, 2, 3)$, and then define the following map:

$$H^{\phi,\delta} : M_1 \times M_2 \times M_3 \to \mathbb{R}^6; \quad H^{\phi,\delta} = (F_1^{\phi,\delta}, F_2^{\phi,\delta}, F_3^{\phi,\delta}, G_1^\phi, G_2^\phi, G_3^\phi).$$

As for F, since the manifolds are orientable, the normal vectors may be chosen smoothly, so this can be regarded as defining a continuous map

$$H : \hat{\Gamma}_3 \times \mathbb{R}^3_+ \to C^\infty(M_1 \times M_2 \times M_3, \mathbb{R}^6); \quad (\phi, \delta) \mapsto H^{\phi,\delta}.$$

Finally, let $\rho^\phi, \sigma^\phi : M_1 \times M_2 \times M_3 \to \mathbb{R}$ be given by

$$\begin{aligned} \rho^\phi(X) &= \mathbf{n}_1(x_1).(\mathbf{n}_2(x_2) \times \mathbf{n}_3(x_3)), \\ \sigma^\phi(X) &= \mathbf{v}_1(x_1).(\mathbf{v}_2(x_2) \times \mathbf{v}_3(x_3)). \end{aligned}$$

Lemma 4.1. *Given a regular 3-point motion defined by $\phi \in \hat{\Gamma}_3$ and $\delta \in \mathbb{R}^3_+$, as above, a point $X \in M_1 \times M_2 \times M_3$ is a configuration of the motion and has:*

1. *a type IIA^0 screw system at X if and only if $X \in (H^{\phi,\delta})^{-1}(0)$ and $\rho^\phi(X) \neq 0$;*
2. *a type IIB^0 screw system at X if and only if $X \in (H^{\phi,\delta})^{-1}(0)$ and $\sigma^\phi(X) \neq 0$.*

Proof. Suppose that $H^{\phi,\delta}(X) = 0$. Then X is certainly a configuration since $F^{\phi,\delta}(X) = 0$.

It is a standard result of line geometry (see for example [10, 15]) that two lines L_1, L_2 intersect, possibly at infinity (i.e., are parallel), if and only if $Q_0(L_1, L_2) = 0$. Thus, since $G_k^\phi(X) = 0$, $k = 1, 2, 3$, the three surface normals either intersect or are parallel, pairwise.

Suppose now that $\rho^{\phi}(X) \neq 0$. It follows that the three direction vectors of the lines are not coplanar and, in particular, no two lines are parallel. Hence the normals intersect pairwise and, to avoid coplanarity, the three points of intersection must coincide at a single point so the corresponding screw system is type IIA^0.

Conversely, if X is a configuration and if the screw system there is type IIA^0, then the relevant conditions all hold. This proves (a).

By duality, the condition $\sigma^{\phi}(X) \neq 0$ corresponds to the three lines not being coincident. Hence they either intersect pairwise in distinct points, in which case the lines are coplanar and the screw system type IIB^0, or at least two are parallel. If all three were parallel then the motion would be singular by Corollary 3.3, contrary to hypothesis. So at most two are parallel and the third must intersect each of them, again resulting in a coplanar system of lines. □

Theorem 4.2. *Given a 3-point motion defined by $\phi \in \hat{\Gamma}_3$ and a coupler triangle with parameters $\delta = (d_{23}, d_{31}, d_{12}) \in \mathbb{R}^3_+$, suppose X is a configuration at which the screw type is IIA^0 (resp. IIB^0). If $H^{\phi,\delta} \pitchfork \{0\}$ then there are open neighborhoods U' of $\phi \in \hat{\Gamma}_3$ and V' of $\delta \in \mathbb{R}^3_+$ such that for any $\phi' \in U'$ and $\delta' \in V'$, the corresponding 3-point motion is regular and possesses a configuration at which the screw type is IIA^0 (resp. IIB^0).*

Proof. By a standard result of transversality (see for example [11]), since $\{0\} \subset \mathbb{R}^6$ is closed, the set of maps

$$\{f \in C^{\infty}(M_1 \times M_2 \times M_3, \mathbb{R}^6) \mid f \pitchfork \{0\}\}$$

is open. Moreover the inequalities in Lemma 4.1 also define open sets in $C^{\infty}(M_1 \times M_2 \times M_3, \mathbb{R}^6)$. The theorem now follows from the lemma and the continuity of H. □

5. Applications

While Theorem 4.2 establishes sufficient conditions for stability of these screw types, we wish to establish whether they hold for the specific examples of 3-point mechanisms discussed in Section 1.

5.1. Darboux motions

Let N_i, $i = 1, 2, 3$, to be three planes in general position. We may assume the planes intersect at the origin and denote unit vectors along the line of intersection of N_i, N_j by $\mathbf{r}_{ij}$ for $(i, j) = (1, 2), (2, 3), (3, 1)$. Then each N_i may be parametrised by

$$\phi_i(u_i, v_i) = u_i \mathbf{r}_{ki} + v_i \mathbf{r}_{ij}$$

where (i, j, k) is a cyclic permutation of $(1, 2, 3)$. A normal vector at any point of N_i is then $\mathbf{n}_i = \mathbf{r}_{ki} \times \mathbf{r}_{ij}$.

Let $a_{ij} = \cos\theta_{ij} = \mathbf{r}_{ik}.\mathbf{r}_{kj}$, where θ_{ij} is the angle between the planes N_i, N_j. These three numbers are effectively design parameters for the motion, in addition to the side lengths d_{ij} of the coupler triangle. The feasible region for these parameters is bounded by $\theta_{12} + \theta_{23} + \theta_{31} = 2\pi$; taking cosines and applying multiple angle formulae results in the boundary condition:

$$a_{12}^2 + a_{23}^2 + a_{31}^2 - 2a_{12}a_{23}a_{31} = 1.$$

The normal line L_i, at a point $\mathbf{y}_i = \phi_i(u_i, v_i) \in N_i$, can be expressed in motor coordinates by $(\mathbf{n}_i, \mathbf{y}_i \times \mathbf{n}_i)$. Hence

$$Q_0(L_i, L_j) = \mathbf{n}_i.(\mathbf{y}_j \times \mathbf{n}_j) + \mathbf{n}_j.(\mathbf{y}_i \times \mathbf{n}_i) = (\mathbf{y}_i - \mathbf{y}_j).(\mathbf{n}_i \times \mathbf{n}_j).$$

Note that $\mathbf{n}_i \times \mathbf{n}_j$ must lie in the intersection $N_i \cap N_j$, so is a non-zero multiple of $\mathbf{r}_{ij}$. Since we are interested in the zeroes of the function $H^{\phi,\delta}$ of Lemma 4.1, we may safely ignore the constant and assume $G_k^{\phi} = (\mathbf{y}_i - \mathbf{y}_j).\mathbf{r}_{ij}$. $H^{\phi,\delta} = (F^{\phi,\delta}, G^{\phi})$ may now be expressed in terms of the 6 design parameters a_{ij}, d_{ij} and the 6 internal variables (u_i, v_i), $i = 1, 2, 3$.

$$\begin{aligned} F_k^{\phi,\delta}(u_1, v_1, u_2, v_2, u_3, v_3) &= (u_i^2 + v_i^2) + (u_j^2 + v_j^2) + 2a_{jk}u_iv_i - 2a_{jk}u_iu_j \\ &\quad - 2a_{ij}u_iv_j - 2u_jv_i - 2a_{ki}v_iv_j + 2a_{ki}u_jv_j; \\ G_k^{\phi}(u_1, v_1, u_2, v_2, u_3, v_3) &= a_{jk}u_i + v_i - u_j + a_{ki}v_j, \end{aligned}$$

for $k = 1, 2, 3$ and (i, j, k) cyclic.

The determinant of the Jacobian may be calculated (most easily using computer algebra software) and has the form:

$$8(1 - a_{12}^2 - a_{23}^2 - a_{31}^2 + 2a_{12}a_{23}a_{31})^2(u_1u_2u_3 + v_1v_2v_3).$$

The repeated factor represents the boundary of the feasible region of design parameters established above. The last factor is proportional to the volume of the tetrahedron whose vertices are the origin and the contact points $\mathbf{y}_i = \phi_i(u_i, v_i)$, $i = 1, 2, 3$:

$$(\mathbf{y}_1 \times \mathbf{y}_2).\mathbf{y}_3 = (u_1u_2u_3 + v_1v_2v_3)[(\mathbf{r}_{12} \times \mathbf{r}_{23}).\mathbf{r}_{31}].$$

It follows that $H^{\phi,\delta}$ is a local diffeomorphism at a configuration with screw type IIA^0 unless the "coupler tetrahedron" collapses to become planar. This represents the boundary in the space of coupler triangles, for which there exist IIA^0 screw systems. For example, in the special case considered by Bottema and Roth, where the contact surfaces are the coordinate planes, the coupler triangle must be right-angled or acute for $H^{\phi,\delta} = 0$, and transversality fails only for the right-angled triangles (see [5]).

5.2. RCC device

A general analysis of the case of three contact spheres has so far proved intractable. We therefore concentrate on the local situation in the standard case described in the Introduction. We may assume that the centres $\mathbf{c}_i$, $i = 1, 2, 3$, are at $(1, 0, 0)$, $(-\frac{1}{2}, \pm\frac{\sqrt{3}}{2}, 0)$. Let the spheres have radius R and suppose the vertices of the coupler triangle lie on a circle of radius $r < 1$. Then the sides of the coupler triangle have length $\sqrt{3}r$. The spheres may by parametrised by:

$$\phi_i(u_i, v_i) = R(\cos v_i \cos u_i, \cos v_i \sin u_i, \sin v_i) + \mathbf{c}_i, \quad i = 1, 2, 3.$$

The home configuration of the device, in which the screw system is type IIA^0, is such that the contact points are at $r.\mathbf{c}_i + (0, 0, \sin v)$, where v is the common value of the parameters v_1, v_2, v_3 for which the triangle is horizontal. A simple trigonometric argument shows that $\cos v = (1-r)/R$. Evaluating the determinant of the Jacobian of the associated map $H^{\phi,\delta}$ shows that it is non-zero, except for the special values $r = 1 \pm R$. (In fact, for these values the motion is singular as the normals are coplanar as well as coincident.) Otherwise, for any local perturbation of the spheres or the coupler triangle, there remains a type IIA^0 screw system for the perturbed motion.

5.3. HVRam device

In this case, it is the presence of a IIB^0 screw system that confers mechanical advantage. The three contact planes may be assumed to intersect in the z-axis and hence to be parametrised by:

$$\phi_i(u_i, v_i) = u_i(a_i, b_i, 0) + v_i(0, 0, 1), \quad i = 1, 2, 3,$$

where we may take $a_i^2 + b_i^2 = 1$. One establishes easily that

$$G_k^{\phi}(u_1, v_1, u_2, v_2, u_3, v_3) = (a_i b_j - a_j b_i)(v_i - v_j)$$

for any cyclic permutation (i, j, k) of $(1, 2, 3)$. It follows that the last three rows of the Jacobian of $H^{\phi,\delta}$, for any choice of coupler triangle, will have rank 2 only, so transversality fails. Indeed, it is clear that the coupler triangle can be translated vertically from the given configuration and will retain screw type IIB^0. In this case, small perturbations of the device may not possess its special property.

Acknowledgment

The authors are indebted to Dr. P. Andreae and Professor C. Wampler for introducing us to the remote centre compliance device.

References

[1] Ball, R.S., *The Theory of Screws*, Cambridge University Press, Cambridge, 1900.

[2] Bottema, O. and Roth, B., *Theoretical Kinematics*, Dover Publications, New York, 1990.

[3] Carretero, J.A., Nahon, M., Buckham, B. and Gosselin, C.M., *Kinematics Analysis of a Three-DoF Parallel Mechanism for Telescope Applications* in Proc. ASME Design Engineering Technical Conf., (Sacramento, 1997), ASME, 1997.

[4] Carretero, J.A., Podhorodeski, R.P. and Nahon M., *Architecture Optimization of a Three-DoF Parallel Mechanism* in Proc. ASME Design Engineering Technical Conf., (Atlanta, 1998) ASME, 1998.

[5] Cocke, M.W., *Natural Constraints on Euclidean Motions.* PhD Thesis, Department of Mathematical Sciences, University of Liverpool, 1998.

[6] Cocke, M.W., Donelan, P.S. and Gibson, C.G., *Instantaneous Singular Sets Associated to Spatial Motions* in Real and Complex Singularities, (São Carlos, 1998), eds. F. Tari and J.W. Bruce, Res. Notes Math., **412**, Chapman and Hall/CRC, 2000, 147–163.

[7] Donelan, P.S. and Gibson, C.G., *First-Order Invariants of Euclidean Motions.* Acta Applicandae Mathematicae **24** (1991), 233–251.

[8] Donelan, P.S. and Gibson, C.G., *On the Hierarchy of Screw Systems.* Acta Applicandae Mathematicae **32** (1993), 267–296.

[9] Gibson, C.G. and Hobbs, C.A., *Local Models for General One-Parameter Motions of the Plane and Space.* Proc. Royal Soc. Edinburgh **125A** (1995), 639–656.

[10] Gibson, C.G. and Hunt, K.H., *Geometry of Screw Systems.* Mech. Machine Theory **25** (1990), 1–27.

[11] Golubitsky, M. and Guillemin, V. *Stable Mappings and Their Singularities.* Springer Verlag, 1973.

[12] Hunt, K.H. *Kinematic Geometry of Mechanisms.* Clarendon Press, 1978.

[13] *Manipulating industrial robots – vocabulary,* ISO 8373, 1994.

[14] Nevins, J.L. and Whitney, D.E., *Assembly Research.* Automation **16** (1980), 595–613.

[15] Pottmann, H. and Wallner, J., *Computational Line Geometry*, Springer Verlag, 2001.

[16] Watson, P.C., *A Multidimensional System Analysis of the Assembly Process as Performed by a Manipulator* presented at 1st North American Robot Conf. (Chicago, 1976).

[17] Whitney, D.E. and Nevins, J.L., *What is the RCC and what can it do?* in Robot Sensors, Tactile and Non-Vision, ed. A.Pugh, IFS Publications, 1986, 3–15.

Matthew W. Cocke
Department of Mathematical Sciences
University of Liverpool
PO Box 147
Liverpool L69 3BX, United Kingdom

Peter Donelan
School of Mathematics, Statistics and Computer Science
Victoria University of Wellington
PO Box 600
Wellington, New Zealand
e-mail: peter.donelan@vuw.ac.nz

Christopher G. Gibson
Department of Mathematical Sciences
University of Liverpool
PO Box 147
Liverpool L69 3BX, United Kingdom
e-mail: c.g.gibson@liverpool.ac.uk

Real and Complex Singularities
Trends in Mathematics, 71–97

Vertices and Inflexions of Plane Sections of Surfaces in $\mathbb{R}^3$

André Diatta and Peter Giblin

Abstract. We discuss the behavior of vertices and inflexions of one-parameter families of plane curves which include a singular member. These arise as sections of smooth surfaces by families of planes parallel to the tangent plane at a given point. We cover all the generic cases, namely elliptic, umbilic, hyperbolic, parabolic and cusp of Gauss points. This work is preliminary to an investigation of symmetry sets and medial axes for these families of curves, reported elsewhere.

Mathematics Subject Classification (2000). 53A04; 53A05; 53A55; 51L15; 34C23; 68U05; 14H50; 14Q05; 14Q10; 51N05.

Keywords. Isophote curve, symmetry set, medial axis, skeleton, vertex, inflexion, plane curve, shape analysis.

1. Introduction

Let M be a smooth surface, and $\mathbf{p}$ be a point of M. We shall consider the intersection of M with a family of planes parallel to the tangent plane at $\mathbf{p}$. This family of plane curves contains a singular member, when the plane is the tangent plane itself; generically the other members of the family close to the tangent plane are nonsingular curves.

The motivation for this work comes from computer vision, where the surface is the intensity surface $z = f(x, y)$ corresponding to the intensity function f of a two-dimensional image, and the plane curves are level sets of this function, that is, isophotes. A great deal of information about the shape of these level sets and the way they evolve through the singular level set is contained in the family of so-called *symmetry sets* and *medial axes* of the level sets (see for example [9]).

This work is a part of the DSSCV project supported by the IST Programme of the European Union (IST-2001-35443.) The first author was supported and the second author partially supported by this grant. We are also very grateful to Terry Wall, Bill Bruce and Vladimir Zakalyukin for helpful suggestions.

These sets in turn take some of their structure from the pattern of vertices and inflexions (curvature extrema and zeros) of the level set.

In this article we concentrate on the vertices and inflexions, and apply this and other results to the study of symmetry sets in articles to appear elsewhere [6, 7]. Besides the patterns of vertices and inflexions we also study the limiting curvatures at the vertices as the level set approaches the singular member of the family.

The contact between a surface and its tangent plane at **p** is an *affine invariant* of the surface. Likewise the inflexions on the intersections with nearby planes are affine invariants, but we are also interested in the curvature extrema on these sections, and these are *Euclidean* invariants. For a generic surface M, the contact between the surface and its tangent plane at a point **p**, as measured by the height function in the normal direction at **p**, can be of the following types. See for example [11] for the geometry of these situations, and [4, 5, 10] for an extensive discussion of the singularity theory.

- The contact at **p** is ordinary ('A_1 contact'), at an elliptic point or at a hyperbolic point (occupying *regions* of M). The intersection of M with its tangent plane at **p** is locally an isolated point or a pair of transverse smooth arcs. (As regards contact there is no distinction between 'ordinary' elliptic points and umbilics, where the principal curvatures coincide. But as we shall see there is a great deal of difference when we consider vertices of the plane sections.)
- The contact is of type A_2 at parabolic points (generically forming *smooth curves* on M), where the asymptotic directions coincide. The intersection of M with its tangent plane at **p** is locally a cusped curve.
- The contact is of type A_3 at a cusp of Gauss, where the parabolic curve is tangent to the asymptotic direction at **p** (these are *isolated points* of M). There are two types, the elliptic cusp and the hyperbolic cusp. The intersection with the tangent plane is locally an isolated point or a pair of tangential arcs.

Other authors have considered the A_1 cases, using different techniques and with slightly different motivations from ours. Vertices in the A_1 case are studied by Uribe-Vargas in [14] and inflexions in the same case by Garay in [8], using more sophisticated techniques of singularity theory aimed at finding normal forms up to an appropriate equivalence. Our very detailed results on the other hand combine vertices and inflexions and apply to all three cases A_1, A_2, A_3 above. They are obtained by direct calculation: our motivation, as above, is to facilitate investigation of the symmetry sets of surface sections, and we do not as yet know how to fit this into a more general theory.

Here is a simple example. Consider a round torus in 3-space, obtained by rotating a circle about an axis in the plane of the circle but not intersecting it. This consists of elliptic and hyperbolic points, separated by two circles of parabolic points along the 'top' and 'bottom' of the torus. (The parabolic curves are far from generic but we shall stay clear of them.) We can take sections by planes parallel to the axis of rotation, as in Figure 1. The sections pass from a connected curve through a nodal curve (a 'figure eight') to two ovals. In the figure we have drawn the

FIGURE 1. Two plane sections of a torus close to a singular section, together with their evolutes. One connected component becomes two ovals, and four inflexions disappear while, locally, two vertices become six.

evolutes of the nonsingular sections: these have cusps at the centres of curvature of the vertices. As the connected curve splits, two vertices (maxima of curvature) come into coincidence at the crossing. After the transition, when there are two components of the curve, *three* vertices (one maximum and two minima of curvature) emerge from the crossing on each component. This transition, two local vertices becoming six local vertices, is written '1+1 $\leftrightarrow$ 3+3'. As regards transitions on vertices this is one of two generic situations at a hyperbolic point on a surface. However as regards inflexions it is special since the figure-eight level curve itself has an inflexion on each branch at the crossing point (in the terminology of [11, p.282] the crossing point is a flecnode for both asymptotic directions). This allows a transition on the inflexions of the plane sections whereby 2+2 $\leftrightarrow$ 0+0: two inflexions on each branch becomes none. Taking into account both vertices and inflexions this becomes one of the types '$\mathbf{H}_7$' below. If we take a hyperbolic point on the torus at which the tangent plane is *not* parallel to the axis of rotation, then it can be shown that neither branch of the nodal curve has an inflexion. While the transition on vertices remains as 1+1 $\leftrightarrow$ 3+3, the inflexions become 2+0 $\leftrightarrow$ 1+1 or 1+1 $\leftrightarrow$ 2+0; this is one of the '$\mathbf{H}_1$' cases in the notation below, which occur in regions of the surface.

The paper is organized as follows. In §2 we state the main results in the various cases. In §3 we describe the various patterns which arise in the hyperbolic case, and the limiting curvatures. The hyperbolic region of our surface M is divided into subregions according to the possible patterns, separated by a set which we call the vertex transition (VT) set. This set consists of those hyperbolic points $\mathbf{p}$ in M for which one of the smooth local components of the intersection between M and the tangent plane at $\mathbf{p}$ has a vertex. The VT set is difficult to calculate in particular cases but in §3.3 we give some examples and explain how the VT set approaches the parabolic curve on M. In §4 we turn to the elliptic case, concentrating on umbilic points since the general elliptic case is very simple. In §5 and §6 we cover the remaining cases, parabolic point and cusp of Gauss respectively. Finally in §7 we summarize and add some remarks on the material of the paper.

2. The vertex and inflexion sets

We always assume that our surface is locally given by an equation $z = f(x, y)$ for some smooth function f, with the tangent plane at the origin given by $z = 0$. Thus our family of curves is $f(x, y) = k$ for constants k close to 0, and (x, y) close to $(0, 0)$. (In some cases the set $f(x, y) = k$ is non-empty only for one sign of k.) We also take the x and y axes to be in principal directions at the origin, so that the surface M assumes the local Monge form

$$\begin{aligned} f(x,y) = {} & \tfrac{1}{2}(\kappa_1 x^2 + \kappa_2 y^2) + b_0 x^3 + b_1 x^2 y + b_2 x y^2 + b_3 y^3 \\ & + c_0 x^4 + c_1 x^3 y + c_2 x^2 y^2 + c_3 x y^3 + c_4 y^4 \\ & + d_0 x^5 + d_1 x^4 y + d_2 x^3 y^2 + d_3 x^2 y^3 + d_4 x y^4 + d_5 y^5 + \text{ h.o.t.} \end{aligned} \tag{1}$$

where κ_1, κ_2 are the principal curvatures at $\mathbf{p}$. We often scale the surface (multiply x, y and z by the same nonzero constant) so that $\kappa_1 = 2$ and the coefficient of x^2 is therefore 1.

We use subscripts to denote partial derivatives: $f_x = \frac{\partial f}{\partial x}$ etc.

To such f we assign two functions V_f and I_f whose zero-level sets $V_f = 0$ and $I_f = 0$ are respectively the sets of all vertices and inflexions of the plane curves $f(x, y) = k$ for constants k. We shall consider both the 'vertex function' V_f and the 'vertex set' $V_f = 0$. In fact for each of the generic cases of elliptic, hyperbolic, parabolic and cusp of Gauss points of M we shall go through the following steps.

- Calculate $V_f = 0$ and $f = 0$ and their Taylor expansions at the origin. In each case there will be several branches, some of which may be singular. (The same also applies to I_f.)
- Decide the possible relative positions of the branches of $f = 0$ and $V_f = 0$ (and $I_f = 0$). These can be indicated on diagrams.
- For k small, the level sets $f = k$ are close to the zero level set $f = 0$. We can read off the pattern of vertices (and inflexions) from the diagrams above.
- Calculate the limiting curvature at vertices of the section $f = k$, when $k \to 0$.

To obtain the function V_f we argue as follows. We want to find the vertices on a *smooth* curve $f(x, y) = k$. For this purpose we may assume that locally the curve is given by $y = h(x)$ for a smooth h, that is $f(x, h(x)) = k$ is an identity. Then the vertex condition is simply $\kappa'(x) = 0$ where $\kappa(x) = \frac{h''(x)}{(1+(h'(x))^2)^{3/2}}$ is the curvature of $y = h(x)$. Working out the derivatives of h in terms of those of f and clearing denominators we arrive at the following. The vertices of any smooth curve $f(x, y) = k$ will be at the intersections with the set $V_f = 0$, where

$$\begin{aligned} V_f = {} & (f_x^2 + f_y^2)(-f_y^3 f_{xxx} + 3 f_x f_y^2 f_{xxy} - 3 f_x^2 f_y f_{xyy} + f_x^3 f_{yyy}) \\ & + 3 f_x f_y (f_y^2 f_{xx}^2 + (f_x^2 - f_y^2) f_{xx} f_{yy} - f_x^2 f_{yy}^2) \\ & + 6 f_x f_y f_{xy}^2 (f_x^2 - f_y^2) \\ & + 3 f_{xy} (f_{xx} f_y^4 - 3 f_x^2 f_y^2 (f_{xx} - f_{yy}) - f_{yy} f_x^4). \end{aligned} \tag{2}$$

The square of the curvature, κ^2, of the curve $f(x, y) = k$ at (x, y) is

$$\kappa^2 = \frac{(f_{xx}f_y^2 - 2f_{xy}f_xf_y + f_{yy}f_x^2)^2}{(f_x^2 + f_y^2)^3}, \tag{3}$$

so that the *inflexion condition* is $I_f = 0$ where

$$I_f(x, y) = f_{xx}f_y^2 - 2f_{xy}f_xf_y + f_{yy}f_x^2 \tag{4}$$

is the usual *Hessian determinant* of f.

The following result gives the number of intersections of the level set $f(x, y) = k$ with $V_f = 0$ and $I_f = 0$, as k passes through 0.

Theorem 2.1. *Let $f = k$ be a section of a generic surface M by a plane close to the tangent plane at $\mathbf{p}$, $k = 0$ corresponding with the tangent plane itself. Then for every sufficiently small open neighborhood U of $\mathbf{p}$ in M, there exists $\varepsilon > 0$ such that $f = k$ has exactly $v(\mathbf{p})$ vertices and $i(\mathbf{p})$ inflexions lying in U, for every $0 < |k| \leq \varepsilon$, where $v(\mathbf{p})$ and $i(\mathbf{p})$ satisfy the following equalities. We also use $\leftrightarrow$ to indicate the numbers of vertices or inflexions on either side of a transition, local to the singular point on $f = 0$, when $f = k$ has two branches. The notation $m + n$ indicates the numbers of vertices or inflexions on the two branches.*

(E) *If $\mathbf{p}$ is an elliptic point, then for one sign of k the section is locally empty; in the non-umbilic case, for the sign of k yielding a locally nonempty intersection we have $v(\mathbf{p}) = 4$, $i(\mathbf{p}) = 0$. Likewise if $\mathbf{p}$ is a generic[1] umbilic point, then $v(\mathbf{p}) = 6$, $i(\mathbf{p}) = 0$. (This is already well known: see for example* [13, §15.3].)

(H) *If $\mathbf{p}$ is a hyperbolic point $v(\mathbf{p})$ satisfies one of the following.*
For $\mathbf{p}$ lying in open regions of M we have
$2 + 2 \leftrightarrow 2 + 2$ *or* $1 + 1 \leftrightarrow 3 + 3$.
In other cases, occurring along curves or at isolated points of M, we can have in addition
$3 + 2 \leftrightarrow 2 + 1$ *or* $3 + 1 \leftrightarrow 2 + 2$.
See §3.1 *for an explanation of the different cases.*

Also using the same notation, $i(\mathbf{p})$ satisfies: $1 + 1 \leftrightarrow 0 + 2$ *or* $1 + 2 \leftrightarrow 0 + 1$*; the full list is in Table* 2.

(P) *If $\mathbf{p}$ is a parabolic point but not a cusp of Gauss, $v(\mathbf{p}) = 3$, $i(\mathbf{p}) = 2$.*

(ECG) *If $\mathbf{p}$ is an elliptic cusp of Gauss, $v(\mathbf{p}) = 4$, $i(\mathbf{p}) = 2$ for one sign of k, and the level set is empty for the other.*

(HCG) *If $\mathbf{p}$ is a hyperbolic cusp of Gauss, we have:*
$v(\mathbf{p}) : 1 + 3 \leftrightarrow 4 + 4$ *or* $2 + 2 \leftrightarrow 4 + 4$,
and for each of these, we can have any of
$i(\mathbf{p}) : 1 + 1 \leftrightarrow 0 + 0$ *or* $2 + 2 \leftrightarrow 0 + 2$ *or* $1 + 1 \leftrightarrow 0 + 4$

[1]The genericity assumption can be stated explicitly: the quadratic terms of f should not divide the cubic terms. See §4.

We split the proof of Theorem 2.1 into different cases discussed in the relevant sections, in which we also carry out a closer investigation of the geometry of the sets $V_f = 0$ and $I_f = 0$.

3. Hyperbolic case

Recall that at a hyperbolic point $\mathbf{p}$ of a surface, the principal curvatures κ_1, κ_2 are not zero and have opposite signs. After scaling, f can be taken in (1) to have quadratic part $x^2 - a^2y^2$ where $a > 0$. We shall write V_h for V_f in this case, and likewise I_h for I_f.

3.1. Patterns of vertices and inflexions on the level sets

Proposition 3.1.

(i) *The vertex set* $V_h = 0$ *has exactly four smooth branches* VH_1, VH_2, VH_3, VH_4 *through* $(0, 0)$*, where* VH_1 *is tangent to the principal direction* $x = 0$*,* VH_2 *is tangent to the principal direction* $y = 0$*,* VH_3 *is tangent to the asymptotic direction* $x - ay = 0$ *and* VH_4 *is tangent to the asymptotic direction* $x + ay = 0$.

(ii) *The level sets* $f = 0$ *and* $I_h = 0$ *have exactly two smooth branches in a neighborhood of* $(0, 0)$*, one of them being tangent to* $x - ay = 0$ *and the other one to* $x + ay = 0$.

The proof for V_h can be done in several ways. We can use the technique exemplified in §5.1, that is, blowing up combined with the implicit function theorem, or, in the present case, we can even prove that V_h is $\mathcal{R}$-equivalent as a function to its lowest terms, which are $192a^4(1 + a^2)xy(x - ay)(x + ay)$. The functions f and I_h are Morse functions, hence equivalent to their quadratic parts.

In order to verify the conclusions of Theorem 2.1 in the hyperbolic case we need to determine the relative positions of the branches of $f = 0$ and $V_f = 0$ (and $I_f = 0$) which are tangent to one another at the origin. To do this we need the higher terms of the Taylor expansions of those branches with the same tangents. The branches VH_1 and VH_2 present no problems since they are always transverse to the branches of the level set $f = 0$. For the branches VH_3 and VH_4 we use Proposition 3.1 and substitute for example $x = ay + x_2y^2 + x_3y^3 +$ higher terms, into the expression of the vertex set V_h, for the branch VH_3.

Notation. Certain expressions occur often in our formulae so we introduce some notation for them.

$f^{(n)}(a)$ means the result of substituting $x = a,\ y = 1$ in the homogeneous part of degree n in the Taylor expansion of f. (We write this rather than the more precise $f^{(n)}(a, 1)$.)

For example,
$f^{(3)}(a) = b_0a^3 + b_1a^2 + b_2a + b_3$, and $f^{(4)}(a) = c_0a^4 + c_1a^3 + c_2a^2 + c_3a + c_4$.

Similarly $f^{(n)}(-a)$ is the result of substituting $x = -a, y = 1$ in the same homogeneous polynomial of degree n.

Proposition 3.2.

(i) *The branches VH_3, VH_4 of the vertex set have the following 3-jets:*

$$VH_3: \quad x = ay - \frac{1}{2a}f^{(3)}(a)y^2 + \frac{1}{4a^3(1+a^2)}\Big(f^{(3)}(a)(3b_0a^5 + b_1a^4 + (5b_0 - b_2)a^3 + (3b_1 - 3b_3)a^2 + b_2a - b_3) - 4a^2(1+a^2)f^{(4)}(a)\Big)y^3$$

$$= ay + x^+_{2v}y^2 + x^+_{3v}y^3 \ \textit{say},$$

$$VH_4: \quad x = -ay + x^-_{2v}y^2 + x^-_{3v}y^3 (\textit{obtained by replacing } a \textit{ with } -a \textit{ in the above.})$$

(ii) *The branches of $f = 0$ have the following 3-jets:*

$$x = ay - \frac{1}{2a}f^{(3)}(a)y^2 + \frac{1}{8a^3}\Big(f^{(3)}(a)(5b_0a^3 + 3b_1a^2 + b_2a - b_3) - 4a^2f^{(4)}(a)\Big)y^3$$

$$= ay + x^+_{2v}y^2 + x^+_{3f}y^3 \ \textit{say, and}$$

$$x = -ay + x^-_{2v}y^2 + x^-_{3f}y^3 \ \textit{obtained by replacing } a \textit{ with } -a \textit{ in the above.}$$

It is evident, from (i) and (ii) of Proposition 3.2, that the branches of vertex set and those of the curve $f = 0$ have at least 3-point contact at the origin: their Taylor expansions agree up to order two. This also means that they have the same osculating circle (circle of curvature) at the origin. The condition for them to have (at least) 4-point contact is that the terms in y^3 agree also. After some manipulation, this 4-point contact condition comes to the following.

Proposition 3.3. Four-point contact condition *The condition for the vertex branch VH_3 to have (at least) 4-point contact with the corresponding branch of $f = 0$ at the origin is*

$$f^{(3)}(a)\left(b_0a^5 - b_1a^4 + (5b_0 - 3b_2)a^3 - (5b_3 - 3b_1)a^2 + b_2a - b_3\right) - 4a^2(1+a^2)f^{(4)}(a) = 0. \quad (5)$$

The condition for VH_4 to have (at least) 4-point contact with the corresponding branch of $f = 0$ is obtained by replacing a by $-a$:

$$f^{(3)}(-a)\left(b_0a^5 + b_1a^4 + (5b_0 - 3b_2)a^3 + (5b_3 - 3b_1)a^2 + b_2a + b_3\right) + 4a^2(1+a^2)f^{(4)}(-a) = 0. \quad (6)$$

For a generic surface M, (5) or (6) then imposes one condition on the point $\mathbf{p}$ and can therefore be expected to hold for points $\mathbf{p}$ along one or more *curves* on M. We call this the *vertex transition set* (VT set) on M.

Remarks 3.4.

(1) The apparently rather complicated conditions in Proposition 3.3 actually state that *one or other of the branches of the curve* $f = 0$ *itself – the intersection between the surface* $M : z = f(x, y)$ *and its tangent plane – has a vertex.*

In fact we have the general result:

> *For any* f *giving a hyperbolic point at the origin, a branch of the curve* $f = 0$ *and the corresponding branch of the vertex set have the same order of contact with their common osculating circle.*

Thus at a point of the VT set, the corresponding branches of $f = 0$ and of the vertex set both have vertices. We shall not use this fact here, but discuss the result and its consequences elsewhere.

(2) Note in particular that (5) holds if $x - ay$ is a factor of both the cubic and quartic terms of the expansion of f. This is a *biflecnode* in the terminology of Koenderink [11, p.296]. As a special case, one of (5), (6) will hold at every point of a ruled surface, since the whole line in one of the asymptotic directions lies on the surface. From the point of view of the VT set, both ruled surfaces and surfaces of revolution (see §3.3) are highly non-generic.

Note that the 4-point contact condition can be regarded as a formula for $f^{(4)}(\pm a)$ in terms of the lower degree coefficients of the expansion of f. It can therefore be regarded as a formula for any of the degree 4 coefficients c_i in terms of the other c_j and lower degree coefficients of f. In a similar way we can write down the additional condition for VH_3 or VH_4 and the corresponding branch of $f = 0$ to have 5-point contact. This can be written in the form $f^{(5)}(\pm a) =$ a polynomial in the lower degree coefficients, but it is complicated and we shall not display it here. (As noted above, this is equivalent to the branch of $f = 0$, or of the vertex set, having a higher vertex.)

Analyzing in a similar way the Taylor expansions of the inflexion function I_h we find the following.

Proposition 3.5. *The branches of the inflexion curve* $I_h = 0$ *have the following 3-jets:*

$$\begin{aligned} x &= ay + \frac{1}{8a^3}\Big(-3f^{(3)}(a)(3b_0a^3 + b_1a^2 - b_2a - 3b_3) + 8a^2 f^{(4)}(a)\Big)y^3 \\ &= ay + x_{3i}^{+}y^3 \ \textit{say}, \end{aligned}$$

and

$$x = -ay + x_{3i}^{-}y^3 \ \textit{obtained by replacing } a \textit{ with } -a \textit{ in the above.}$$

Symbol	$x = ay$ branch	$x = -ay$ branch	'codim'	Comment
$\mathbf{H}_1$	$\mathbf{V}_1\mathbf{I}_1$	$\mathbf{V}_1\mathbf{I}_1$	0	the most generic case
$\mathbf{H}_2$	$\mathbf{V}_2\mathbf{I}_1$	$\mathbf{V}_1\mathbf{I}_1$	1	along curves in the VT set
$\mathbf{H}_3$	$\mathbf{V}_2\mathbf{I}_1$	$\mathbf{V}_2\mathbf{I}_1$	2	self-intersections of the VT set
$\mathbf{H}_4$	$\mathbf{V}_3\mathbf{I}_1$	$\mathbf{V}_1\mathbf{I}_1$	2	isolated points of the VT set
$\mathbf{H}_5$	$\mathbf{V}_1\mathbf{I}_2$	$\mathbf{V}_1\mathbf{I}_1$	1	curves in the hyperbolic region
$\mathbf{H}_6$	$\mathbf{V}_2\mathbf{I}_1$	$\mathbf{V}_1\mathbf{I}_2$	2	isolated points
$\mathbf{H}_7$	$\mathbf{V}_1\mathbf{I}_2$	$\mathbf{V}_1\mathbf{I}_2$	2	isolated points
$\mathbf{H}_8$	$\mathbf{V}_2\mathbf{I}_3$	$\mathbf{V}_1\mathbf{I}_1$	2	$\mathbf{V}_2\mathbf{I}_2$ and $\mathbf{V}_1\mathbf{I}_3$ do not occur

TABLE 1. The possibilities for contact between $f = 0$ and the vertex and inflexion curves. See Propositions 3.3 and 3.6, and Lemma 3.7 for further information.

Note that there are no quadratic terms in these expansions: the branches of the inflexion curve $I_h = 0$ *themselves have inflexions* at the origin. Accordingly the branches of $I_h = 0$ and $f = 0$ tangent to $x = ay$, say, have 2-point contact unless the branch of $f = 0$ also has an inflexion (that is, $f^{(3)}(a) = 0$).

Altogether the possibilities for contact between branches of the vertex and inflexion sets and the branches of $f = 0$ in the present hyperbolic cases are as follows.

Notation

- $\mathbf{V}_1$ A branch of $f = 0$ and of $V_h = 0$ have the minimum 3-point contact,
- $\mathbf{V}_2$ A branch of $f = 0$ and of $V_h = 0$ have 4-point contact; see (5) or (6),
- $\mathbf{V}_3$ A branch of $f = 0$ and of $V_h = 0$ have 5-point contact,
- $\mathbf{I}_1$ A branch of $f = 0$ and of $I_h = 0$ have the minimum 2-point contact,
- $\mathbf{I}_2$ A branch of $f = 0$ and of $I_h = 0$ have 3-point contact.
- $\mathbf{I}_3$ A branch of $f = 0$ and of $I_h = 0$ have 4-point contact.

The possible ways of combining these at the two branches of $f = 0$ tangent to $x = \pm ay$ are therefore as shown in Table 1. Here 'codim' refers to the codimension of the locus of these points in the hyperbolic region.

For the 'most generic' case $\mathbf{H}_1$, we give in Figure 2 the three possible ways (up to rotation or reflection of the diagram) in which the different elements can intersect. We use the notation $2+2 \leftrightarrow 2+2$ and $1+1 \leftrightarrow 3+3$ to indicate the numbers of vertices on the pair of branches of $f=k$ for small k first of one sign and then of the other. The inflexions in the first case follow the pattern $2+0 \leftrightarrow 1+1$ whereas in the second case the two patterns $2+0 \leftrightarrow 1+1$ and $1+1 \leftrightarrow 2+0$ occur. Examining cases we find the following.

Proposition 3.6.

(i) *In the case* $\mathbf{H}_1$, *the vertex transition* $2+2 \leftrightarrow 2+2$ *occurs when the left-hand sides of* (5) *and* (6) *have opposite signs.*

(ii) *The vertex transition* $1+1 \leftrightarrow 3+3$ *occurs when the left-hand sides of* (5) *and* (6) *have the same sign.*

 (iia) *The inflexion transition* $1+1 \leftrightarrow 2+0$ *occurs when, in addition to* (ii), *'left-hand sides both* <0*' is accompanied by* $f^{(3)}(a)f^{(3)}(-a)>0$ *and 'left-hand sides both* >0*' by* $f^{(3)}(a)f^{(3)}(-a)<0$.

 (iib) *The inflexion transition* $2+0 \leftrightarrow 1+1$ *occurs when, in addition to* (ii), *'left-hand sides both* <0*' is accompanied by* $f^{(3)}(a)f^{(3)}(-a)<0$ *and 'left-hand sides both* >0*' by* $f^{(3)}(a)f^{(3)}(-a)>0$.

The other cases also require an analysis of the order of branches of $f=0, V_h=0, I_h=0$ around each of the lines $x=\pm ay$. In Figure 3 the principal cases for a single branch tangent to $x=ay$ are illustrated. By putting these together with similar information at $x=-ay$, and including the other branches of $V_h=0$ tangent to the two coordinate axes (see Proposition 3.1) we arrive at the classification in Table 2.

Here is an indication of the calculations which allow us to draw the cases in Figure 3.

Lemma 3.7.

(i) *The condition for* $\mathbf{V}_1\mathbf{I}_2$ *on the branch tangent to* $x=ay$ *is* $f^{(3)}(a)=0$, $f^{(4)}(a)\neq 0$ *and the configuration of* $f=0, V_h=0$, *and* $I_h=0$ *is determined by the sign of* $f^{(4)}(a)$, *as in Figure* 3.

(ii) *The conditions for* $\mathbf{V}_2\mathbf{I}_2$ *or* $\mathbf{V}_1\mathbf{I}_3$ *on the branch tangent to* $x=ay$ *are* $f^{(3)}(a)=f^{(4)}(a)=0, f^{(5)}(a)\neq 0$, *and this situation is in fact* $\mathbf{V}_2\mathbf{I}_3$. *The configuration of* $f=0, V_h=0$ *and* $I_h=0$ *is determined by the sign of* $f^{(5)}(a)$, *as in Figure* 3.

Proof For (i), note that we require the branches of $I_h=0$ and $f=0$ tangent to $x=ay$ to have the same 2-jet, and using the formulae of Propositions 3.2 and 3.5 this requires $f^{(3)}(a)=0$. They have different 3-jets provided $f^{(4)}(a)\neq 0$, since the coefficients of y^3 in the two Taylor series are then $\frac{1}{a}f^{(4)}(a)$ and $-\frac{1}{2a}f^{(4)}(a)$ respectively. Since the coefficient of y^3 in the Taylor series of V_h is $-\frac{1}{a}$ we find the two orderings of the branches depicted in Figure 3.

For (ii) we use the same Propositions, noting that $\mathbf{I}_2$ together with $\mathbf{V}_2$ imply $f^{(3)}(a)=f^{(4)}=0$ which in turn imply that the 3-jets of the branches of $V_h=0$ and $f=0$ agree. Further calculations then show that the coefficients of y^4 in

Symbol	Vertex transitions	Inflexion transitions	Comment
$\mathbf{H}_1$ (i)	$2+2 \leftrightarrow 2+2$	$2+0 \leftrightarrow 1+1$	Figure 2(i)
(iia) (iib)	$1+1 \leftrightarrow 3+3$	$1+1 \leftrightarrow 2+0$ $2+0 \leftrightarrow 1+1$	Figure 2(iia) Figure 2(iib)
$\mathbf{H}_2$	$3+2 \leftrightarrow 2+1$	$1+1 \leftrightarrow 2+0$ $2+0 \leftrightarrow 1+1$ $0+2 \leftrightarrow 1+1$ $1+1 \leftrightarrow 0+2$	
$\mathbf{H}_3$	$3+1 \leftrightarrow 2+2$	$2+0 \leftrightarrow 1+1$ $1+1 \leftrightarrow 2+0$ $0+2 \leftrightarrow 1+1$	
$\mathbf{H}_4$			As for $\mathbf{H}_1$
$\mathbf{H}_5$ (i)	$2+2 \leftrightarrow 2+2$	$1+2 \leftrightarrow 1+0$	
(ii)	$1+1 \leftrightarrow 3+3$	$1+2 \leftrightarrow 1+0$	
$\mathbf{H}_6$	$3+2 \leftrightarrow 2+1$	$1+0 \leftrightarrow 2+1$	
$\mathbf{H}_7$ (i)	$2+2 \leftrightarrow 2+2$	$1+1 \leftrightarrow 1+1$	
(ii)	$1+1 \leftrightarrow 3+3$	$2+2 \leftrightarrow 0+0$	Torus, Figure 1
$\mathbf{H}_8$	$3+2 \leftrightarrow 2+1$	$1+1 \leftrightarrow 0+2$ $0+2 \leftrightarrow 1+1$	

TABLE 2. The transitions on vertices and inflexions in the hyperbolic case.

the three branch expansions are $f = 0 : -\frac{1}{2a}f^{(5)}(a)$, $V_h = 0 : -\frac{5}{2a}f^{(5)}(a)$ and $I_h = 0 : \frac{5}{2a}f^{(5)}(a)$ from which the results now follow.

3.2. Extrema of curvature and limiting curvature

In order to analyze the vertices further, we need to decide which vertices correspond to maxima and which to minima of curvature on the curve. (This is of significance when we apply the results to the symmetry set and the medial axis, since only minima – indeed absolute minima – can contribute to the latter.) We proceed as follows. The different branches of the vertex set locally divide the plane into regions where the derivative κ' of the curvature κ (with respect to any regular parametrisation of the curve) has a constant sign, the vertex branches being the loci of points where this derivative vanishes. Note that the sign of κ' does *not* depend on the orientation of the curve. However κ' has the same sign as the vertex condition $V_h(x, y)$.

To decide the sign of κ', for instance in the (local) region between the vertex branches tangent to $y = 0$ and $x - ay = 0$, let us then check the sign of $V_h(x, y)$

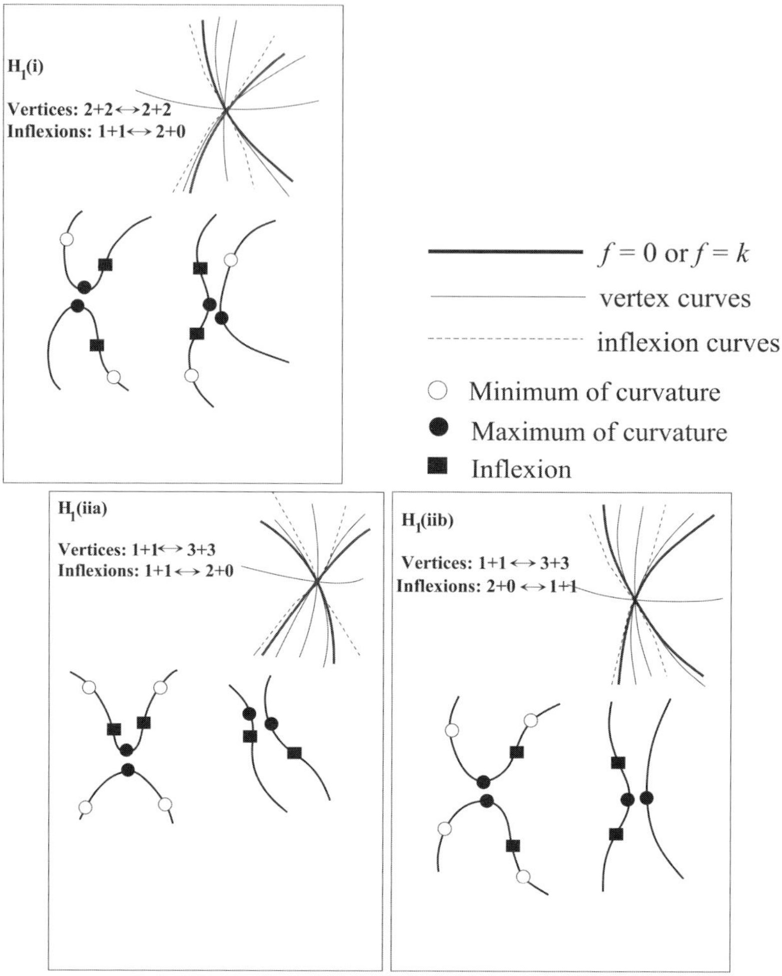

FIGURE 2. Arrangements of vertices and inflexions on the level sets of f, hyperbolic case $\mathbf{H}_1$ (see Table 2). In each case, we show, above, the vertex and inflexion curves – that is, the loci of vertices and inflexions on the level sets of f – and, below, a sketch of the level curves for $f < 0$, $f > 0$, showing the positions of these vertices and inflexions. The orientation chosen for the branches of $f = k$ is shown in Figure 4.

along the line $x = 2ay$, which is inside this region. Along this line, the sign of the vertex condition is positive, at least for y small, as the Taylor expansion of the vertex condition is: $V_h(2ay, y) = 1152(a^7 + a^9)y^4 + O(y^5)$ and $a > 0$.

We can complete the sign of κ' in all other regions by just alternating it before and after vertex branches. This completely describes the growth of κ on

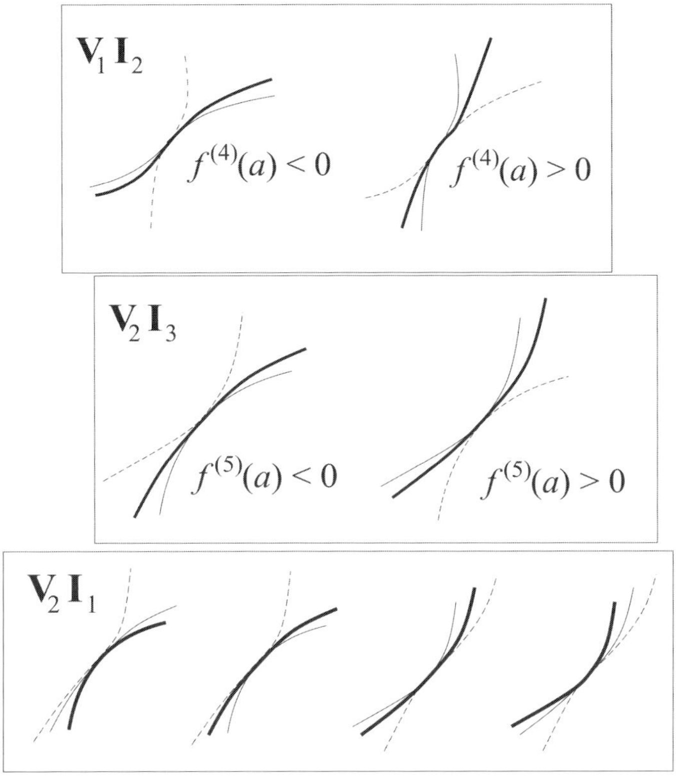

FIGURE 3. The arrangements of branches tangent to $x = ay$: thick line $f = 0$, thin solid line the vertex curve $V_h = 0$ and dashed line the inflexion curve $I_h = 0$. Three cases are illustrated here, the notation being that of Table 1.

the level sets of f in the plane. We shall always orient the branches of $f = k$ as indicated in Figure 4, and in this orientation κ will have a definite maximum or a minimum at a given vertex.

Proposition 3.8. *In the notation of Proposition* 3.1, *the limiting curvature of the level curves* $f = k$, $k \to 0$ *at vertices on the various branches is, up to sign,*

- *infinite, along* VH_1 *and* VH_2
- $f^{(3)}(a)/a(1+a^2)^{3/2}$, *along* VH_3
- $-f^{(3)}(-a)/a(1+a^2)^{3/2}$, *along* VH_4.

To prove this, we use the Taylor expansions of the branches of the vertex set, given above in Proposition 3.2, and the formula (3) for the square of the curvature of a plane curve. For the branch VH_1 we find the numerator and denominator of

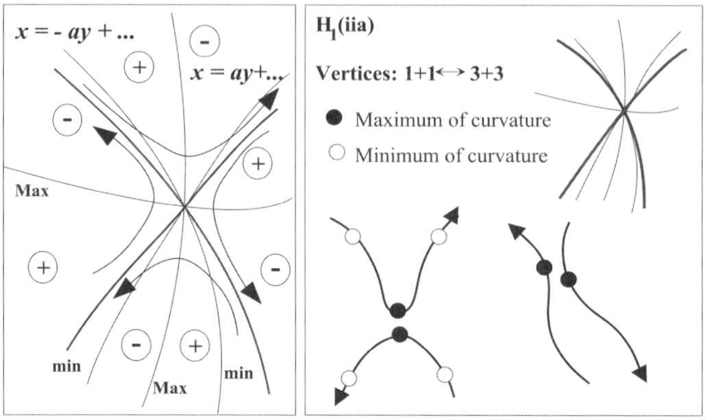

FIGURE 4. Case $\mathbf{H}_1$(iia) (compare Figure 2). The sign of κ': following the indicated orientations on $f = k$, before the curve $f = k$ intersects a vertex branch "Max", the derivative κ' of κ is positive, then vanishes at the vertex branch and becomes negative afterwards. So the intersections of the vertex branches "Max" and the curves $f = k$ are the vertices on $f = k$ where κ reaches a local maximum; likewise the 'min' describes the patterns of the local minima of curvature of $f = k$ when k goes through zero. The diagram on the right takes into account inflexions on $f = k$.

κ^2 come to $64a^8y^4 + O(y^5)$ and $64a^{12}y^6 + O(y^7)$ respectively, so that as $y \to 0$ the limiting curvature is infinite. The situation for VH_2 is similar.

For VH_3 the numerator and denominator come to $64a^4f^{(3)}(a)^2y^6 + O(y^7)$ and $64a^6(1+a^2)^3y^6 + O(y^7)$, which gives the required result. Note that this limiting curvature is zero precisely for a *flecnodal* point, at which the quadratic and cubic terms have a common factor $x - ay$. The limiting curvatures for both branches VH_3 and VH_4 are zero when the whole of the quadratic terms are a factor of the cubic terms, that is for the intersection of two flecnodal curves corresponding to different asymptotic directions on the surface M.

3.3. The vertex transition (VT) set

Given a generic surface M, we can apply our analysis to any point $\mathbf{p}$ of the surface: we are then looking at the family of plane sections of the surface close to the tangent plane section. The '4-point contact condition' (5) or (6) is generically expected to hold for points $\mathbf{p}$ along a set of *curves* on M, the vertex transition (VT) set. Of course the VT set lies entirely in the hyperbolic region, though it may have limit points on the parabolic set (see below); it separates those points where the family of sections parallel to the tangent plane exhibits behavior $\mathbf{H}_1$(i) in Table 2 from those exhibiting $\mathbf{H}_1$(ii).

It is clearly of interest to determine, for a given surface M, the subregions into which the hyperbolic region is separated by the VT set. This set can self-intersect, when *both* local branches of $f = 0$ have 4-point contact with the corresponding local branches of the vertex set $V_h = 0$: this is $\mathbf{H}_3$ in Table 1. Also there are special points on the VT set where a branch of the vertex set and $f = 0$ have 5-point contact: this is $\mathbf{H}_4$ in Table 1. Although the local conditions are quite easy to calculate – see the above formulae – it is not so easy to take a global surface and determine the VT set. We consider below the case of a surface of revolution M, which turns out to be non-generic in the sense that a point of M lies on both branches of the VT set or on neither. We also consider the limit points of the VT set on the parabolic curve of a general surface M.

Torus and surface of revolution. Consider a torus of revolution M in $\mathbb{R}^3$, obtained by rotating a circle about a line in its plane, not intersecting the circle. Naturally the VT set will be one or more circular 'latitude parallels' of the torus in view of the circular symmetry. In fact, for a circle of radius r rotating so that its centre describes a circle C of radius $R > r$, the two latitude parallels in the hyperbolic region of the torus making an angle $\cos^{-1}(r/R)$ with the plane of C lie on both branches of the VT set. Thus for points $\mathbf{p}$ on these two latitude parallels, *both* branches of the local intersection of M with its tangent plane have 4-point contact with the corresponding branches of the vertex set at $\mathbf{p}$ (or equivalently both branches of the intersection of M with its tangent plane have a vertex at $\mathbf{p}$). At other hyperbolic points of M *neither* branch has these properties. Crossing the VT set we therefore cross it twice, so that, apart from points $\mathbf{p}$ on the VT set itself, the pattern of vertices on sections of M parallel to the tangent plane at $\mathbf{p}$ is always the same. In fact we find that, in the expansion of the torus in Monge form at any hyperbolic point, the coefficients b_1, b_3, c_1, c_3 are all zero. It is clear that, in this situation, the two expressions in Proposition 3.3 become identical so that, in the case $\mathbf{H}_1$ of Theorem 2.1, only (ii) is possible. Thus all hyperbolic points away from the VT set exhibit the same pattern of vertices. Interestingly, when we consider inflexions, then both possibilities in Table 2 occur. In fact let $\mathbf{p}$ be a point of the torus of the form $(r \sin t, 0, R + r \cos t)$ (where the axis of rotation is the x-axis and we can without loss of generality take $\mathbf{p}$ to be in the xz-plane). Then using Proposition 3.6 we find that if $-r/R < \cos t < 0$ then the inflexion transition is $1 + 1 \leftrightarrow 2 + 0$ but if $-1 < \cos t < -r/R$ it is $2 + 0 \leftrightarrow 1 + 1$. Note that $\cos t < 0$ since $\mathbf{p}$ is hyperbolic, and $t = \pi$ gives the symmetrical case $\mathbf{H}_7$(ii) of Table 2 and Figure 1.

The same happens in fact for any surface of revolution generated by rotating a plane curve, say in the x, z-plane, about the z-axis. We find that b_1, b_3, c_1, c_3 are all zero and the conclusion follows as before. If we rotate the curve $y = 0$, $x = a + bz + cz^2 + dz^3 + ez^4 + \cdots$ about the z-axis then the condition for the point $(a, 0, 0)$ to be hyperbolic is $ac > 0$ and the condition for this point to lie on the VT set determines e uniquely in terms of a, b, c, d. For example, the curve $x = a + cz^2 - (c^2/2a)z^4$ has the latter property, as does $x = 4 - 2z + 2z^2 + z^3$.

The VT set and the parabolic curve. The analysis of sections parallel to the tangent plane at a parabolic point and at a cusp of Gauss is given in §§5, 6. Here we are concerned with the hyperbolic region near a parabolic point and we ask which type, $\mathbf{H}_1$(i) or $\mathbf{H}_1$(ii), the points of this region can be.

Suppose we consider a sequence of hyperbolic points tending to a parabolic point $\mathbf{p}$ of M. If we let $a \to 0$ in (5) and (6), the left-hand sides both tend to $-b_3^2$, since $f^{(3)}(0) = b_3$. Hence, if $b_3 \neq 0$ at a point $\mathbf{p}$ of the parabolic curve, then all hyperbolic points sufficiently close to $\mathbf{p}$ are of type $\mathbf{H}_1$(iia), by Proposition 3.6(ii). In particular the VT set cannot have a limit point on the parabolic curve except where $b_3 = 0$, that is at the cusps of Gauss. It is possible to calculate the local form of the VT set at cusps of Gauss; we find the following.

- At an *elliptic* cusp of Gauss $\mathbf{p}$ (in (1) $\mathbf{p} = (0,0,0)$ and $\kappa_2 = 0$, $b_2^2 < 2\kappa_1 c_4$ or, scaling κ_1 to 2, $b_2^2 < 4c_4$), there is locally no VT set.
- At a *hyperbolic* cusp of Gauss $\mathbf{p}$ (the previous inequalities are reversed), there is either locally no VT set, or locally a VT set consisting of two curves tangent to the parabolic curve at $\mathbf{p}$ and having inflexional contact with each other (equivalent by a change of coordinates in the parameter plane of M to $(x - y^3)(x + y^3) = 0$). The criterion separating these cases is the sign of a polynomial in coefficients of the Monge form of M at $\mathbf{p}$ of order ≤ 4, together with d_5. When $d_5 = 0$ a VT set exists if and only if c_4 lies between 0 and $20b_2^2 c_3(b_1 b_2 - c_3)/(4b_1 b_2 + c_3)^2$. There is a similar, slightly more complicated formula, for general d_5.

4. Elliptic points

We sketch this case for completeness; the chief interest for us lies in the symmetry set and medial axis in the umbilic case as in [6].

At an elliptic point, say $\mathbf{p} = (0,0,0)$ on a surface $z = f(x,y)$, the two principal curvatures are of the same sign, say positive: $\kappa_1 > 0, \kappa_2 > 0$. Using (1), the function f, after scaling of the variables x, y, z, is of the form $f_e(x,y) = x^2 + a^2y^2 + b_0x^3 + b_1x^2y + b_2xy^2 + b_3y^3 +$ h.o.t., where we may assume $a > 0$. We can distinguish two cases here: the generic case where $a \neq 1$ and the case $a = 1$ of umbilic points, where the principal curvatures are equal. Umbilic points are isolated points in the elliptic region of a surface. See Figure 5 for the vertex set and some level curves $f_e(x,y) = k$ in the umbilic case.

4.1. Proof of Theorem 2.1: elliptic case

The results on vertices in this case are well known; to deduce them from the function V_e note that it has 4-jet

$$-192a^4xy(a^2-1)(x^2+a^2y^2)$$

so that, when $a \neq 1$, there can be only two real branches of $V_e = 0$, with tangents $x = 0$ and $y = 0$.

The inflexion condition I_e has 2-jet $8a^2(x^2 + a^2y^2)$, hence the set $I_e = 0$ contains no real points apart from the origin.

The sections $f_e = k$ will therefore have four vertices for small $k > 0$, just as in the case of an ellipse.

The umbilic case. Let us consider the case $a = 1$. The vertex set $V_u = 0$ is now given by

$$\begin{aligned}\frac{1}{192}V_u &= px^5 - 3qx^4y - 2px^3y^2 - 2qx^2y^3 - 3pxy^4 + qy^5 + \text{ h.o.t.},\\ &= (x^2+y^2)(px^3 - 3qx^2y - 3pxy^2 + qy^3) + \text{ h.o.t.},\end{aligned}$$

where $p = b_3 - b_1,\ q = b_2 - b_0$. The discriminant of the form of degree 3 is

$$108(p^2+q^2)^2,$$

so that, unless $p = q = 0$ (which amounts to saying that $x^2 + y^2$ is a factor of the cubic terms), the discriminant is > 0 and the branches of $V_u = 0$ through the origin are distinct and exactly three of them are real. It follows that there are always six vertices on the section $f_u = k$ for small $k > 0$. (Compare [13, §15.3].) Not surprisingly, there are no inflexions on $f_u = k$. The inflexion set has equation $I_u = 0$, which has the form $8(x^2 + y^2)$+h.o.t.

Naturally, the curvature at the vertices tends to infinity as $k \to 0$ through positive values; in fact the curvature behaves like that of a circle of radius $\sqrt{k}$.

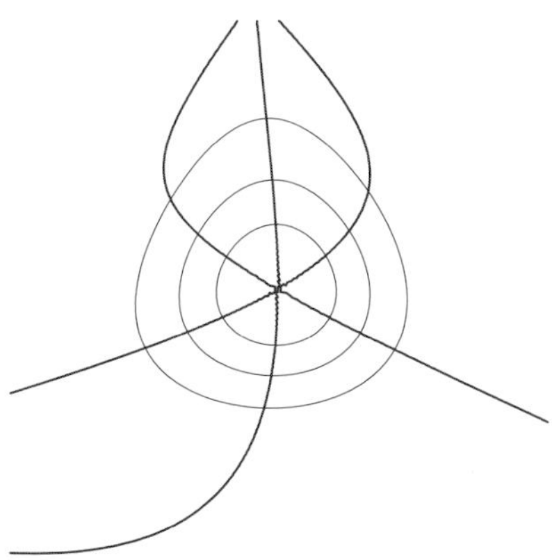

FIGURE 5. Loci of vertices in a 1-parameter family of level sets $f = k$ (closed curves), in the umbilic case. The vertex curve has three branches through the origin, giving rise to six vertices on the level set for all small k.

5. Parabolic case

At a parabolic point $\mathbf{p}$ the contact of the surface M with its tangent plane is of type A_2 at least; we consider the case of ordinary parabolic points where the contact is exactly A_2 in this section. One of the principal curvatures vanishes. After scaling of the variables x, y, z in (1) f can be written

$$f_p(x, y) = x^2 + b_0x^3 + b_1x^2y + b_2xy^2 + b_3y^3 + \text{higher order terms} \qquad (7)$$

where $b_3 \neq 0$. (The case $b_3 = 0$ is that of a cusp of Gauss; see §6.)

Proposition 5.1.
(i) *The vertex set $V_p = 0$ has three branches, one being smooth and the other two having ordinary cusps.*
(ii) *The inflexion set $I_p = 0$ has two branches, one smooth and one having an ordinary cusp.*
(iii) *The zero level set $f_p = 0$ has one branch, having an ordinary cusp.*

See §5.1 for the proof.

By the same method as in Proposition 3.5 we can show the following.

Proposition 5.2. *Suppose that $b_3 > 0$ (see the Remark below for the contrary case $b_3 < 0$).*
(i) *The smooth branch VP_1 of the vertex set has the following 3-jet:*
- VP_1: $(-\frac{1}{2}b_2t^2 + \frac{b_2(b_1-3b_3)-c_3}{2}t^3, t)$.

The two cusped branches of the vertex set have the following 4-jets:
- VP_2: $(x_3't^3 - \frac{1}{2}b_2t^4, -t^2)$,
- VP_3: $(x_3''t^3 - \frac{1}{2}b_2t^4, -t^2)$,

where $2x_3' = \sqrt{9+3\sqrt{3}}\sqrt{b_3}$, and $2x_3'' = \sqrt{9-3\sqrt{3}}\sqrt{b_3}$.

(ii) *The branches of the inflexion set can be parametrized as*
- $(3b_3t, -b_2t + \cdots)$ *(recall $b_3 \neq 0$)*
- $(\frac{1}{2}\sqrt{3b_3}t^3 - \frac{3}{8}b_2t^4 + \cdots, -t^2)$.

(iii) *The level set $f_p = 0$ has the following 5-jet:*
- $(\sqrt{b_3}t^3 - \frac{1}{2}b_2t^4 + \frac{b_2^2+4b_1b_3-4c_4}{8\sqrt{b_3}}t^5, -t^2)$.

Comparing the coefficients of the t^3-terms of the cuspidal branches in (i), (ii) and (iii) we have $\frac{1}{2}\sqrt{3b_3} < x_3'' < \sqrt{b_3} < x_3'$. It follows that the branch of $f_p = 0$ is always between the two cusped branches of the vertex set, and also the cusped branch of the inflexion set is inside all these three cusps. See Figure 6.

Hence each level curve $f_p = k$ has only three vertices, near the origin, for small k. Thus, when k passes though 0, the number of vertices of the curves $f_p = k$ remains unchanged: $3 \leftrightarrow 3$, as claimed in Theorem 2.1. The number of inflexions does not change as k passes through 0: each curve $f = k$ has two inflexions near the origin. Hence the transition of inflexions is $2 \leftrightarrow 2$.

Remark. If $b_3 < 0$, then in Proposition 5.2 we use $y = t^2$ instead of $y = -t^2$ and replace $\sqrt{b_3}$ by $\sqrt{-b_3}$ wherever it occurs. The two cases $'$ and $''$ in (i) are then reversed.

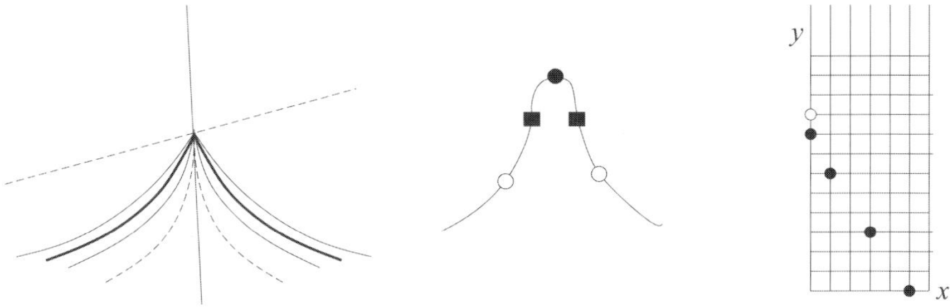

FIGURE 6. Left: a schematic picture of the vertex set $V_p = 0$ (thin solid line), the inflexion set $I_p = 0$ (dashed line) and the zero level set $f_p = 0$ (thick line) in the parabolic case. The vertex set has two cuspidal branches and one smooth branch, the inflexion set has one cuspidal branch and one smooth branch, and $f_p = 0$ has one cuspidal branch. The level set $f_p = k$ then evolves so that the number of vertices remains as 3 and the number of inflexions as 2 for both signs of k, with k small. Centre: a sketch of the level curve $f_p = k$ for $k \neq 0$, marking vertices (circles) and inflexions (squares). Right: the Newton polygon for the parabolic case; see §5.1.

5.1. Proof of Proposition 5.1

In this section we show briefly how we are able to deduce that the vertex and inflexion sets have branches as claimed above. We do so by looking at the Newton polygon and then applying the well-known techniques of blowing-up combined with the implicit function theorem. We give this example in detail; all the other cases encountered in this article can be dealt with similarly.

The Newton polygon for the function V_p contains the following monomials with coefficients: $192b_3x^5 + 864b_3^2x^3y^3 + 648b_3^3xy^6 + 324b_2b_3^3y^8$. Since $b_3 \neq 0$ all but the last term are definitely present. The last term is absent when $b_2 = 0$, which means that the parabolic curve is tangent to the *other* principal direction at the origin. When this is the case, there is a term $324b_3^3c_3y^9$, which will be present unless $c_3 = 0$. Generically $b_2 = c_3 = 0$ will not happen anywhere on our surface M. Thus the Newton polygon has terms x^5, x^3y^3, xy^6 and either y^8 or y^9; see Figure 6.

Let us write the above as $g(x, y) = ax^5 + bx^3y^3 + cxy^6 + dy^8$ and consider the case where $d \neq 0$. Note that a, b, c and $b^2 - 4ac = 248832b_3^4$ are all nonzero. The function V_p will then be of the form g + terms above the Newton polygon; we can think of the latter as linear combinations of monomials x^my^n where $3m + 2n > 15$ and $(m, n) \neq (0, 8)$. We first blow up by $x = ty$, so that the 'blow-down' transformation is $(t, y) \to (ty, y)$ and $y = 0$ is the exceptional divisor. The result after cancelling y^5 is

$$at^5 + bt^3y + cty^2 + dy^3 + \text{ linear combination of monomials } t^my^{m+n-5}.$$

Note that $m+n>5$ for all monomials above the Newton polygon. Hence intersecting with $y=0$ gives five coincident points at the origin $t=y=0$ (and there are no points sent to infinity, that is $y=tx$ produces no points on the exceptional divisor, using $a\neq 0$).

For the second blow-up we use $y=ut$, with blow-down map $(t,u)\to(t,ut)$; we find after cancelling t^3

$$at^2+but+cu^2+du^3+\text{ linear combination of monomials } u^{m+n-5}t^{2m+n-8},$$

and $2m+n-8>0$ for all monomials above the Newton polygon. This meets $t=0$ in $cu^2+du^3=0$, that is a cusp at the origin and a transverse crossing of the u-axis at $u=-c/d$, since $d\neq 0$. The transverse crossing provides us with a smooth branch of the blown-up curve $V_p=0$, parametrized by t, using the implicit function theorem, and by blowing-down we obtain one of the branches of our curve $V_p=0$ (in fact also a smooth branch). No further points are obtained from the alternative blow-up $t=uy$.

Blowing up the origin a third time, using $t=uw$, with blow-down map $(w,u)\to(uw,u)$, we obtain after cancelling u^2,

$$aw^2+bw+c+du+\text{ linear combination of monomials } u^{3m+2n-15}w^{2m+n-8},$$

and again $3m+2n-15>0$ for all points above the Newton polygon. Finally this meets the exceptional divisor $u=0$ in distinct points, since $b^2\neq 4ac$, each of which gives a transverse crossing of $u=0$ so that the two branches of the blown-up curve can be locally parametrized smoothly by u, using the implicit function theorem. These branches blow-down to the remaining two branches (actually ordinary cusps) of $V_p=0$.

5.2. The limiting curvature of $f_p=k$, at vertices

Here again, we would like to evaluate the curvature κ of $f_p=k$ at a vertex. Then we will take the limit of κ, as one approaches the parabolic point $\mathbf{p}=(0,0,0)$, along that vertex branch.

Proposition 5.3. *The limiting curvature of the level curves $f=k$ is infinite as $k\to 0$, at vertices on any of the branches of the vertex set.*

We substitute the parametrizations of the branches of $V_p=0$ given in Proposition 5.2 into the expression for κ^2 given in (3). The result is (in all cases using $b_3\neq 0$) for the branch VP_1, $\kappa^2\sim t^{-4}$, while for VP_2 and VP_3, $\kappa^2\sim t^{-2}$. The result follows.

6. Non-degenerate cusps of Gauss

For a non-degenerate cusp of Gauss the Monge form (1) can be written, after scaling the variables, as

$$f_g=x^2+b_0x^3+b_1x^2y+b_2xy^2+c_0x^4+c_1x^3y+c_2x^2y^2+c_3xy^3+c_4y^4+\text{ h.o.t}$$

where $b_2^2 - 4c_4 \neq 0$, that is, the lowest degree terms in the weighted sense, namely the x^2, xy^2 and y^4 terms, are non-degenerate. Since cusps of Gauss are isolated we can assume generically that other conditions on the coefficients are avoided. By changing the sign of x if necessary we can assume $b_2 > 0$.

There are two broad cases:

Elliptic cusp of Gauss: $b_2^2 - 4c_4 < 0$. Then the curve $f_g = k$ is locally a closed loop for $k > 0$ and empty for $k < 0$.

Hyperbolic cusp of Gauss: $b_2^2 - 4c_4 > 0$. Then $f_g = k$ has two local branches for $k \neq 0$ and two tangential branches for $k = 0$.

Note that the principal direction $x = 0$ is tangent to the parabolic curve at a cusp of Gauss. See [1] for an extensive discussion of cusps of Gauss, and [11, pp. 245, 276] for further geometrical information.

6.1. Vertices and inflexions on level sets at a cusp of Gauss

In a neighborhood of a cusp of Gauss, the vertex condition now reads

$$V_g = 192(c_3 - b_1b_2)x^6 + 192(4c_4 - b_2^2)x^5y + \text{h.o.t.}$$

Since $b_2^2 - 4c_4 \neq 0$, there will be a nonzero coefficient of x^5y here.

Proposition 6.1.

(i) *In the case of an elliptic cusp of Gauss, i.e.,* $b_2^2 - 4c_4 < 0$ *(the closed curve intersection), there are two smooth real branches of the vertex set* $V_g = 0$ *through the origin, one of which is tangent to the axis* $x = 0$*, and the other one to* $(b_2^2 - 4c_4)y = (c_3 - b_1b_2)x$*.*

(ii) *In the case* $b_2^2 - 4c_4 > 0$ *(hyperbolic cusp of Gauss), the vertex set has six smooth real branches* VG_i *for* $i = 1, \ldots, 6$*. All except* VG_6 *are tangent to* $x = 0$ *while* VG_6 *is tangent to* $(b_2^2 - 4c_4)y = (c_3 - b_1b_2)x$*.*

(iii) *If in addition to* $b_2^2 - 4c_4 > 0$*, we have* $b_2^2 - 8c_4 > 0$*, then the inflexion set has three smooth branches (see Figures 9 and 10), whereas when* $b_2^2 - 8c_4 < 0$*, there is only one smooth branch (see Figure 8).*

The claimed number of branches can be deduced from the Newton polygon in the same way as §5.1; the present case is easier. The Newton polygon for V_g is illustrated in Figure 7, right. The terms on the Newton polygon are

$$\begin{aligned}&192(c_3 - b_1b_2)x^6 - 192(b_2^2 - 4c_4)x^5y - 480b_2(b_2^2 - 4c_4)x^4y^3\\ &-48(b_2^2 - 4c_4)(7b_2^2 + 12c_4)x^3y^5 - 24b_2(b_2^2 - 4c_4)(b_2^2 + 36c_4)x^2y^7\\ &+24(b_2^2 - 4c_4)(b_2^4 - 10b_2^2c_4 - 16c_4^2)xy^9 + 24b_2c_4(b_2^2 - 4c_4)(b_2^2 - 8c_4)y^{11}.\end{aligned}$$

The key fact is this: ignoring the first term and then cancelling y, the remaining terms form a quintic polynomial in x and y^2 which has *distinct* roots; in fact it factorizes as

$$(b_2^2 - 4c_4)(2x + b_2y^2)(x^2 + b_2xy^2 + c_4y^4)(4x^2 + 4b_2xy^2 - (b_2^2 - 8c_4)y^4).$$

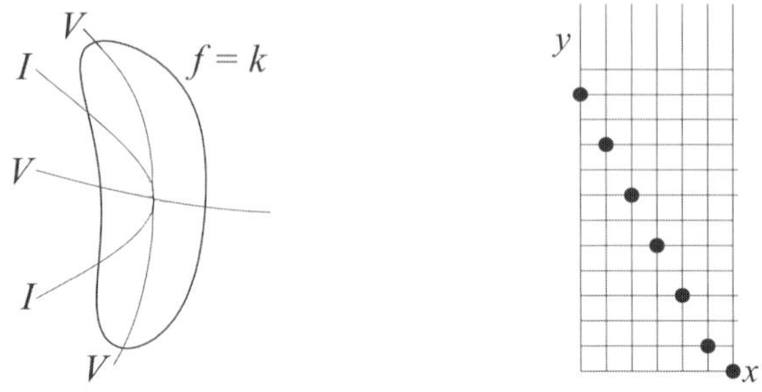

FIGURE 7. Left: Vertices and inflexions in the case of an elliptic cusp of Gauss: the curves marked V are the vertex set, those marked I are the inflexion set and $f = k$ is one level set of f. As k increases through 0, the curve passes from empty to one with four vertices and two inflexions. Right: the Newton polygon for a hyperbolic cusp of Gauss; compare §6.1.

The discriminant is a nonzero constant times $(b_2^2 - 4c_4)^{18}$ and the number of real roots is 1 for an elliptic cusp and 5 for a hyperbolic cusp. Two blow-ups $x = ty$ and $y = tu$ suffice to find the real branches of the singular point $V_g = 0$.

We can parametrize the branches of the inflexion set as follows. Substitute $x = x_1y + x_2y^2 + \dots$ in the inflexion condition $I_g = 0$; this gives the solution $x_1 = 0$, implying that the inflexion branches are all tangent to the y-axis. Then the coefficient z of y^2 is a solution of a cubic equation $I(z) = 0$ where

$$I(z) = -4c_4(b_2^2 - 8c_4) - 6b_2(b_2^2 - 8c_4)z + 48c_4z^2 + 8b_2z^3 \tag{8}$$

with discriminant $D = 6912(b_2^2 - 8c_4)(b_2^2 - 4c_4)^4$. So $D > 0$, giving 3 solutions for z, if and only if $b_2^2 - 8c_4 > 0$. However, if $b_2^2 < 4c_4$ (elliptic cusp) then $c_4 > 0$ so automatically $b_2^2 < 8c_4$ and the 3 solutions case applies only to hyperbolic cusps.

Once we know the number of smooth branches we can find their Taylor expansions using the same method of substitution of power series that was used in the previous cases. We find the following.

Proposition 6.2. *When $b_2^2 - 4c_4 > 0$ (hyperbolic cusp of Gauss), the branches of the vertex set, tangent to the principal direction $x = 0$, can be parametrized as follows:*

$$VG_1: \quad x = -\frac{1}{2}\left(b_2 - \sqrt{b_2^2 - 4c_4}\right)y^2 + h.o.t.,$$

$$VG_2: \quad x = -\frac{1}{2}\left(b_2 + \sqrt{b_2^2 - 4c_4}\right)y^2 + h.o.t.,$$

$$VG_3: \quad x = -\frac{1}{2}\left(b_2 - \sqrt{2b_2^2 - 8c_4}\right)y^2 + h.o.t.,$$

$$VG_4: \quad x = -\frac{1}{2}\left(b_2 + \sqrt{2b_2^2 - 8c_4}\right) y^2 + h.o.t.$$

$$VG_5: \quad x = -\frac{1}{2}b_2 y^2 - \frac{1}{2}(c_3 - b_1 b_2)y^3 + h.o.t.$$

The level set $f = 0$ has two branches which can be parametrized as:

$$FG_1: \quad x = -\frac{1}{2}\left(b_2 - \sqrt{b_2^2 - 4c_4}\right) y^2 + h.o.t.,$$

$$FG_2: \quad x = -\frac{1}{2}\left(b_2 + \sqrt{b_2^2 - 4c_4}\right) y^2 + h.o.t.$$

This Proposition implies in particular that the vertex branch VG_1 and the branch FG_1 of $f = 0$ have at least 3-point contact at the origin. The same holds for VG_2 and FG_2.

The conditions for 4-point contact are given below; since cusps of Gauss are isolated on a generic surface, only the signs of the expressions below will be of significance.

Proposition 6.3. *The vertex branch VG_1 and the branch FG_1 of $f_g = 0$ have at least 4-point contact at the origin if and only if $D_1 = 0$ where*

$$D_1 = -b_1 b_2^2 + b_1 b_2\sqrt{b_2^2 - 4c_4} + 2b_1 c_4 + b_2 c_3 - c_3\sqrt{b_2^2 - 4c_4} - 2d_5.$$

The same holds for VG_2 and FG_2 if and only if $D_2 = 0$ where

$$D_2 = b_1 b_2^2 + b_1 b_2\sqrt{b_2^2 - 4c_4} - 2b_1 c_4 - b_2 c_3 - c_3\sqrt{b_2^2 - 4c_4} + 2d_5.$$

The signs of the D_i determine the relative positions of the branches VG_i and FG_i. More precisely, $D_1 > 0$ if and only if, above the x-axis, the curve VG_1 is to the right of FG_1. (Below the x-axis this is reversed, since they have 3-point contact at the origin.) Similarly, $D_2 > 0$ if and only if, above the x-axis, VG_2 is to the right of FG_2. Note that both $D_1 > 0$ and $D_2 > 0$ can be regarded as conditions on the coefficient d_5.

6.2. Hyperbolic cusp of Gauss

Let x_{2i}, $i = 1, \ldots, 5$ be the coefficient of y^2 in the expansion of the branch VG_i as in Proposition 6.2, and let z_0 or $z_1 < z_2 < z_3$ denote the real roots of (8), as appropriate. Thus the z_i are the coefficients of y^2 in the expansion(s) of the branch(es) of the inflexion set: $x = z_i y^2 + \cdots$. Recall that we also assume $b_2 > 0$. The following is obtained from the expressions in Proposition 6.2 and the sign of the polynomial I in (8) at the values x_{2i}.

Proposition 6.4.

(a) *Suppose $b_2^2 - 8c_4 > 0$.*
   (1) *If $c_4 > 0$ then $x_{24} < x_{22} < z_1 < x_{25} < x_{21} < z_2 < 0 < x_{23} < z_3$.*
   (2) *If $c_4 < 0$ then $x_{24} < x_{22} < z_1 < x_{25} < 0 < z_2 < x_{21} < x_{23} < z_3$.*

(b) *Suppose $b_2^2 - 8c_4 < 0$. Then $x_{24} < x_{22} < z_0 < x_{25} < x_{21} < x_{23} < 0$.*

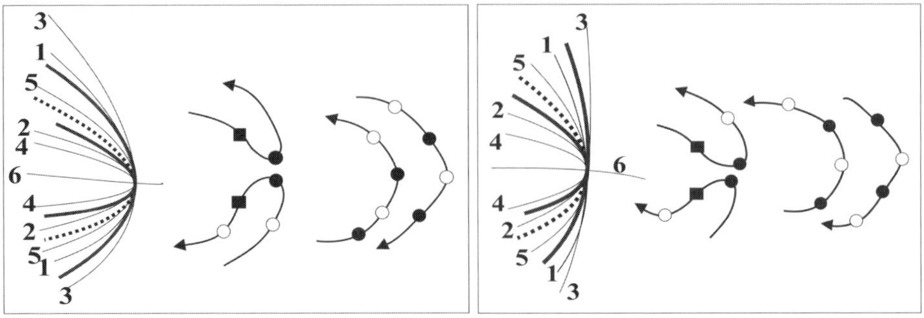

FIGURE 8. Sketch of the hyperbolic cusp of Gauss case, $b_2 > 0,\ 4c_4 < b_2^2 < 8c_4$ (so $c_4 > 0$); see Proposition 6.1. The left box has D_1 and D_2 as in Proposition 6.3 of the same sign (negative) and the right box of opposite signs ($D_1 > 0$). The thick lines are $f = 0$ on the left of each figure and $f = k$ for the two signs of small nonzero k on the right. The thin lines are the VG_i of Proposition 6.2, labelled by i, and the dashed line is the single branch of the inflexion set. As before, solid circles are maxima and open circles are minima of curvature, for the orientations indicated, and squares are inflexions.

These are illustrated in Figures 8-10. The sign of the derivative of curvature is determined as for the hyperbolic case; this determines the pattern of maxima and minima of curvature. The statements of Theorem 2.1, case (HCG), follow from these diagrams.

7. Conclusion

In this article, we have derived detailed results on the pattern of vertices and inflexions on families of plane curves of the form $f(x, y) = k$, which can be interpreted as the parallel plane sections of a generic surface close to the tangent plane at a given point $\mathbf{p}$. This is part of an investigation of the symmetry sets and medial axes of 1-parameter families of plane curves which evolve through a singular member. The symmetry set of a nonsingular plane curve γ is the closure of the locus of centres of circles tangent to γ in more than one place ('bitangent circles'). It has endpoints in the cusps of the evolute, that is at the centres of curvature of the vertices of γ. Thus the pattern of vertices has a strong influence on the branches of the symmetry set. Inflexions have a direct effect on the evolute – it goes to infinity – and, through the associated double tangents, an indirect effect on the symmetry set, which has a point at infinity for every double tangent (a bitangent circle of infinite radius). The limiting curvatures at vertices, as $k \to 0$, determines the limiting position of the endpoints of the symmetry set as the plane section becomes singular.

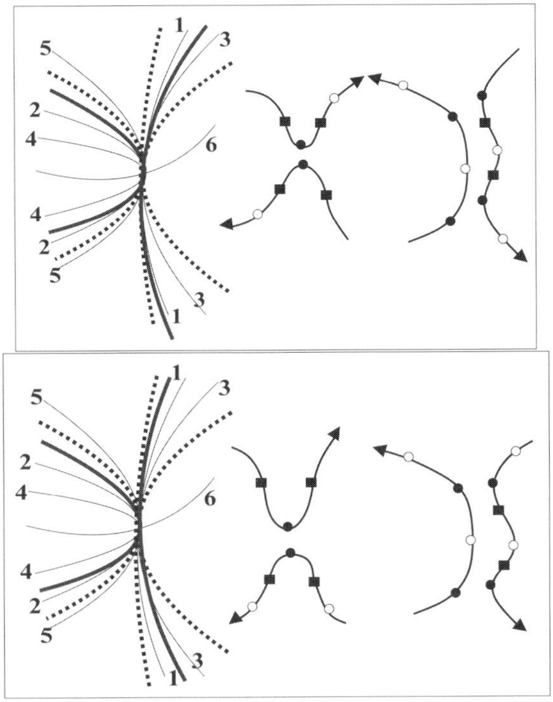

FIGURE 9. Sketch of the hyperbolic cusp of Gauss case, $b_2 > 0$, $c_4 < 0$ (hence $b_2^2 > 8c_4$); see Proposition 6.1. The left box has D_1 and D_2 as in Proposition 6.3 of the same sign (negative) and the right box of opposite signs ($D_1 > 0$). The thick lines are $f = 0$ on the left of each figure and $f = k$ for the two signs of small nonzero k on the right. The thin lines are the VG_i of Proposition 6.2, labelled by i, and the dashed lines are the three branches of the inflexion set. As before, solid circles are maxima and open circles are minima of curvature, for the orientations indicated, and squares are inflexions.

The investigation of symmetry sets involves many other factors, such as an investigation of circles which are tangent in three places to γ (these produce triple crossings on the symmetry set) and circles which are circles of curvature at one point of γ and tangent elsewhere (these produce cusps on the symmetry set). These and other matters are reported elsewhere, beginning with [6].

We conclude with some remarks and questions about the material of this article.

1. Is it possible to calculate the VT curve for classes of global examples where the two branches do not coincide? Compare §3.3.
2. Can the parabolic and cusp of Gauss cases be approached by more general methods of singularity theory, as in [8, 14]?

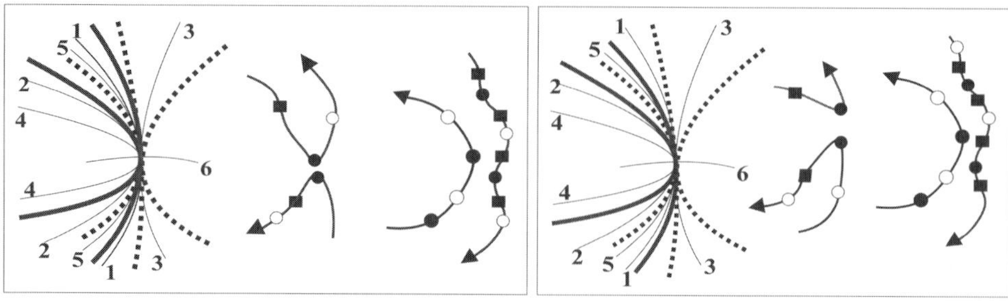

FIGURE 10. As for Figure 9 except that $b_2 > 0$, $b_2^2 > 8c_4$, $c_4 > 0$; see Proposition 6.1.

3. For the purpose of plotting symmetry sets it is much more convenient to have a parametrized curve rather than a level set $f(x, y) = k$. A method of parametrizing the level sets to arbitrarily high accuracy is given in [7].

Acknowledgements

This work is a part of the DSSCV project supported by the IST Programme of the European Union (IST-2001-35443). The first author was supported and the second author partially supported by this grant. We are also very grateful to Terry Wall, Bill Bruce and Vladimir Zakalyukin for helpful suggestions.

References

[1] T.F. Banchoff, T. Gaffney and C. McCrory, *Cusps of Gauss Mappings*, Pitman Research Notes in Mathematics, 55, 1982.

[2] J.W. Bruce and P.J. Giblin, *Curves and Singularities* Cambridge University Press, 2nd ed. (1992).

[3] J.W. Bruce and P.J. Giblin, 'Growth, motion and one-parameter families of symmetry sets', *Proc. Royal Soc.Edinburgh* 104A (1986), 179–204.

[4] J.W. Bruce, P.J. Giblin and F. Tari, 'Parabolic curves of evolving surfaces,' *Int. J. Computer Vision.* 17 (1996), 291–306.

[5] J.W. Bruce, P.J. Giblin and F. Tari, 'Families of surfaces: height functions, Gauss maps and duals', in *Real and Complex Singularities,* W.L. Marar (ed.), *Pitman Research Notes in Mathematics*, Vol. 333 (1995), 148–178.

[6] A. Diatta, P.J. Giblin, 'Geometry of isophote curves', *Scale Space Theory and PDE Methods in Computer Vision* 2005. *Lecture Notes in Computer Science* 3459 (2005), 50–61.

[7] A. Diatta, P.J. Giblin, B. Guilfoyle and W. Klingenberg, 'Plane sections of surfaces and applications to symmetry sets', *Mathematics of Surfaces XI, Lecture Notes in Computer Science* 3604 (2005), 147–160.

[8] M.D. Garay, 'On vanishing inflection points of plane curves', *Ann. Inst. Fourier, Grenoble* 52 (2002), 849–880.

[9] P.J. Giblin, 'Symmetry sets and medial axes in two and three dimensions', *The Mathematics of Surfaces IX*, Roberto Cipolla and Ralph Martin (eds.), Springer-Verlag 2000, pp. 306–321.

[10] P.L. Hallinan, G.G. Gordon, A.L. Yuille, P. Giblin and D. Mumford, *Two and three dimensional patterns of the face*, viii+262 pages, Natick, Massachusetts: A.K. Peters 1999.

[11] J.J. Koenderink, *Solid Shape*, M.I.T. Press (1990)

[12] R.J. Morris, *Liverpool Surface Modelling Package*, also known as *SingSurf*, http://www.singsurf.org/singsurf/SingSurf.html
See also R.J. Morris, 'The use of computer graphics for solving problems in singularity theory', in *Visualization in Mathematics*, H.-C.Hege & K.Polthier, Heidelberg: Springer-Verlag (1997), 173–187.

[13] I.R. Porteous, *Geometric Differentiation*, Cambridge University Press, 1994 and 2001.

[14] R. Uribe-Vargas, 'On the stability of bifurcation diagrams of vanishing flattening points', *Functional Analysis and its Applications* 37 (2003), 236–240.

André Diatta and Peter Giblin
University of Liverpool
Liverpool L69 3BX, England
e-mail: `adiatta@liv.ac.uk`
e-mail: `pjgiblin@liv.ac.uk`

Real and Complex Singularities
Trends in Mathematics, 99–111

Local Topology of Reducible Divisors

Alexandru Dimca and Anatoly Libgober

Abstract. We show that the universal abelian cover of the complement to a germ of a reducible divisor on a complex space Y with isolated singularity is $(\dim Y - 2)$-connected provided that the divisor has normal crossings outside of the singularity of Y. We apply this result to obtain a vanishing property for the cohomology of local systems of rank one and also study vanishing in the case of local systems of higher rank.

1. Introduction

The topology of holomorphic functions near an isolated singular point is a classical subject (cf. [23], [4]). Among the main results are the existence of Milnor fibration and the connectivity of the Milnor fiber yielding a very simple picture for the latter: it has the homotopy type of a wedge of spheres. Starting with the case of a germ of holomorphic function on $\mathbb{C}^N$ considered by Milnor ([23]), these results were eventually extended to the germs of holomorphic functions on analytic spaces (cf. [12], [19]).

In [22], it was shown that if the divisor of a holomorphic function on $\mathbb{C}^N$ is *reducible* then the results on the connectivity of Milnor fibers (cf. [23], [17]) can be refined. This refinement is based on the observation that the Milnor fiber is homotopy equivalent to the infinite cyclic cover of the total space of the Milnor fibration. So the classical connectivity results by Milnor and Kato-Matsumoto can be restated in terms of the connectivity of this cyclic cover.

In the case when the divisor of a holomorphic function is reducible, it is the associated universal abelian cover which has interesting connectivity properties generalizing the connectivity properties in the cyclic cover case, see Theorem 3.2 below. The present paper studies the case of reducible divisors on *arbitrary* isolated singularities.

More precisely the situation we consider is the following. Let $(Y, 0) \subset (\mathbb{C}^N, 0)$ be the intersection of a ball of a sufficiently small radius about the origin 0 with a $(n+1)$-dimensional irreducible complex analytic space with an isolated singularity at 0. Let $(D_j, 0) \subset (Y, 0)$ for $j = 1, \ldots, r$ be r irreducible Cartier divisors on $(Y, 0)$. We set $X = \cup_{i=1,r} D_i$, $M = Y \setminus X$ and regard M as the complement of the hypersurface arrangement $\mathcal{D} = (D_j)_{j=1,r}$. Since $(Y, 0)$ is irreducible, the complement M is connected and so we can unambiguously talk about the fundamental group $\pi_1(M)$ without mentioning a base point.

In this paper we investigate the topology of this complement M. In Section 2 we generalize a case of the Lê-Saito result in [18] asserting that if $(Y, 0)$ is a smooth germ and X is an isolated non-normal crossing divisor (see the definition below), then the fundamental group $\pi_1(M)$ is abelian. Our proof is based on an idea used in [24] in the global case and is much shorter than the proof in [18].

In Section 3 we consider the case when the hypersurface arrangement $\mathcal{D}$ is an arrangement based on a hyperplane arrangement $\mathcal{A}$ in the sense of Damon [3]. We show that the (co)homology of M is determined up-to degree $(n-1)$ by the hyperplane arrangement $\mathcal{A}$. The key fact here is the functoriality of the Gysin sequence and the splitting of the Gysin sequence associated to a triple $(\mathcal{A}, \mathcal{A}', \mathcal{A}'')$ of hyperplane arrangements into short exact sequences. Note that the proof of Theorem 3.1 can be done *only in this paper more general setting*, i.e., the deletion and restriction argument cannot be performed in the setting of [22] when $(Y, 0)$ is a smooth germ.
Combining the results above and following the approach in [22], we show that the universal abelian cover $\tilde{M}$ of M is homotopically a bouquet of spheres of dimension n which is the refinement of [23] and [12] we mentioned earlier.

In the last two sections we prove vanishing results for the (co)homology of the complement M with coefficients in a local system $\mathcal{L}$ on M. The case when the rank of $\mathcal{L}$ is equal one is treated in Section 4 and in this context we give a description for the dimension of the non zero homology groups $H_*(M, \mathcal{L})$. The general case when rank $\mathcal{L} \geq 1$ is treated in Section 5 where we allow a more general setting for the ambient space $(Y, 0)$ and for the divisor $(X, 0)$. The vanishing result in this case follows the general philosophy in [7], but the use of perverse sheaves as in [2] is unavoidable. Note that in our case the space M may be singular so one cannot use the technique of integrable connections to get vanishing results. A new point in our proof is the need to use the *interplay between constructible complexes of sheaves on real and complex spaces.* Indeed, real spaces occur in the picture in the form of links of singularities.

2. Fundamental group of the complements to INNC

Let $(Y, 0) \subset (\mathbb{C}^N, 0)$ be as above an $(n+1)$-dimensional irreducible complex analytic space germ with an isolated singularity at the origin. Let $(D_j, 0) \subset (Y, 0)$ for $j = 1, \ldots, r$ be r irreducible Cartier divisors on $(Y, 0)$, i.e., each D_j is given (with its reduced structure) as the zero set of a holomorphic function germ f_j : $(Y, 0) \to (\mathbb{C}, 0)$. When the local ring $\mathcal{O}_{Y,0}$ is factorial, then any hypersurface germ in $(Y, 0)$ is Cartier. This is the case for instance when $(Y, 0)$ is smooth or an isolated complete intersection singularity (ICIS for short) with dim $Y \geq 4$, see [11]. See also Example 2.1. Here and in the sequel we identify germs with their (good) representatives.

In particular the local homotopy groups of Y and $M = Y \setminus X$ are well defined as $\pi_j(L_Y)$ and $\pi_j(L_Y \setminus L_X)$, where the links L_Y and L_X are as defined below. We assume in this section that the following condition holds.

(**C1**) The divisor $X = \cup_{i=1}^{i=r} D_i$ has only normal crossing singularities on Y except possibly at the origin. We say in this case that X is an isolated non normal crossing divisor (for short INNC) on $(Y, 0)$.

In particular each germ $(D_j, 0)$ has an isolated singularity at the origin as well. Since the $(r+1)$-tuple $(Y, D_1, \ldots, D_r)$ has a conical structure (cf. [9]) we have an isomorphism:

$$\pi_1(L_Y \setminus L_X) = \pi_1(M \cap \partial B_\epsilon) \to \pi_1(M) \tag{1}$$

where L_Y (resp. L_X) denotes the link of Y (resp. of X), i.e., the intersection of Y (resp. of X) with the boundary ∂B_ϵ of a small ball B_ϵ about 0. In particular we get an epimorphism

$$\pi_1(M) = \pi_1(L_Y \setminus L_X) \to \pi_1(L_Y)$$

induced by the inclusion $L_Y \setminus L_X \to L_Y$.

Theorem 2.1. *For $n \geq 2$, the kernel of the surjection $\pi_1(M) \to \pi_1(L_Y)$ is contained in the center of the group $\pi_1(M)$. In particular, if L_Y is simply connected, then the fundamental group $\pi_1(M)$ is abelian.*

Proof. First notice that if dim$Y > 3$ and if H is a generic linear subspace passing through 0 such that the codimension of H in $\mathbb{C}^N$ is dim$Y - 3$, then, by Lefschetz hyperplane section theorem (cf. [9], p. 26 and p. 155), we have an isomorphism

$$\pi_1(M \cap H) \to \pi_1(M). \tag{2}$$

Hence it is enough to consider the case dim$Y = 3$ only (though the arguments below work for any dimension ≥ 3).

Next notice that $\kappa = \text{Ker}(\pi_1(L_Y \setminus L_X) \to \pi_1(L_Y))$ is the normal subgroup spanned by the set of elements in the fundamental group $\pi = \pi_1(L_Y \setminus L_X)$ represented by the loops δ_i each of which is the boundary of a fiber over a non singular point of a small closed tubular neighborhood $T(D_i)$ of the submanifold $D_i \cap L_Y$ in the manifold L_Y. Indeed a loop representing an element γ in the kernel κ is the image of the boundary of a 2-disk under a map $\phi : D^2 \to L_Y$ which is isotopic to an embedding (since dim$L_Y \geq 5$) and which we may assume to be transversal to all the submanifolds $D_i \cap L_Y$. Now δ_i are the ϕ-images of loops in D^2 each of which is composed of a path α_i going from the point in D^2 corresponding to the base point $p \in L_Y$ to the vicinity of a point $y \in D^2$ corresponding to a point in $\phi(D^2) \cap D_i$, a small loop about y and back along α_i^{-1}. So it is enough to show that all these loops δ_i (note that there may be several of them for a given i) belong to the center of $\pi_1(L_Y \setminus L_X)$.

Let $T(D_i)$ be a tubular neighborhood of $D_i \cap L_Y$ in L_Y as above. We claim that for any i ($i = 1, \ldots, r$) there is a surjection:

$$\pi_1(T(D_i) \setminus L_X) \to \pi_1(L_Y \setminus L_X). \tag{3}$$

Notice that assuming the surjectivity in (3) we can conclude the proof as follows. Since the divisors D_i's have normal intersections in L_Y, the space $T(D_i) \setminus L_X$ is homotopy equivalent to the total space of a locally trivial circle fibration over $D_i \setminus \cup_{j \neq i} D_j$. The fiber δ_i' of this fibration, which is a loop based at a point p', is in the center of $\pi_1(T(D_i) \setminus L_X, p')$.
Indeed, if $\alpha : S^1 \to T(D_i) \setminus L_X$ is any loop, we set $\beta = \pi \cdot \alpha$ with $\pi : T(D_i) \setminus L_X \to D_i \setminus \cup_{j \neq i} D_j$ the the corresponding projection. Then the commutativity $\alpha\delta' = \delta'\alpha$ follows from the triviality of the pull-back of the normal bundle $\pi : T(D_i) \setminus L_X \to D_i \setminus \cup_{j \neq i} D_j$ under β. This triviality in turn follows from the triviality of any complex line bundle over a circle S^1.
Therefore the surjectivity in (3) yields that the class of δ' commutes with any element in $\pi_1(L_Y \setminus L_X, p')$ and hence with any element in $\pi_1(L_Y \setminus L_X, p)$.

To show the surjectivity (3), let us consider a generic holomorphic function g on Y sufficiently close to f_i so that $L_Y \cap \{g = 0\} \subset T(D_i)$. We have the decomposition

$$\begin{array}{ccc} \pi_1(L_Y \cap \{g=0\} \setminus L_X) & & \\ \downarrow & \searrow & \\ \pi_1(T(D_i) \setminus L_X) & \to & \pi_1(L_Y \setminus L_X) \end{array} \tag{4}$$

corresponding to the factorization of the embeddings. This yields that the horizontal map is surjective provided the map:

$$\pi_1(L_Y \cap (g=0) \setminus L_X) \to \pi_1(L_Y \setminus L_X) \tag{5}$$

is surjective. But this follows from [14]. □

Note that this result in the case when $Y = \mathbb{C}^{n+1}$ is a consequence of a theorem of Lê Dung Trang and K.Saito (cf. [18]).

Example 2.1.
(i) If $(Y, 0)$ is an ICIS with dim $(Y, 0) \geq 3$, then it follows from [12] that the link L_Y is simply-connected.

(ii) If $V \subset \mathbb{P}^m$ is a locally complete intersection such that $n = \dim V > \operatorname{codim} V$, then the morphism $\pi_2(V) \to \pi_2(\mathbb{P}^m) = \mathbb{Z}$ induced by the inclusion $V \to \mathbb{P}^m$ is an epimorphism by the generalized Barth Theorem, see [9], p. 27. It follows that the associated affine cone $(Y, 0) = (CV, 0)$ has a simply-connected link.

If we assume that V is smooth and that $n = \dim V > \operatorname{codim} V + 1$, then the divisor class group $C\ell(\mathcal{O}_{Y,0})$ is trivial, i.e., any divisor on this germ $(Y, 0)$ is Cartier. This follows from the exact sequence in [15], Exercise II.6.3 comparing the divisor class groups in the local and the global settings, the usual isomorphism $C\ell(V) =$

$H^1(V, \mathcal{O}_V^*)$, see [15], II.6.12.1 and II.6.16 and the GAGA results allowing to use the exponential sequence, see [15], Appendix B, to relate topology to $H^1(V, \mathcal{O}_V^*)$.

If $E = \cup_{j=1,r} E_j$ is a normal crossing divisor on the smooth variety V, then the associated cone $X = \cup_{j=1,r} CE_j$ is an INNC divisor on the cone $(Y, 0)$. The epimorphism $\pi_1(M) \to \pi_1(V \setminus E)$ can then be used to show that this last fundamental group is abelian.

3. Homology of the complements to reducible divisors

Assume in this section that the germ $(Y, 0)$ is an ICIS with $\dim Y = n+1$ and let $\mathcal{A} = \{H_i\}_{i=1,\dots,r}$ be a central hyperplane arrangement in $\mathbb{C}^m$. Suppose given an analytic map germ $f : (Y, 0) \to (\mathbb{C}^m, 0)$ such that the following condition holds.

(**C2**) For any edge $L \in L(\mathcal{A})$ with codim $L = c$, the (scheme-theoretic) pull-back $D_L = f^{-1}(L)$ is an ICIS in $(Y, 0)$ of codimension exactly c for $c \leq n$ and $D_L = \{0\}$ for $c \geq n+1$.

This condition (**C2**) is equivalent to asking that $f : (Y, 0) \to (\mathbb{C}^m, 0)$ is transverse to $\mathcal{A}$ off 0 in the sense of Damon, see [3], Definition 1.2. In his language, $X = \cup_{i=1,r} D_i$ is a nonlinear arrangement of hypersurfaces based on the central arrangement $\mathcal{A}$, with $D_i = f^{-1}(H_i)$. Note that for $n \geq 2$ all the germs D_j are irreducible by [12], but on the other hand the condition (**C1**) may well fail in this setting.

Consider the complements $M = Y \setminus X$ and $N = \mathbb{C}^m \setminus \cup_{i=1,r} H_i$, and note that there is an induced mapping $f : M \to N$. Our result is the following

Theorem 3.1. *With this notation,*

$$f_* : H_j(M) \to H_j(N)$$

is an isomorphism for $j < n$ and an epimorphism for $j = n$. Similarly

$$f^* : H^j(N) \to H^j(M)$$

is an isomorphism for $j < n$ and a monomorphism for $j = n$. In particular, the algebra $H^*(M)$ is spanned by $H^1(M)$ up-to degree $(n-1)$.

Proof. For $n = 1$ everything is clear, so we can assume in the sequel $n > 1$.

The proof is by induction on r. For $r = 1$ the result follows since M can be identified to the total space of the Milnor fibration, whose Milnor fiber is a bouquet of n-dimensional spheres by work of Hamm, see [12].

Assume now that $r > 1$ and apply the deletion and restriction trick, see more on this in [25], p. 4. Namely, let $\mathcal{A}' = \{H_i\}_{i=2,\dots,r}$ and $\mathcal{A}'' = \{H_1 \cap H_i\}_{i=2,\dots,r}$. Then both $\mathcal{A}'$ and $\mathcal{A}''$ are central arrangements with at most $(r-1)$ hyperplanes.

Since $L(\mathcal{A}') \subset L(\mathcal{A})$, it is clear that $f : (Y, 0) \to (\mathbb{C}^m, 0)$ satisfies the condition **C2** with respect to $\mathcal{A}'$. Moreover, $L(\mathcal{A}'') \subset L(\mathcal{A})$ (with a 1-shift in codimensions) and $f : (D_1, 0) \to (\mathbb{C}^{m-1}, 0)$ satisfies the condition **C2** with respect to $\mathcal{A}''$.

Let M', M'', N', N'' be the corresponding complements. Since M'' (resp. N'') is a smooth hypersurface in M' (resp. N') we have the following ladder of Gysin sequences.

$$\begin{array}{ccccccccc}
\cdots \longrightarrow & H_{j-1}(M'') & \longrightarrow & H_j(M) & \longrightarrow & H_j(M') & \longrightarrow & H_{j-2}(M'') & \longrightarrow \cdots \\
 & \downarrow f_* & & \downarrow f_* & & \downarrow f_* & & \downarrow f_* & \\
\cdots \longrightarrow & H_{j-1}(N'') & \longrightarrow & H_j(N) & \longrightarrow & H_j(N') & \longrightarrow & H_{j-2}(N'') & \longrightarrow \cdots
\end{array}$$

It is a standard fact in hyperplane arrangement theory that the bottom-right morphism $H_j(N') \to H_{j-2}(N'')$ is zero. Usually this result is stated for cohomology, see [25], p.191, but since both homology and cohomology of hyperplane arrangement complements are torsion free, the vanishing holds for homology as well.

An easy diagram chasing, using the induction hypothesis, shows that for $j \leq n$ the top-right morphism $H_j(M') \to H_{j-2}(M'')$ is zero as well. This implies the claim, again by an easy diagram chasing and using the induction hypothesis.

The proof for the cohomological result is completely dual. □

Using Theorems 2.1 and 3.1 and Example 2.1,(i) we get the following result.

Corollary 3.1. *Assume that $n \geq 2$ and that the divisor X satisfies the condition* **C1**. *Then*

$$\pi_1(M) = H_1(M) = H_1(N)$$

is a free abelian group of rank $r = |\mathcal{A}|$.

Remark 3.1. Note that M is a Stein manifold, and hence $H_j(M) = 0$ for $j > n+1$ by [13]. It follows that there are only two Betti numbers $b_j(M)$ to compute, namely for $j = n, n+1$. Indeed, for $j < n$ the Betti number $b_j(M) = b_j(N)$ is known by the results in [25], Theorem 5.93. Moreover by the additivity of Euler characteristics, see [8], it follows that $\chi(M) = \chi(Y) - \chi(X) = 0$. This gives a relation between the two top unknown Betti numbers of M.

Similarly to [22], the above results yield the following.

Theorem 3.2. *Let $(Y, 0)$ be a an isolated complete intersection singularity of dimension $n+1 \geq 3$. Let $X = \cup_{i=1,r} D_i$ be a union of Cartier divisors of Y which have normal crossing outside of the origin. Then the universal abelian cover $\tilde{M}$ of $M = Y \setminus X$ has the homotopy type of a bouquet of spheres of dimension n.*

Proof. The proof is similar to the proof of Thm. 2.2 in [22]. Firstly, let us consider the exact homotopy sequence corresponding to the map $f : M \to \mathbb{C}^{*r}$, obtained by using the equations $f_i = 0$ for the divisors D_i. The isomorphism $\pi_1(M) \to \pi_1(\mathbb{C}^{*r})$ which follows from Theorems 2.1 and 3.1 and the known fact $\pi_j(\mathbb{C}^{*r}) = 0$ for $j > 1$ yield that $\pi_2(\mathbb{C}^{*r}, M) = 0$ and $\pi_j(\mathbb{C}^{*r}, M) = \pi_{j-1}(M)$ for $j > 2$ (here we assume that f is replaced by an embedding, which is of course possible up-to homotopy type). Moreover we can show exactly as in [22] that the action of $\pi_1(M)$

on $\pi_j(\mathbb{C}^{*r}, M)$ is trivial. Hence we can apply the relative Hurewich theorem to the pair $(\mathbb{C}^{*r}, M)$ and note that we have a vanishing of the relative homology of this pair as a consequence of the previous theorem. Looking now at $f : M \to \mathbb{C}^{*r}$ as a homotopy fibration with fiber $\tilde{M}$, we get the vanishing of the homotopy groups of the universal abelian cover $\tilde{M}$ of M up to dimension $n-1$. On the other hand, the existence of the Milnor fibration of $g = f_1 \dots f_r : (Y, 0) \to (\mathbb{C}, 0)$ (theorem of Hamm in [12]) yields that M admits a cyclic cover which has the homotopy type of a CW complex of dimension n (i.e., the Milnor fiber F of the hypersurface X in Y). Hence the universal abelian cover $\tilde{M}$, which is the universal abelian cover of this Milnor fiber F has the homotopy type of an n-complex. Therefore the universal abelian cover $\tilde{M}$ is homotopy equivalent to the wedge of spheres S^n. □

4. Homology of local systems (rank one case)

Let $(Y, 0)$ be a germ of an isolated complete intersection singularity and let $X = \bigcup_{i=1,r} D_i$ be a divisor which has normal crossings outside of the origin, i.e., we place ourselves again in the setting of Theorem 3.2. The above notation is still used here.

Let $\rho : \pi_1(M) \to \mathbb{C}^*$ be a character of the fundamental group or equivalently a local system $\mathcal{L}$ of rank one on M. The space M, being a Stein space of dimension $(n+1)$, has the homotopy type of an $(n+1)$ complex and hence $H_j(M, \mathcal{L}) = H_j(M, \rho) = 0$ for $j > n+1$.

The main result of this section is the following:

Theorem 4.1. *Let $\rho : \pi_1(M) \to \mathbb{C}^*$ be a non trivial character and let $\mathcal{L}$ be the associated rank one local system on M. Then:*

(i) $H_j(M, \mathcal{L}) = 0$ *for* $j \neq n, n+1$;

(ii) dim $H_n(M, \mathcal{L})$ *is the largest integer k such that ρ belongs to the zero set V_k of the kth Fitting ideal of the $\mathbb{C}[\pi_1(M)]$- module $\pi_n(M) \otimes_{\mathbb{Z}} \mathbb{C}$.*

(iii) *The largest integer k such that the trivial character of $\pi_1(M)$ belongs to V_k is equal to*

$$\dim \operatorname{Ker} (\Lambda^{n+1} H^1(M) \to H^{n+1}(M)) + \dim H_n(M) - \binom{r}{n}.$$

Proof. Recall the spectral sequence for the cohomology of local systems (cf. [1]) Thm. 8.4. Let $C_*^\rho(\tilde{M})$ be the chain complex on which $H_1(M, \mathbf{Z})$ acts from the right via $g(x) = \rho(g)xg$ $(g \in H_1(M, \mathbb{Z}), x \in C_*(\tilde{M}, \mathbb{C}))$ where $x \to xg$ is the action via deck transformations. Let $H_q^\rho(\tilde{M})$ be the homology of this complex. We have a spectral sequence:

$$E_{p,q}^2 = H_p(H_1(M, \mathbb{Z}), H_q^\rho(\tilde{M})) \Rightarrow H_{p+q}(M, \rho).$$

Recall that M has the homotopy type of an $(n+1)$-complex and $\tilde{M}$ has the homotopy type of a bouquet of spheres of dimension n, see 3.2.

The group $H_q^\rho(\tilde{M})$ carries the canonical structure of $H_1(M,\mathbb{Z})$-module coming from the corresponding module structure on chains. We have the isomorphism: $H_0^\rho(\tilde{M}) = \mathbb{C}_\rho$ where $\mathbb{C}_\rho$ is the one-dimensional representation of $H_1(M,\mathbb{Z})$ given by ρ. Indeed, if x is a generator of $C_0^\rho(\tilde{M})$ as $H_1(M,\mathbb{Z})$-module then $x - x\cdot g = 0$ in $H_0^\rho(\tilde{M})$. On the other hand $x - x\cdot g = x - \rho(g)g^{-1}x$. Hence $gx = \rho(g)x$ in $H_0^\rho(\tilde{M})$.

Since for $\rho \neq 1$ one has $H_p(H_1(M,\mathbb{Z}),\mathbb{C}_\rho) = 0$, it follows from the vanishing theorem in the last section that the term E_2 has only one horizontal row: $q = n$. This yields the claim (i). We have

$$H_0(H_1(M,\mathbb{Z}), H_n^\rho(\tilde{M})) = H_n^\rho(\tilde{M})_{Inv} = H_n(\tilde{M}) \otimes_{H_1(M,\mathbb{Z})} \mathbb{C}_\rho$$

On the other hand, taking tensor product with $\mathbb{C}_\rho$ in the resolution $\Phi : \Lambda^s \to \Lambda^t \to H_n(\tilde{M}) \to 0$ we obtain the resolution of $H_n(\tilde{M}) \otimes_{H_1(M,\ Z)} \mathbb{C}_\rho$ in which the matrix of $\Phi_\rho : \Lambda^s \otimes \mathbb{C}_\rho \to \Lambda^t \otimes \mathbb{C}_\rho$ is obtained from the matrix of Φ by replacing its entries by values of the entries at ρ. Hence if ρ belongs to the set of zeros of the kth Fitting ideal (and k is maximal with this property), then the corank of Φ_ρ is k. This yields the second claim.

Let us consider the exact sequence:

$$\begin{aligned} H_{n+1}(M) \to H_{n+1}(H_1(M,\mathbb{Z})) \to H_n(\tilde{M}) \otimes_{H_1(M,\mathbb{Z})} \mathbb{C} \\ \to H_n(M) \to H_n(H_1(M,\mathbb{Z})) \to 0 \end{aligned} \tag{6}$$

corresponding to the spectral sequence:

$$H_p(H_1(M,\mathbb{Z}), H_q(\tilde{M})) \Rightarrow H_{p+q}(M) \tag{7}$$

We have

$$\begin{aligned} &\dim \operatorname{Coker}(H_{n+1}(M) \to H_{n+1}(H_1(M,\mathbb{Z}))) \\ &= \dim \operatorname{Ker} H_{n+1}(H_1(M,\mathbb{Z}))^* \to H_{n+1}(M)^*) \end{aligned} \tag{8}$$

The latter kernel (using Kronecker pairing identification $H_i^* = H^i$ over $\mathbb{C}$) is isomorphic to $\dim \operatorname{Ker} H^{n+1}(H_1(M)) \to H^{n+1}(M)$. Since $H_1(M,\mathbb{Z}) = \mathbb{Z}^r$ we have $\Lambda^i(H_1(M)) = H^i(H_1(M))$ with the isomorphism provided by the cup product. Hence the dimension in (8) is equal to

$$\dim \operatorname{Ker}(\Lambda^{n+1}(H_1(M)) \to H^{n+1}(M)).$$

Therefore, using the sequence (6) and the equality $\dim H_n(H_1(M)) = \binom{r}{n}$, we obtain

$$\begin{aligned} &\dim H_n(\tilde{M}) \otimes_{H_1(M,\mathbb{Z})} \mathbb{C} \\ &= \dim \operatorname{Ker}(\Lambda^{n+1}(H_1(M)) \to H^{n+1}(M)) + \dim H_n(M) - \binom{r}{n}. \end{aligned}$$

This yields the last claim in Theorem 4.1. □

Notice that the claim 4.1 (iii) is a generalization of a result in [20] and that the space of local systems with non vanishing cohomology in the cases when $Y = \mathbb{C}^2$ and $Y = \mathbb{C}^{n+1}$ was studied in [21] and [22] respectively.

Remark 4.1. The morphism $\Lambda^{n+1}H^1(M) \to H^{n+1}(M)$ in the above Theorem is surjective in the following two cases.

(i) $n = 1$ and $(Y, 0) = (\mathbb{C}^2, 0)$, see [4], Corollary 2.20 ;

(ii) $(Y, 0) = (\mathbb{C}^{n+1}, 0)$ and X a central hyperplane arrangement, see [25], Corollary 5.88.

The following result is a generalization of Example (6.1.8) in [5], where Y was assumed to be a smooth germ and the proof uses properties of the vanishing cycle functor and a generalization of Prop. 4.6 from [22] where the case of X smooth was treated.

Corollary 4.1. *Let F be the Milnor fiber of the reduced germ $g : (Y, 0) \to (\mathbb{C}, 0)$ which defines the divisor X in Y. With the above assumptions, the monodromy action $h^j : H^j(F, \mathbb{C}) \to H^j(F, \mathbb{C})$ is trivial for $j \leq n - 1$.*

Proof. Let ρ_a be the representation sending each elementary loop to the same complex number $a \in \mathbb{C}^*$. Then it is well-known, see for instance [5], Corollary 6.4.9, that

$$\dim H^q(M, \rho_a) = \dim \operatorname{Ker} (h^q - aId) + \dim \operatorname{Ker} (h^{q-1} - aId).$$

The unipotence follows by applying this equality to $a \neq 1$.

To obtain triviality of the monodromy action, notice that due to the Milnor's fibration theorem, the Milnor fiber F is homotopy equivalent to the infinite cyclic cover of M. Hence, it is a quotient of the universal abelian cover $\tilde{M}$ by the action of the kernel of $\pi : \pi_1(M) \to \mathbf{Z}$ where π sends each elementary loop to 1. Let us consider the corresponding spectral sequence:

$$H^p(\operatorname{Ker}\pi, H^q(\tilde{M}, \mathbb{C})) \Rightarrow H^{p+q}(F, \mathbb{C}) \tag{9}$$

for this action of the group $\operatorname{Ker}\pi = \mathbb{Z}^{r-1}$ on the universal abelian cover $\tilde{M}$. Notice that this is a spectral sequence of $\mathbb{C}[t, t^{-1}]$ modules where the action on $H^p(\operatorname{Ker}\pi, H^q(\tilde{M}))$ is the standard action of the generator of $\pi_1/\operatorname{Ker}\pi$ and the action of t on cohomology of the Milnor fiber is the monodromy action. Since, by Theorem 3.2, $H^q(\tilde{M}) = 0$ for $1 \leq q < n$ we have $n - 1$ zero-rows in the term E_2 and hence the isomorphism $H^j(F, \mathbb{C}) = H^j(\operatorname{Ker}\pi, H^0(\tilde{M}))$ for $1 \leq j \leq n-1$. Since the map of the classifying spaces $(S^1)^r \to S^1$ corresponding to the homomorphism π has trivial monodromy, the action of $\pi_1/\operatorname{Ker}\pi$ on $H^j(\operatorname{Ker}\pi, \mathbb{C})$ is trivial for any j in the range $0 \leq j \leq n - 1$ and the claim follows. □

Remark 4.2. One can obtain the triviality of the monodromy action also using mixed Hodge theory, at least for $j < n - 1$. See for details [6], Theorem 0.2. Note that the above proof shows that $\dim H^j(F) = \binom{r-1}{j}$ for $j \leq n - 1$ (cf. [22]).

5. Homology of local systems (higher rank case)

In this section we work with weaker assumptions on the germs $(Y, 0)$ and $(D_j, 0)$ as above. Indeed, we simply need that M has only locally complete intersection singularities (which is weaker than asking $(Y, 0)$ to be an isolated singularity) and that there is a $\mathbb{Q}$-Cartier divisor, say D_1, among the divisors D_j such that the INNC condition for X holds only along D_1. In particular, the singularities of the divisors $D_j \setminus D_1$ for $j > 1$ can be arbitrary.

To start, note that if $D_1 \setminus \{0\}$ is contained in the smooth part of the space $Y \setminus \{0\}$, then one has an elementary loop δ_1 which goes once about the irreducible divisor D_1 (in a transversal at a smooth point). It follows that a rank m local system $\mathcal{L}$ on $Y \setminus X$ which corresponds to a representation

$$\rho : \pi_1(Y \setminus X) \to GL_m(\mathbb{C})$$

gives rise to a monodromy operators $T_1 = \rho(\delta_1)$. Of course, both δ_1 and T_1 are well-defined only up-to conjugacy. The following result should be compared to the vanishing part of Theorem 0.2 in [6].

Theorem 5.1. *Let $\mathcal{L}$ be a local system on M such that*

(i) *M is a locally complete intersection and D_1 is an irreducible $\mathbb{Q}$-Cartier divisor, i.e., there is an integer m such that mD_1 is the zero set of a holomorphic germ;*

(ii) *$D_1 \setminus \{0\}$ is contained in the smooth part of the space $Y \setminus \{0\}$ and X has only normal crossings along $D_1 \setminus \{0\}$;*

(iii) *the corresponding monodromy operator T_1 has not 1 as an eigenvalue.*

Then $H^k(M, \mathcal{L}) = 0$ for $k < n$ and for $k > n + 1$.

Proof. For this proof we assume that the (good) representatives for our germs $Y, D_j, \ldots$ exist as closed analytic subspaces in an open ball B of radius 2ϵ centered at the origin. This implies in particular that Y is a Stein space, as well as $Y \setminus X$, which is the complement of the zero set of a holomorphic function on Y. Such a Stein space has the homotopy type of a CW complex of dimension at most $(n+1)$ by [13], and this already gives $H^k(Y \setminus X, \mathcal{L}) = 0$ for $k > n + 1$.

These representatives are good in the sense that all the germs $Y, D_j, \ldots$ have a conic structure inside the ball B such that the corresponding retractions are the same for all these germs. We represent the links $L_Y, L_X, L_{D_j}, \ldots$ as the intersections of these representatives inside B with a sphere S of radius ϵ. In such a way we have an inclusion $i_\epsilon : L_Y \to Y$ and a retraction $r_\epsilon : Y^* \to L_Y$, with $Y^* = Y \setminus \{0\}$, which induces inclusions and retractions for the other germs.

The main tool for the proof below is the theory of constructible (resp. perverse) sheaves. For all necessary background material on this subject we refer to [16] and [5].

Let $i : Y \setminus X \to Y \setminus D_1$ be the inclusion and set $\mathcal{F}^* = Ri_*\mathcal{L} \in D^b_c(Y \setminus D_1)$, $\mathcal{F}_1^* = \mathcal{F}^*|(L_Y \setminus L_{D_1})$. The constructible sheaf complex $\mathcal{F}^*$ has constant cohomology sheaves along the fibers of the retraction r_ϵ (since the topology is constant along such a fiber). It follows, as in Lemma 2.7.3 in [16], that

$$H^k(Y \setminus X, \mathcal{L}) = \mathbb{H}^k(Y \setminus D_1, \mathcal{F}^*) = \mathbb{H}^k(L_Y \setminus L_{D_1}, \mathcal{F}_1^*).$$

Let $j_1 : L_Y \setminus L_{D_1} \to L_Y$ be the inclusion and note that

$$Rj_{1*}\mathcal{F}_1^* = Rj_{1!}\mathcal{F}_1^*$$

exactly as in [7] and [2], the key points being the assumptions (ii) and (iii) in the above statement. Since the link L_Y is compact, it plays the role of the compact algebraic variety in [7] and [2], and we get

$$\mathbb{H}^k(L_Y \setminus L_{D_1}, \mathcal{F}_1^*) = \mathbb{H}^k_c(L_Y \setminus L_{D_1}, \mathcal{F}_1^*)$$

for any integer k.

The new difficulty we encounter here is that $L_Y \setminus L_{D_1}$ is not a Stein space (not even properly homotopically equivalent to a Stein space as the retraction $r_\epsilon : Y \setminus D_1 \to L_Y \setminus L_{D_1}$ is not proper!), hence the vanishing for the last hypercohomology group is not obvious.

We proceed as follows. We apply first Poincaré-Verdier Duality on the real semialgebraic set $L_Y \setminus L_{D_1}$ and get

$$\mathbb{H}^k_c(L_Y \setminus L_{D_1}, \mathcal{F}_1^*)^\vee = \mathbb{H}^{-k}(L_Y \setminus L_{D_1}, D_{\mathbb{R}}\mathcal{F}_1^*).$$

Here $D_{\mathbb{R}}\mathcal{F}_1^*$ is the dual sheaf of $\mathcal{F}_1^*$ in this real setting. Note that we can also consider the (complex) dual sheaf $D\mathcal{F}^* \in D^b_c(Y \setminus D_1)$. It follows that

$$D\mathcal{F}^*|(L_Y \setminus L_{D_1}) = D_{\mathbb{R}}\mathcal{F}_1^*[1]$$

since the inclusion $L_Y \setminus L_{D_1} \to Y \setminus D_1$ is normally nonsingular in the sense of [10] (this is what corresponds to a non-characteristic embedding in the sense of [16], Definition 5.4.12 in the case of singular spaces).

This yields the following isomorphisms

$$\begin{aligned}\mathbb{H}^{-k}(L_Y \setminus L_{D_1}, D_{\mathbb{R}}\mathcal{F}_1^*) &= \mathbb{H}^{-k}(L_Y \setminus L_{D_1}, D\mathcal{F}^*[-1]|(L_Y \setminus L_{D_1}))\\ &= \mathbb{H}^{-k-1}(L_Y \setminus L_{D_1}, D\mathcal{F}^*|(L_Y \setminus L_{D_1})).\end{aligned}$$

Here we are again in the presence of a constructible sheaf complex, namely $D\mathcal{F}^*$, whose cohomology sheaves are constant along the fibers of the retraction r_ϵ. This implies that

$$\begin{aligned}\mathbb{H}^{-k-1}(L_Y \setminus L_{D_1}, D\mathcal{F}^*|(L_Y \setminus L_{D_1})) &= \mathbb{H}^{-k-1}(Y \setminus D_1, D\mathcal{F}^*)\\ &= \mathbb{H}^{k+1}_c(Y \setminus D_1, \mathcal{F}^*)\end{aligned}$$

the last isomorphism coming from Poincaré-Verdier Duality on the algebraic variety $Y \setminus D_1$.

Now it is time to note that the shifted local system $\mathcal{L}[n+1]$ is a perverse sheaf on the locally complete intersection variety M and hence $\mathcal{F}^*[n+1]$ is a perverse sheaf on the variety $Y \setminus D_1$ since the morphism i is Stein and quasi-finite. It follows that

$$\mathbb{H}_c^{k+1}(Y \setminus D_1, \mathcal{F}^*) = \mathbb{H}_c^{k-n}(Y \setminus D_1, \mathcal{F}^*[n+1]) = 0$$

for any $k < n$ by Artin's Vanishing Theorem in the Stein setting, see [16] Proposition 10.3.3 (iv) and Theorem 10.3.8 (ii). □

References

[1] H. Cartan, S. Eilenberg, Homological Algebra. Princeton University Press, Princeton, NJ, 1956.

[2] D. Cohen, A. Dimca and P. Orlik, Nonresonance conditions for arrangements, Ann. Institut Fourier 53(2003), 1883–1896.

[3] J. Damon, Critical points of affine multiforms on the complements of arrangements, Singularity Theory, ed. J. W. Bruce and D. Mond, London Math. Soc. Lect. Notes 263 (1999), CUP, 25–53.

[4] A. Dimca, Singularities and Topology of Hypersurfaces, Universitext, Springer Verlag, 1992.

[5] A. Dimca, Sheaves in Topology, Universitext, Springer Verlag, 2004.

[6] A. Dimca, M. Saito, Some consequences of perversity of vanishing cycles, Ann.Inst. Fourier, Grenoble 54(2004), 1769-1792.

[7] H. Esnault, E. Viehweg, *Logarithmic de Rham complexes and vanishing theorems*, Invent. Math. **86** (1986), 161–194.

[8] W. Fulton, *Introduction to Toric Varieties*, Annals of Math, Studies 131, Princeton University Press, 1993.

[9] M. Goresky, R. MacPherson, Stratified Morse Theory. Springer Verlag.

[10] M. Goresky, R. MacPherson, Intersection homology II, Invent. Math. 71 (1983), 77–129.

[11] A. Grothendieck, Cohomologie locale des faisceaux cohérents et Théorèmes de Lefschetz locaux et globaux, SGA2, North-Holland, 1968.

[12] H. Hamm, Lokale topologische Eigenschaften komplexer Räume, Math. Ann. 191 (1971), 235–252.

[13] H. Hamm, Zum Homotopietyp Steinscher Räume, J. Reine Angew. Math. 338 (1983), 121–135.

[14] H. Hamm, Lê Dung Trang, Local generalization of Lefschetz-Zariski theorem, J. Reine Angew. Math. 389 (1988), 157–189.

[15] R. Hartshorne, *Algebraic Geometry*, GTM 52, Springer 1977.

[16] M. Kashiwara, P. Schapira, *Sheaves on Manifolds*, Grundlehren Math. Wiss., vol. 292, Springer-Verlag, Berlin, 1994.

[17] M. Kato, Y. Matsumoto, On the connectivity of the Milnor fiber of a holomorphic function at a critical point. Manifolds – Tokyo 1973 (Proc. Internat. Conf., Tokyo, 1973), pp. 131–136. Univ. Tokyo Press, Tokyo, 1975.

[18] Lê Dung Trang, K. Saito, The local π_1 of the complement to a hypersurface with normal crossings in codimension 1 is abelian. Ark. Mat. 22 (1984), no. 1, 1–24.

[19] Lê, D.T., Complex analytic functions with isolated singularities, J. Algebraic Geometry 1 (1992), 83–100.

[20] A. Libgober, On the homology of finite abelian coverings. Topology Appl. 43 (1992), no. 2, 157–166.

[21] A. Libgober, Hodge decomposition of Alexander invariants. Manuscripta Math. 107 (2002), no. 2, 251–269.

[22] A. Libgober, Isolated non normal crossings, in Real and Complex Singularities, eds. M. Ruas and T. Gaffney, Contemporary Mathematics, **354**, 2004, 145–160.

[23] J. Milnor, Singular points of complex hypersurfaces, Annals of Mathematical Studies, 61, Princeton University Press, Princeton, 1968.

[24] M. Nori, Zariski conjecture and related problems, Ann. Sci. Ecole Norm. Sup. (4), 16, (1983), 305–344.

[25] P. Orlik, H. Terao, *Arrangements of Hyperplanes*, Grundlehren Math. Wiss., vol. 300, Springer-Verlag, Berlin, 1992.

Alexandru Dimca
Laboratoire J. A. Dieudonné
UMR CNRS 6621
Université de Nice Sophia-Antipolis
Parc Valrose
F-06108 Nice Cedex 2, France

Anatoly Libgober
Department of Mathematics
University of Illinois at Chicago
Chicago, IL 60607, USA

Real and Complex Singularities
Trends in Mathematics, 113–126

The Geometry of Flecnodal Pairs

Daniel Dreibelbis

Abstract. We generalize the definition of a flecnode on a surface in $\mathbb{R}^3$ to a definition for a general immersed manifold in Euclidean space. Instead of considering a flecnode as a point on the manifold, we consider it as a pair of a normal vector and a tangent vector, called the flecnodal pair. The structure of this set is considered, as well as its connection to binormals and A_3 singularities in the set of height functions. The specific case of a surface immersed in $\mathbb{R}^4$ is studied in more detail, with the generic singularities of the flecnodal normals and the flecnodal tangents classified. Finally, the connection between the flecnodals and bitangencies are studied, especially in the case where the dimension of the manifold equals the codimension.

Mathematics Subject Classification (2000). Primary 58K30, 14E20; Secondary 53A07, 58K05.

Keywords. Flecnode, differential geometry.

1. Introduction

Given a surface immersed in $\mathbb{R}^3$, one of the main third-order invariants of the surface is known as the *flecnodal curve*. This is the set of points on the surface where there exists a tangent line with at least four point (A_3) contact with the surface ([10]). Flecnodes have several important geometric properties, but the most important one to the author is that they are one of the boundary sets for the set of line bitangencies. We wish to generalize the idea of flecnodes to immersed manifolds of any dimension and any codimension.

Unfortunately, the concept of a tangent line with four-point contact does not generalize well. The notion of n-point contact is supposed to be a generalization of a transversal intersection, and we cannot really talk about it unless the codimension is equal to one. Furthermore, if we look at the set of points on a manifold with A_3 contact with a tangent line, it is possible for every point on the manifold to have such a tangent line, and so every point would be considered a flecnode. To generalize the notion of a flecnode, we need to include the tangent vector into the definition, and it will turn out that a corresponding normal vector will also be necessary. Thus we will get our generalization of the flecnode, which we will call the *flecnodal pair*.

In Section 1 we give the definition of a flecnodal pair, a geometric description of the concept, and a calculation of the dimension of the set of flecnodal pairs. In Section 2 we introduce the definitions of binormals and A_3 points ([2, 5]), and we show the connection between these objects and the flecnodal pair. In particular, we see for surfaces immersed in a low codimension, the A_3 points completely determine the singularity set of the Gauss map on the flecnodal normals.

In Section 3 we focus on the particular case of surfaces immersed in $\mathbb{R}^4$. In particular, we look at the tangent vector field and the normal surface defined by the flecnodal pair. We study the type of singularities possible for both, and then provide an example. Finally, in Section 4, we show the link between these flecnodal pairs and bitangencies (the concept of flecnodal pairs first came out when studying bitangencies). In particular, we describe how the concept of flecnodal pairs can lead us to a general counting formula for bitangencies on n-manifolds immersed in $\mathbb{R}^{2n}$.

2. Definitions

Through out research on bitangencies ([4]), the following concept became important to us: given an immersion of a manifold M into Euclidean space, let p be a point on M and let $\mathbf{n}$ be a normal vector at p. We can project M into the subspace spanned by the tangent plane at p and the normal vector $\mathbf{n}$. The result will be a local immersion of M in a codimension one Euclidean space. When the codimension is one, we have a good definition of a flecnode: a point p is a flecnode if there exists a line with at least A_3 contact with the manifold at p. Hence we can say that $\mathbf{n}$ is a flecnodal normal if p is a flecnode of the projected manifold, and we will say the spanning vector of the line with high contact is the flecnodal tangent. If we have a parametrization for the immersion of M and a local frame field of the normal space, then we can explicitly describe a parametrization of the projected manifold. Then we can take this parametrization and determine the conditions for A_3 contact with a line. We will use the resultant equations as our official definition of a flecnodal pair:

Definition 2.1. Let $\mathbf{s} : M^n \to \mathbb{R}^{n+m}$ be an immersion of an n-manifold into $(n+m)$-dimensional Euclidean space. Then a point $(p, \mathbf{n}, \mathbf{v}) \in (UN \times UT)M$ is a *flecnodal pair* if $\mathbf{n}$ and $\mathbf{v}$ satisfy the following two equations at p:

$$\mathbf{n} \cdot d^2\mathbf{s}\mathbf{v}^2 = 0 \tag{2.1}$$

$$(d^2\mathbf{n} \cdot d\mathbf{s} - d\mathbf{n} \cdot d^2\mathbf{s} + 3\sum_{k=1}^{m}(\mathbf{e}_k \cdot d\mathbf{n})(\mathbf{e}_k \cdot d^2\mathbf{s}))\mathbf{v}^3 = 0 \tag{2.2}$$

where $\{\mathbf{e}_1, \ldots, \mathbf{e}_m\}$ is a locally defined orthonormal frame field of the normal bundle. The normal vector of a flecnodal pair is a *flecnodal normal*, while the tangent vector is a *flecnodal tangent*.

As written, it is not clear that flecnodal pairs are well defined. In particular, the definition relies on the parametrization of $\mathbf{s}$, it relies on the derivatives of $\mathbf{n}$ (even though flecnodal pairs are supposed to be a pointwise definition), and it relies on the choice of local orthonormal frame. We need to show that the definition is invariant of all of these variables.

Theorem 2.2. *The equations in Definition* 2.1 *are invariant of parametrization.*

Proof. First we need to show that the definition is invariant of the parametrization of the immersion. So let $\mathbf{r} = \mathbf{s} \circ h$, where h is the change in coordinates. Also, let $\mathbf{w}$ be the vector where $\mathbf{v} = dh(\mathbf{w})$, let $\mathbf{m} = \mathbf{n} \circ h$ and $\mathbf{f}_k = \mathbf{e}_k \circ h$. Then:

$$\begin{aligned}\mathbf{m} \cdot d^2\mathbf{r}\mathbf{w}^2 &= (\mathbf{n} \circ h) \cdot d^2(\mathbf{s} \circ h)\mathbf{w}^2 \\ &= (\mathbf{n} \circ h) \cdot (d^2\mathbf{s}(dh)^2 + d\mathbf{s}d^2h)\mathbf{w}^2 \\ &= \mathbf{n} \cdot d^2\mathbf{s}\mathbf{v}^2\end{aligned}$$

and

$$\begin{aligned}&(d^2\mathbf{m} \cdot d\mathbf{r} - d\mathbf{m} \cdot d^2\mathbf{r} + 3\sum_k (\mathbf{f_k} \cdot d\mathbf{m})(\mathbf{f}_k \cdot d^2\mathbf{r}))\mathbf{w}^3 \\ &\quad = (d^2(\mathbf{n} \circ h) \cdot d(\mathbf{s} \circ h) - d(\mathbf{m} \circ h) \cdot d^2(\mathbf{s} \circ h) \\ &\qquad + 3\sum_k ((\mathbf{e}_k \circ h) \cdot d(\mathbf{n} \circ h))((\mathbf{e}_k \circ h) \cdot d^2(\mathbf{s} \circ h)))\mathbf{w}^3 \\ &\quad = ((d^2\mathbf{n}dh^2 + d\mathbf{n}d^2h) \cdot (d\mathbf{s}dh) - (d\mathbf{n}dh) \cdot (d^2\mathbf{s}dh^2 + d\mathbf{s}d^2h) \\ &\qquad + 3\sum_k ((\mathbf{e}_k \circ h) \cdot (d\mathbf{n}dh))(\mathbf{e}_k \circ h) \cdot (d^2\mathbf{s}dh^2 + d\mathbf{s}d^2h))\mathbf{w}^3 \\ &\quad = ((d^2\mathbf{n}dh^2) \cdot (d\mathbf{s}dh) - (d\mathbf{n}dh) \cdot (d^2\mathbf{s}dh^2) \\ &\qquad + 3\sum_k (\mathbf{e}_k \cdot d\mathbf{n}dh)(\mathbf{e}_k \circ h) \cdot (d^2\mathbf{s}dh^2))\mathbf{w}^3 \\ &\quad = (d^2\mathbf{n} \cdot d\mathbf{s} - d\mathbf{n} \cdot d^2\mathbf{s} + 3\sum_k (\mathbf{e}_k \cdot d\mathbf{n})(\mathbf{e}_k \cdot d^2\mathbf{s}))\mathbf{v}^3\end{aligned}$$

Also, since the definition of the flecnodal pair is supposed to be a pointwise definition, but Equation 2.2 involves the derivatives of $\mathbf{n}$, we need to show that the definition does not depend on how $\mathbf{n}$ is defined away from p. To this effect, let $\mathbf{n} = \sum a_k\mathbf{e}_k$. We need to show that Equation 2.2 does not depend on the values of da_k or d^2a_k:

$$\begin{aligned}&(d^2\mathbf{n} \cdot d\mathbf{s} - d\mathbf{n} \cdot d^2\mathbf{s} + 3\sum_k (\mathbf{e}_k \cdot d\mathbf{n})(\mathbf{e}_k \cdot d^2\mathbf{s}))\mathbf{v}^3 \\ &\quad = (d^2(\sum_k a_k\mathbf{e}_k) \cdot d\mathbf{s} - d(\sum_k a_k\mathbf{e}_k) \cdot d^2\mathbf{s} \\ &\qquad + 3\sum_k (\mathbf{e}_k \cdot d(\sum_j a_j\mathbf{e}_j))(\mathbf{e}_k \cdot d^2\mathbf{s}))\mathbf{v}^3\end{aligned}$$

$$= \sum_k ((d^2 a_k \mathbf{e}_k + 2da_k d\mathbf{e}_k + a_k d^2\mathbf{e}_k) \cdot d\mathbf{s} - (da_k \mathbf{e}_k + a_k d\mathbf{e}_k) \cdot d^2\mathbf{s}$$

$$+ 3(\sum_j \mathbf{e}_k \cdot (da_j \mathbf{e}_j + a_j d\mathbf{e}_j)(\mathbf{e}_k \cdot d^2\mathbf{s}))\mathbf{v}^3$$

$$= \sum_k (a_k d^2\mathbf{e}_k \cdot d\mathbf{s} - a_k d\mathbf{e}_k \cdot d^2\mathbf{s} + 3(\mathbf{e}_k \cdot d^2\mathbf{s}) \sum_j a_j \mathbf{e}_k \cdot d\mathbf{e}_j)\mathbf{v}^3$$

The last reduction needed the relations $\mathbf{e}_k \cdot d\mathbf{s} = 0$, $d\mathbf{e}_k \cdot d\mathbf{s} + \mathbf{e}_k \cdot d^2\mathbf{s} \equiv 0$, and $\mathbf{e}_k \cdot \mathbf{e}_j = \delta_{ij}$. The final equation does not depend on the derivatives of the a_k's, and so our equations are independent of how we extend $\mathbf{n}$.

Finally, we need to show the definition is independent of our choice of orthonormal frame field. This follows from the fact that the summation in Equation 2.2 is just the trace of the quadratic form $Q : N_pM \times N_pM \to \mathbb{R}$ defined by $Q(\mathbf{a}, \mathbf{b}) = (\mathbf{a} \cdot d\mathbf{n}\mathbf{v})(\mathbf{b} \cdot d^2\mathbf{s}\mathbf{v}^2)$. □

Notes

- In the definition, we restricted $\mathbf{n}$ and $\mathbf{v}$ to be unit vectors. But since both Equation 2.1 and Equation 2.2 are homogeneous in both $\mathbf{n}$ and $\mathbf{v}$, we can use non-unit vectors if it is more convenient. This is useful for computational purposes.
- In fact, because the equations are homogeneous, we could define our flecnodal pair to be an element of $(NP^{m-1} \times TP^{n-1})M$, i.e., the projectified normal and tangent bundles. While there are some advantages to doing this, the unit bundles are more natural to the geometry of our objects, and so we will use them.
- If $m = 1$ (so we have a hypersurface), the trace term in Equation 2.2 disappears and the formula becomes considerably easier.
- If p is a parabolic point (meaning all of its second fundamental forms have a common root) and $\mathbf{v}$ is the common root, then again the trace term disappears from Equation 2.2 and the formula again becomes easier.

The concept of flecnodal pairs came from the study of bitangencies ([4]). Specifically, we needed to project surfaces in $\mathbb{R}^4$ into a 3-space to get a surface with a flecnode at a particular point p. Obviously a particular tangent vector would project to the line with A_3 contact, but it also turned out that the direction of projection needed to be perpendicular to a particular normal vector. This concept can be generalized for any immersed manifold, no matter what the dimension or codimension are.

Theorem 2.3. *A point $(p, \mathbf{n}, \mathbf{v})$ is a flecnodal pair if and only if $\mathbf{v}$ spans a line with at least A_3 contact to the manifold $\pi_{\mathbf{n}}(M)$, where $\pi_{\mathbf{n}}$ is the projection into the subspace spanned by $\mathbf{n}$ and $d\mathbf{s}$ at p.*

Proof. Since our previous theorem showed that our definition is invariant of parametrization, we can do all of our work in Monge form. So, let $\mathbf{s}(x_1, \ldots, x_n)$ be the function

$$\mathbf{s}(x_1, \dots x_n) = (x_1, \dots, x_n, f_1(x_1, \dots x_n), \dots, f_m(x_1, \dots, x_n))$$

where f_k are polynomials in the x_j's with no linear or constant terms. We will assume that e_1 and e_{n+1} (i.e., the first and $n+1$ coordinate vector of $\mathbb{R}^{n+m}$) are the flecnodal tangent and the flecnodal normal at the origin. The condition for this is that e_1 needs to be a root of both the quadratic part and the cubic part of f_1. Our projection will be

$$\pi_{e_{n+1}}\mathbf{s} = (x_1, \dots, x_n, f_1(x_1, \dots, x_n))$$

and the condition for e_1 to span a line with A_3 contact is precisely that e_1 is a root of both the quadratic and cubic parts of f_1. □

The projection in the previous proof could be replaced with $\pi_{\mathbf{n}+\mathbf{w}}$, where $\mathbf{w}$ is an arbitrary tangent vector. The important part is the normal component.

Definition 2.4. Define the sets $FP \in (UN \times UT)M$, $FN \in UNM$ and $FT \in UTM$ as the set of flecnodal pairs, flecnodal normals, and flecnodal tangents, respectively.

Proposition 2.5. *In general, the set FP is an embedded smooth submanifold of $(UN \times UT)M$ of dimension $m + 2n - 4$.*

Proof. The manifold $(UN \times UT)M$ has dimension $m + 2n - 2$, and FP is defined as the intersection of the zero sets of two equations on $(UN \times UT)M$. It remains to show that the zero sets are transversal. We can describe our surface in Monge form, as we did in the proof of Theorem 2.3, and then explicitly write down the formulas for Equations 2.1 and 2.2. We take the gradients of both equations and determine when they are parallel. The condition for them to be parallel reduces to $m + 2n - 1$ conditions on the second, third, and fourth-order terms of the parametrization $\mathbf{s}$. The conditions have no relation among them, and we only have n degrees of freedom, so we have no solutions as long as $m + 2n - 1 > n$, i.e., $m + n > 1$, which of course is always true. □

Examples

- $n = 1$, $m = 1$: Planar curves. In general, this set has dimension -1, and so we would not expect a flecnodal pair on a general planar curve, but we would expect them in a one-parameter family of curves. These are just points with A_3 contact with their tangent line.
- $n = 1$, $m = 2$: Space curves. In general, this set has dimension 0. It corresponds to the points on the curve where the torsion is zero. The flecnodal normal is the binormal vector B of the Frenet frame.
- $n = 2$, $m = 1$: Surfaces in $\mathbb{R}^3$. This is the original definition. In general, the set of points gives a curve on the surface, and the corresponding tangent vector is the asymptotic vector with higher order contact.
- $n = 2$, $m = 2$: Surfaces in $\mathbb{R}^4$. In general, FP will be two-dimensional. This situation will be studied in detail in Section 4. We will show that every point on the surface has at least one corresponding point on FP.

3. A_3 and binormals

Closely linked to the set of flecnodes is the set of binormals and the set of A_3 points. These points have been studied in [2] and [5], but we will describe them in a slightly different form.

Definition 3.1. Let $\mathbf{s} : M^n \to \mathbb{R}^m$ be an immersion. A point $(p, \mathbf{n}, \mathbf{v}) \in (UN \times UT)M$ is a *binormal pair* if

$$\mathbf{n} \cdot d^2\mathbf{s}\mathbf{v} = 0$$

A point $(p, \mathbf{n}, \mathbf{v})$ is an A_3 *pair* if

$$\mathbf{n} \cdot d^2\mathbf{s}\mathbf{v} = 0, \qquad \mathbf{n} \cdot d^3\mathbf{s}\mathbf{v}^3 = 0$$

We will call the set of binormal pairs BP, the set of A_3 pairs A_3P, and the corresponding restrictions to the unit normal bundle and the unit tangent bundle BN, A_3N, BT and A_3T.

The set BP (resp. A_3P) correspond to height functions on the manifold with A_2 (resp. A_3) singularities, where $\mathbf{n}$ is the direction of the height function, p is the point of the singularity, and $\mathbf{v}$ is the probe structure vector (see [5]). The set BN is the singularity set of the Gauss map $\Gamma : UNM \to S^{n+m-1}$, and the set A_3P is the singularity set of the Gauss map $\Gamma : BN \to S^{n+m-1}$.

Proposition 3.2. *Generically, the sets BP and A_3P are $n+m-2$ and $n+m-3$ embedded submanifolds of $(UN \times UT)M$.*

Proof. The condition $\mathbf{n} \cdot d^2\mathbf{s}\mathbf{v} = 0$ is actually n different equations. In general, the zero sets of these equations are transverse, and so the dimension of BP is the dimension of $(UN \times UT)M$ minus n, which gives us $n+m-2$. The conditions for A_3P adds one more equation, which is still in general transverse to all the others. So the dimension of A_3P is one less, or $n+m-3$. □

Note that the sets FP and BP are not transverse. In fact their intersection will have dimension $n+m-3$, where a transverse intersection would have dimension $n+m-4$. It turns out that A_3P is the intersection set.

Proposition 3.3. *For an immersed manifold $\mathbf{s} : M^n \to \mathbb{R}^{n+m}$, $A_3P = BP \cap FP$.*

Proof. We need to show that the conditions to be a point in A_3P are equivalent to the union of the conditions needed to be in BP and FP. The quadratic condition for FP is clearly contained in the condition for BP, which in turn is contained in the condition for A_3P. It remains to show that if $\mathbf{n} \cdot d^2\mathbf{s}\mathbf{v} = 0$, then the cubic for FP and the cubic for A_3P are equivalent.

Assume that $\mathbf{n} \cdot d^3\mathbf{s}\mathbf{v}^3 = 0$. By differentiating the relation $\mathbf{n} \cdot d\mathbf{s} \equiv 0$ twice, we will get the relation $\mathbf{n} \cdot d^3\mathbf{s} + 2d\mathbf{n} \cdot d^2\mathbf{s} + d^2\mathbf{n} \cdot d\mathbf{s} \equiv 0$, and so $(2d\mathbf{n} \cdot d^2\mathbf{s} + d^2\mathbf{n} \cdot d\mathbf{s})\mathbf{v}^3 = 0$. Plugging this into the flecnode's cubic gives us:

$$3(d\mathbf{n} \cdot d^2\mathbf{s})\mathbf{v}^3 = 3\left(\sum (\mathbf{e}_k \cdot d\mathbf{n})(\mathbf{e}_k \cdot d^2\mathbf{s})\right)\mathbf{v}^3$$

Since $\mathbf{n} \cdot d^2\mathbf{s}\mathbf{v} = 0$, we can rewrite this as $d\mathbf{n}\mathbf{v} \cdot d\mathbf{s} = 0$, which implies that $d\mathbf{n}\mathbf{v}$ is a normal vector, and so $d\mathbf{n}\mathbf{v} = \sum \alpha_k \mathbf{e}_k$ for some values α_k.

Hence our equation reduces to

$$3\left(\sum \alpha_k \mathbf{e}_k\right) \cdot d^2\mathbf{s}\mathbf{v}^2 = 3\sum(\alpha_k)(\mathbf{e}_k \cdot d^2\mathbf{s})\mathbf{v}^2$$

which shows that $A_3P \subset BP \cap FP$. Following the argument backwards will give you the other inclusion. □

We can use the result of Proposition 3.3 to connect the Gauss maps of FN and BN, at least for low dimensions.

Proposition 3.4. *If M^1 is a curve, then $FP = A_3P$. In particular, $FN = A_3N$, and so FN is the singularity set of the Gauss map of BN.*

Proof. Since $n = 1$, the dimension of FP and A_3P are both equal to $m - 2$. More to the point, since $d^2\mathbf{s}$ is just $\mathbf{s}''$, the quadratic conditions for A_3P and FP are the same, and it follows (as in the proof of of Proposition 3.3) that the cubic conditions will be the same. □

This phenomenon can be seen in the case of a space curve: here the set of flecnodal normals is just the Frenet binormal B when the torsion is equal to zero, and this is exactly the condition for the binormal curve B to have a cusp.

We can get an equivalent condition for surfaces also, but only for low codimension.

Proposition 3.5. *If M^2 is a generic surface and $m \leq 5$, then the projection $FP \to FN$ is an immersion. In this case, the singularity set of the Gauss map $\Gamma : FN \to S^{m+1}$ is the set A_3N, which is also the singularity set of the Gauss map $\Gamma : BN \to S^{m+1}$.*

Proof. Simply by comparing dimensions, we cannot expect the projection $FP \to FN$ to be an immersion unless the dimension of FP is less than the dimension of UNM, i.e., $2n + m - 4 < n + m - 1$. This implies $n < 3$, so curves and surfaces are the only manifolds for which this projection is an immersion (FP can be immersed into UNM for curves no matter what the codimension). For a surface, the only way the projection can have a singularity is if there is a point p which has a flecnodal normal that can be paired with every tangent vector. The first time such a point occurs generically on a surface is when the codimension is 6 (at such a point, the projection will give us a cone point in FN).

Now let us assume that FN is immersed in UNM. In this case, it is only possible for Γ restricted to FN to be singular when FN is on the singularity set of the Gauss map, which is precisely BN. As already seen, $FP \cap BP = A_3P$, so $FN \cap BN = A_3N$. Hence the singularity set of FN is A_3N, which is the singularity set of BN. □

We can see this situation for surfaces in $\mathbb{R}^3$. In general, the flecnodal curve and the parabolic curve (which corresponds to our binormals) will only meet at a cusp of Gauss ([1]), which are the points with an A_3 singularity of the height function. If we look at the Gauss map of both the parabolic curve and the flecnodal curve, both of them will have a cusp at the cusp of Gauss.

4. Surfaces in $\mathbb{R}^4$

Now we want to look at case of surfaces in $\mathbb{R}^4$ in more detail. We will assume our surface is in Monge form:

$$\begin{aligned}\mathbf{s}(x,y) &= (x,y,f(x,y),g(x,y))\\ &= (x,y,\sum_{i+j\geq 0} a_{ij}x^iy^j, \sum_{i+j\geq 0} b_{ij}x^iy^j)\end{aligned}$$

and we will work out the conditions for a flecnodal pair at the origin. By setting $\mathbf{n} = a(-f_x,-f_y,1,0)+b(-g_x,-g_y,0,1)$ and $\mathbf{v} = c\mathbf{s}_x + d\mathbf{s}_y$, our equations for a flecnodal pair at the origin reduce to

$$a(a_{20}c^2 + a_{11}cd + a_{02}d^2) + b(b_{20}c^2 + b_{11}cd + b_{02}d^2) = 0$$

$$a(a_{30}c^3 + a_{21}c^2d + a_{12}cd^2 + a_{03}d^3) + b(b_{30}c^3 + b_{21}c^2d + b_{12}cd^2 + b_{03}d^3) = 0$$

We can solve the first equation for (a,b) and plug the result into the second equation to get the following fundamental quintic:

$$\begin{aligned}&(a_{30}b_{20} - a_{20}b_{30})c^5 + (a_{30}b_{11} + a_{21}b_{20} - a_{20}b_{21} - a_{11}b_{30})c^4d\\ &+(a_{30}b_{02} + a_{21}b_{11} - a_{20}b_{12} + a_{12}b_{20} - a_{11}b_{21} - a_{02}b_{30})c^3d^2\\ &+(a_{21}b_{02} - a_{20}b_{03} + a_{12}b_{11} - a_{11}b_{12} + a_{03}b_{20} - a_{02}b_{21})c^2d^3\\ &+(a_{12}b_{02} - a_{11}b_{03} + a_{03}b_{11} - a_{02}b_{12})cd^4 + (a_{03}b_{02} - a_{02}b_{03})d^5 = 0\end{aligned} \tag{4.1}$$

So the set of flecnodal tangents at the origin (and hence the number of flecnodal pairs) is given by the number of real solutions to this quintic equation. In particular, every point on the surface has at least one associated flecnodal pair, and in general no point has more than five flecnodal pairs (in a nongeneric case, we can have a one-parameter set of normals associated with a single tangent).

Equation 4.1 can be considered as a quintic differential equation on the surface, its solutions being the integral curves of the flecnodal tangent vector field. This vector field cannot have any singularities (a singularity would require the quintic to reduce to the zero polynomial, which cannot happen on a generic surface). We can expect points where the quintic has a double root, which occur when the discriminant of Equation 4.1 is zero. These are the points where the projection map $FP \to M$ is singular, and we will call them the *flecnodal edge*. The condition for the origin to be on the flecnodal edge is a 16 degree homogeneous equation in the second and third order derivatives of $\mathbf{s}$, which is an 8 degree homogeneous equation in the a_{ij}'s, the b_{ij}'s, the second order terms and the third order terms (it's formula is too large to include here). If we normalize to make $a_{20} = a_{30} = 0$ (thus guaranteeing that $((0,0),e_1,e_3)$ is a flecnodal pair), then the origin is on the flecnodal edge when $a_{21}b_{20} = a_{11}b_{30}$.

In general, the flecnodal edge will have ordinary cusps when the quintic has a triple root. Again, the general equation for this is too large to include, but in the case where $a_{20} = a_{30} = 0$, the flecnodal edge will have a cusp at the origin when $a_{21}b_{20} = a_{11}b_{30}$ and $a_{21}b_{11} + a_{12}b_{20} = a_{11}b_{21} + a_{02}b_{30}$.

Finally, if the flecnodal edge meets the parabolic curve, the corresponding flecnodal tangent and asymptotic vector matching, then the point of intersection will occur on the A_3 curve, and all three points will be tangent at this point (we will see that a similar thing happens with the flecnodal normals, the binormals, and the A_3 normals).

Proposition 4.1. *The flecnodal edge is tangent to the parabolic curve at A_3 points. At such a point, both curves are tangent to the A_3 curve, and the corresponding flecnodal tangent (which is also the asymptotic vector) is tangent to the parabolic curve.*

Proof. We begin by assuming our surface is given in Monge form with $a_{20} = a_{30} = 0$, hence ensuring $((0,0), e_1, e_3)$ is a flecnodal pair. This point will be on the edge if $a_{21}b_{20} = a_{11}b_{30}$. We are making the additional assumption that the origin is a parabolic point with e_1 the asymptotic vector, so $b_{20} = 0$. Our condition to be on the edge reduces to either $b_{30} = 0$, which gives us a full flecnodal point (see [6]), which does not occur on a generic surface, or our condition reduces to $a_{11} = 0$, which gives us an A_3 parabolic point. Now we can calculate the gradient to the flecnodal edge and the gradient to the parabolic curve, and we find with our restrictions on the coefficients, these two vectors are parallel. The A_3 curve is tangent to the parabolic curve whenever they meet ([3]), so all three curves are tangent. □

There is a similar quintic polynomial whose solutions give the flecnodal normals, but its formula is too big to include here. While the quintic is more complicated, its discriminant is the same as the discriminant of Equation 4.1. As seen in the previous section, the Gauss map of the flecnodal normals has the same singularity set as the Gauss map of the binormal vectors, namely the A_3 points. However, there is a difference in the types of singularities that occur, namely at the inflection points.

Theorem 4.2. *For a general surface in $\mathbb{R}^4$, the Gauss map $\Gamma : FN \to S^3$ can have the following singularities:*

- *A cuspidal edge at the A_3 points.*
- *A swallowtail at the A_4 points (points where the height function has an A_4 singularity).*
- *A cuspidal pinch point at an inflection point.*

Furthermore, three cuspidal pinch points will meet at an elliptic umbilic (i.e., an inflection where three distinct A_3 curves meet), and the three corresponding cuspidal edges will be tangent at the meeting point.

Proof. The majority of this theorem follows from the fact that the singularity set of the binormals is known ([5]) and the singularity set of the flecnodal normals is the same as the the singularity set of the binormals. The only claim that still needs to be shown is that the flecnodal normals will have a cuspidal pinch point at an inflection point. As usual, we do the computations in Monge form. We set up

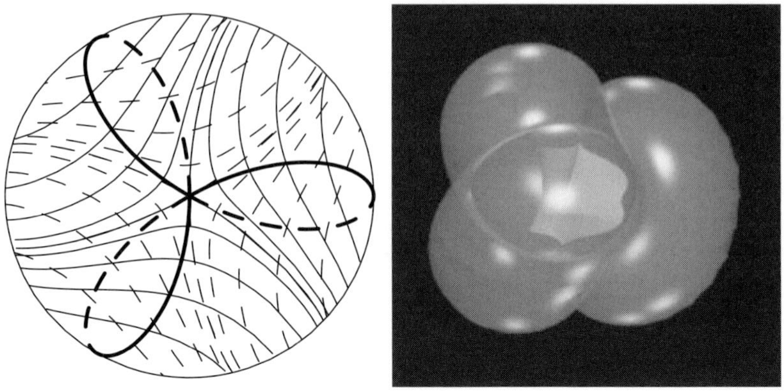

FIGURE 1. Left: The parabolic curve, A_3 curve, and asymptotic curves. Right: The Gauss map of the binormal vectors (both pictures from [5]).

the surface so that it has an inflection point at the origin with the A_3 pair given by e_1 (for the tangent) and e_3 (for the normal). We can calculate the flecnodal normals for this surface, calculate its Gauss map, and then look at the singularity type at the origin. Generically, the result is a cuspidal pinch point. □

Example. Perturbed $z^3/3$.

Consider the surface

$$\mathbf{s}(x, y) = (x, y, x^2 + y^2 + x^3/3 - xy^2, x^2y - y^3/3)$$

This is the function graph of the complex function $z^3/3$, modified by a quadratic term. We used this example in [5] because it has an easy parabolic curve (the unit circle) and a relatively simple, though interesting, binormal surface. With some effort, we can draw the flecnodal tangents and the Gauss map of the flecnodal normal.

Figure 1 shows the structure of the parabolic curve (the circle), the A_3 curve (the three-leafed rose) and the asymptotic curves. It also shows the the Gauss map of the binormal vectors after a stereographic projection (Both pictures are from [5]). Note in particular the three parabolic A_3 points, the inflection point at the origin, and the structure of the singularity set on the binormal vectors.

Figure 2 shows the flecnodal tangent vector field. Figure 3 shows the parameter space colored by the number of flecnodal tangents per point: the brightest color stands for five tangents, the middle color stand for 3 tangents, the darker color stands for a single tangent. This picture gives a good idea of the flecnodal edge: the exact equation for the flecnodal edge of this surface is the zero set of a sixty-eight degree polynomial. Note the various cusps, corresponding to points with a triple tangent, and note that the three points where the flecnodal edge is tangent to the

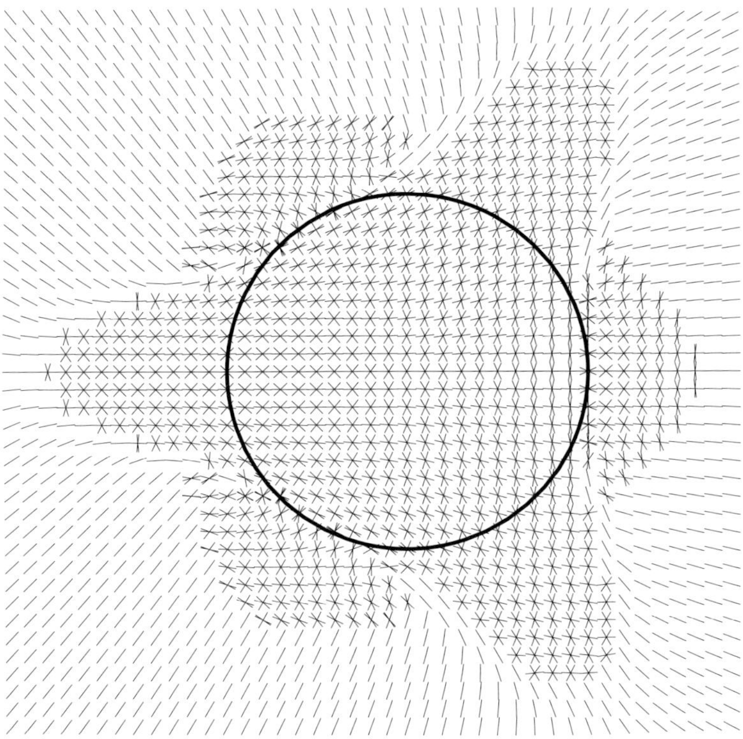

FIGURE 2. The flecnodal tangent vector field.

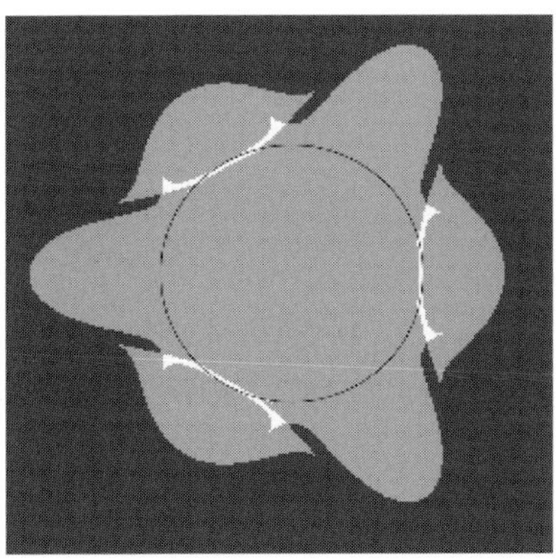

FIGURE 3. The flecnodal edge.

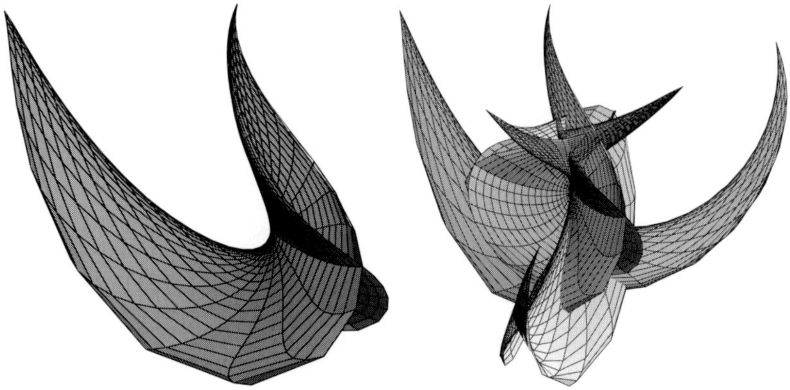

FIGURE 4. Left: An individual sheet of the flecnodal normals.
Right: All three sheets at the same time.

parabolic curve are the three parabolic A_3 points. For the other places where the flecnodal edge is transverse to the parabolic curve, the corresponding flecnodal tangent and the asymptotic vector are different.

Finally, Figure 4 shows part of the Gauss map of the flecnodal normals. The left picture shows one of the six sheets over the unit disk (there are three sheets and their negatives over the unit disk). In particular, you can see the cuspidal edge (corresponding to the A_3 points) and the cuspidal pinch point (corresponding to the inflection point at the origin). The right picture shows the three sheets that meet over the inflection point. The three cuspidal edges are tangent at this point, but obviously this is difficult to see. The cuspidal edges match up with the cuspidal edges on the bottom half of the binormal vectors pictured in Figure 1.

5. Flecnodes and bitangencies

A *line bitangency* for an immersed surface is a pair of points $(p,q) \in M \times M$, $p \neq q$ and $\mathbf{s}(p) \neq \mathbf{s}(q)$, with $\mathbf{s}(p) - \mathbf{s}(q) \in T_pM$ and $\mathbf{s}(p) - \mathbf{s}(q) \in T_qM$. Bitangencies and flecnodes are closely related: a bitangency is a line with 2-point contact at two points, while a flecnode is a single point with 4-point contact. For surfaces in $\mathbb{R}^3$, flecnodes are the boundary set of bitangencies as $p \to q$.

Proposition 5.1. *Let* $\mathbf{s}$ *be an immersion* $\mathbf{s} : M^2 \to \mathbb{R}^3$. *If* (p_t, q_t) *is a one parameter family of bitangencies with* $\lim_{t\to 0} p_t = \lim_{t\to 0} q_t = p_0$, *then* p_0 *is a flecnode, and its corresponding flecnodal tangent is the limit of* $(\mathbf{s}(p_t) - \mathbf{s}(q_t))/|\mathbf{s}(p_t) - \mathbf{s}(q_t)|$.

The original definition of the flecnodal normal came about in the study of bitangencies on surfaces in $\mathbb{R}^4$. In particular, we can restrict our attention to the parabolic curve $\mathcal{P}$. Each point on the curve has a unique asymptotic vector, and corresponding to that vector is a unique normal vector which is the flecnodal normal (unique up to sign, of course). The number of bitangencies on a closed, oriented surface turned out to be related to the structure of this flecnodal normal vector field along the parabolic curve.

Theorem 5.2 ([7]). *Let* $\mathbf{s}$ *be an immersion of a closed, oriented surface into* $\mathbb{R}^4$. *Let* $\mathcal{P}$ *be the parabolic curve on this surface, and let* F *be the flecnodal normal vector field along* $\mathcal{P}$. *Then number of* F *about* $\mathcal{P}$ *is equal to*

$$B + D - \frac{1}{2}\nu(\mathbf{s}(M))^2$$

where B *is the number of bitangencies, counted with sign,* D *is the number of double points, counted without sign, and* $\nu(\mathbf{s}(M))$ *is the normal Euler number of the immersion.*

In general, we can look at the case of a closed oriented manifold M^n immersed in $\mathbb{R}^{2n}$. In such a case, there are (in general) a finite number of line bitangencies on the immersion, and so we should be able to have a counting formula. As in the case of the surface in $\mathbb{R}^4$, the counting formula will reduce to a topological invariant equal to $B + D - 1/2\nu(\mathbf{s}(M))^2$. Evidence suggests that we can describe this invariant in terms of the flecnodal normals at parabolic points.

For our immersed surface, we will define $\mathcal{P}$ as the set of points p such that there is a vector $\mathbf{v}$ with $\mathbf{n} \cdot d^2\mathbf{s}\mathbf{v}^2 = 0$ for all normal vectors $\mathbf{n}$ at p. Note that at such a point, the quadratic condition for a flecnodal pair is satisfied for any normal vector $\mathbf{n}$, and the trace term in the cubic condition disappears. The dimension of $\mathcal{P}$ is $n - 1$, and the set of flecnodal normals at each point form a $n - 2$ sphere. In particular, at each point of $\mathcal{P}$, there exists a unique (up to sign) unit normal vector which is perpendicular to all of the flecnodal normals. We call this vector field E. Evidence supports the idea that structure of E is the proper topological invariant to use in the bitangency equation. In the case of surfaces, we looked at the structure of F instead of E. But since the vector fields are perpendicular at all points of $\mathcal{P}$, the linking number of one is equal to the linking number of the other. So the theorem is still correct, only it does not generalize well to higher dimensions. The generalization will appear in a later paper.

6. Conclusions and further research

Considering surfaces in $\mathbb{R}^4$, perhaps the most interesting structure of the flecnodal pairs occur when the tangent is the asymptotic vector at a parabolic point. It is the structure of this set that is connected with the number of bitangencies, and because it uses the asymptotic vectors, it connects bitangencies to the binormals. In particular, we should be able to extend the flecnodal vector field from the parabolic curve in such a way that it is smooth everywhere except at the inflections. Doing this would connect the bitangencies to the inflections. Examples support the idea of this connection, but the work has not been done.

References

[1] T. Banchoff, T. Gaffney, and C. McCrory. *Cusps of Gauss Mappings*, volume 55 of *Research Notes in Mathematics.* Pittman, 1982.

[2] J.W. Bruce and A.C. Nogueira. Surfaces in $\mathbb{R}^4$ and duality. *Quart. J. Math.*, 2(49):433–443, 1998.

[3] J.W. Bruce and F. Tari. Families of surfaces in $\mathbb{R}^4$. *Proc. Edinburgh Math. Soc.*, 45:181–203, 2002.

[4] D. Dreibelbis. A bitangency theorem for surfaces in four-dimensional Euclidean space. *Q. J. Math.*, 52:137–160, 2001.

[5] D. Dreibelbis. Singularities of the guass map and the binormal surface. *Adv. Geom.*, 3:453–468, 2003.

[6] D. Dreibelbis. Birth of bitangencies in a family of surfaces in $\mathbb{R}^4$. *Submitted, Differential Geometry and its Applications*, 2004.

[7] D. Dreibelbis. Invariance of the diagonal contribution in a bitangency formula. In *Real and Complex Singularities*, volume 354 of *Contemporary Mathematics*, pages 45–55. American Mathematical Society, 2004.

[8] R.A. Garcia, D.K.H. Mochida, M.C. Romero Fuster, and M.A.S. Ruas. Inflection points and topology of surfaces in 4-space. *Trans. Amer. Math. Soc.*, 352(7):3029–3043, 2000.

[9] D. Mochida, M.C. Romero Fuster, and M. Ruas. The geometry of surfaces in 4-space from a contact viewpoint. *Geom. Dedicata*, 54:323–332, 1995.

[10] I.R. Porteous. *Geometric Differentiation.* Cambridge University Press, 1994.

Daniel Dreibelbis
University of North Florida
4567 St. John's Bluff Rd.
Jacksonville, FL, 32224, USA
e-mail: ddreibel@unf.edu

Real and Complex Singularities
Trends in Mathematics, 127–141

Path Formulation for $Z_2 \oplus Z_2$-equivariant Bifurcation Problems

João Carlos Ferreira Costa and Angela Maria Sitta

Abstract. M. Manoel and I. Stewart ([10]) classify $\mathbb{Z}_2 \oplus \mathbb{Z}_2$-equivariant bifurcation problems up to codimension 3 and 1 modal parameter, using the classical techniques of singularity theory of Golubistky and Schaeffer [8]. In this paper we classify these same problems using an alternative form: the path formulation (Theorem 6.1). One of the advantages of this method is that the calculates to obtain the normal forms are easier. Furthermore, in our classification we observe the presence of only one modal parameter in the generic core. It differs from the classical classification where the core has 2 modal parameters. We finish this work comparing our classification to the one obtained in [10].

Mathematics Subject Classification (2000). Primary 58K70; Secondary 37G40.

Keywords. Path formulation, equivariant bifurcation problems, $Z_2 \oplus Z_2$-symmetry, classification.

1. Introduction

The symmetry group of a rectangle, $\mathbb{Z}_2 \oplus \mathbb{Z}_2$, appears in many bifurcation problems: in PDE problems with rectangular domains, like in the buckling of a rectangular plate [14], or in some Hopf-Hopf mode interaction [8]. More generally, when there is an interaction of two modes in symmetric systems where the normalizer of the isotropy subgroups of the two modes have Weyl group $\mathbb{Z}_2$ and are "independent". We refer to the introduction of the paper of Manoel and Stewart [10] for a substantive list of references with such equivariant problems. The classification of $\mathbb{Z}_2 \oplus \mathbb{Z}_2$-equivariant bifurcation problems of corank two up to codimension topological codimension two appears in [10] using the classical techniques of singularity theory of Golubistky and Schaeffer [8]. In this paper we classify these problems using an alternative form: the path formulation. The basic idea of path formulation

This work was partially supported by CAPES and FAPESP.

was suggested by Golubitsky and Schaeffer in [7] where they related bifurcation problems in one state variable (without symmetry) with a path through a miniversal unfolding of the cuspoid x^{m+1}. This idea has been extended and applied to more complex situations (cf. [4], [12], [1]). The main idea is to consider the bifurcation problem $g(x, \lambda) = 0$ as an unfolding (perturbation) with parameter λ of the core $g_0(x) = g(x, 0)$. If g_0 is of finite codimension, with respect to some equivalence of maps relevant to our problem (cf. Section 2), we have a miniversal unfolding G_0 (with parameter $\alpha \in \mathbb{R}^k$ where k is the codimension of g_0) of g_0 such that g is equivalent (in the previous sense) to a pull-back $\bar{\alpha}^*G_0$ where $\bar{\alpha} : (\mathbb{R}^l, 0) \to (\mathbb{R}^k, 0)$ is the path associated with g_0 (given the core g_0). Then we can compare paths and determine their miniversal unfoldings (cf. Section 3 thereafter for more precision). The path formulation differentiates between the singular behavior attributable to the core and to the paths. Moreover we can discuss efficiently multiparameter situations and forced symmetry breaking (cf. [4]).

The main goal of this paper is to use the path formulation as an alternative method to obtain the classification of $\mathbb{Z}_2 \oplus \mathbb{Z}_2$-equivariant bifurcation problems: Theorem 6.1. To do this, we first classify the normal forms to the cores: Theorem 4.1 and Theorem 4.2. We observe the presence of only one modal parameter in the generic core. It differs from the classical classification where the core has two modal parameters (cf. [8], [10]). As a byproduct of our approach we get a set-up that can be easily generalized to multiple bifurcation parameters, even with some additional complex internal structure (cf. Section 3). Moreover we also get new information on the structure of vector fields liftable over the projection onto the parameter space of a miniversal unfolding of singularities in the equivariant case (cf. Sections 5 and 6).

We assume that the reader has some familiarity with Mather's approach to singularity theory as generalized in Damon [3]. In Section 2 we recall the basic ingredients and results we need to classify the bifurcation problems and their perturbations modulo changes of coordinates following the now classical approach of Golubitsky and Schaeffer [8]. In Section 3 we introduce the basic results of the alternative approach of path formulation theory. Then we apply those ideas to classify $\mathbb{Z}_2 \oplus \mathbb{Z}_2$-equivariant bifurcation problems in corank 2 using this formulation. To that effect we discuss the cores up to codimension 3 in Section 4, followed, in Section 5, by the main technical ingredient: the module of the vector fields liftable over the projection onto the parameter space of their miniversal unfoldings. We can then achieve the classification in Section 6 before finishing this work with some comments and comparing our classification to the one obtained in [10].

2. $\mathbb{Z}_2 \oplus \mathbb{Z}_2$-equivariant bifurcation problems and their equivalence

We consider the usual action of $\mathbb{Z}_2 \oplus \mathbb{Z}_2$ on the plane given by $(\epsilon, \delta) \cdot (x, y) = (\epsilon x, \delta y)$ with $\epsilon^2 = \delta^2 = 1$. The action on the bifurcation and other parameters is always trivial. We denote (x, y) by z and by $GL_l(\mathbb{R})$ the set of real invertible $l \times l$-matrices with identity denoted by I_l. Derivatives are denoted by subscripts, for example, f_x is $\frac{\partial f}{\partial x}$, and the superscript ^{o} denotes the value of any function at the origin, for example $f^o = f(0)$, $f_x^o = f_x(0)$. For any variable, or set of variables, $a \in \mathbb{R}^n$, we denote by $\mathcal{E}_a$ the ring of germs $f : (\mathbb{R}^n, 0) \to \mathbb{R}$ and by $\mathcal{M}_a$ its maximal ideal. For $b \in \mathbb{R}^m$, we denote by $\vec{\mathcal{E}}_{a,b}$ the $\mathcal{E}_a$-module of smooth germs $g : (\mathbb{R}^n, 0) \to \mathbb{R}^m$ and by $\vec{\mathcal{M}}_{a,b}$ the $\mathcal{E}_a$-submodule of germs vanishing at the origin. When b is clear from the context, we denote $\vec{\mathcal{E}}_{a,b}$ by $\vec{\mathcal{E}}_a$ and $\vec{\mathcal{M}}_{a,b}$ by $\vec{\mathcal{M}}_a$. If R is some ring, we denote by $\langle g_1, \ldots, g_k \rangle_R$ the R-module generated by $\{g_i\}_{i=1}^k$.

2.1. $\mathbb{Z}_2 \oplus \mathbb{Z}_2$-equivariant map germs

The following is well known (for instance [7]). The ring $\mathcal{E}_z^{\mathbb{Z}_2 \oplus \mathbb{Z}_2}$ of smooth $\mathbb{Z}_2 \oplus \mathbb{Z}_2$-invariant germs is generated by $u = x^2$ and $v = y^2$. The module $\vec{\mathcal{E}}_z^{\mathbb{Z}_2 \oplus \mathbb{Z}_2}$ of smooth $\mathbb{Z}_2 \oplus \mathbb{Z}_2$-equivariant maps $f : (\mathbb{R}^2, 0) \to \mathbb{R}^2$ is generated over $\mathcal{E}_z^{\mathbb{Z}_2 \oplus \mathbb{Z}_2}$ by $(x, 0)$ and $(0, y)$.

Hence, any bifurcation germ $g : (\mathbb{R}^2 \times \mathbb{R}^l, 0) \to \mathbb{R}^2$ (with parameters $\lambda \in (\mathbb{R}^l, 0)$) has the form

$$g(z, \lambda) = (p(u, v, \lambda)x, q(u, v, \lambda)y),$$

with $p, q \in \mathcal{E}_{(z,\lambda)}^{\mathbb{Z}_2 \oplus \mathbb{Z}_2}$. We use the notation $g = [p, q]$.

The zero-set of $g = 0$ is a stratified set composed of four pieces, each depending on the isotropy of the solutions. With maximal isotropy we have $\mathcal{S}_0$ of solution $z = 0$. With each copy of Z_2, we have $(x, 0, \lambda) \in \mathcal{S}_x$ of equation $p(x^2, 0, \lambda) = 0$ and $(0, y, \lambda) \in \mathcal{S}_y$ of equation $q(0, y^2, \lambda) = 0$. Finally, with the trivial isotropy, we have $(z, \lambda) \in \mathcal{S}_z$ of equation $p(z, \lambda) = q(z, \lambda) = 0$.

2.2. $\mathcal{K}_\lambda^{\mathbb{Z}_2 \oplus \mathbb{Z}_2}$-equivalence

The classification of bifurcation problems is via contact equivalences preserving the $\mathbb{Z}_2 \oplus \mathbb{Z}_2$-symmetry. Let $f, g \in \vec{\mathcal{E}}_{(z,\lambda)}^{\mathbb{Z}_2 \oplus \mathbb{Z}_2}$, f is $\mathbb{Z}_2 \oplus \mathbb{Z}_2$-*equivalent* to g if there exists a map $S : (\mathbb{R}^{2+l}, 0) \to GL_2(\mathbb{R})$ and a diffeomorphism $(z, \lambda) \mapsto (Z(z, \lambda), L(\lambda))$ such that

$$f(z, \lambda) = S(z, \lambda)\, g(Z(z, \lambda), L(\lambda))$$

where $(Z^o, L^o) = (0, 0)$, $\det(L_\lambda^o) > 0$, $Z \in \vec{\mathcal{E}}_{(z,\lambda)}^{\mathbb{Z}_2 \oplus \mathbb{Z}_2}$, $L \in \vec{\mathcal{E}}_\lambda$ and

$$S((\epsilon, \delta) \cdot (x, y, \lambda)) \begin{pmatrix} \epsilon & 0 \\ 0 & \delta \end{pmatrix} = \begin{pmatrix} \epsilon & 0 \\ 0 & \delta \end{pmatrix} S(x, y, \lambda). \tag{2.1}$$

In addition, we require S^o and Z_z^o to be (diagonal) matrices with positive entries. The set of $\mathbb{Z}_2 \oplus \mathbb{Z}_2$-equivalences (S, Z, L) has a semidirect product group structure by composition, and is denoted by $\mathcal{K}_\lambda^{\mathbb{Z}_2 \oplus \mathbb{Z}_2}$.

2.3. Unfolding theory

The perturbations of any $g \in \vec{\mathcal{E}}_{(z,\lambda)}^{\mathbb{Z}_2\oplus\mathbb{Z}_2}$ are described by k-parameter unfoldings of g which are $\mathbb{Z}_2 \oplus \mathbb{Z}_2$-equivariant map-germs $G : (\mathbb{R}^{2+l+k}, 0) \to \mathbb{R}^2$ such that $G(z, \lambda, 0) = g(z, \lambda)$. We extend in a straightforward manner the previous definitions to the k-parametrized version for unfoldings. The unfolding and finite determinacy theorems follow the general theory of Damon [3] as our group is a geometric subgroup. Let $G(z, \lambda, \alpha)$ and $F(z, \lambda, \beta)$ be unfoldings of a germ $g \in \vec{\mathcal{E}}_{(z,\lambda)}^{\mathbb{Z}_2\oplus\mathbb{Z}_2}$ with $\alpha \in (\mathbb{R}^k, 0)$ and $\beta \in (\mathbb{R}^s, 0)$. We say that F *maps into* G, or F *factors through* G, if there exist a β-unfolding (S, Z, L) of the identity in $\mathcal{K}_\lambda^{\mathbb{Z}_2\oplus\mathbb{Z}_2}$ and a map $A : (\mathbb{R}^s, 0) \to (\mathbb{R}^k, 0)$ such that

$$F(z, \lambda, \beta) = S(z, \lambda, \beta)\, G(Z(z, \lambda, \beta), L(\lambda, \beta), A(\beta)).$$

The unfolding G is called *versal* if any unfolding F of g maps into G and *miniversal* if it has the minimal number of parameters necessary to be versal. That number is given by the $\mathcal{K}_\lambda^{\mathbb{Z}_2\oplus\mathbb{Z}_2}$-codimension $c(g)$ of g that is calculated as the real dimension of the extended normal space.
The *extended normal space* of $g = [p, q] \in \vec{\mathcal{E}}_{(z,\lambda)}^{\mathbb{Z}_2\oplus\mathbb{Z}_2}$ is defined by

$$\mathcal{N}_e\mathcal{K}_\lambda^{\mathbb{Z}_2\oplus\mathbb{Z}_2}(g) = \vec{\mathcal{E}}_{(z,\lambda)}^{\mathbb{Z}_2\oplus\mathbb{Z}_2} / \mathcal{T}_e\mathcal{K}_\lambda^{\mathbb{Z}_2\oplus\mathbb{Z}_2}(g),$$

where

$$\mathcal{T}_e\mathcal{K}_\lambda^{\mathbb{Z}_2\oplus\mathbb{Z}_2}(g) = \langle [p, 0], [vq, 0], [0, up], [0, q], [up_u, uq_u], [vp_v, vq_v] \rangle_{\mathcal{E}_{(z,\lambda)}^{\mathbb{Z}_2\oplus\mathbb{Z}_2}} + \langle [p_\lambda, q_\lambda] \rangle_{\mathcal{E}_\lambda},$$

is the *extended tangent space* of g (cf. [7]) which is a module over the system of rings $\{\mathcal{E}_{(z,\lambda)}^{\mathbb{Z}_2\oplus\mathbb{Z}_2}, \mathcal{E}_\lambda\}$.

Moreover, if $\{d_i\}_{i=1}^{c(g)}$ is a basis for $\mathcal{N}_e\mathcal{K}_\lambda^{\mathbb{Z}_2\oplus\mathbb{Z}_2}(g)$, then a miniversal unfolding of g is

$$G(z, \lambda, \alpha) = g(z, \lambda) + \sum_{i=1}^{c(g)} \alpha_i d_i(z, \lambda).$$

2.4. Determinacy and recognition theories

Let $g \in \vec{\mathcal{E}}_{(z,\lambda)}^{\mathbb{Z}_2\oplus\mathbb{Z}_2}$. Another consequence of finite codimension is that g is *finitely determined*, that is, there exists an integer $k \geq 1$ such that every germ with the same k^{th}-jet as g is $\mathcal{K}_\lambda^{\mathbb{Z}_2\oplus\mathbb{Z}_2}$-equivalent to g. The *recognition problem* seeks conditions under which a germ $g \in \vec{\mathcal{E}}_{(z,\lambda)}^{\mathbb{Z}_2\oplus\mathbb{Z}_2}$ is $\mathcal{K}_\lambda^{\mathbb{Z}_2\oplus\mathbb{Z}_2}$-equivalent to a given normal form. To solve a particular recognition problem means to explicitly characterize a $\mathcal{K}_\lambda^{\mathbb{Z}_2\oplus\mathbb{Z}_2}$-equivalence class in terms of a finite number of polynomial equalities and inequalities to be satisfied by the Taylor coefficients of the elements of that class. For this we need further ideas and results. A subspace $M \subset \vec{\mathcal{E}}_{(z,\lambda)}^{\mathbb{Z}_2\oplus\mathbb{Z}_2}$ is *intrinsic* if it contains the $\mathcal{K}_\lambda^{\mathbb{Z}_2\oplus\mathbb{Z}_2}$-orbit of all its elements. If $V \subset \vec{\mathcal{E}}_{(z,\lambda)}^{\mathbb{Z}_2\oplus\mathbb{Z}_2}$ then the *intrinsic part* of V, denoted by $\operatorname{itr} V$, is the largest intrinsic subspace of $\vec{\mathcal{E}}_{(z,\lambda)}^{\mathbb{Z}_2\oplus\mathbb{Z}_2}$ contained in V. In [10] it is shown that the the ideals generated by powers of $u = x^2$, $v = y^2$ or

λ are intrinsic. Moreover a $\mathcal{E}^{\mathbb{Z}_2\oplus\mathbb{Z}_2}_{(z,\lambda)}$-module $M = [\mathcal{I}, \mathcal{J}] = \{[p,q] \mid p \in \mathcal{I},\, q \in \mathcal{J}\}$ is intrinsic if and only if $\mathcal{I}, \mathcal{J}$ are intrinsic ideals such that $v\mathcal{J} \subset \mathcal{I}$ and $u\mathcal{I} \subset \mathcal{J}$. Let $g \in \vec{\mathcal{E}}^{\mathbb{Z}_2\oplus\mathbb{Z}_2}_{(z,\lambda)}$, the "perturbation term" $w \in \vec{\mathcal{E}}^{\mathbb{Z}_2\oplus\mathbb{Z}_2}_{(z,\lambda)}$ is of *higher order* with respect to g if $f+w$ is $\mathcal{K}^{\mathbb{Z}_2\oplus\mathbb{Z}_2}_{\lambda}$-equivalent to g for every f in the $\mathcal{K}^{\mathbb{Z}_2\oplus\mathbb{Z}_2}_{\lambda}$-orbit to g. By definition such a perturbation cannot enter into a solution of the recognition problem for g. We denote by $\mathcal{P}(g)$ the set of all higher order terms of g. Then, for each $g \in \vec{\mathcal{E}}^{\mathbb{Z}_2\oplus\mathbb{Z}_2}_{(z,\lambda)}$ the set $\mathcal{P}(g)$ is an intrinsic $\mathcal{E}^{\mathbb{Z}_2\oplus\mathbb{Z}_2}_{(z,\lambda)}$-submodule of $\vec{\mathcal{E}}^{\mathbb{Z}_2\oplus\mathbb{Z}_2}_{(z,\lambda)}$ (cf. [8]). To evaluate $\mathcal{P}(g)$ we introduce a subgroup $\mathcal{UK}^{\mathbb{Z}_2\oplus\mathbb{Z}_2}_{\lambda}$ of $\mathcal{K}^{\mathbb{Z}_2\oplus\mathbb{Z}_2}_{\lambda}$ of *unipotent equivalences* represented by equivalences whose linear part is unipotent. The unipotent tangent space of $g = [p,q] \in \vec{\mathcal{E}}^{\mathbb{Z}_2\oplus\mathbb{Z}_2}_{(z,\lambda)}$ is given by

$$\begin{aligned} \mathcal{TUK}^{\mathbb{Z}_2\oplus\mathbb{Z}_2}_{\lambda}(g) = {} & \langle [p,0], [0,q], [up_u, uq_u], [vp_v, vq_v] \rangle_{\mathcal{M}^{\mathbb{Z}_2\oplus\mathbb{Z}_2}_{(z,\lambda)}} \\ & + \langle [vq,0], [0,up] \rangle_{\mathcal{E}^{\mathbb{Z}_2\oplus\mathbb{Z}_2}_{(z,\lambda)}} + \langle [p_\lambda, q_\lambda] \rangle_{\mathcal{M}^2_\lambda}. \end{aligned}$$

Following the proof of Theorem 1.17 ([6], p. 108) we know that

$$\text{itr } \mathcal{TUK}^{\mathbb{Z}_2\oplus\mathbb{Z}_2}_{\lambda}(g) \subset \mathcal{P}(g).$$

We finalize this section with two theorems that present the normal forms that classify the $Z_2 \oplus Z_2$-equivariant bifurcation problems using classical techniques of singularity theory.

Theorem 2.1. ([8]) *Let $g = [p,q] \in \vec{\mathcal{E}}_{x,y,\lambda}(Z_2 \oplus Z_2)$ a bifurcation problem with $\lambda \in (\mathbb{R}, 0)$. If*

$$p_u^o, \quad q_v^o, \quad p_\lambda^o, \quad q_\lambda^o, \quad p_u^o q_v^o - p_v^o q_u^o, \quad p_u^o q_\lambda^o - p_\lambda^o q_u^o, \quad q_v^o p_\lambda^o - p_v^o q_\lambda^o$$

are all nonzero at the origin, then g is $Z_2 \oplus Z_2$-equivalent to

$$h_1 = [\varepsilon_1 u + m\, v + \varepsilon_2 \lambda, \eta u + \varepsilon_3 v + \varepsilon_4 \lambda],$$

$\epsilon_1 = \text{sign}(p_u^o)$, $\epsilon_2 = \text{sign}(p_\lambda^o)$, $\epsilon_3 = \text{sign}(q_v^o)$, $\epsilon_4 = \text{sign}(q_\lambda^o)$ *and modal parameters*

$$m = \left| \frac{q_\lambda^o}{q_v^o p_\lambda^o} \right| p_v^o, \qquad \eta = \left| \frac{p_\lambda^o}{p_u^o q_\lambda^o} \right| q_u^o.$$

Moreover, the moduli μ and η satisfy the conditions

$$m \neq \epsilon_2 \epsilon_3 \epsilon_4, \qquad \eta \neq \epsilon_1 \epsilon_2 \epsilon_4, \qquad m\eta \neq \epsilon_1 \epsilon_3.$$

Also, it is shown in [7] that the normal form h_1 has $c(h_1) = 3$ and miniversal unfolding

$$H_1(x, y, \lambda, \tilde{m}, \tilde{\eta}, \sigma) = [\varepsilon_1 u + \tilde{m} v + \varepsilon_2 \lambda, \tilde{\eta} + u + \varepsilon_3 v + \varepsilon_4(\lambda - \sigma)]$$

where $(\tilde{m}, \tilde{\eta}, \sigma)$ varies on a neighborhood of $(m, \eta, 0)$.

The table with the other normal forms that complete the classification of $\mathbb{Z}_2 \oplus \mathbb{Z}_2$-equivariant bifurcation problems is given by the theorem to follow:

Theorem 2.2. ([10]) *If a germ* $g = [p, q] \in \vec{\mathcal{E}}_{x,y,\lambda}(Z_2 \oplus Z_2)$, $\lambda \in (\mathbb{R}, 0)$, *satisfies the recognition conditions in following table, then* g *is* $Z_2 \oplus Z_2$*-equivalent to* h_j, $j = 2, \ldots, 8$.

Normal Forms

$$\begin{aligned}
h_2 &= [\varepsilon_1 u + \varepsilon_4 v + \varepsilon_2 \lambda + \varepsilon_5 u^2, \varepsilon_1\varepsilon_3\varepsilon_4 u + \varepsilon_3 v + \kappa\lambda] \\
h_3 &= [\varepsilon_1 u + \mu v + \varepsilon_2 \lambda, \varepsilon_1\varepsilon_2\varepsilon_4 u + \varepsilon_3 v + \varepsilon_4 \lambda + \varepsilon_5 \lambda^2] \\
h_4 &= [\varepsilon_1 u + \varepsilon_2\varepsilon_3\varepsilon_4 v + \varepsilon_2 \lambda, \eta u + \varepsilon_3 v + \varepsilon_4 \lambda + \varepsilon_5 \lambda^2] \\
h_5 &= [\varepsilon_1 u^2 + \mu v + \varepsilon_2 \lambda, \varepsilon_5 u + \varepsilon_3 v + \varepsilon_4 \lambda] \\
h_6 &= [\varepsilon_1 u + \varepsilon_5 v + \varepsilon_2 \lambda, \eta u + \varepsilon_3 v^2 + \varepsilon_4 \lambda] \\
h_7 &= [\varepsilon_1 u + \varepsilon_5 v + \varepsilon_2 \lambda^2, \eta u + \varepsilon_3 v + \varepsilon_4 \lambda] \\
h_8 &= [\varepsilon_1 u + \mu v + \varepsilon_2 \lambda, \varepsilon_5 u + \varepsilon_3 v + \varepsilon_4 \lambda^2]
\end{aligned}$$

Normal Form	*Recognition Conditions*	ε_1	ε_2	ε_3	ε_4	ε_5	*Modal Parameter*	*Unfolding Terms*
h_2	$p^o_u q^o_v - p^o_v q^o_u = 0$ $p^o_u, q^o_v, p^o_v, p^o_\lambda, q^o_\lambda, \rho_2,$ $p^o_u q^o_\lambda - p^o_\lambda q^o_u \neq 0$	p^o_u	p^o_λ	q^o_v	p^o_v	$\rho_2 q^o_v$	$\kappa = \lvert\frac{p^o_v}{p^o_\lambda q^o_v}\rvert q^o_\lambda$	$[u, 0]$ $[0, \lambda]$ $[0, 1]$
h_3	$p^o_\lambda q^o_u - p^o_u q^o_\lambda = 0$ $p^o_u, q^o_v, p^o_\lambda, q^o_\lambda, \rho_3,$ $p^o_u q^o_v - p^o_v q^o_u \neq 0$	p^o_u	p^o_λ	q^o_v	q^o_λ	$\rho_3 p^o_\lambda$	$\mu = \lvert\frac{q^o_\lambda}{p^o_\lambda q^o_v}\rvert p^o_v$	$[v, 0]$ $[0, u]$ $[0, 1]$
h_4	$p^o_v q^o_\lambda - p^o_\lambda q^o_v = 0$ $p^o_u, q^o_v, p^o_\lambda, q^o_\lambda, \rho_4,$ $p^o_u q^o_v - p^o_v q^o_u \neq 0$	p^o_u	p^o_λ	q^o_v	q^o_λ	$\rho_4 p^o_\lambda$	$\eta = \lvert\frac{p^o_\lambda}{p^o_u q^o_\lambda}\rvert q^o_u$	$[v, 0]$ $[0, u]$ $[0, 1]$
h_5	$p^o_u = 0$ $p^o_\lambda, q^o_v, q^o_\lambda, p^o_{uu},$ $p^o_v q^o_\lambda - p^o_\lambda q^o_v \neq 0$	p^o_{uu}	p^o_λ	q^o_v	q^o_λ	q^o_u	$\mu = \lvert\frac{q^o_\lambda}{p^o_\lambda q^o_v}\rvert p^o_v$	$[v, 0]$ $[u, 0]$ $[0, 1]$
h_6	$q^o_v = 0$ $p^o_u, p^o_\lambda, q^o_\lambda, p^o_v, q^o_u, q^o_{vv},$ $p^o_u q^o_\lambda - p^o_\lambda q^o_u \neq 0$	p^o_u	p^o_λ	q^o_{vv}	q^o_λ	p^o_v	$\eta = \lvert\frac{p^o_\lambda}{p^o_u q^o_\lambda}\rvert q^o_u$	$[0, v]$ $[0, u]$ $[1, 0]$
h_7	$p^o_\lambda = 0$ $p^o_u, p^o_v, q^o_v, q^o_\lambda, p^o_{\lambda\lambda},$ $p^o_u q^o_v - p^o_v q^o_u \neq 0$	p^o_u	$p^o_{\lambda\lambda}$	q^o_v	q^o_λ	p^o_v	$\eta = \lvert\frac{p^o_v}{p^o_u q^o_v}\rvert q^o_u$	$[\lambda, 0]$ $[0, u]$ $[1, 0]$
h_8	$q^o_\lambda = 0$ $p^o_u, p^o_\lambda, q^o_u, q^o_v, q^o_{\lambda\lambda}$ $p^o_u q^o_v - p^o_v q^o_u \neq 0$	p^o_u	p^o_λ	q^o_v	$q^o_{\lambda\lambda},$	q^o_u	$\mu = \lvert\frac{q^o_u}{p^o_u q^o_v}\rvert p^o_v$	$[0, \lambda]$ $[v, 0]$ $[1, 0]$

$$\begin{aligned}
\rho_2 &= q^o_{uv} p^o_u (p^o_v)^2 - q^o_{vv} (p^o_u)^2 p^o_v - q^o_{uu} (p^o_v)^3 - p^o_{uv} q^o_v p^o_u p^o_v + p^o_{vv} q^o_v (p^o_u)^2 + p^o_{uu} q^o_v (p^o_v)^2 \\
\rho_3 &= q^o_{uu} (p^o_\lambda)^3 - p^o_{uu} (p^o_\lambda)^2 q^o_\lambda - q^o_{u\lambda} (p^o_\lambda)^2 p^o_u + q^o_{\lambda\lambda} p^o_\lambda (p^o_u)^2 + p^o_{u\lambda} p^o_\lambda q^o_\lambda p^o_u - p^o_{\lambda\lambda} (p^o_u)^2 q^o_\lambda \\
\rho_4 &= q^o_{\lambda\lambda} p^o_\lambda (q^o_v)^2 - p^o_{vv} p^o_\lambda (q^o_\lambda)^2 - q^o_{v\lambda} p^o_\lambda q^o_\lambda q^o_v - p^o_{vv} (q^o_\lambda)^3 - p^o_{\lambda\lambda} (q^o_v)^2 q^o_\lambda + p^o_{v\lambda} (q^o_\lambda)^2 q^o_v
\end{aligned}$$

3. Path formulation

This section is divided into two parts. First we explain how we associate a path to each bifurcation problem. Second we recall the equivalence on paths that corresponds to the contact equivalence for the bifurcation diagrams.

3.1. Cores and paths

Let $g \in \vec{\mathcal{E}}_{(z,\lambda)}^{\mathbb{Z}_2 \oplus \mathbb{Z}_2}$. The germ $g_0 \in \vec{\mathcal{E}}_z^{\mathbb{Z}_2 \oplus \mathbb{Z}_2}$ defined by $g_0(z) = g(z, 0)$ is called the *core* of g. When $\lambda = 0$, the group $\mathcal{K}_\lambda^{\mathbb{Z}_2 \oplus \mathbb{Z}_2}$ simplifies to $\mathcal{K}^{\mathbb{Z}_2 \oplus \mathbb{Z}_2}$, the classical group of $\mathbb{Z}_2 \oplus \mathbb{Z}_2$-equivalences without parameters. A germ g is of *finite core* if g_0 is of finite $\mathcal{K}^{\mathbb{Z}_2 \oplus \mathbb{Z}_2}$-codimension, say k. Consider now g as an unfolding of g_0 with l parameters. From the $\mathcal{K}^{\mathbb{Z}_2 \oplus \mathbb{Z}_2}$-theory of unfoldings, if G_0 is a miniversal unfolding of g_0 (of codimension k, say), then g factors through G_0. That is, there exists changes of coordinates S, Z such that

$$g(z, \lambda) = S(z, \lambda)\, G_0(Z(z, \lambda), \bar{\alpha}(\lambda)), \tag{3.1}$$

where $\bar{\alpha} : (\mathbb{R}^l, 0) \to (\mathbb{R}^k, 0)$. We say that $\bar{\alpha}$ is the *path* associated to g. Thus $\bar{\alpha}$ induces a new bifurcation problem defined by

$$\bar{\alpha}^\star G_0(z, \lambda) = G_0(z, \bar{\alpha}(\lambda)).$$

The *space of paths* will be denoted by $\vec{\mathcal{P}}_\lambda = \{\bar{\alpha} : (\mathbb{R}^l, 0) \to (\mathbb{R}^k, 0)\}$.

From (3.1), g and the pull-back $\bar{\alpha}^\star G_0$ are $\mathcal{K}^{\mathbb{Z}_2 \oplus \mathbb{Z}_2}$-equivalent (with (S, Z, I_l) providing the equivalence).

3.2. Path equivalence and its tangent spaces

We can now define an equivalence between two paths with the same core. That is, we say that $\bar{\alpha}, \bar{\beta} : (\mathbb{R}^l, 0) \to (\mathbb{R}^k, 0)$ are *path equivalent* if

$$\bar{\alpha}(\lambda) = H(\lambda, \bar{\beta}(L(\lambda))) \tag{3.2}$$

where $L : (\mathbb{R}^l, 0) \to (\mathbb{R}^l, 0)$ is an orientation-preserving diffeomorphism and $H : (\mathbb{R}^{l+k}, 0) \to (\mathbb{R}^k, 0)$ is a λ-parametrized family of local orientation-preserving diffeomorphism on $(\mathbb{R}^k, 0)$ that lifts to $\mathbb{Z}_2 \oplus \mathbb{Z}_2$-equivariant diffeomorphism on $G_0^{-1}(0)$. More precisely, there exists a λ-family of $\mathbb{Z}_2 \oplus \mathbb{Z}_2$-equivariant diffeomorphisms $\Phi : (\mathbb{R}^{2+k+l}, 0) \to (\mathbb{R}^{2+k}, 0)$ preserving $G_0^{-1}(0)$ such that $H \circ \pi_{G_0} = \pi_{G_0} \circ \Phi$ on $G_0^{-1}(0)$ where $\pi_{G_0} : G_0^{-1}(0) \to (\mathbb{R}^k, 0)$ is the restriction to $G_0^{-1}(0)$ of the natural projection $(\mathbb{R}^{2+k}, 0) \to (\mathbb{R}^k, 0)$. In this case we shall see in Section 5 that this definition is equivalent to the definition in [12].

For a fixed core g_0, the group of path equivalences, denoted by $\mathcal{K}_{\Delta^{G_0}}$, is a geometric subgroup which acts on the space of paths, hence the general theory of [3] applies. Note that we cannot in general simplify H in (3.2) to a λ-parametrized matrix like with the usual contact-equivalence. An explicit description of the diffeomorphisms H is in general very hard, if not impossible. But the tangent space of $\bar{\alpha}$ can be determined explicitly. Let $\operatorname{Derlog}^\star(\Delta^{G_0})$ be the module of *liftable vector fields* ξ satisfying

$$S(z, \alpha)\, G_0(z, \alpha) = (\mathrm{d}G_0)_z(z, \alpha)\, Z(z, \alpha) + (\mathrm{d}G_0)_\alpha(z, \alpha)\, \xi(\alpha), \quad \alpha \in \mathbb{R}^k. \tag{3.3}$$

The tangent space at a path $\bar{\alpha}$ is the $\mathcal{E}_\lambda$-module of $\vec{\mathcal{P}}_\lambda$ given by

$$\mathcal{T}_e\mathcal{K}_{\Delta^{G_0}}(\bar{\alpha}) = \langle \bar{\alpha}_\lambda \rangle_{\mathcal{E}_\lambda} + \bar{\alpha}^\star \mathrm{Derlog}^\star(\Delta^{G_0}). \tag{3.4}$$

Also we define the codimension of path $\bar{\alpha}$ by $\mathrm{cod}_{\mathcal{K}_{\Delta^{G_0}}}(\bar{\alpha}) = \dim_{\mathbb{R}} \vec{\mathcal{P}}_\lambda / \mathcal{T}_e\mathcal{K}_{\Delta^{G_0}}(\bar{\alpha})$.

In Section 5 we discuss the geometric interpretation of the path equivalence showing that H is also exactly the discriminant preserving contact equivalences, hence the notation with Δ^{G_0}. In the mean time one has the following result about path and contact equivalences.

Theorem 3.1.

1. *Let $g \in \vec{\mathcal{E}}^{\mathbb{Z}_2\oplus\mathbb{Z}_2}_{(z,\lambda)}$ be a finite $\mathcal{K}^{\mathbb{Z}_2\oplus\mathbb{Z}_2}_\lambda$-codimension. If g has a core of finite $\mathcal{K}^{\mathbb{Z}_2\oplus\mathbb{Z}_2}$-codimension, then there exists a path $\bar{\alpha}$ such that g is $\mathcal{K}^{\mathbb{Z}_2\oplus\mathbb{Z}_2}_\lambda$-equivalent to $\bar{\alpha}^\star G_0$, where G_0 is a miniversal unfolding of the core g_0 of g.*
2. *The $\mathcal{K}_{\Delta^{G_0}}$-codimension of $\bar{\alpha}$ is finite if and only if the $\mathcal{K}^{\mathbb{Z}_2\oplus\mathbb{Z}_2}_\lambda$-codimension of $\bar{\alpha}^\star G_0$ is finite, as correspondents normal spaces are isomorphics.*
3. *Let $\bar{\alpha}$, $\bar{\beta}$ be two paths in $\vec{\mathcal{P}}_\lambda$. Then, $\bar{\alpha}$ is $\mathcal{K}_{\Delta^{G_0}}$-equivalent to $\bar{\beta}$ if and only if $\bar{\alpha}^\star G_0$ is $\mathcal{K}^{\mathbb{Z}_2\oplus\mathbb{Z}_2}_\lambda$-equivalent to $\bar{\beta}^\star G_0$ for finite codimension problems.*

The proof may be adapted to the correspondent theorem in [5].

4. $\mathbb{Z}_2 \oplus \mathbb{Z}_2$-equivariant cores

In this section we discuss the cores and their universal unfolding of low codimension. The Theorem 4.1 and Theorem 4.2 classify the normal forms for the cores. We observe the presence of only one modal parameter in the generic core h_1^c. It differs from the classical classification where the core has two modal parameters (see the core of h_1 in the Theorem 2.1). To simplify the description of their recognition problem we define the following quantities: $\epsilon_1 = \mathrm{sign}(p_u^o)$, $\epsilon_2 = \mathrm{sign}(p_v^o)$, $\epsilon_3 = \mathrm{sign}(q_u^o)$, $\epsilon_4 = \mathrm{sign}(q_v^o)$, $\epsilon_5 = \mathrm{sign}(p_{uu}^o)$ and $\epsilon_6 = \mathrm{sign}(q_{vv}^o)$.

The generic core and its miniversal unfolding is given in the following result whose proof follows from a simple re-scaling and a straightforward calculation of the unipotent tangent space.

Theorem 4.1. *Let $g \in \vec{\mathcal{E}}^{\mathbb{Z}_2\oplus\mathbb{Z}_2}_z$ be given by $g(x,y) = (p(u,v)x, q(u,v)y) = [p,q]$. If ϵ_i, $1 \le i \le 4$, and $p_u^o q_v^o - p_v^o q_u^o$ are all nonzero at the origin, then g is $\mathbb{Z}_2 \oplus \mathbb{Z}_2$-equivalent to*

$$h_1^c = [\epsilon_1 u + mv, \epsilon_3 u + \epsilon_4 v]$$

with the modal parameter $m = \left| \frac{q_u^o}{p_u^o q_v^o} \right| p_v^o$ $(m \neq \epsilon_1\epsilon_3\epsilon_4)$. A miniversal unfolding of the core h_1^c is

$$H_1^c = [\epsilon_1 u + (m+\alpha_3)v + \alpha_1, \epsilon_3 u + \epsilon_4 v + \alpha_2].$$

Proof. Consider the change of coordinates given by $S(z)g(Z(z)) = [\bar{p}, \bar{q}]$, $Z = [a, b] \in \vec{\mathcal{E}}_z^{\mathbb{Z}_2 \oplus \mathbb{Z}_2}$, $S = \begin{pmatrix} s_1 & s_2 \\ s_3 & s_4 \end{pmatrix}$ satisfying (2.1) (without λ). It follows

$$\bar{p}_u^0 = p_u^0 a_0^3 s_1^0, \quad \bar{p}_v^0 = p_v^0 a_0 b_0^2 s_1^0, \quad \bar{q}_u^0 = q_u^0 a_0^2 b_0 s_4^0, \quad \bar{q}_v^0 = q_v^0 b_0^3 s_4^0,$$

where $a_0 > 0$, $b_0 > 0$, $s_1^0 > 0$, $s_4^0 > 0$ are the coefficients in the Taylor expansion of the functions a, b, s_1 and s_4, respectively.

Normalizing these coefficients we get the formula to h_1^c and the respective parameters ϵ_i's and m.

Now, to find the miniversal unfolding, first we calculate that the quadratic terms are in the unipotent tangent space of h_1^c to justify that they can be ignored. Then we use the normal space to find the unfolding terms. If $m \neq \epsilon_1\epsilon_3\epsilon_4$, then $d_1 = [1, 0]$, $d_2 = [0, 1]$ and $d_3 = [v, 0]$ are generators of the extended normal space of h_1^c, thus H_1^c is a versal unfolding of h_1^c. □

The generic core is of smooth codimension 3 but topological codimension 2. The remaining cores of codimension 3 are given by degenerating the previous conditions in Theorem 4.1. There are 5 of them but they can be grouped in 3 types by interchanging x and y (as well then as p and q). Define

$$\rho_2 = q_{uv}^o p_u^o (p_v^o)^2 - q_{vv}^o (p_u^o)^2 p_v^o - q_{uu}^o (p_v^o)^3 - p_{uv}^o q_v^o p_u^o p_v^o + p_{vv}^o q_v^o (p_u^o)^2 + p_{uu}^o q_v^o (p_v^o)^2.$$

Theorem 4.2. *Let $g \in \vec{\mathcal{E}}_z^{\mathbb{Z}_2 \oplus \mathbb{Z}_2}$ be given by $g = [p, q]$.*

(a) *If $p_u^o q_v^o - p_v^o q_u^o = 0$ with ϵ_i, $1 \leq i \leq 4$, and ρ_2 all nonzero, then g is $\mathbb{Z}_2 \oplus \mathbb{Z}_2$-equivalent to*

$$h_2^c = [\epsilon_1 u + \epsilon_2 v + \hat{\epsilon}_5 u^2, \epsilon_1\epsilon_2\epsilon_4 u + \epsilon_4 v],$$

with miniversal unfolding $H_2^c = [(\epsilon_1 + \alpha_3)u + \epsilon_2 v + \hat{\epsilon}_5 u^2 + \alpha_1, \epsilon_1\epsilon_2\epsilon_4 u + \epsilon_4 v + \alpha_2]$ and $\hat{\epsilon}_5 = sign(q_v^o \rho_2)$.

(b) *If $p_u^o = 0$ with ϵ_i, $2 \leq i \leq 5$, all nonzero, then g is $\mathbb{Z}_2 \oplus \mathbb{Z}_2$-equivalent to*

$$h_5^c = [\epsilon_5 u^2 + \epsilon_2 v, \epsilon_3 u + \epsilon_4 v]$$

with miniversal unfolding $H_5^c = [\epsilon_5 u^2 + \alpha_3 u + \epsilon_2 v + \alpha_1, \epsilon_3 u + \epsilon_4 v + \alpha_2]$. When $q_v^0 = 0$, interchanging x and y, we obtain $h_6^c = [\epsilon_1 u + \epsilon_2 v, \epsilon_3 u + \epsilon_6 v^2]$ of miniversal unfolding $H_6^c = [\epsilon_1 u + \epsilon_2 v + \alpha_1, \epsilon_3 u + \epsilon_6 v^2 + \alpha_3 v + \alpha_2]$.

(c) *If $q_u^o = 0$ with $\epsilon_1, \epsilon_3, \epsilon_4$ all nonzero, then g is $\mathbb{Z}_2 \oplus \mathbb{Z}_2$-equivalent to*

$$h_9^c = [\epsilon_1 u + \epsilon_2 v, \epsilon_4 v]$$

with miniversal unfolding $H_9^c = [\epsilon_1 u + \epsilon_2 v + \alpha_1, \alpha_3 u + \epsilon_4 v + \alpha_2]$. When $p_v^0 = 0$, interchanging x and y, we obtain $h_{10}^c = [\epsilon_1 u, \epsilon_3 u + \epsilon_4 v]$ with miniversal $H_{10}^c = [\epsilon_1 u + \alpha_3 v + \alpha_1, \epsilon_3 u + \epsilon_4 v + \alpha_2]$.

Proof. We proceed as before with more complicated calculations. This is done with details in [2]. □

5. Derlogs

In this section we calculate here the Derlogs of the cores of Theorems 4.1 and 4.2. First we discuss an important geometric notion linked with the liftable vector fields.

5.1. Discriminants and Derlogs

The *discriminant* Δ^{G_0} of G_0 is the local bifurcation set of G_0, that is, the set of $\alpha \in (\mathbb{R}^k, 0)$ where G_0 is singular. Here the liftability condition on H can be replaced by preserving the discriminant Δ^{G_0} of G_0 in the sense that $H(\lambda, \Delta^{G_0}) \subset \Delta^{G_0}$ for all $\lambda \in (\mathbb{R}^l, 0)$. This is a weaker condition because any liftable vector field must be tangent to the discriminant. In the non equivariant case it is well known that both notion coincide (cf. [9]). Here both modules are also equal and also free. The proof is similar to the one in [5].

The discriminants of the miniversal unfoldings of the cores are formed of the following local bifurcation varieties:

(a) $\mathcal{P}_x, \mathcal{P}_y$ of equations $q(0,0,\alpha) = 0$, resp. $p(0,0,\alpha) = 0$, representing the bifurcations of the branches $\mathcal{S}_x$, resp. $\mathcal{S}_y$, from the trivial branch,

(b) $\mathcal{P}_{y,z}$ of equation $p(0,v,\alpha) = q(0,v,\alpha) = 0$ representing the bifurcation of the branches $\mathcal{S}_z$ from $\mathcal{S}_y$, $\mathcal{B}_y$ of equation $p(0,v,\alpha) = p_u(0,0,\alpha) = 0$ representing turning points on the $\mathcal{S}_y$-branches. In a similar fashion we define $\mathcal{B}_x$ and $\mathcal{P}_{x,z}$.

(c) $\mathcal{B}_z$ of equation $p = q = p_u q_v - p_v q_u = 0$ representing fold points in $\mathcal{S}_z$.

To fully exploit methods from algebraic geometry we need to complexify our situation. Nothing will be lost in finite codimension because we can work with germs equivalent to polynomials and we take care to preserve the real and complex algebras. Our results are valid for real germs viewed as real slices of the holomorphic objects. Let $G_0^{\mathbb{C}}$ be the complexification of the miniversal unfolding G_0 (chosen as a polynomial from finite determinacy). The discriminant $\Delta^{G_0^{\mathbb{C}}}$ of $G_0^{\mathbb{C}}$ is the set of singular values of the projection $\pi_{G_0^{\mathbb{C}}} : (G_0^{\mathbb{C}})^{-1}(0) \to \mathbb{C}^k$. The real slice of $\Delta^{G_0^{\mathbb{C}}}$ defines the discriminant Δ^{G_0} instead of the equivalent formula for G_0. We actually define $\mathrm{Derlog}^\star(\Delta^{G_0})$ as the submodule of the real vector fields of the module $\mathrm{Derlog}^\star(\Delta^{G_0^{\mathbb{C}}})$ of liftable vector fields. Let $I(\Delta^{G_0^{\mathbb{C}}})$ be the ideal of germs vanishing on $\Delta^{G_0^{\mathbb{C}}}$. The module of vector fields tangent to $\Delta^{G_0^{\mathbb{C}}}$, called $\mathrm{Derlog}(\Delta^{G_0^{\mathbb{C}}})$, is given by

$$\mathrm{Derlog}(\Delta^{G_0^{\mathbb{C}}}) = \{\xi : (\mathbb{C}^k, 0) \to \mathbb{C}^k \mid \xi \cdot g_\alpha \in I(\Delta^{G_0^{\mathbb{C}}}),\ \forall g \in I(\Delta^{G_0^{\mathbb{C}}})\}.$$

5.2. Liftable vector fields

The discriminant of H_1^c is

$$\Delta^{H_1^c} = \alpha_1\alpha_2(\epsilon_1\alpha_1 - \epsilon_3\alpha_2)(\alpha_1 - \epsilon_4\alpha_2(m + \alpha_3)).$$

To find the vectors of $\mathrm{Derlog}^*(\Delta^{H_1^c})$ we solve the expression (3.3). The coordinates of $\mathrm{Derlog}^*(\Delta^{H_1^c})$, with respect to the basis $[1,0]$, $[0,1]$ and $[v,0]$ of unfolding terms, are

$$\xi_1 = \begin{pmatrix} \alpha_1 \\ \alpha_2 \\ 0 \end{pmatrix}, \quad \xi_2 = \begin{pmatrix} -\epsilon_1\alpha_1^2 \\ -\epsilon_3\alpha_2^2 \\ (-\epsilon_1\alpha_1 + \epsilon_3\alpha_2)(m+\alpha_3) \end{pmatrix}, \quad \xi_3 = \begin{pmatrix} 0 \\ 0 \\ \alpha_1 - \epsilon_4\alpha_2(m+\alpha_3) \end{pmatrix}.$$

Note that the determinant of the matrix formed by these vectors is equal to the discriminant $\Delta^{H_1^c}$ which shows that Derlog$^*(\Delta^{H_1^c})$ is free and equal to Derlog$(\Delta^{H_1^c})$ using Saito's criterion (cf. [13]).

The discriminant of H_{10}^c is

$$\Delta^{H_{10}^c} = \alpha_1\alpha_2(\epsilon_1\alpha_1 - \epsilon_3\alpha_2)(\alpha_1 - \epsilon_4\alpha_2\alpha_3).$$

The generators of Derlog$^*(\Delta^{H_{10}^c})$ are

$$\xi_1 = \begin{pmatrix} \alpha_1 \\ \alpha_2 \\ 0 \end{pmatrix}, \quad \xi_2 = \begin{pmatrix} \epsilon_3\alpha_1\alpha_2 - \epsilon_1\alpha_1^2 \\ 0 \\ -\epsilon_1\alpha_1\alpha_3 + \epsilon_3\alpha_2\alpha_3 \end{pmatrix}, \quad \xi_3 = \begin{pmatrix} 0 \\ 0 \\ \alpha_1 - \epsilon_4\alpha_2\alpha_3 \end{pmatrix},$$

and the module is still free.

The discriminant of H_2^c is $\Delta^{H_2^c} = \alpha_1\alpha_2\,(\epsilon_4\alpha_1 - \epsilon_2\alpha_2)\,(\epsilon_1\epsilon_2\epsilon_4\alpha_2\alpha_3^3 + 3\epsilon_2\epsilon_4\alpha_2\alpha_3^2 + 3\epsilon_1\epsilon_2\epsilon_4\alpha_2\alpha_3 + \epsilon_2\epsilon_4\alpha_2 - \alpha_1\alpha_3^2 - 2\epsilon_1\alpha_1\alpha_3 - \alpha_1 - \epsilon_5\alpha_2^2\alpha_3^2 - 2\epsilon_1\epsilon_5\alpha_2^2\alpha_3 - \epsilon_5\alpha_2^2 - 4\epsilon_1\epsilon_2\epsilon_4\epsilon_5\alpha_1\alpha_2\alpha_3 + 4\epsilon_5\alpha_1^2 - 4\epsilon_2\epsilon_4\epsilon_5\alpha_1\alpha_2 + 4\alpha_1\alpha_2^2)\,(\epsilon_2\epsilon_4\epsilon_5\alpha_3^2 - 4\epsilon_2\epsilon_4\alpha_1 + 4\alpha_2)$. The coordinates of Derlog$^*(\Delta^{H_2^c})$, with respect to the basis $[1,0]$, $[0,1]$ and $[u,0]$ of unfolding terms, are

$$\xi_1 = \begin{pmatrix} 4\alpha_1(\epsilon_4\alpha_1 - \epsilon_2\alpha_2) + A\alpha_3 \\ 2\alpha_2(\epsilon_4\alpha_1 - \epsilon_2\alpha_2) + B\alpha_3 \\ 2\epsilon_1(\epsilon_4\alpha_1 - \epsilon_2\alpha_2) + C\alpha_3 \end{pmatrix},$$

with $A = \epsilon_1\epsilon_4\epsilon_5\alpha_1 + \epsilon_4\epsilon_5\alpha_1\alpha_3 - 2\epsilon_1\epsilon_2\alpha_1\alpha_2$, $B = \epsilon_1\epsilon_4\epsilon_5\alpha_2 + \epsilon_4\epsilon_5\alpha_2\alpha_3 - 2\epsilon_1\epsilon_2\alpha_2^2$, $C = -3\epsilon_2\alpha_2 + 4\epsilon_4\alpha_1 - \epsilon_1\epsilon_2\alpha_2\alpha_3$,

$$\xi_2 = \begin{pmatrix} 2\epsilon_1\alpha_1(\epsilon_4\epsilon_5\alpha_1 - \epsilon_2\epsilon_5\alpha_2 - 2\epsilon_2\alpha_1\alpha_2 + 2\epsilon_4\alpha_2^2) + D\alpha_3 \\ 2\epsilon_1\alpha_2(\epsilon_4\epsilon_5\alpha_1 - \epsilon_2\epsilon_5\alpha_2 - 2\epsilon_2\alpha_1\alpha_2 + 2\epsilon_4\alpha_2^2) + E\alpha_3 \\ -6\epsilon_2\alpha_1\alpha_2 + 4\epsilon_4\alpha_1^2 + 2\epsilon_4\alpha_2^2 + F\alpha_3 \end{pmatrix},$$

with $D = -5\epsilon_2\epsilon_5\alpha_1\alpha_2 + 4\epsilon_4\epsilon_5\alpha_1^2 - 3\epsilon_1\epsilon_2\epsilon_5\alpha_1\alpha_2\alpha_3 + 4\epsilon_4\alpha_1\alpha_2^2$, $E = -5\epsilon_2\epsilon_5\alpha_2^2 + 4\epsilon_4\epsilon_5\alpha_1\alpha_2 - 3\epsilon_1\epsilon_2\epsilon_5\alpha_2^2\alpha_3 + 4\epsilon_4\alpha_2^3$, $F = -4\epsilon_1\epsilon_2\alpha_1\alpha_2 - 2\epsilon_2\epsilon_5\alpha_2\alpha_3 + 4\epsilon_1\epsilon_4\alpha_2^2 + \epsilon_1\epsilon_4\epsilon_5\alpha_1 + \epsilon_4\epsilon_5\alpha_1\alpha_3 - \epsilon_1\epsilon_2\epsilon_5\alpha_2 - \epsilon_1\epsilon_2\epsilon_5\alpha_2\alpha_3^2 + 2\epsilon_4\alpha_2^2\alpha_3$, and

$$\xi_3 = \begin{pmatrix} 2\epsilon_1\alpha_1(3\alpha_2 - 2\epsilon_2\epsilon_4\alpha_1 - 4\epsilon_2\epsilon_4\epsilon_5\alpha_2^2) + G\alpha_3 \\ 2\epsilon_1\alpha_2(3\alpha_2 - 2\epsilon_2\epsilon_4\alpha_1 - 4\epsilon_2\epsilon_4\epsilon_5\alpha_2^2) + H\alpha_3 \\ -2\epsilon_2\epsilon_5(-\epsilon_2\epsilon_5\alpha_2 + \epsilon_4\epsilon_5\alpha_1 - 2\epsilon_2\alpha_1\alpha_2 + 2\epsilon_4\alpha_2^2) + I\alpha_3 \end{pmatrix}$$

with $G = -\epsilon_2\epsilon_4\epsilon_5\alpha_1 - \epsilon_1\epsilon_2\epsilon_4\epsilon_5\alpha_1\alpha_3 + 8\alpha_1\alpha_2$, $H = -\epsilon_2\epsilon_4\epsilon_5\alpha_2 - \epsilon_1\epsilon_2\epsilon_4\epsilon_5\alpha_2\alpha_3 + 8\alpha_2^2$ and $I = -4\epsilon_1\epsilon_2\epsilon_4\alpha_1 + 5\epsilon_1\alpha_2 + 3\alpha_2\alpha_3 - 4\epsilon_1\epsilon_2\epsilon_4\epsilon_5\alpha_2^2$. The module is free.

The discriminant of H_5^c is $\Delta^{H_5^c} = \alpha_1\alpha_2(\epsilon_2\alpha_1 - \epsilon_4\alpha_2)(-\epsilon_1 - \epsilon_1\alpha_3^2 - 4\epsilon_2\epsilon_4\alpha_2 + 2\epsilon_2\epsilon_3\epsilon_4\epsilon_1\alpha_3 + 4\alpha_1)(\epsilon_3\alpha_2\alpha_3^3 - \alpha_1\alpha_3^2 - \epsilon_1\alpha_2^2\alpha_3^2 - 4\epsilon_3\epsilon_1\alpha_1\alpha_2\alpha_3 + 4\epsilon_1\alpha_1^2 + 4\alpha_1\alpha_2^2)$. The generators of $\mathrm{Derlog}^*(\Delta^{H_5^c})$ are listed below

$$\xi_1 = \begin{pmatrix} -\epsilon_2\epsilon_3\epsilon_1\alpha_1\alpha_3 + \epsilon_4\epsilon_1\alpha_1\alpha_3^2 - 2\epsilon_3\epsilon_4\alpha_1\alpha_2\alpha_3 + 4\epsilon_4\alpha_1^2 - 2\epsilon_2\alpha_1\alpha_2 \\ -\epsilon_2\epsilon_3\epsilon_1\alpha_2\alpha_3 + \epsilon_4\epsilon_1\alpha_2\alpha_3^2 - 2\epsilon_3\epsilon_4\alpha_2^2\alpha_3 + 2\epsilon_4\alpha_1\alpha_2 \\ -2\epsilon_2\epsilon_3\alpha_1 + 4\epsilon_4\alpha_1\alpha_3 - \epsilon_3\epsilon_4\alpha_2\alpha_3^2 - \epsilon_2\alpha_2\alpha_3 \end{pmatrix},$$

$$\xi_2 = \begin{pmatrix} -4\epsilon_3\epsilon_4\alpha_1^2\alpha_2 - 2\epsilon_2\epsilon_3\epsilon_1\alpha_1^2 + A_1\alpha_3 \\ -4\epsilon_3\epsilon_4\alpha_1\alpha_2^2 - 2\epsilon_2\epsilon_3\epsilon_1\alpha_1\alpha_2 + A_2\alpha_3 \\ 4\epsilon_4\alpha_1^2 - 2\epsilon_2\alpha_1\alpha_2 + A_3\alpha_3 \end{pmatrix},$$

with $A_1 = \epsilon_2\epsilon_1\alpha_1\alpha_2 - 3\epsilon_3\epsilon_4\epsilon_1\alpha_1\alpha_2\alpha_3 + 4\epsilon_4\alpha_1\alpha_2^2 + 4\epsilon_4\epsilon_1\alpha_1^2$, $A_2 = \epsilon_2\epsilon_1\alpha_2^2 + 4\epsilon_4\epsilon_1\alpha_1\alpha_2 - 3\epsilon_3\epsilon_4\epsilon_1\alpha_2^2\alpha_3 + 4\epsilon_4\alpha_2^3$, $A_3 = -\epsilon_2\epsilon_3\epsilon_1\alpha_1 + \epsilon_4\epsilon_1\alpha_1\alpha_3 - 4\epsilon_3\epsilon_4\alpha_1\alpha_2 + \epsilon_2\epsilon_1\alpha_2\alpha_3 + 2\epsilon_4\alpha_2^2\alpha_3 - \epsilon_3\epsilon_4\epsilon_1\alpha_2\alpha_3^2$, and

$$\xi_3 = \begin{pmatrix} -4\epsilon_3\alpha_1^2 - \epsilon_3\epsilon_1\alpha_1\alpha_3^2 + \epsilon_2\epsilon_4\epsilon_1\alpha_1\alpha_3 + 8\alpha_1\alpha_2\alpha_3 - 8\epsilon_3\epsilon_1\alpha_1\alpha_2^2 - 2\epsilon_2\epsilon_3\epsilon_4\alpha_1\alpha_2 \\ -4\epsilon_3\alpha_1\alpha_2 - \epsilon_3\epsilon_1\alpha_2\alpha_3^2 + \epsilon_2\epsilon_4\epsilon_1\alpha_2\alpha_3 + 8\alpha_2^2\alpha_3 - 8\epsilon_3\epsilon_1\alpha_2^3 - 2\epsilon_2\epsilon_3\epsilon_4\alpha_2^2 \\ -4\epsilon_3\alpha_1\alpha_3 + 2\epsilon_2\epsilon_4\alpha_1 + 3\alpha_2\alpha_3^2 + 4\epsilon_1\alpha_1\alpha_2 - 4\epsilon_3\epsilon_1\alpha_2^2\alpha_3 - \epsilon_2\epsilon_3\epsilon_4\alpha_2\alpha_3 \end{pmatrix}.$$

5.3. Modal spaces

We observe that there is a modal parameter in our classification of the generic core instead of two modal parameters in the core of bifurcation problem in Theorem 2.1. Because of the equivalence between contact and path equivalence we expect the values of the modal space to be left *pointwise* invariant by path equivalence. This means that the liftable vector fields must vanish along the modal space. In general, this is a mechanism by which $\mathrm{Derlog}^*(\Delta)$ may be strictly smaller than $\mathrm{Derlog}(\Delta)$ because vector fields in the latest only need to be tangent to the discriminant, so we need to select those that actually vanish on the modal space. Such examples occur for the corank two representation of the dihedral group $\mathbb{D}_4$ (cf. [5]). Surprisingly, in the present case, all vector fields tangent to the discriminant also vanish on the modal space. In all our cases the modal space is the same, given by $m = \alpha_3$, $\alpha_1 = \alpha_2 = 0$. By inspection, each generator of the Derlogs vanishes on that space confirming that both Derlogs are the same.

6. Classification of $\mathbb{Z}_2 \oplus \mathbb{Z}_2$-equivariant bifurcation problems via path formulation

The classification of the paths up to topological codimension 2 is at follows. First we define a few quantities (when they exist): $\delta_1 = \mathrm{sign}(p_\lambda^o)$, $\delta_2 = \mathrm{sign}(q_\lambda^o)$, $\delta_3 = \mathrm{sign}(p_{\lambda\lambda}^o)$, $\delta_4 = \mathrm{sign}(q_{\lambda\lambda}^o)$, $\delta_{51} = \mathrm{sign}(p_\lambda^o\rho_3)$ and $\delta_{52} = \mathrm{sign}(p_\lambda^o\rho_4)$ with

$$\rho_3 = q_{uu}^o(p_\lambda^o)^3 - p_{uu}^o(p_\lambda^o)^2q_\lambda^o - q_{u\lambda}^o(p_\lambda^o)^2p_u^o + q_{\lambda\lambda}^op_\lambda^o(p_u^o)^2 + p_{u\lambda}^op_\lambda^oq_\lambda^op_u^o - p_{\lambda\lambda}^o(p_u^o)^2q_\lambda^o,$$

and

$$\rho_4 = q_{\lambda\lambda}^op_\lambda^o(q_v^o)^2 - p_{vv}^op_\lambda^o(q_\lambda^o)^2 - q_{v\lambda}^op_\lambda^oq_\lambda^oq_v^o - p_{vv}^o(q_\lambda^o)^3 - p_{\lambda\lambda}^o(q_v^o)^2q_\lambda^o + p_{v\lambda}^o(q_\lambda^o)^2q_v^o.$$

Finally, the modal parameters are $\chi = \left|\frac{p_u^o}{q_u^op_\lambda^o}\right| q_\lambda^o$, $\kappa = \left|\frac{p_v^o}{q_v^op_\lambda^o}\right| q_\lambda^o$.

Second we define the following paths α_i's (with their miniversal unfoldings A_i's of unfolding parameters ν_j, $j = 1, 2, 3$):

- $\bar{\alpha}_1(\lambda) = (\delta_1\lambda, \chi\lambda, 0)$ with $A_1(\lambda, \nu) = (\delta_1\lambda, (\chi + \nu_2)\lambda + \nu_1, \nu_3)$ of topological codimension 1,
- $\bar{\alpha}_2(\lambda) = (\delta_1\lambda, \kappa\lambda, 0)$ with $A_2(\lambda, \nu) = (\delta_1\lambda, (\kappa + \nu_2)\lambda + \nu_1, \nu_3)$ of topological codimension 1,
- $\bar{\alpha}_3(\lambda) = (\delta_3\lambda^2, \delta_2\lambda, 0)$ with $A_3(\lambda, \nu) = (\delta_3\lambda^2 + \nu_1 + \nu_2\lambda, \delta_2\lambda, \nu_3)$ of topological codimension 2,
- $\bar{\alpha}_4(\lambda) = (\delta_1\lambda, \delta_4\lambda^2, 0)$ with $A_4(\lambda, \nu) = (\delta_1\lambda, \delta_4\lambda^2 + \nu_1 + \nu_2\lambda, \nu_3)$ of topological codimension 2 and
- $\bar{\alpha}_{5i}(\lambda) = (\delta_1\lambda, \delta_2\lambda + \delta_{5i}\lambda^2, 0)$, $i = 1, 2$, with $A_{5i}(\lambda, \nu) = (\delta_1\lambda, (\delta_2 + \nu_2)\lambda + \delta_{5i}\lambda^2 + \nu_1, \nu_3)$ of topological codimension 2.

Theorem 6.1 (Main Theorem). *Classification of $\mathbb{Z}_2 \oplus \mathbb{Z}_2$-equivariant bifurcation problems using path formulation up to topological codimension two and one bifurcation parameter:*

(a) *With the core H_1^c, the paths are $\bar{\alpha}_1$ (when $\chi \neq \delta_1\epsilon_1\epsilon_3$), $\bar{\alpha}_3$ (when $\delta_1 = 0$), $\bar{\alpha}_4$ (when $\delta_2 = 0$) and $\bar{\alpha}_{51}, \bar{\alpha}_{52}$ (when $\chi = \delta_1\epsilon_1\epsilon_3$).*

(b) *With the core H_{10}^c the paths are $\bar{\alpha}_1$ (when $\chi \neq \delta_1\epsilon_1\epsilon_3$), $\bar{\alpha}_4$ (when $\delta_2 = 0$) and $\bar{\alpha}_{51}$ (when $\chi = \delta_1\epsilon_1\epsilon_3$). With the core H_9^c the path are $\bar{\alpha}_2$ (when $\kappa \neq \delta_1\epsilon_2\epsilon_4$), $\bar{\alpha}_3$ (when $\delta_1 = 0$), and $\bar{\alpha}_{52}$ (when $\kappa - \delta_1\epsilon_2\epsilon_4$).*

(c) *With the core H_6^c the path is $\bar{\alpha}_1 = (\delta_1\lambda, \chi\lambda, 0)$ (when $\chi \neq \delta_1\epsilon_1\epsilon_3$), and with the core H_5^c the path is $\bar{\alpha}_2 = (\delta_1\lambda, \kappa\lambda, 0)$ (when $\kappa \neq \delta_1\epsilon_2\epsilon_4$).*

(d) *With the core H_2^c the path is $\bar{\alpha}_2 = (\delta_1\lambda, \kappa\lambda, 0)$ when $\kappa \neq \delta_1\epsilon_2\epsilon_4$.*

Proof. The proof of each case follows the same pattern. We show the first case with some details. Let $\bar{\alpha} : (\mathbb{R}, 0) \to (\mathbb{R}^3, 0)$ be a path such that $\bar{\alpha}(\lambda) = (\delta_1\lambda, \chi\lambda, 0) + \vec{\mathcal{M}}_\lambda^2$. From (3.4), the tangent space of $\bar{\alpha}$ is

$$\mathcal{T}_e\mathcal{K}_{\Delta^{H_1^c}}(\bar{\alpha}) = \langle\bar{\alpha}_\lambda\rangle_{\mathcal{E}_\lambda} + \bar{\alpha}^\star \mathrm{Derlog}(\Delta^{H_1^c}).$$

At their lowest order the generators of $\mathcal{T}_e\mathcal{K}_{\Delta^{H_1^c}}(\bar{\alpha})$ are:

$$v_1 = \begin{pmatrix} \delta_1 \\ \chi \\ 0 \end{pmatrix}, \quad v_2 = \begin{pmatrix} \delta_1\lambda \\ \chi\lambda \\ 0 \end{pmatrix}, \quad v_3 = \begin{pmatrix} -\epsilon_1\lambda^2 \\ -\epsilon_3\chi^2\lambda^2 \\ (-\epsilon_1\delta_1 + \epsilon_3\chi)m\lambda \end{pmatrix},$$

$$v_4 = \begin{pmatrix} 0 \\ 0 \\ (\delta_1 - \epsilon_4\chi m)\lambda \end{pmatrix}.$$

If $\delta_1 - \epsilon_4\chi m \neq 0$ the generator $(0, 0, \lambda)$ is in the tangent space $\mathcal{T}_e\mathcal{K}_{\Delta^{H_1^c}}(\bar{\alpha})$. If $\chi \neq \epsilon_1\epsilon_2\epsilon_3$ and $\chi \neq 0$, we get the generator $(0, \lambda^2, 0)$. We find that the normal space $\mathcal{N}_e\mathcal{K}_{\Delta^{H_1^c}}(\bar{\alpha})$ is generated by $[0, 1]$, $[v, 0]$ and $[0, \lambda]$, or in vector notation $(0, 1, 0)$, $(0, 0, 1)$ and $(0, \lambda, 0)$. In fact, consider the following table.

generators	$[1,0]$	$[\lambda,0]$	$[v,0]$	$[0,1]$	$[0,\lambda]$
v_1	δ_1	0	0	χ	0
v_2	0	δ_1	0	0	χ
$[0,1]$	0	0	0	1	0
$[v,0]$	0	0	1	0	0
$[0,\lambda]$	0	0	0	0	1

Since this 5×5 matrix has maximum rank, we obtain the generators of $\mathcal{N}_e\mathcal{K}_{\Delta^{H_1^c}}(\bar{\alpha})$. The miniversal unfolding $H_1^\star$ of $h_1^\star$ is given by

$$H_1^\star = [\epsilon_1 u + \tilde{m}v + \delta_1\lambda, \epsilon_3 u + \epsilon_4 v + \tilde{\chi}\lambda + \nu_1]$$

where $\tilde{m} = m + \nu_3$ and $\tilde{\chi} = \chi + \nu_2$. Hence the topological codimension is 1 but the smooth codimension is 3.

For the other cases we proceed similarly. □

6.1. Comparison with the classical theory

The generic bifurcation diagram in [10, 8] is $h_1 = [\epsilon_1 u + \mu v + \delta_1\lambda, \eta u + \epsilon_4 v + \delta_2\lambda]$. As expected it has two modal parameters but they live in the core. Our analysis shows that actually only one modal parameter is associated with the core, the other is linked with the path. More explicitly, the link between the two sets of modal parameters are $\mu = \delta_1\chi m$, $\eta = \delta_1\epsilon_3\chi^{-1}$.

Another point to make is that the classification of [10] contains 8 cases they denote h_1 to h_8. In our classification theorem we have 14 cases. Actually a more detailed comparison indicates that some of the cases in [10] correspond to several of our cases. Explicitly,

- h_1 corresponds to 3 pull-backs: $\bar{\alpha}_1^* H_1^c$, $\bar{\alpha}_2^* H_9^c$ or $\bar{\alpha}_1^* H_{10}^c$,
- the next 4 to two pull-backs: $h_{3,8}$ to $\bar{\alpha}_{51,4}^* H_1^c$ or $\bar{\alpha}_{51,4}^* H_{10}^c$ and $h_{4,7}$ to $\bar{\alpha}_{52,3}^* H_1^c$ or $\bar{\alpha}_{51,4}^* H_9^c$, and
- the final 3 to only one pull-back: h_2 to $\bar{\alpha}_2^* H_2^c$, h_5 to $\bar{\alpha}_2^* H_5^c$ and h_6 to $\bar{\alpha}_1^* H_6^c$.

Acknowledgment

The authors would like to thank Professor Ian Stewart and Dr. Jacques Elie Furter.

References

[1] T.J. Bridges, J.E. Furter, *Singularity theory and equivariant symplectic maps.* Springer Lectures Notes **1558**, Heidelberg, 1993.

[2] J.C.F. Costa, *Formulação por caminhos para problemas de bifurcação* $\mathbb{Z}_2 \oplus \mathbb{Z}_2$*-equivariantes.* Dissertação de Mestrado, Ibilce-Unesp, São José do Rio Preto, 2002.

[3] J. Damon, *The unfolding and determinacy theorems for subgroups of* $\mathcal{A}$ *and* $\mathcal{K}$. Memoirs of the American Mathematical Society **306** (AMS, Providence RI), 1984.

[4] J.E. Furter, A.M. Sitta, I. Stewart, *Algebraic path formulation for equivariant bifurcation problems.* Math. Proc. Camb. Phil. Soc. **124** (1998) 275–304.

[5] J.E. Furter, A.M. Sitta, *Path formulation for equivariant bifurcation problem: the* $\mathbb{D}_3$*-classification.* To appear.

[6] T. Gaffney, *Some new results in the classification theory of bifurcation problems*, in *Multiparameter bifurcation theory.* (Eds. Golubitsky, M. and Guckenheimer, J.), Contemporary Mathematics **56**, Amer. Math. Soc., Providence RI (1986) 97–116.

[7] M. Golubitsky, D.G. Schaeffer, *Singularities and groups in bifurcation theory.* Vol. I, Applied Mathematics and Sciences **51**, Springer-Verlag, New York, 1985.

[8] M. Golubitsky, I. Stewart, D.G. Schaeffer, *Singularities and groups in bifurcation theory.* Vol. II, Applied Mathematics and Sciences **69**, Springer-Verlag, New York, 1988.

[9] E.J.N. Looijenga, *Isolated singular points of complete intersections.* LMS Lecture Notes, Vol. **17**, Cambridge University Press, 1984.

[10] M.G. Manoel, I. Stewart, *Degenerate bifurcations with* $\mathbb{Z}_2 \oplus \mathbb{Z}_2$*-symmetry.* International Journal of Bifurcation and Chaos, Vol. 9, No. **8** (1999) 1653–1667.

[11] J. Mather, *Stability of* $\mathcal{C}^\infty$*-mappings, III: finitely determined map germs.* Publ. Math. I.H.E.S. **35** (1969), 127–156.

[12] J. Montaldi, *The path formulation of bifurcation theory*, in *Dynamics, bifurcation and symmetry: new trends and new tools.* (Ed. P. Chossat), Nato ASI Series **437**, Kluwer, Dordrecht (1994) 259–278.

[13] K. Saito, *Theory of logarithmic differential forms and logarithmic vectorfields.* J. Fac. Sci. Univ. Tokyo Sec 1A **27** (1980), 265–291.

[14] D.G. Schaeffer, M. Golubitsky, *Boundary conditions and mode jumping in the buckling of a rectangular plate.* Commun. Math. Phys. **69** (1979) 209–236.

João Carlos Ferreira Costa and Angela Maria Sitta
Depto. Matemática - IBILCE-UNESP
Rua Cristóvão Colombo, 2265, CEP 15054-000
São José do Rio Preto-SP, Brazil
e-mail: `jcosta@ibilce.unesp.br`
`angela@ibilce.unesp.br`

Real and Complex Singularities
Trends in Mathematics, 143–168

The Multiplicity of Pairs of Modules and Hypersurface Singularities

Terence Gaffney

Abstract. This paper applies the multiplicity polar theorem to the study of hypersurfaces with non-isolated singularities. The multiplicity polar theorem controls the multiplicity of a pair of modules in a family by relating the multiplicity at the special fiber to the multiplicity of the pair at the general fiber. It is as important to the study of multiplicities of modules as the basic theorem in ideal theory which relates the multiplicity of an ideal to the local degree of the map formed from the generators of a minimal reduction. In fact, as a corollary of the theorem, we show here that for M a submodule of finite length of a free module F over the local ring of an equidimensional complex analytic germ, that the number of points at which a generic perturbation of a minimal reduction of M is not equal to F, is the multiplicity of M.

Specifically, we apply the multiplicity polar theorem to the study of stratification conditions on families of hypersurfaces, obtaining the first set of invariants giving necessary and sufficient conditions for the A_f condition for hypersurfaces with non-isolated singularities.

Introduction

In [10] the author introduced the notion of the multiplicity of a pair of modules as a way of working with modules of non-finite colength. Some applications of this notion to equisingularity problems were described in [11]. The invariants introduced using this tool have the advantage that they must be independent of the parameters in the family when the stratification condition they describe holds. These invariants provide a framework for studying the equisingularity conditions W, W_f and A_f for very general families of spaces and functions. In this paper we will illustrate the use of these invariants in the study of families of functions with non-isolated singularities and show how the invariants arise naturally in the work of Pellikaan ([27], [28])and Zaharia ([32], [33]). Pellikaan studied functions f whose singular set was an isolated complete intersection singularity (ICIS) of dimension 1, Zaharia those of of dimension 2. The principal tool for connecting the multiplicity of the pair with geometry is the multiplicity polar theorem (Theorem 1.1) which we review in Section 1. This theorem is used to relate multiplicity information at the special fiber of a family with information at the generic fiber. As an illustration of the theorem we use it to give a geometric interpretation of the multiplicity of a module (Theorem 1.2). This interpretation is then used in Remark 1.3 to connect the multiplicity of a module with Fulton's kth degeneracy class. In Section 2 we show how the multiplicity of the pair $(J(f), I)$ appears naturally in the work of Pellikaan and give two formulas for it. The first formula relates this multiplicity to the number of D_∞ and A_1 points appearing in a deformation of f. The second formula shows that the multiplicity of the pair $(J(f), I)$, if defined, is actually the length of $I/J(f)$. This length is Pellikaan's invariant $j(f)$. Both formulas are contained in Theorem 2.3 and its proof. These formulas are used to give a new formula for the Lê number of dimension 0 (Proposition 2.4). (Cf. [24] for details on the Lê numbers.)

In Section 3, we extend the results of Pellikaan for singular sets of dimension 1 to ICIS of dimension d, then use these results to prove extensions of the theorems of section 2. The computation of the formula for the Lê number of dimension 0 uses Zaharia's computation of the homology of the Milnor fiber. These formulae suggest in general that the Lê number of dimension 0 is the sum of the invariant which controls the A_f condition, (which in turn is related the multiplicity of a pair of ideals), and invariants of dimension 0 related to the other singularity types in the singular set of f. Section 3 also shows that the condition that $j(f)$ is finite imposes strong restrictions on f – there must exist a set of generators of I, $\{g_1, \ldots, g_p\}$ such that $f = \sum g_i^2$. This implies that every such function is the composition of a function h with a Morse singularity at the origin and the map G whose components are generators of I. In particular, all of the germs of type $D(d, p)$, with $d > 1$, studied by Pellikaan have $j(f) = \infty$, contrary to assertions made in Remark 5.3 on page 52 of [27] and in Remark 5.4 on page 373 of [28].

In Section 4 we then use the multiplicity of the pair to give a necessary and sufficient condition for the A_f condition to hold for a family of functions f_y (Theorem 4.5). The proof of this result involves a new trick which is used to pass information from strata in the singular set of f to the ambient geometry of f_0. This enables us to drop the hypothesis that the "natural" stratification of the singular set of f satisfies Whitney A.

In the case that the singular locus of f_0 is an ICIS of dimension 1, we use the relation between our invariant and the Lê numbers, to show that a strong form of the A_F condition also implies that the Lê numbers are constant as well (Corollary 4.7). This is used to show that in this situation the strong form of the A_f condition implies the triviality of the Milnor fibrations (Corollary 4.8). In Example 4.9, by modifying the example of Briancon-Speder we show that both the A_f condition and topological triviality of the family may hold, yet the Lê numbers may not be constant. It remains open whether the strong form of the A_f condition implies the Lê number of dimension 0 is constant in general, or if the strong form of A_f is needed if the dimension of $S(f) = 1$.

We then discuss the W_f condition for the situation of Theorem 4.5. Here we show that the independence from parameter of a single invariant is all that is required for a W_f-Whitney stratification of a family of functions, which implies the topological triviality of the family (Theorem 4.10). This invariant is then related to the relative polar multiplicities of the members of the family and the multiplicity of the pair that is used to control the A_f condition (Corollary 4.13). In turn, this implies that the A_f condition combined with the independence from parameter of the relative polar multiplicities implies that we have a W_f-Whitney stratification (Corollary 4.14).

The application of the multiplicity of the pair to equisingularity problems grew out of a long series of conversations with Steven Kleiman; the author thanks him for his encouragement. The author also thanks David Massey and James Damon for helpful conversations, and the referee for his careful reading of the paper, and helpful suggestions.

1. The multiplicity polar theorem

In this paper we work with complex analytic sets and maps. Let $\mathcal{O}_X$ denote the structure sheaf on a complex analytic space X. If a module M has finite colength in $\mathcal{O}^p_{X,x}$, it is possible to attach a number to the module, its Buchsbaum-Rim multiplicity ([3]). We can also define the multiplicity of a pair of modules $M \subset N$, M of finite colength in N, as well, even if N does not have finite colength in $\mathcal{O}^p_X$. We recall how to construct these numbers following the approach of Kleiman and Thorup ([20]). Given a submodule M of a free $\mathcal{O}_X$ module F of rank p, we can associate a subalgebra $\mathcal{R}(M)$ of the symmetric $\mathcal{O}_X$ algebra on p generators. This is known as the Rees algebra of M. If $(m_1, \ldots, m_p)$ is an element of M then $\sum m_i T_i$ is the corresponding element of $\mathcal{R}(M)$. Then Projan$(\mathcal{R}(M))$, the projective analytic spectrum of $\mathcal{R}(M)$ is the closure of the projectivised row spaces of M at points where the rank of a matrix of generators of M is maximal. Denote the projection to X by c, or by c_M where there is ambiguity. If M is a submodule of N or h is a section of N, then h and M generate ideals on Projan $\mathcal{R}(N)$; denote them by $\rho(h)$ and $\rho(\mathcal{M})$. If we can express h in terms of a set of generators $\{n_i\}$ of N as $\sum g_i n_i$, then in the chart in which $T_1 \neq 0$, we can express a generator of $\rho(h)$ by $\sum g_i T_i / T_1$. Having defined the ideal sheaf $\rho(\mathcal{M})$, we blow up by it. On the blowup $B_{\rho(\mathcal{M})}(\text{Projan}\,\mathcal{R}(N))$ we have two tautological bundles, one the pullback of the bundle on Projan $\mathcal{R}(N)$, the other coming from Projan $\mathcal{R}(M)$; denote the corresponding Chern classes by l_M and l_N, and denote the exceptional divisor by $D_{M,N}$. Suppose the generic rank of N (and hence of M) is e. Then the multiplicity of a pair of modules M, N is:

$$e(M,N) = \sum_{j=0}^{d+e-2} \int D_{M,N} \cdot l_M^{d+e-2-j} \cdot l_N^j.$$

The multiplicity of the pair is well defined as long as the set of points where N is not integrally dependent on M is isolated ([20]). If the pair is M and $\mathcal{O}^p_X$, then this condition implies that $M = \mathcal{O}^p_X$ except at isolated points so $M \subset \mathcal{O}^p_X$ is of finite colength and the multiplicity of M is the multiplicity of the pair $(M, \mathcal{O}^p_X)$. Later in this section we will give a new geometric interpretation of this number based on polar methods. If $\mathcal{O}_{X^d,x}$ is Cohen-Macauley, and M has $d+p-1$ generators then there is a useful relation between M and its ideal of maximal minors; the multiplicity of M is the colength of M, is the colength of the ideal of maximal minors, by some theorems of Buchsbaum and Rim [3], 2.4 p. 207, 4.3 and 4.5 p. 223. In Section 2 we will see a first generalization of this result to pairs of modules. We next develop the notion of polar varieties which is the other term in the multiplicity polar theorem. Assume we have a module M which is a submodule of a free module on X^d, an equidimensional, analytic space, reduced off a nowhere dense subset of X, and the generic rank of M is e on each component of X. The hypothesis on the equidimensionality of X and on the rank of M ensures that $\operatorname{Projan}\mathcal{R}(M)$ is equidimensional of dimension $d+e-1$. Note that $\operatorname{Projan}\mathcal{R}(M)$ can be embedded in $X \times \mathbf{P}^{r-1}$, provided we can chose a set of generators of M with r elements.

The *polar variety of codimension* k of M in X denoted $\Gamma_k(M)$ is constructed by intersecting $\operatorname{Projan}\mathcal{R}(M)$ with $X \times H_{e+k-1}$ where H_{e+k-1} is a general plane of codimension $e+k-1$, then projecting to X. This notion was developed by Teissier in the case where $M = JM(F)$, $X = F^{-1}(0)$ ([31]). Think of H as the projectivised row space of a linear submersion π. Then $\Gamma_k(JM(F))$ consists of the set of points where the matrix formed from $D\pi$ and DF has less than maximal rank, hence greater than minimal kernel rank. These are the points where the restriction to X of π is singular. In general, think of $\Gamma_k(M)$ as the set of points where the module whose matrix of generators consists of the matrix of generators of M augmented by the rows of the linear submersion π, has less than maximal rank $n-k+1$. When we consider M as part of a pair of modules M, N, where the generic rank of M is the same as the generic rank of N, then other polar varieties become interesting as well. In brief, we can intersect $B_{\rho(\mathcal{M})}(\operatorname{Projan}\mathcal{R}(N)) \subset X \times \mathbf{P}^{N-1} \times \mathbf{P}^{p-1}$ with a mixture of hyperplanes from the two projective spaces which are factors of the space in which the blowup is embedded. We can then push these intersections down to $\operatorname{Projan}\mathcal{R}(N)$ or X as is convenient, getting mixed polar varieties in $\operatorname{Projan}\mathcal{R}(N)$ or in X. These mixed varieties play an important role in the proof of the multiplicity-polar theorem, the theorem we next describe.

Setup: We suppose we have families of modules $M \subset N$, M and N submodules of a free module F of rank p on an equidimensional family of spaces with equidimensional fibers $\mathcal{X}^{d+k}$, $\mathcal{X}$ a family over a smooth base Y^k. We assume that the generic rank of M, N is $e \leq p$. Let $P(M)$ denote $\operatorname{Projan}\mathcal{R}(M)$, π_M the projection to $\mathcal{X}$. let $C(M)$ denote the locus of points where M is not free, i.e., the points where the rank of M is less than e, $C(\operatorname{Projan}\mathcal{R}(M))$ its inverse image under π_M, $C(\mathcal{M})$ the cosupport of $\rho(\mathcal{M})$ in $P(\operatorname{Projan}\mathcal{R}(N))$.

We will be interested in computing the change in the multiplicity of the pair (M, N), denoted $\Delta(e(M, N))$. We will assume that the integral closures of M and N agree off a set C of dimension k which is finite over Y, and assume we are working on a sufficiently small neighborhood of the origin, that every component of C contains the origin in its closure. Then $e(M, N, y)$ is the sum of the multiplicities of the pair at all points in the fiber of C over y, and $\Delta(e(M, N))$ is the change in this number from 0 to a generic value of y. If we have a set S which is finite over Y, then we can project S to Y, and the degree of the branched cover at 0 is $\text{mult}_y S$. (Of course, this is just the number of points in the fiber of S over our generic y.) We can now state our theorem.

Theorem 1.1. *Suppose in the above setup we have that* $\overline{M} = \overline{N}$ *off a set* C *of dimension* k *which is finite over* Y. *Suppose further that* $C(\text{Projan}\,\mathcal{R}(M))(0) = C(\text{Projan}\,\mathcal{R}(M(0)))$ *except possibly at the points which project to* $0 \in \mathcal{X}(0)$. *Then, for* y *a generic point of* Y,

$$\Delta(e(M, N)) = \text{mult}_y \Gamma_d(M) - \text{mult}_y \Gamma_d(N).$$

Proof. The proof in the ideal case appears in [11]; the general proof will appear in [12]. □

Now we describe an application of the result to the simple case where N is free. The following geometric interpretation of the multiplicity of an ideal is well known. Given an ideal I of finite colength in $\mathcal{O}_{X,x}$, X^d equidimensional, choose d elements $(f_1, \ldots, f_d)$ of I which generate a reduction of I. (Recall that if M is a submodule of N, then M is *reduction* of N if they have the same integral closure.) Then the multiplicity of I is the degree at x of F where F is the branched cover defined by the map-germ with components $(f_1, \ldots, f_d)$, or the number of points in a fiber of F over a regular value close to 0. We wish to give a similar interpretation of the multiplicity of a module.

Theorem 1.2. *Given* M *a submodule of* $\mathcal{O}^p_{X,x}$, X^d *equidimensional, choose* $d+p-1$ *elements which generate a reduction* K *of* M. *Denote the matrix whose columns are the* $d+p-1$ *elements by* $[K]$; $[K]$ *induces a section of* $\text{Hom}\,(\mathbf{C}^{d+p-1}, \mathbf{C}^p)$ *which is a trivial bundle over* X. *Stratify* $\text{Hom}\,(\mathbf{C}^{d+p-1}, \mathbf{C}^p)$ *by rank. Let* $[\epsilon]$ *denote a* $p\times(d+p-1)$ *matrix, whose entries are small, generic constants. Then, on a suitable neighborhood* U *of* x *the section of* $\text{Hom}\,(\mathbf{C}^{d+p-1}, \mathbf{C}^p)$ *induced from* $[K] + [\epsilon]$ *has at most kernel rank 1, is transverse to the rank stratification, and the number of points where the kernel rank is 1 is* $e(M)$.

Proof. The first step is to explain by construction what we mean by "generic constants". Consider the family of maps G_a from X^d, parametrised by $\mathbf{C}^{p(d+p-1)}$ to $\text{Hom}\,(\mathbf{C}^{d+p-1}, \mathbf{C}^p)$ defined by $G_a(x) = G(x, a) = [K(x)] + [A]$, where $[A]$ is the $p\times(d+p-1)$ matrix whose entries are coordinates $a_{i,j}$ on $\mathbf{C}^{p(d+p-1)}$. Let $\tilde{X}$ be a resolution of X, so we have an induced family of maps $\tilde{G}$ on $\tilde{X}$. Since the map $\tilde{G}(x, a)$ is a submersion, it follows that for a Z-open subset V of $\mathbf{C}^{p(d+p-1)}$, that for $a \in V$, the map $\tilde{G}_a$ is transverse to the rank stratification. We claim that the points of V

are the generic constants in the theorem. Note that the points of Hom $(\mathbf{C}^{d+p-1}, \mathbf{C}^p)$ of kernel rank 1 have codimension $1 \cdot ((d+p-1)-(p-1)) = d$; so since $\tilde{G}_a$ is transverse it can only hit points of the rank stratification of kernel rank 1, and only if $D\tilde{G}_a$ has maximal rank at such points which implies X is smooth at the projection of such points. Let $\tilde{K}$ be the submodule of $\mathcal{O}^p_{X\times\mathbf{C}^{p(d+p-1)}}$ defined by the matrix $[K(x)]+[A]$. Now apply the multiplicity-polar theorem to $X \times \mathbf{C}^{p(d+p-1)}$, thought of as a family parametrised by $\mathbf{C}^{p(d+p-1)}$, and $(\tilde{K}, \mathcal{O}^p_{X\times\mathbf{C}^{p(d+p-1)}})$. Use a point of V as the generic parameter value ϵ. Then $\mathcal{O}^p_{X\times\mathbf{C}^{p(d+p-1)}}$ has no polar, because it is free, $\tilde{K}$ has no polar, because $\tilde{K}$ is generated by $d+p-1$ elements. Choose U a neighborhood of $x \times \mathbf{C}^{p(d+p-1)}$ sufficiently small such that every component of the cosupport of $\tilde{K}$ which meets U has $(x,0)$ in its closure. Now at ϵ the cosupport of $\tilde{K}_\epsilon$ is just the points where $[K]+[\epsilon]$ has less than maximal rank. At such points $e(\tilde{K}_\epsilon)$ is 1, because since we are at a smooth point of X, the local ring of X is Cohen-Macaulay, so $e(\tilde{K}_\epsilon)$ is just the colength, which is 1. Hence $e(M) = e(K) = e(\tilde{K}_0) = e(\tilde{K}_\epsilon)$, which is the number of points where the kernel rank of $[K]+[\epsilon]$ is 1. □

Remark 1.3. In [5] p. 254 Fulton describes the kth degeneracy class associated to σ a homomorphism of vector bundles over X^d. The support of the class is the set of points where the rank of σ is less than or equal to k. Suppose $\sigma : E \to F$ where the rank of E is e and the rank of F is f, $e \geq f$, $e-f+1 = d$. Then the $f-1$ degeneracy class is supported at isolated points. Fulton shows that if X is Cohen-Macaulay at x, the contribution to the class at x is the colength of the ideal of maximal minors of the matrix of σ at x for some suitable local trivializations of E and F. Note that this is just the Buchsbaum-Rim multiplicity of the module generated by the columns of the matrix associated to σ. Theorem 1.2 shows that in this situation if X is pure dimensional, the contribution to the degeneracy locus is always the Buchsbaum-Rim multiplicity associated to σ at x, the Cohen-Macaulay hypothesis is unnecessary. (Just use the proof of 1.2 to construct a rational equivalence to go back to Fulton's case close to x.)

2. Hypersurface singularities with 1-dimensional singular locus

In his thesis ([27]) Pellikaan studied non-isolated hypersurface singularities. This is the setup for his work. He assumed that $f : \mathbf{C}^{n+1} \to \mathbf{C}$, f had a 1-dimensional singular locus Σ, which is a complete intersection curve defined by an ideal I. He assumed that $f \in I^2$. This ensured that $J(f)$, the jacobian ideal of f was in I as well. (In fact for the singular locus a complete intersection Pellikaan proved that if f and its partials were in I then f was in I^2.) One of the key invariants of f was

$$j(f) = \dim_{\mathbf{C}} \frac{I}{J(f)}$$

which plays the same role in Pellikaan's work as the dimension of $\frac{\mathcal{O}_{n+1}}{J(f)}$ does in the case of isolated singularities. Two important examples of non-isolated singularities are germs of type A_∞ which have the normal form $f(z_1,\ldots,z_{n+1}) = \sum_{i=1}^{n} z_i^2$ and germs of type D_∞ which have normal form $f(z_1,\ldots,z_{n+1}) = z_1z_2^2 + \sum_{i=3}^{n+1} z_i^2$. Note that if n=2 then D_∞ is just a Whitney umbrella. For A_∞ germs $j(f) = 0$ while for D_∞ germs $j(f) = 1$. Using these building blocks, Pellikaan was able to give a nice geometric description of $j(f)$.

Theorem 2.1. *Suppose f is as above and $j(f)$ finite. Then f has a deformation F such that F_y has Σ_y as singular locus for generic y where Σ_y is the Milnor fiber of Σ, with only A_1 singularities off Σ_y and only A_∞ singularities at points of Σ_y, except for isolated D_∞ points. Moreover*

$$j(f) = \#\{D_\infty(F_y)\} + \#\{A_1(F_y)\}.$$

Proof. Cf. [27] p. 87 Proposition 7.20. □

In applying the theory of integral closure to ambient stratification conditions like A_f or W_f in Pellikaan's situation, we see that there are three strata – the open stratum, $\Sigma - 0$ and the origin. So, there are two pairs of ideals $(I, \mathcal{O}_{n+1})$ and $(J(f), I)$ that we are interested in. We wish to give a geometric interpretation of $(J(f), I)$ using Pellikaan's theorem and the multiplicity-polar theorem. First we look at our building block germs.

Proposition 2.2. *If f is a germ of type A_∞, then $e(I, J(f)) = 0$, if f is a germ of type D_∞, then $e(I, J(f)) = 1$.*

Proof. If f is a germ of type A_∞, then $I = J(f)$, so $e(J(f), I) = 0$. So suppose f is a germ of type D_∞. We may assume f is in normal form, as changes of coordinates do not affect the multiplicity of the pair. We have to compute a sum of intersection numbers:

$$e(J(f), I) = \sum_{j=0}^{n} \int D_{J(f),I} \cdot l_{J(f)}^{n-j} \cdot l_I^j.$$

Consider the part of the sum of form:

$$\sum_{j=1}^{n} \int D_{J(f),I} \cdot l_{J(f)}^{n-j} \cdot l_I^j = \sum_{j=0}^{n-1} \int (D_{J(f),I} \cdot l_I) \cdot l_{J(f)}^{n-1-j} \cdot l_I^j.$$

This is $e(J(f), I)$ where both ideals are restricted to the codimension 1 polar variety of I. Consider the family of candidate polar varieties defined by $z_2 = \sum_{i=3}^{n+1} a_i z_i$. Since this a Z-open subset of all potential polar varieties, if we show that for a Z-open subset of them that the multiplicity of the pair of the restriction of the ideals to each candidate in the set is zero then we will have shown that all of terms in this second sum are zero and all these candidates are actually polars. Now it is obvious from the normal form of f that when we restrict our two ideals to any element of this set the two ideals become equal so all of the terms in the second sum are zero. It remains to compute $\int D_{J(f),I} \cdot l^n_{J(f)}$. Our approach is to choose a Z-open set of candidate polar curves of $J(f)$, then show that each candidate gives the same value for the computation of the desired intersection number. Consider the family of curves defined by ideals $J_{a,b,c} = (b_1(z_1z_2) + c_1z_2^2 + \sum_{i=3}^{n+1} a_{1,i}z_i, \dots, b_n(z_1z_2) + c_nz_2^2 + \sum_{i=3}^{n+1} a_{n,i}z_i)$. For a Z-open set of coefficients, we can re-write the ideals defining these curves as

$$J_{a,b,c} = (z_1z_2 + cz_2^2, \dots, z_i + b_iz_1z_2, \dots)$$

where $3 \leq i \leq n+1, c \neq 0$. Each curve in this family has two components; one of which (given by $z_2 = 0$) lies in $V(J(f))$. The other component is the candidate polar curve. So we get the family of parmeterizations $\phi(t) = (-ct, t, \dots, b_ict^2, \dots)$ for the candidate polar curves. Now the intersection number we want is just the multiplicity of the pair restricted to a polar curve; by the additivity of the multiplicity ([20]) this is just $e(J(f)) - e(I)$ restricted to the polar curve; given a parameterization this is just the order of vanishing of $\phi^*(J(f))$ less the order of vanishing of $\phi^*(I)$. Now $\phi^*(J(f)) = (t^2)$ and $\phi^*(I) = (t)$ for all parameterizations, so the value of this intersection number is $2 - 1 = 1$, so $e(J(f), I) = 1$. □

For our basic building block germs we have seen that $j(f) = e(J(f), I)$. The next theorem shows that this is true in general. If F depends on coordinates (y, z), let $J_z(F)$ denote the ideal generated by the partials of F with respect to z.

Theorem 2.3. *Suppose $f : \mathbf{C}^{n+1} \to \mathbf{C}$, f has a 1-dimensional singular locus Σ, which is a complete intersection curve defined by an ideal I, $f \in I^2$ and $j(f)$ finite. Then*

$$j(f) = \dim_{\mathbf{C}} \frac{I}{J(f)} = e(J(f), I).$$

Proof. Let F be the deformation of Theorem 2.1. Denote the parameter space by Y^k. The singular set of F is given by a complete intersection $\tilde{I}$. We are interested in the family of pairs of ideals given by $(J_z(F), \tilde{I})$ as these restrict to $(J(f), I)$ at $y = 0$. Since $\tilde{I}$ defines a complete intersection it has no polar variety of dimension k. Since $J_z(F)$ is generated by $n+1$ generators it has no polar of dimension k either. This means that the multiplicity of the pair at the origin is same as the sum of the multiplicities over a generic parameter value by the multiplicity-polar formula. Pick a generic y. We have $(J_z(F))_y = J(F_y)$, so the cosupport of $(J_z(F))_y$ consists of A_1 points off Σ_y, isolated D_∞ points on Σ_y and A_∞ points. Off Σ_y, $(\tilde{I})_y = \mathcal{O}_{n+1}$, so off Σ_y, at A_1 points, $e(J(F_y), \tilde{I}_y) = e(J(F_y), \mathcal{O}_{n+1}) = 1$ and 0 elsewhere off Σ_y. On Σ_y, $e(J(F_y), \tilde{I}_y) = 1$ at D_∞ points, otherwise it is 0 by proposition 2.2. So the sum of the $e(J(F_y), \tilde{I}_y), z)$ at points where it is non-zero is just $\#\{\mathrm{D}_\infty(F_y)\} + \#\{\mathrm{A}_1(F_y)\}$.

Then, by the multiplicity-polar formula we know that

$$e(J(f), I) = \#\{\mathrm{D}_\infty(F_y)\} + \#\{\mathrm{A}_1(F_y)\}$$

which proves the theorem. □

If R is Cohen-Macaulay of dimension d, M a submodule of a rank p free module F of finite colength, then by a theorem of Buchsbaum and Rim ([3]), $e(M,F)$, which is $e(M)$, is just the colength of M if M has $d+p-1$ generators. Theorem 3 can be viewed as a first step in generalizing this result to pairs of modules.

Using some other results of Pellikaan, we can link $e(J(f),I)$ and the Lê numbers introduced by Massey. In the situation of Theorem 2.3 there are two Lê numbers – $\lambda^0(f)$ and $\lambda^1(f)$; denote the number of D_∞ points of f by $\delta(f)$.

Proposition 2.4. *Assume the hypotheses of Theorem 2.3, then*

$$\lambda^0(f) = e(J(f),I) + e(JM(\Sigma)) + \delta(f)$$

Proof. If F is the Milnor fiber of f, we have that

$$\chi(F) = 1 + (-1)^{n-1}\lambda^1(f) + (-1)^n\lambda^0(f) = 1 + (-1)^n(j(f) + \delta_f + \mu(\Sigma) - 1).$$

The first equality is due to Massey ([24]), while the second is due to Pellikaan ([27], p113, proposition 10.11). In the present situation, since the transverse Milnor number is 1, $\lambda^1(f) = \text{mult }(\Sigma)$, while $e(JM(\Sigma)) = \mu(\Sigma) + \text{mult }(\Sigma) - 1$. Therefore, substituting and cancelling we get

$$\lambda^0(f) = e(J(f),I) + e(JM(\Sigma)) + \delta(f). \qquad \square$$

Now we turn to the extension of these ideas to hypersurface singularities with a higher-dimensional singular locus.

3. Hypersurface singularities with d-dimensional singular locus

In this section we assume that $I = (g_1,\ldots,g_p) \subset \mathcal{O}_n$ defines a complete intersection of dimension $d > 1$, and $S(f) = V(I)$, hence we can write f as $f = \sum_{i,j}^p h_{i,j}g_ig_j$, where $h_{i,j} = h_{j,i}$, for some $h_{i,j}$. Let $[H]$ denote the symmetric matrix with entries $h_{i,j}$. We will want to study those germs f for which $j(f) < \infty$. Basic examples of such germs are those of type $A(d)$. For these germs up to a change of coordinates, $I = (z_1,\ldots,z_{n-d})$, $f = \sum_{i=1}^{n-d} z_i^2$, z_i part of a coordinate system on $\mathbf{C}^n$. It turns out that the condition that $j(f) < \infty$ is much more restrictive than in the case where dimension of $V(I) = 1$. Pellikaan already showed that $j(f) < \infty$ implies I defines an ICIS. The next proposition gives a further restriction.

Proposition 3.1. *Suppose f, I as above, then if $[H]$ has less than maximal rank at the origin, the set of points on $V(I)$ where the singularity type is not $A(d)$ is of codimension 1 in $V(I)$, hence $j(f)$ is not finite.*

Proof. If f has an $A(d)$ singularity at $x \in V(I)$ then $V(I)$ is smooth at x and the matrix $[H(x)]$ must have rank $n-d$. But the points where $\det[H] = 0$ defines a non-empty hypersurface in $V(I)$, since $\det[H(0)] = 0$ and the dimension of $V(I) > 1$. Hence, at these points f does not have an $A(d)$ singularity. Since at these points $I \neq J(f)$, it follows that $j(f) = \infty$. $\square$

There are two types of Lê cycles; those which are the images in $\mathbf{C}^n$ of components of the exceptional divisor of the Jacobian blow up, called fixed cycles, and the polar varieties of the fixed cycles called moving cycles.

Corollary 3.2. *Suppose f, I, $[H]$ as above. Then $V(I)$ contains a fixed Lê cycle of dimension $d-1$.*

Proof. We can deform f so that the D^∞ points are dense in the zero set of $\det[H]=0$. These points are clearly the image of a component of the exceptional divisor by Proposition 2.4, and by the properties of the Lê numbers. Then when we specialize, the component of E will specialize as well. □

In [27] and in [28] Pellikaan defines the singularities of type $D(d,p)$; here d is the dimension of $S(f)$, while p is the dimension of the kernel rank of $[H]$ at the point in question. Then $f:\mathbf{C}^n,x\to\mathbf{C},0$ has type $D(d,p)$ at x if local coordinates can be chosen so that f has the local form

$$f=z_1^2+\cdots+z_q^2+\sum_{1\leq i\leq j\leq p} x_{i,j}y_iy_j$$

where z,x,y are part of a coordinate system on $\mathbf{C}^n$ at x, $n-d=q+p$. From 3.1 it follows that if f has singularity type $D(d,p)$ at the origin, and $d>1$, then, since $\det[H(0)]=0$, it follows that $j(f)=\infty$, contrary to remark 5.3 of [27] and Remark 5.4 of [28]. This shows that $j(f)$ fails to be finite in what seems to be the next most simple case to the $A(d)$ singularities when $d>1$. Instead, the structure of $S(f)$ seems more like a discriminant, in that the non-generic points appear in codimension 1.

In the next lemma we begin to look at those germs where $[H]$ has maximal rank, so we can characterize those germs where $j(f)<\infty$.

Lemma 3.3. *Suppose $f=\sum_{i,j}^p h_{i,j}g_ig_j$, $\det[H(0)]\neq 0$, $I=(g_1,\dots,g_p)\subset\mathcal{O}_n$. Then one can chose a set of generators $(g_1',\dots,g_p')$ of I such that $f=\sum_i^p(g_i')^2$.*

Proof. The proof is standard, so we just sketch the details. Given an invertible matrix $[R]$ with entries in $\mathcal{O}_n$, it is clear that if

$$[g]=[R][g'],$$

where $[g]$ is the column vector whose entries are the g_i, $[g']$ another column vector, that the entries of $[g']$ are also a set of generators of I. Given

$$[f]=[g]^t[H][g] \qquad \text{and} \qquad [g]=[R][g']$$

it follows that

$$[f]=[g']^t([R]^t[H][R])[g'].$$

Hence, we need to show that by choice of $[R]$ we can reduce $[H]$ to the identity matrix. This is done in two steps – first we can chose $[R]\in Gl(p,\mathbf{C})$ so that we can assume $[H(0)]=I$. (This follows because the action of $Gl(p,\mathbf{C})$ clearly preserves rank, the orbits of $Gl(p,\mathbf{C})$ are connected constructible sets, and the orbits of non-singular matrices are open, by a tangent space calculation.) For the second step we assume $[H(0)] = I$, consider the linear homotopy from I to $[H]$; this stays inside the set of invertible symmetric matrices. The congruence transformation gives an action of the group $\mathcal{C}$ of invertible $p \times p$ matrices with entries in $\mathcal{O}_n$ on the $p \times p$ symmetric matrices. Applying the techniques of Mather-Damon produces a homotopy in $\mathcal{C}$ which trivializes our linear homotopy, which finishes the proof. □

Lemma 3.3 also appears as a remark without proof in [32] (see page 87). Given a set of generators $\{g_1, \ldots, g_p\}$ for an ideal, we can form the function G whose components are the g_i. If $\{g_1, \ldots, g_p\}$ define an ICIS, then the map G is said to be of *finite singularity type.*

Corollary 3.4. *Suppose* $f = \sum_{i,j}^{p} h_{i,j} g_i g_j$, $\det[H(0)] \neq 0$, $I = (g_1, \ldots, g_p) \subset \mathcal{O}_n$, $\{z_i\}$ *coordinates on* $\mathbf{C}^p$ *then generators* $(g'_1, \ldots, g'_p)$ *of* I *can be chosen so that*

$$f = \sum_{i=1}^{p} z_i^2 \circ G'$$

Proof. By Lemma 3.3 we have there exists generators $(g'_1, \ldots, g'_p)$ of I such that

$$f = \sum_{i=1}^{p} (g'_i)^2 = \sum_{i=1}^{p} z_i^2 \circ G'. \qquad \square$$

Thus the study of functions with $j(f) < \infty$ is intimately tied up with the study of functions on the discriminant of a map germ of finite singularity type as we shall see below. We wish to describe a condition which will ensure that the pullback by G of a function on $\mathbf{C}^p$ with a Morse singularity at the origin gives a function on $\mathbf{C}^n$ with $j(f) < \infty$ for the ideal defined by the components of G. This completes our geometric description of the meaning of $j(f)$ finite. Our condition is based on the intersection of the levels of the Morse function in the target with the discriminant, $\Delta(G)$, of G. At this point we assume that I defines an ICIS. This implies that if G comes from a minimal set of generators of I, then $G|S(G)$ is a finite map.

We can partition $S(G)$ by the $S_i(G)$ which denotes points of $S(G)$ where the kernel rank of G is i. We can also partition $\Delta(G)$ as follows. For each point z of $\Delta(G)$, list the points S_z of $S(G)$ mapped to z. The points z and z' are in the same element of the partition if there is a bijection between S_z and $S_{z'}$ which which preserves components of the $S_i(G)$. It is easy to see that the elements of this partition are constructible sets since $G|S(G)$ is finite. Given an element of the partition of $\Delta(G)$, we now associate a collection of systems of linear sub spaces of $T\mathbf{C}^p$ over the underlying set P of the partition element. Since G has constant rank on each $S_i(G)$, $D(G)|_{S_i(G)\cap G^{-1}(P)}(T\mathbf{C}^n|S_i(G) \cap G^{-1}(P))$ is a well-defined sub bundle of $G^*T\mathbf{C}^p$ over $S_i(G) \cap G^{-1}(P)$. Since the restriction of G to each component of $G^{-1}(P)$ is a homeomorphism or finite cover, the push forward by G of these sub bundles gives the desired collection of systems of linear spaces. We call the partition of $\Delta(G)$ together with the collection of linear spaces on each element of the partition an *enriched partition.* A smooth subset V of $\mathbf{C}^p$ is *enriched transverse* to the enriched partition if at every point of intersection with the elements of the partition the tangent space of V is transverse to each of the linear spaces we have associated to the element of the partition at that point. Since the restriction of G to each component of $G^{-1}(P)$ is a homeomorphism or finite cover, all of the linear spaces at a smooth point in a partition element contain the tangent space to the partition element. So if V is transverse to each element of the partition it is enriched transverse. The next proposition describes a situation in which transversality and enriched transversality are equivalent.

Proposition 3.5. *Suppose there exists an element P of the partition which is a Z-open subset of $\Delta(G)$ whose pre-images lie in the Z-open subset $S_{n-p+1}(G)$ on which G is immersive. Then all of the systems of linear spaces associated to P are just the tangent bundle to P.*

Proof. Suppose $y \in P$, z a preimage in $S_{n-p+1}(G)$. Since G restricted to $S_{n-p+1}(G)$ is immersive at z, the dimension of $DG(TS_{n-p+1}(G))$ is $p-1$ which is the dimension of $D(G)(z)T\mathbf{C}^n$, so these spaces are equal; further $DG(TS_{n-p+1}(G))$ is the tangent space to $\Delta(G)$ at y, which is the tangent space to P at y. □

Now we give our condition for $j(f)$ finite.

Theorem 3.6. *Suppose $I = (g_1, \ldots, g_p) \subset \mathcal{O}_n$ defines an ICIS of dimension $d > 1$, G the mapgerm whose components are the g_i, $h : \mathbf{C}^p, 0 \to \mathbf{C}, 0$ a function with an isolated singularity at the origin, $f = h \circ G$. Then*

$$j(f, G^*(J(h))\mathcal{O}_n) := \dim_{\mathbf{C}} \frac{G^*(J(h))\mathcal{O}_n}{J(f)} < \infty$$

if and only if $h^{-1}(0)$ is enriched transverse to the enriched partition of $\Delta(G)$ except possibly at the origin.

Proof. Suppose $j(f, G^*(J(h))\mathcal{O}_n)$ finite. Then, except possibly at the origin, $J(f) = G^*(J(h))\mathcal{O}_n$. If the enriched transversality condition fails, there must be a curve $\phi : \mathbf{C} \to \Delta G$, such that the image of ϕ lies in an element of the partition, and the tangent space to $h^{-1}(0)$ contains one of the systems of linear spaces along the partition element. This implies that contained system is in the kernel of Dh along ϕ. Then ϕ has a lift to the component of $S_i(G)$ associated to the contained system, denoted ψ. Along the image of ψ we have

$$Df \circ \psi(T\mathbf{C}^n) = Dh \circ (G \circ \psi) DG \circ \psi(T\mathbf{C}^n) = 0$$

Hence $V(J(f)) \supset \mathrm{im}\psi$, while $V(G^*(J(h))\mathcal{O}_n) = V(I)$ which is a contradiction. Suppose enriched transversality holds. If $j(f, G^*(J(h))\mathcal{O}_n)$ is not finite, there exists a curve ψ whose image properly contains the origin in $\mathbf{C}^n$, such that $J(f) \neq G^*(J(h))\mathcal{O}_n$ along ψ. At points of $\mathbf{C}^n$ off $S(G)$, G is a submersion, hence $J(f) = G^*(J(h))\mathcal{O}_n$. If ψ lies in $S(G)$, then the image of ψ lies in $S(f)$ since $G^*(J(h))\mathcal{O}_n = \mathcal{O}_n$ at such points. Then ψ lies in the zero set of F, hence $G \circ \psi$ lies in the zero set of h. Then enriched transversality fails along $G \circ \psi$. □

Corollary 3.7. *Suppose h has a Morse singularity at the origin in the set-up of Theorem 3.6, then $j(f)$ is finite if and only if $h^{-1}(0)$ is enriched transverse to the enriched partition of $\Delta(G)$ except possibly at the origin.*

Proof. If h has a Morse singularity, then $G^*(J(h))\mathcal{O}_n = I$. □

Corollary 3.8. *Suppose $I = (g_1, g_2)$ in the setup of Theorem 3.6.*
Then $j(f, G^*(J(h))\mathcal{O}_n)$ is finite iff $f^{-1}(0) \cap S(G)$ is the origin.

Proof. If $p = 2$, then $\Delta(G)$ is a curve, and G restricted to each branch of $S(G)$ is an immersion except at the origin. Then enriched transversality becomes ordinary transversality, so $h^{-1}(0)$ must miss $\Delta(G)$ off the origin, so $f^{-1}(0) \cap S(G)$ is the origin. □

Theorem 3.6 introduces an interesting class of functions. Given an ICIS, by using appropriate h we can construct examples of non-isolated singularities in which the singular locus is the ICIS, but the transverse singularity type is constant and is that of h. In studying the equisingularity of families of such examples, the key invariant is the multiplicity of the pair $J(f), G^*(J(h))\mathcal{O}_n$. This number should also be linked to the way $h^{-1}(0)$ meets the discriminant of G at the origin.

Now we show that such functions with $j(f)$ finite are plentiful.

Proposition 3.9. *Suppose $G : \mathbf{C}^n, 0 \to \mathbf{C}^p, 0$, $G^{-1}(0)$ an ICIS, $p > 1$. Then if $h_a(x) = \sum a_i x_i^2$, for $a \in U$, U a Z-open subset of $\mathbf{C}^p$, $h^{-1}(0)$ is transverse to the enriched partition of $\Delta(G)$ except perhaps at the origin.*

Proof. Consider $H(a, z) = \sum a_i x_i^2 \circ G(z)$. We have

$$DH = \langle \dots, x_i^2 \circ G, \dots, 2a_i x_i \circ G, \dots \rangle$$

This implies that H is a submersion except along $\mathbf{C}^p \times G^{-1}(0)$. Denote by π the projection of $H^{-1}(0)$ to $\mathbf{C}^p$. By Sard's lemma for varieties (Prop. 3.7 p. 42 [25]) there exists a Z-open subset $U \subset \mathbf{C}^p$ such that π is smooth at $z \in H^{-1}(0) \cap \pi^{-1}(U)/\mathbf{C}^p \times G^{-1}(0)$. This implies that the fiber of π, which is the fiber of $h_a \circ G$ over 0 is smooth at z; in addition since π maps $T_z(H^{-1}(0)) = \ker DH_z$ surjectively to $\mathbf{C}^p$, the $\ker DH_z$ does not contain $\mathbf{C}^n$, thus $h_a \circ G$ is a submersion at z as well, hence h_a is enriched transverse to the enriched partition of $\Delta(G)$, except perhaps at the origin. □

Now that we know that it is worth proving results about functions with $j(f)$ finite for $V(I)$ an ICIS of dimension > 1, we prove the analogue of 2.3. To do this we first study a special deformation of $f = \sum_1^p z_i^2$. We call the following pair of deformations a smoothing of f.

$$F(u, b, z) = \sum_i (1 + \sum_j b_{i,j} z_j)(g_i - u_i)^2$$

$$\tilde{G}(u, z) = (g_1(z) - u_1, \dots, g_p(z) - u_p)$$

This is called a smoothing because of the following lemma:

Lemma 3.10. *For a Z-open subset U of $\mathbf{C}^p \times \mathbf{C}^{pn}$, $f_{u,b}$ has only A_1 singularities off $G^{-1}(u)$, $G^{-1}(u)$ is smooth and $f_{u,b}$ has only $A(d)$ singularities on $G^{-1}(u)$.*

Proof. let $V \subset \mathbf{C}^p$ be the complement of $\Delta(G)$ in $\mathbf{C}^p$, then $G^{-1}(u)$ is smooth for $u \in V$.

We claim $D_z F(u, b, z)$ is a submersion off $\mathbf{C}^{pn} \times \Gamma(G)$, where $\Gamma(G) \subset \mathbf{C}^n \times \mathbf{C}^p$ denotes the graph of G. Let e_i, where $1 \leq i \leq n$ denote the unit vectors in $\mathbf{C}^n$. Then we have

$$\frac{\partial D_z(F)}{\partial b_{i,j}} = (g_i - u_i)^2 e_j.$$

This implies $D_z F(u, b, z)$ has maximal rank when some $g_i - u_i$ is not zero which proves the claim. Now consider $D_z F(u, b, z)^{-1}(0)$. The claim shows this is smooth off $\mathbf{C}^{pn} \times \Gamma(G)$. As in the proof of 3.9 we consider the projection of this set to $\mathbf{C}^p \times \mathbf{C}^{pn}$, let W be the Z-open subset of the base over which π is smooth off $\mathbf{C}^{pn} \times \Gamma(G)$. Now the tangent space to $D_z F(u, b, z)^{-1}(0)$ at a point x is just the kernel at x of $D(D_z F(u, b, z))$, which has dimension $p+pn$ and which surjects to $\mathbf{C}^p \times \mathbf{C}^{pn}$. Hence $D_z^2 F(u, b, z)$ has maximal rank, so $f_{u,b}$ has only Morse singularities off $G = u$. Let $U = W \cap \mathbf{C}^{pn} \times V$, then for $(u, b) \in U$, we have g_u has a smooth fiber over zero. Since the set of points where the matrix H with entries $h_{i,i} = \sum_j 1 + b_{i,j} z_j$, $h_{i,j} = 0$ $i \neq j$ has maximal rank on some Z-open subset of $\mathbf{C}^{pn} \times \mathbf{C}^n \times \mathbf{C}^p$ which contains zero, we can ensure that each of the $f_{u,b}$ has only $A(d)$ singularities on some fixed neighborhood of the origin in $\mathbf{C}^n$ on $g_u = 0$. □

Remark 3.11. It was pointed out to the author by the referee that this lemma also follows from the statement and proof of Theorem 1 of [4]

Now we extend Theorem 2.3 to ICIS of dimension greater than 1.

Theorem 3.12. *Suppose $f : \mathbf{C}^{n+1} \to \mathbf{C}$, f has a d-dimensional singular locus Σ, $d > 1$, which is an ICIS defined by an ideal I, $f \in I^2$ and $j(f)$ finite. Then*

$$j(f) = \dim_{\mathbf{C}} \frac{I}{J(f)} = e(J(f), I) = \#A_1(f).$$

where $\#A_1(f)$ is the number of A_1 singularities appearing in a smoothing of f.

Proof. The proof is similar to that of 2.3. By [27] Theorem 3.1 p. 145 the quotient of the ideals $(g_1(z) - u_1, \ldots, g_p(z) - u_p)/J_z(F)$ is perfect, where F is part of a smoothing of f, hence the length of the quotients $(g_u)/J(f_{u,b})$ is independent of parameter, and for generic parameter value is just $\#A_1(f)$. Meanwhile, $J_z(F)$ and $(g_1(z) - u_1, \ldots, g_p(z) - u_p)$ have no polar varieties of dimension $p + p(n+1)$, so as in Theorem 2.3, the multiplicity polar theorem implies that $e(J(f_{u,b}, (g_u))$ is independent of parameter, so again is $\#A_1(f)$, hence the theorem follows. □

Now we wish to extend Proposition 2.4 to ICIS of dimension greater than 1.

In [32], Prop 5.5.5, p. 86, (cf. also [26]), Zaharia computed the homology of the Milnor fiber, $\hat{f}$, of a function germ f defined on $\mathbf{C}^{n+1}$ whose singular set Σ

was an ICIS of codimension p such that $j(f) < \infty$. His result was:

$$H_*(\hat{f}) = \begin{cases} Z, \text{ if } * = 0, p-1 \\ Z^{\mu_\Sigma + \sigma} \text{ if } * = n \\ 0, \text{ otherwise} \end{cases}$$

Here σ is the number of A_1 points appearing in a smoothing, which we have shown is $j(f)$.

Proposition 3.13. *Assume the hypotheses of Theorem* 3.11, *then*

$$\lambda^0(f) = e(J(f), I) + e(JM(\Sigma))$$

Proof. By Massey ([24]) we have that

$$\chi(\hat{f}) = 1 + \sum_{i=0}^{d}(-1)^{n-i}\lambda^i(f) = 1 + (-1)^n\lambda^0(f) + \sum_{i=1}^{d}(-1)^{n-i}\lambda^i(f)$$

Now, for $i > 0$, since Σ is the only fixed Lê cycle of dimension greater than 0, and f has transverse Milnor number 1, since the type of f is $A(d)$ generically on Σ,

$$\lambda^i(f) = m_{d-i}(\Sigma),$$

where $m_{d-i}(\Sigma)$ is the $d-i$ polar multiplicity of the ICIS Σ. In turn, $m_{d-i}(\Sigma) = \mu^{d-i}(\Sigma) + \mu^{d-i-i}(\Sigma)$ ([9]) where $\mu^{d-i}(\Sigma)$ is the Milnor number of $\Sigma \cap H_i$ where H_i is a generic plane of codimension i, and where $\mu^{-1} = 1$. Substituting, the sum telescopes to:

$$\chi(\hat{f}) = 1 + (-1)^n\lambda^0(f) + (-1)^{n-d} + (-1)^{n-1}\mu^{d-1}(\Sigma).$$

Calculating $\chi(\hat{f})$ from the homology calculation of [32] we get:

$$1 + (-1)^n\lambda^0(f) + (-1)^{n-d} + (-1)^{n-1}\mu^{d-1}(\Sigma) = 1 + (-1)^{n-d} + (-1)^n(\mu_\Sigma + \sigma),$$

Hence

$$\lambda^0(f) = \sigma + \mu_\Sigma + \mu(\Sigma \cap H_1) = e(J(f), I) + e(JM(\Sigma)).$$ □

Remark 3.14. There are two other general calculations of the homology of the Milnor fiber in [32] (Theorem 5.5.4 and Proposition 5.5.6). (Note, however the typo in the formula of 5.5.4 – the coefficients of μ_Δ and μ_Σ should be exchanged.)

Using these calculations, it is possible to prove by the same methods as 3.12, two other formulas for $\lambda^0(f)$. In the first case, assume $V(I) = \Sigma$ is an ICIS of dimension 2, write $[f] = [g]^t[H][g]$ as we did earlier, let H denote the ideal generated by I^2 and the entries of $[H][g]$, assume $\dim_{\mathbf{C}} H/J(f)$ is finite, $V(\det[H]) \cap \Sigma = \Delta$, where Δ is an ICIS of dimension 1. We can consider the smoothing used by Zaharia to study this situation, and the ideal H extends to $\tilde{H}$ in a natural way, to the space of the smoothing. Then the polar of $\tilde{H}$ may be non-empty if the kernel rank of $[H]$ is > 2. Call the multiplicity of the polar of $\tilde{H}$ over the base $m(\Gamma(\tilde{H}))$. Then the multiplicity polar theorem applied to the smoothing gives $e(J(f), H) + m(\Gamma(\tilde{H})) = \#(A_1(f))$ and hence,

$$\lambda^0(f) = e(J(f), H) + m(\Gamma(\tilde{H})) + e(JM(\Sigma)) + 2e(JM(\Delta)).$$

In the second case, assume $V(I)$ has dimension $d > 1$, assume the rank of $[H(0)]$ is $p-1$ (one less than maximal). Then, as Zaharia remarks ([32] p. 87), generators $(g_1, \ldots, g_p)$ for I can be found so that $f = \det([H])g_1^2 + g_2^2 + \cdots + g_p^2$. Then the ideal H of the last paragraph is just $(\det([H])g_1, g_1^2, g_2, \ldots, g_p)$. Since H has only $p+1$ generators as does $\tilde{H}$ the polar of $\tilde{H}$ is empty and

$$\lambda^0(f) = e(J(f), H) + e(JM(\Sigma)) + 2e(JM(\Delta)).$$

The form of these formulae makes it likely that they are special cases of a more general theorem. It has long been known that in cases like those considered here, that the independence from parameter of the Lê numbers implies that the families $\Sigma(t)$ and $\Delta(t)$ are Whitney equisingular. (See, for example, [13] Prop 4.6, for the case where $I = J(f)$, and use the fact that the components of the exceptional divisor of the blowup of $\mathbf{C}^{n+1}$ by $J(f)$ which project to Σ and Δ are the conormals of Σ and Δ.) Thus, a relation between the Lê numbers and the invariants used to control the Whitney equisingularity of Σ and Δ is not unexpected. That the formulae relate λ_0 so simply to the zero dimensional invariants of the strata and to the A_f invariant is surprising.

Now we develop some results which shows how well $e(J(f), I)$ is linked to the A_f and W_f conditions.

4. Conditions A_f and W_f

In this section, we'll study Thom's Condition A_f, and Henry, Merle and Sabbah's Condition W_f , which concern limiting tangent hyperplanes at a singular point of a complex analytic space. First we recall the notions of integral dependence and strict dependence. Let $(X, 0)$ be the germ of a complex analytic space, and $\mathcal{E} := \mathcal{O}_X^p$ a free module of rank p at least 1. Let M be a coherent submodule of $\mathcal{E}$, and h a section of $\mathcal{E}$. Given a map of germs $\varphi\,(\mathbf{C}, 0) \to (X, 0)$, denote by $h \circ \varphi$ the induced section of the pullback $\varphi^*\mathcal{E}$, or $\mathcal{O}_{\mathbf{C}}^p$, and by $M \circ \varphi$ the induced submodule. Call h *integrally dependent* (resp., *strictly dependent*) on M at 0 if, for every φ, the section $h \circ \varphi$ of $\varphi^*\mathcal{E}$ is a section of $M \circ \varphi$ (resp., of $m_1(M \circ \varphi)$, where m_1 is the maximal ideal of 0 in $\mathbf{C}$). The submodule of $\mathcal{E}$ generated by all such h will be denoted by $\overline{M}$, resp., by $M^\dagger$. In the context of hypersurface singularities, given a family of map-germs $F(y, z)$ parametrised by $Y = \mathbf{C}^k$, where $F : \mathbf{C}^k \times \mathbf{C}^{n+1}, \mathbf{C}^k \times 0, 0 \to \mathbf{C}, 0, 0$ Thom's A_f condition holds for the pair $(\mathbf{C}^k \times \mathbf{C}^{n+1} - S(F), \mathbf{C}^k \times 0)$ at $y \in Y$ if and only if every limit of tangent hyperplanes to the fibers of F on $\mathbf{C}^k \times \mathbf{C}^{n+1} - S(F)$ contains TY at y. The condition holds for the pair if it holds for the pair at every y. Although this condition looks like it says nothing about strata other than the open stratum, this can be deceiving, as we shall see.

Proposition 4.1. *Suppose $F : \mathbf{C}^k \times \mathbf{C}^{n+1}, \mathbf{C}^k \times 0, 0 \to \mathbf{C}, 0, 0$ then the following are equivalent:*

1) *The A_F condition holds for the pair $(\mathbf{C}^k \times \mathbf{C}^{n+1} - S(F), \mathbf{C}^k \times 0)$ at 0.*

2) *The fiber over 0 of the exceptional divisor E of the blowup of $\mathbf{C}^k \times \mathbf{C}^{n+1}$ by $J(F)$, denoted $B_{J(F)}(\mathbf{C}^k \times \mathbf{C}^{n+1})$ is contained in $C(Y)$, the conormal of Y.*
3) $\frac{\partial F}{\partial y_i} \in J(F)^\dagger$ *for* $1 \le i \le k$.
4) $\frac{\partial F}{\partial y_i} \in J_z(F)^\dagger$ *for* $1 \le i \le k$.

Proof. The fiber over 0 of the exceptional divisor E of $B_{J(F)}(\mathbf{C}^k \times \mathbf{C}^{n+1})$ is exactly the set of limiting tangent hyperplanes at 0 to the fibers of F on $\mathbf{C}^k \times \mathbf{C}^{n+1} - S(F)$; saying that this fiber lies in the conormal of Y just says that each limit contains the tangent space to Y at 0. This shows 1) and 2) are equivalent. The equivalences of 1) and 3) and 4) can be found in [14]. □

The W_F condition holds for the pair $(\mathbf{C}^k \times \mathbf{C}^{n+1} - S(F), \mathbf{C}^k \times 0)$ at 0 if there exist a (Euclidean) neighborhood U of 0 in $\mathbf{C}^k \times \mathbf{C}^{n+1}$ and a constant $C > 0$ such that, for all y in $U \cap Y$ and all x in $U \cap (\mathbf{C}^k \times \mathbf{C}^{n+1} - S(F))$, we have

$$\operatorname{dist}\big(T_yY(F(y)), T_x(\mathbf{C}^k \times \mathbf{C}^{n+1})(F(x))\big) \le C \operatorname{dist}(x, Y)$$

where $T_yY(F(y))$ and $T_x(\mathbf{C}^k \times \mathbf{C}^{n+1})(F(x))$ are the tangent spaces to the indicated fibers of F and the restriction $F|Y$.

Proposition 4.2. *Suppose* $F : \mathbf{C}^k \times \mathbf{C}^{n+1}, \mathbf{C}^k \times 0, 0 \to \mathbf{C}, 0, 0$ *then the following are equivalent:*

1) *The* W_F *condition holds for the pair* $(\mathbf{C}^k \times \mathbf{C}^{n+1} - S(F), \mathbf{C}^k \times 0)$ *at* 0.
2) $\frac{\partial F}{\partial y_i} \in \overline{\mathbf{m}_Y J(F)}$ *for* $1 \le i \le k$.
3) $\frac{\partial F}{\partial y_i} \in \overline{\mathbf{m}_Y J_z(F)}$ *for* $1 \le i \le k$.

Proof. This follows from Proposition 1.1 of [15] □

Now we want to look at the connection between the multiplicity of the pair, $e(J(f), I)$, and the A_F condition. At this point we no longer assume that I defines a curve singularity. We do need two simple lemmas first.

Lemma 4.3. *Suppose I is an ideal generated by d elements in an equidimensional local ring R of dimension n such that R/I has dimension $n - d$. Suppose $J \subset I$ is a reduction of I. Then $J = I$.*

Proof. The proof is by induction on d. Assume $d = 1$, denote the generator of I by p_1. Let $J = (f_1 p_1, \dots, f_k p_1)$. If some f_i is a unit, then we are done. Suppose no f_i is, and denote the ideal they generate by F. If p_1 satisfies a relation of integral dependence, then we get

$$(p_1)^k + \sum_{i=0}^{k-1} g_i p_1^i = 0$$

where $g_i \in J^{k-i}$. Then $g_i \in F^{k-i}(p^{k-i})$, so the equation of integral dependence implies that there exists a unit u such that $u \cdot p^k = 0$ which is a contradiction. Assume I is generated by d elements; work on $R' = R/(p_1)$, then applying the induction hypothesis to the homomorphic images of J and I in R' we have that these images are equal, hence $p_i = g_i + r_i p_1$ where $g_i \in J$. Notice that $\{p_1, p_2 - r_2 p_1, p_3, \ldots, p_d\}$ is a set of generators for I. Now mod out by $p_2 - r_2 p_1 = g_i$, and again apply the induction hypothesis. This shows that $\{p_1, p_3, \ldots, p_d\}$ are in J hence I is in J since the missing generator of I is already in J. □

Note that it in the above proof it is not necessary for I to be radical.

We say that $f : \mathbf{C}^{n+1}, x \to \mathbf{C}, 0$ has singularity of type A(d) at x, if local coordinates $(z_1, \ldots, z_d, w_1, \ldots, w_r)$ can be found such that

$$f(z, w) = w_1^2 + \cdots + w_r^2.$$

If f has singularity of type A(d) at x then $S(f) = V(w_1, \ldots, w_r) = J(f)$ so $j(f) = 0$. There is a partial converse.

Lemma 4.4. *Suppose $f : \mathbf{C}^{n+1}, 0 \to \mathbf{C}, 0$. Suppose I defines a complete intersection Σ^d at 0 with reduced structure, and suppose $j(f) = 0$. Then f has a singularity of type $A(d)$ at 0.*

Proof. If $d = 0$ the hypothesis implies that $J(f) = m_{n+1}$, and the result is implied by the Morse lemma. Suppose $d > 0$, then Theorem 5.14 p. 59 of [27] implies that Σ is an ICIS, and f is A(d) except perhaps at 0. Further, the formula of 5.14 implies that the Tjurina number of Σ is 0, hence Σ is smooth at the origin. Then proposition 3.13 p. 35 of [27], the formula cited above, and remark 5.3 on p. 52 imply that f is A(d) at the origin as well. □

Now we are ready to prove our result about A_f.

Theorem 4.5. *Suppose $F : \mathbf{C}^k \times \mathbf{C}^{n+1}, \mathbf{C}^k \times 0, 0 \to \mathbf{C}, 0, 0$, suppose the singular set of F, $S(F)$ is $V(I)$ where I defines a family of complete intersections with isolated singularities of fiber dimension d, and every component of $V(I)$ contains $Y = \mathbf{C}^k \times 0$. Suppose further that $J(F) = I$ off Y. Then:*

1) *If the pair $(\mathbf{C}^k \times \mathbf{C}^{n+1} - S(F), \mathbf{C}^k \times 0)$ satisfies the A_F condition then $e(J(f_y), I_y, (y, 0))$ is independent of y.*
2) *If $e(J(f_y), I_y, (y, 0))$ is independent of y, then $\{\mathbf{C}^k \times \mathbf{C}^{n+1} - S(F), V(I) - Y, Y\}$ is an A_F stratification on some neighborhood of Y.*

Proof. To start the proof of 1), assume the A_F condition; this implies that $\frac{\partial F}{\partial y_i} \in J_z(F)^{\dagger}$ for $1 \le i \le k$, by Proposition 3.1. Now

$$e(J(F)(y), I(y), (y, z)) = e(J_z(F)(y), I(y), (y, z)) = e(J(f_y), I(y), (y, z))$$

for all (y,z) in some neighborhood of $(0, 0)$. Since $J(F) = I$ off Y, this implies $e(J(f_y), I(y), (y, z)) = 0$ off Y. Since $\Gamma^k(I) = \Gamma^k(J_z(F)) = \emptyset$, by the multiplicity-polar theorem,

$$e(J(f_0), I(0), (0, 0)) = e(J(f_y), I(y), (y, 0))$$

for all y. Now we prove 2). By hypothesis we have $I = J(F)$ off Y. So by Lemma 3.4 off of Y we have that $V(I)$ is smooth and F has only A($k+d$) singularities. So the pair $\{\mathbf{C}^k \times \mathbf{C}^{n+1} - S(F), V(I) - Y\}$ has the A_F property. Since $e(J(f_y), I_y, (y, 0))$

is independent of y, and I and $J_z(F)$ have no polars of dimension k, it then follows from the multiplicity-polar theorem that $e(J(f_y), I_y, (y,z)) = 0$, for $z \neq 0$. This implies that $\overline{J(f_y)} = \overline{I_y}$. By Lemma 3.3, $J(f_y) = I_y$. In turn this implies by Lemma 3.5 that $V(I_y)$ is smooth off the origin and f has an A(d) singularity at points of $V(I_y)$ off the origin. Now we have that $J_z(F) \subset I$ and at a point (y,z) of Σ off Y,

$$\dim_{\mathbf{C}} J(f_y,z)/(J(f_y,z)\cap m_z^2) = n+1-d \leq \dim J_z(F)/(J_z(F)\cap m_{(y,z)}^2)$$

$$\leq \dim I/(I\cap m_{(y,z)}^2) = n+1-d$$

Hence $J_z(F) = I$ at points of Σ off Y. Using what we have learned about F above, we can describe the components of the exceptional divisor E of $(B_{J_z(F)}(\mathbf{C}^k \times \mathbf{C}^{n+1}), \pi)$; we do this in order to get ready to apply 2) of 3.1, which will finish the proof. Let Σ_i be the ith component of Σ; then there exists a component V_i of E which surjects to Σ_i. Suppose V is a component of E such that $\pi(V)$ is not contained in Y. Let x be a point off Y in $\pi(V)$. Then there is a neighborhood U of x in $\mathbf{C}^k \times \mathbf{C}^{n+1}$ such that on U, $J(F) = J_z(F) = I$, and only one component of Σ intersects U. Hence over U the corresponding blowups are isomorphic; in particular there is only one component of each exceptional divisor which projects to $\Sigma \cap U$. So the V_i are the only components of E whose image does not lie in Y. Suppose W is a component of E whose image lies in Y. Then $W^{n+k} \subset Y^k \times \mathbf{P}^n$, hence $W = Y^k \times \mathbf{P}^n$ if W exists. We have shown that every component of E projects to a set which contains Y in its closure. (This uses the fact that every Σ_i contains Y in its closure.) Since A_F is true generically, there exists a Z-open set U which contains a Z-open subset of Y, and on U we have $\frac{\partial F}{\partial y_i} \in J_z(F)^\dagger$ for $1 \leq i \leq k$. This implies that if we pull back $J_z(F)$ and $J(F)$ to the normalization of $B_{J_z(F)}(\mathbf{C}^k \times \mathbf{C}^{n+1})$, then along every component of the exceptional divisor E_N which meets $\pi_N^{-1}(U)$ in a Z-open set, that $\pi_N^*(J(F)) = \pi_N^*(J_z(F)$. But this is true for all components of E_N, since every component of E of $B_{J_z(F)}(\mathbf{C}^k \times \mathbf{C}^{n+1})$ projects to a set which contains Y in its closure. This implies that $\overline{J_z(F)} = \overline{J(F)}$ at all points of Y ([23]). The last equality implies that E_J, the exceptional divisor of $B_{J(F)}(\mathbf{C}^k \times \mathbf{C}^{n+1})$, is finite over E. The components of E_J which are in $\pi_N^{-1}(Y)$, have dimension $k+n$ and have fiber dimension n, which is the fiber dimension of W, since they are finite over W. Hence they surject onto W, and hence Y. Since A_F holds generically, these components are in $C(Y)$, the conormal of Y, which also has dimension $n+k$, hence they are equal to the conormal, so there is only 1 such component. Over each V_i as we have seen there is only one component of E_J; since A_F holds between the open stratum and these components, a dimension count shows that this unique component is $C(\Sigma_i)$. The proof will be complete if we can show that each component of Σ satisfies Whitney A over Y. (This is also what it means for A_f to hold for the pair (Σ, Y).)

Claim: For every i, $C(\Sigma_i) \cap \pi_N^{-1}(Y \cap U)$ is dense in $C(\Sigma_i) \cap \pi_N^{-1}(Y)$.

Since $C(\Sigma_i) \cap \pi_N^{-1}(Y \cap U)$ lies in $C(Y)$ this will finish the proof by 2) of 3.1. By Lemma 5.7 p230 of [16], we know that each component of $C(\Sigma_i) \cap \pi_N^{-1}(Y)$ has dimension $n+k-1$, that is, must be a hypersurface in $C(\Sigma_i)$. (This uses the fact that I defines a complete intersection.) If the claim fails there must be a component for some i of $C(\Sigma_i)\cap\pi_N^{-1}(Y)$ which does not surject onto Y. Since $E_N|Y$ is finite over $E|Y\subset Y\times\mathbf{P}^n$, this component must map to a subset of Y of dimension $k-1$, and must have constant fiber dimension n. Let C be the fiber of the bad component over 0. Consider $B_{J(F)(0)}(0\times\mathbf{C}^{n+1})$. This must contain C as a component of its exceptional divisor, as C is a subset of $B_{J(F)}(\mathbf{C}^k\times\mathbf{C}^{n+1})\cap 0\times\mathbf{C}^{n+1}\times\mathbf{P}^{n+k}$, and its dimension is too small to be a component of the intersection. Construct a polar variety of $J(F)$ of dimension $k+1$. This is a family of curves over Y; the fiber over 0 contains a curve which is the projection of the intersection of the plane defining the polar with $B_{J(F)(0)}(0\times\mathbf{C}^{n+1})$ forced by the existence of C. Let Γ be the component of our polar which contains this curve. We choose the plane of codimension n of $\mathbf{P}^{n+k}$ so that it misses the points of $C\cap C(Y)$. On some sufficiently small metric neighborhood of the origin in Γ, then we know that Γ intersects Y only at $(0,0)$. Restrict I and $J(F)$ to Γ. Now we apply the multiplicity-polar theorem again. $J(F)$ has no polar, because it is integrally dependent on $J_z(F)$ which has no polar. Over a generic y value, the only points where $J(F)$ has support are on $\Sigma-Y$ hence $e(J(F)(y),I_y)=0$ at such points. We claim that the multiplicity of the pair $(J(F)(0),I_0)$ on $\Gamma(0)$ at $(0,0)$ is not zero. This number has an alternate meaning. It is part of the intersection number $\int D_{J(F)(0),I(0)}\cdot l^n_{J(F)(0)}$, which in turn is part of $e((J(F)(0),I_0),(0,0))$ on $\mathbf{C}^{n+1}$. We know that $B_{\rho(J(F)(0))}(\operatorname{Projan}\mathcal{R}(I_0))$, dominates both $B_{I_0}(\mathbf{C}^{n+1})$ and $B_{J(F)(0)}(\mathbf{C}^{n+1})$; corresponding to C there is a component of the exceptional divisor of $B_{\rho(J(F)(0))}(\operatorname{Projan}\mathcal{R}(I_0))$. The map to $B_{I_0}(\mathbf{C}^{n+1})$ cannot be finite on this component, because the component projects to the origin in $\mathbf{C}^{n+1}$, and the fiber dimension of the exceptional divisor of $B_{I_0}(\mathbf{C}^{n+1})$ over the origin must have dimension less than $n-d<n$, hence this component over C makes a non-zero contribution to $\int D_{J(F)(0),I(0)}\cdot l^n_{J(F)(0)}$, so the multiplicity of the pair $(J(F)(0),I_0)$ on $\Gamma(0)$ at $(0,0)$ is not zero, so the multiplicity-polar theorem gives a contradiction – the change in multiplicity from the special fiber to the generic fiber is positive, but there is no polar variety of dimension k of $J(F)$. So C does not exist, which implies Whitney A holds for $(\Sigma-Y,Y)$ and the theorem is proved. □

Remark 4.6. The key point in the last proof, was the ability to take information about the $k+d$ dimensional strata of the total space, and relate it to the open stratum of f_0. This was possible because we had good control on the conormals of the $k+d$ dimensional strata.

The above proof shows that it is easy to show that a stratification condition implies that the associated invariants are independent of parameter. To prove that the independence from parameter implies the stratification condition requires in general the principle of specialization of integral dependence developed in [12].

As we shall see in general (Remark 4.9) the A_F condition does not imply that the Lê numbers are independent of parameter. We can introduce a stronger notion of A_F which does imply that the Lê numbers are constant if our ICIS is a curve. In the situation of Theorem 4.5 we say the strong A_F condition holds if the A_F condition holds, and for a generic linear function l the A_l condition holds for the pair $V(I) - Y, Y$.

From Theorem 4.5, and the formula for λ^0 in Theorem 2.4, we can now show that the strong A_F condition implies that the Lê numbers are constant in the setup originally considered by Pellikaan.

Corollary 4.7. *Suppose* $F : \mathbf{C}^k \times \mathbf{C}^{n+1}, \mathbf{C}^k \times 0, 0 \to \mathbf{C}, 0, 0$, *suppose the singular set of* F, $S(F)$ *is* $V(I)$ *where* I *defines a family of complete intersection curves with isolated singularities, and every component of* $V(I)$ *contains* $Y = \mathbf{C}^k \times 0$. *Suppose further that* $J(F) = I$ *off* Y. *Suppose the pair* $(\mathbf{C}^k \times \mathbf{C}^{n+1} - S(F), \mathbf{C}^k \times 0)$ *satisfies the* A_F *condition, and the pair* $V(I) - \mathbf{C}^k \times 0, \mathbf{C}^k \times 0$ *satisfies the* A_l *condition for a generic linear function* l, *then the Lê numbers of* f_y *at the origin are independent of* y.

Proof. Theorem 4.5 1) and Theorem 2.3 imply that $j(f_y)$ is constant along Y. The condition that the singular set of F is $V(I)$ implies that F is in I^2 (p. 8, Prop 1.9 [27]), hence $F = \sum_{i,j} h_{i,j} g_i g_j$ where $\{g_i\}$ are a set of generators of I, and $h_{i,j} = h_{j,i}$ ([27], p. 54). Let Δ be the determinant of the matrix with entries $h_{i,j}$. Then the number of D_∞ points at (y, z) is just the colength of (Δ_y) in $\mathcal{O}(V(I_y), z)$ ([27], p. 81 Lemma 7.17). This number is just the local degree at (y, z) of the map with components (Δ, p) where p is projection to the parameter space Y on $V(I)$. Thus if $\delta(f_y)$ varies along Y it must be upper semicontinuous, and if the value for generic y is less than the value over $y = 0$, there must be other points in the fiber over y where $\delta(f_y)$ is non-zero. However as the proof of Theorem 3.5 2) shows off Y f_y has only A_∞ singularities on $V(I_y)$. Hence, $\delta(f_y)$ is constant along Y.

Since the pair $V(I) - \mathbf{C}^k \times 0, \mathbf{C}^k \times 0$ satisfies the A_l condition for a generic linear function l, by Theorem 5.8 p. 232 of [16] the Milnor numbers of $V(I_y)$ and $V(I(y)) \cap V(l)$ are constant. Since l is generic, the sum of these Milnor numbers is just $e(JM(\Sigma_y)$, which is then independent of y. The result for λ^0 now follows from the formula for λ^0 in Proposition 2.4. Since the Milnor number of $V(I(y)) \cap V(l)$ is just the multiplicity of Σ_y, less 1, the multiplicity of $V(I_y)$ is independent of Y. Since the transverse Milnor number is always 1, and the multiplicity of $V(I_y)$ constant, it follows that λ^1 is independent of Y as well. □

Corollary 4.8. *Suppose* $F : \mathbf{C}^k \times \mathbf{C}^{n+1}, \mathbf{C}^k \times 0, 0 \to \mathbf{C}, 0, 0$, *suppose the singular set of* F, $S(F)$ *is* $V(I)$ *where* I *defines a family of complete intersection curves with isolated singularities, and every component of* $V(I)$ *contains* $Y = \mathbf{C}^k \times 0$. *Suppose further that* $J(F) = I$ *off* Y. *Suppose the pairs* $(\mathbf{C}^k \times \mathbf{C}^{n+1} - S(F), \mathbf{C}^k \times 0)$, $V(I) - \mathbf{C}^k \times 0, \mathbf{C}^k \times 0$ *satisfy the strong* A_F *condition at* $(0, 0)$ *then*

1) *The homology of the Milnor fibre of of* f_y *at the origin is independent of* y *for all* y *small. If* $n \geq 3$

2) *The fibre homotopy-type of the Milnor fibrations of f_y at the origin is independent of y for all y small. If $n \geq 4$*
3) *The diffeomorphism-type of the Milnor fibrations of f_y at the origin is independent of y for all y small.*

Proof. Since the strong A_F condition holds, Corollary 4.7 implies that the Lê numbers are constant, then Theorem 9.4 of [24] p. 90 gives the result. (Although Massey states his theorem for the case where the dimension of the parameter stratum is 1, it also applies to the case at hand.) □

This raises the interesting question of whether a strong A_F stratification or an A_F stratification implies the triviality (in the sense of the last corollary) of the Milnor fibrations. The formulae in Proposition 3.12 and Remark 3.13 show that a strong A_F stratification implies that $\lambda^0(f_t)$ is independent of t in these cases.

As the next example shows, in the A_F case, this problem cannot be tackled by hoping that the existence of an A_F stratification implies that the Lê numbers are constant.

Remark 4.9. This example shows that neither the A_f condition nor topological triviality imply that the Lê numbers are constant. Let

$$f_t = z^5 + ty^6z + y^7x + x^{15}.$$

This family of functions was introduced by Briançon-Speder, ([2]) who showed that $\mu^3(f_t) := \mu(f_t) = 364$ for all t, while the Milnor number of a generic hyperplane slice $\mu^2(f_t)$ is 28 when $t = 0$ and 26 otherwise. Historically, this example was important, because it showed that the μ_* constant condition was stronger than topological triviality. Now consider $F_t = f_t^2 + w^2$ where w is a disjoint variable. Then

$$J_z(F) = \left\langle w, 2f_t\frac{\partial f_t}{\partial x}, 2f_t\frac{\partial f_t}{\partial y}, 2f_t\frac{\partial f_t}{\partial z}\right\rangle.$$

So the singular locus of F is defined by $\langle w, f_t\rangle$, hence is a family of complete intersections with isolated singularities. A computation shows that:

$$e(J(F_t), (w, f_t)) = j(F_t) = \mu^3(f_t).$$

Now, the only Lê cycle of dimension 2 is $V(w, f_t)$, so

$$\lambda^2(F_t) = m(X_t) = 5,$$

while

$$\lambda^1(F_t) = m(\Gamma_1^1(X_t, 0)) = \mu^2(f_t) + \mu^1(f_t).$$

Now by 3.12

$$\lambda^0(F_t) = e(J(F_t), (w, f_t)) + e(JM(V(w, f_t))) = \mu^3(f_t) + (\mu^2(f_t) + \mu^3(f_t)).$$

The first equality shows that the A_F condition holds by Theorem 3.5. However $\lambda^0(F_t)$ and $\lambda^1(F_t)$ vary with t. It is not hard to check by a vector field argument that the family of functions F_t are topologically trivial; however this can be seen directly by the following argument which was pointed out to me by J.N. Damon.

We know that there exists a topological trivialization $\phi(z,t) : \mathbf{C}^4 \to \mathbf{C}^3$ of f_t, by [2], so $f_t(\phi(z,t)) = f_0(z)$ Then, we can define $\Phi(z,w,t) = (\phi(z,t),w) : \mathbf{C}^5 \to \mathbf{C}^4$, which gives a topological trivialization of F_t since

$$F_t(\Phi(x,w,t)) = f_t(\phi(x,t)) + w^2 = f_0(x) + w^2 = F_0(x).$$

Now we turn to the W_f condition. It is a paradox, but because this condition is stronger, it is easier to prove results about it.

Theorem 4.10. *Suppose $F : \mathbf{C}^k \times \mathbf{C}^{n+1}, \mathbf{C}^k \times 0, 0 \to \mathbf{C}, 0, 0$, suppose the singular set of F, $S(F)$ is $V(I)$ where I defines a family of complete intersections with isolated singularities of fiber dimension d, and every component of $V(I)$ contains $Y = \mathbf{C}^k \times 0$. Suppose further that $J(F) = I$ off Y. Then:*

1) *If the pair $(\mathbf{C}^k \times \mathbf{C}^{n+1} - S(F), \mathbf{C}^k \times 0)$ satisfies the W_F condition then $e(m_{n+1}J(f_y), I_y, (y,0))$ is independent of y.*
2) *If $e(m_{n+1}J(f_y), I_y, (y,0))$ is independent of y, then the pair $(\mathbf{C}^k \times \mathbf{C}^{n+1} - S(F), \mathbf{C}^k \times 0)$ satisfies the W_F condition, and $\{\mathbf{C}^k \times \mathbf{C}^{n+1} - V(F), V(F) - V(I), V(I) - Y, Y\}$ is a Whitney stratification on some neighborhood of Y.*

Proof. 1) Suppose the pair $(\mathbf{C}^k \times \mathbf{C}^{n+1} - S(F), \mathbf{C}^k \times 0)$ satisfies the W_F condition, then by Theorem 2.1 p. 23 of [8], the dimension of the fiber of the exceptional divisor over Y of $B_{m_Y J(F)}(\mathbf{C}^k \times \mathbf{C}^{n+1})$ is independent of y and is n. This implies that the polar of dimension k of $m_Y J(F)$ is empty; hence by the multiplicity polar theorem $e(m_{n+1}J(f_y), I_y, (y,0))$ is independent of y.

2) Suppose $e(m_{n+1}J(f_y), I_y, (y,0))$ is independent of y. Off Y, $m_{n+1}J(f_y) = J(f_y)$, so off Y by the same arguments found in the proof of 3.6, $J(f_y) = I_y$, so by the multiplicity polar theorem, $\Gamma^k(m_Y J(F))$ is empty, hence the dimension of the fiber of the exceptional divisor over Y of $B_{m_Y J(F)}(\mathbf{C}^k \times \mathbf{C}^{n+1})$ is n, hence is constant over Y. Then by Corollary 2.1, p. 19 of [8], the pair $(\mathbf{C}^k \times \mathbf{C}^{n+1} - S(F), \mathbf{C}^k \times 0)$ satisfies the W_F condition. This implies $V(F) - V(I)$ is Whitney over Y. Since F is of type A_∞ off Y it follows that $V(F) - V(I)$ is Whitney over $V(I) - Y$. It remains to show $V(I) - Y$ is Whitney over Y. Suppose not; then for each C and neighborhood U of the origin there exists a sequence of points $x_i \in U$ on some component of $V(I)$, converging to the origin, and hyperplanes H_i which are tangent hyperplanes to $V(I)$ at x_i such that

$$\operatorname{dist}(Y, H_i) > C \operatorname{dist}(x, Y).$$

From the proof of theorem 3.6, we have $C(V(I)) \subset B_{J(F)}(\mathbf{C}^k \times \mathbf{C}^{n+1})$. This implies we can find points $\tilde{x}_i \in U \cap (\mathbf{C}^k \times \mathbf{C}^{n+1} - S(F))$ and hyperplanes $\widetilde{H}_i$ tangent to the fibers of F at x_i, such that the distance between x_i and $\tilde{x}_i$, H_i and $\widetilde{H}_i$ is as small as desired. Then a similar inequality holds for $\tilde{x}_i$ and $\widetilde{H}_i$, hence W_F fails, which is a contradiction. □

Corollary 4.11. *Suppose in the above setup $e(m_{n+1}J(f_y), I_y, (y,0))$ is independent of y, then the family of functions $\{f_y\}$ is topologically trivial.*

Proof. Since $e(m_{n+1}J(f_y), I_y, (y,0))$ is independent of y, we have the pair $(\mathbf{C}^k \times \mathbf{C}^{n+1} - S(F), \mathbf{C}^k \times 0)$ satisfies the W_F condition, and $\{\mathbf{C}^k \times \mathbf{C}^{n+1} - V(F), V(F) - V(I), V(I) - Y, Y\}$ is a Whitney stratification on some neighborhood of Y. Then we can lift the constant fields over $V(F)$, to the ambient space in such a way that the resulting fields can be integrated to give homeomorphisms. □

There is a nice geometric interpretation of the number $e(m_{n+1}J(f_y))$ which we now describe. We denote the multiplicity of the relative polar variety of f_y of dimension i by $m^i(f_y)$.

Theorem 4.12. *Suppose $f : \mathbf{C}^{n+1}, 0 \to \mathbf{C}, 0$, J any ideal in $\mathcal{O}_{n+1}$ such that $\dim_{\mathbf{C}} J/J(f) < \infty$, then*

$$e(m_{n+1}J(f), J) = e(J(f), J) + 1 + \sum_{i=1}^{n} \binom{n+1}{i} m^i(f_y).$$

Proof. This is exactly the content of the formula in Theorem 9.8 (i) p. 221 [20]. □

Corollary 4.13. *Suppose $f : \mathbf{C}^{n+1}, 0 \to \mathbf{C}, 0$, $S(f)$ is $V(I)$ where I defines a complete intersection with isolated singularities of dimension d, and suppose further that $J(f) = I$ off Y. Then*

$$e(m_{n+1}J(f), I) = e(J(f), I) + 1 + \sum_{i=1}^{n} \binom{n+1}{i} m^i(f_y).$$

Proof. Follows immediately from Theorem 3.12 □

Corollary 4.14. *Suppose $F : \mathbf{C}^k \times \mathbf{C}^{n+1}, \mathbf{C}^k, 0 \to \mathbf{C}, 0, 0$, suppose the singular set of F, $S(F)$ is $V(I)$ where I defines a family of complete intersections with isolated singularities of fiber dimension d, and every component of $V(I)$ contains $Y = \mathbf{C}^k \times 0$. Suppose further that $J(F) = I$ off Y. Then the following are equivalent:*

1) *$e(J(f_y), I_y)$ and the relative polar multiplicities of f_y are independent of y.*
2) *A_F holds for the pair $(\mathbf{C}^k \times \mathbf{C}^{n+1} - V(I), Y)$, and the relative polar multiplicities of f_y are independent of y.*
3) *The pair $(\mathbf{C}^k \times \mathbf{C}^{n+1} - V(I), \mathbf{C}^k \times 0)$ satisfies the W_F condition.*

Proof. 1) and 2) are equivalent by Theorem 3.5, while 2 and 3 are equivalent by Corollary 3.13 and Theorem 3.9. □

References

[1] J. Briançon, P. Maisonobe and M. Merle, *Localisation de systèmes différentiels, stratifications de Whitney et condition de Thom,* Invent. Math. **117** (1994), 531–50

[2] J. Briançon and J.P. Speder, *La trivialité topologique n'implique pas les conditions de Whitney*, C.R. Acad. Sc.Paris, v. 280 365–367 (1975)

[3] D.A. Buchsbaum and D.S. Rim, *A generalized Koszul complex. II. Depth and multiplicity,* Trans. Amer. Math. Soc. **111** (1963), 197–224

[4] J. Fernández de Bobadilla, *Approximations of non-isolated singularities of finite codimension with respect to an isolated complete intersection singularity.* Bull. London Math. Soc. 35 (2003), no. 6, 812–816

[5] W. Fulton, "Intersection Theory," Ergebnisse der Mathematik und ihrer Grenzgebiete, 3. Folge · Band 2, Springer-Verlag, Berlin, 1984

[6] T. Gaffney, *Aureoles and integral closure of modules,* in "Stratifications, Singularities and Differential Equations II", Travaux en Cours **55**, Herman, Paris, 1997, 55–62

[7] T. Gaffney, *Integral closure of modules and Whitney equisingularity,* Invent. Math. **107** (1992), 301–22

[8] T. Gaffney, *Plane sections,* W_f *and* A_f , in "Real and complex singularities (São Carlos, 1998)", Chapman and Hall Res. Notes Math. **412**, 2000, 17–32

[9] T. Gaffney, *Multiplicities and equisingularity of ICIS germs,* Invent. Math. **123** (1996), 209–220

[10] T. Gaffney, *Generalized Buchsbaum-Rim Multiplicities and a Theorem of Rees,* Communications in Algebra, vol 31 #8 pp. 3811–3828, 2003

[11] T. Gaffney,*Polar methods, invariants of pairs of modules and equisingularity,* Real and Complex Singularities (Sao Carlos, 2002), Ed. T.Gaffney and M.Ruas, Contemp. Math.,#354, Amer. Math. Soc., Providence, RI, June 2004, 113–136

[12] T. Gaffney, *The Multiplicity-Polar Formula and Equisingularity,* in preparation

[13] T. Gaffney and R. Gassler, *Segre numbers and hypersurface singularities,* J. Algebraic Geom. **8** (1999), 695–736

[14] T. Gaffney and S. Kleiman, *Specialization of integral dependence for modules* Invent. Math. **137** (1999), 541–574

[15] T. Gaffney and S. Kleiman, W_f *and specialization of integral dependence for modules,* in "Real and complex singularities (São Carlos, 1998)", Chapman and Hall Res. Notes Math. **412**, 2000, 33–45

[16] T. Gaffney and D. Massey, *Trends in equisingularity,* London Math. Soc. Lecture Note Ser. **263** (1999), 207–248

[17] R. Gassler, *Segre numbers and hypersurface singularities.* Thesis, Northeastern University, 1999

[18] J.P.G. Henry and M. Merle, *Conormal Space and Jacobian module. A short dictionary,* in "Proceedings of the Lille Congress of Singularities," J.-P. Brasselet (ed.), London Math. Soc. Lecture Notes **201** (1994), 147–74

[19] G. Jiang, *Functions with non-isolated singularities on singular spaces,* Thesis, Universiteit Utrecht, 1997

[20] S. Kleiman and A. Thorup, *A geometric theory of the Buchsbaum–Rim multiplicity,* J. Algebra **167** (1994), 168–231

[21] S. Kleiman and A. Thorup, *The exceptional fiber of a generalized conormal space,* in "New Developments in Singularity Theory." D. Siersma, C.T.C. Wall and V. Zakalyukin (eds.), Nato Science series, II Mathematics, Physics and Chemistry, Vol. 21 2001 401–404

[22] D.T. Lê and B. Teissier, *Cycles evanescents, sections planes et conditions de Whitney. II,* Proc. Sympos. Pure Math. **40**, part 2, Amer. Math. Soc., 1983, 65–104

[23] M. Lejeune-Jalabert and B. Teissier, *Clôture intégrale des ideaux et equisingularité, chapitre 1* Publ. Inst. Fourier (1974)

[24] D. Massey *Lê Cycles and Hypersurface Singularities,* Springer Lecture Notes in Mathematics 1615, (1995)

[25] D. Mumford, *Algebraic Geometry I Complex Projective Varieties,* Springer-Verlag 1976

[26] A. Nemethi, *Hypersurface singularities with a 2-dimensional critical locus*, J. London Math. Soc. (3), 59, (1999), 922–938

[27] R. Pellikaan, *Hypersurface singularities and resolutions of Jacobi modules,* Thesis, Rijkuniversiteit Utrecht, 1985

[28] R. Pellikaan, *Finite determinacy of functions with non-isolated singularities,* Proc. London Math. Soc. vol. 57, pp. 1–26, 1988

[29] D. Siersma, *Isolated line singularities.* Singularities, Part 2 (Arcata, Calif., 1981), 485–496, Proc. Sympos. Pure Math., 40, Amer. Math. Soc., Providence, RI, 1983

[30] B. Teissier, *The hunting of invariants in the geometry of the discriminant,* in "Real and complex singularities, Oslo 1976," P. Holm (ed.), Sijthoff & Noordhoff (1977), 565–678

[31] B. Teissier, *Multiplicités polaires, sections planes, et conditions de Whitney,* in "Proc. La Rábida, 1981." J.M. Aroca, R. Buchweitz, M. Giusti and M. Merle (eds.), Springer Lecture Notes in Math. **961** (1982), 314–491

[32] A. Zaharia, *A study about singularities with non-isolated critical locus,* Thesis, Rijkuniversiteit Utrecht, 1993

[33] A. Zaharia, *Topological properties of certain singularities with critical locus a 2-dimensional complete intersection,* Topology Appl. 60 (1994), no. 2, 153–171

Terence Gaffney
Department of Mathematics
567 Lake Hall
Northeastern University
Boston, Ma 02115, USA
e-mail: `gaff@neu.edu`

Real and Complex Singularities
Trends in Mathematics, 169–185

Lagrangian and Legendrian Singularities

Victor V. Goryunov and V.M. Zakalyukin

Abstract. These are notes of the introductory courses on the subject we lectured in Trieste in 2003 and Luminy in 2004. The lectures contain basic notions and fundamental theorems of the local theory.

Mathematics Subject Classification (2000). Primary 32S05, 57R45, 58K05; Secondary 58K25, 58K60.

Keywords. Symplectic and contact geometry, Lagrangian and Legendrian submanifolds and singularities, generating family.

Almost all applications of singularity theory are related to wave fronts and caustics: they can be visualized and recognized in many physical models.

Suppose, for example, that a disturbance (such as a shock wave, light, an epidemic or a flame) is propagating in a medium from a given submanifold (called the *initial wave front*). To determine where the disturbance will be at time t, according to the Huygens principle, we must lay a segment of length t along every normal to the initial front. The resulting variety is called an *equidistant* or a wave front.

Along with wave fronts, ray systems may also be used to describe propagation of disturbances. For example, we can consider the family of all normals to the initial front. This family has the envelope, which is called *caustic* – "burning" in Greek – since the light concentrates at it. A caustic is clearly visible on the inner surface of a cup put in the sunshine. A rainbow in the sky is the caustic of a system of rays which have passed through drops of water with the total internal reflection.

Generic caustics in three-dimensional space have only standard singularities. Besides regular surfaces, cuspidal edges and their generic (transversal) intersections, these are: the swallowtail, the 'pyramid' (or 'elliptic umbilic') and the 'purse' (or 'hyperbolic umbilic'). They are a part of R.Thom's famous list of simple catastrophes. It is not so difficult to see that the singularities of a propagating wave front slide along the caustic and trace it out.

The study of singularities of wave fronts and caustics was the starting point of the theory of Lagrangian and Legendrian mappings developed by V.I.Arnold and his school some thirty years ago. Since then the significance of Lagrangian and

Legendrian submanifolds of symplectic and respectively contact spaces has been recognized throughout all mathematics, from algebraic geometry to differential equations, optimization problems and physics.

Symplectic space is essentially the phase space (space of positions and momenta) of classical mechanics, inheriting a rich set of important properties.

It turns out that caustics and wave fronts are the critical value loci of special non-generic mappings either between manifolds of the same dimension or between n- and $(n+1)$-dimensional manifolds. The general definitions of such mappings were introduced by V.I. Arnold in terms of projections of Lagrangian and Legendrian submanifolds embedded into symplectic and contact spaces. These constructions describe many special classes of mappings, such as Gauss, gradient and others.

A Lagrangian or Legendrian mapping is determined by a single family of functions. This crucial feature makes the theory transparent and constructive.

In particular, stable wave fronts and caustics are discriminants and bifurcation diagrams of function singularities. That is why their generic low-dimensional singularities are governed by the famous Weyl groups.

Recently new areas in theory of integrable systems and mathematical physics (for example, Frobenius structures, D-modules etc.) opened up new fields for applications of theory of Lagrangian and Legendrian singularities.

In these lecture notes, we do not touch the fascinating results in symplectic and contact topology, a young branch of mathematics which answers questions on global behavior of Lagrangian and Legendrian submanifolds. An interested reader may be addressed to the book [4] and paper [5] forming a good introduction to that area. Our lectures were designed as an introduction to the original local theory. We hope that they will inspire the reader to do more extensive reading. Items [1, 3, 2] on our bibliography list may be rather useful for this.

1. Symplectic and contact geometry

1.1. Symplectic geometry

A symplectic form ω on a manifold M is a closed 2-form, non-degenerate as a skew-symmetric bilinear form on the tangent space at each point. So $d\omega = 0$ and ω^n is a volume form, $\dim M = 2n$.

Manifold M equipped with a symplectic form is called symplectic. It is necessarily even-dimensional.

If the form is exact, $\omega = d\lambda$, the manifold M is called *exact symplectic.*

Examples

1. Let $K = M = \mathbf{R}^{2n} = \{q_1, \ldots, q_n, p_1, \ldots, p_n\}$ be a vector space, and

$$\lambda = pdq = \sum_{i=1}^{n} p_i dq_i\,, \qquad \omega = d\lambda = dp \wedge dq\,.$$

In these co-ordinates the form ω is constant. The corresponding bilinear form on the tangent space at a point is given by the matrix

$$J = \begin{pmatrix} 0 & -I_n \\ I_n & 0 \end{pmatrix}.$$

NOTICE: for any non-degenerate skew-symmetric bilinear form on a linear space, there exists a basis (called Darboux basis) in which the form has this matrix.

2. $M = T^*N$. Take for λ the *Liouville form* defined in an invariant (co-ordinate-free) way as

$$\lambda(\alpha) = \pi(\alpha)\Big(\rho_*(\alpha)\Big),$$

where

$$\alpha \in T(T^*N), \quad \pi : T(T^*N) \to T^*N \quad \text{and} \quad \rho : T^*N \to N.$$

This is an exact symplectic manifold. If $q_1, \dots, q_n$ are local co-ordinates on the base N, the dual co-ordinates $p_1, \dots, p_n$ are the coefficients of the decomposition of a covector into a linear combination of the differentials dq_i:

$$\lambda = \sum_{i=1}^{n} p_i dq_i .$$

3. On a Kähler manifold M, the imaginary part of its Hermitian structure $\omega(\alpha, \beta) = Im(\alpha, \beta)$ is a skew-symmetric 2-form which is closed.

4. Product of two symplectic manifolds. Given two symplectic manifolds (M_i, ω_i), $i = 1, 2$, their product $M_1 \times M_2$ equipped with the 2-form $(\pi_1)_*\omega_1 - (\pi_2)_*\omega_2$, where the π_i are the projections to the corresponding factors, is a symplectic manifold.

A diffeomorphism $\varphi : M_1 \to M_2$ which sends the symplectic structure ω_2 on M_2 to the symplectic structure ω_1 on M_1,

$$\varphi^*\omega_2 = \omega_1 ,$$

is called a symplectomorphism between (M_1, ω_1) and (M_2, ω_2). When the (M_i, ω_i) are the same, a symplectomorphism preserves the symplectic structure. In particular, it preserves the volume form ω^n.

Symplectic group

For $K = (\mathbf{R}^{2n}, dp \wedge dq)$ of our first example, the group $Sp(2n)$ of linear symplectomorphisms is isomorphic to the group of matrices S such that

$$S^{-1} = -JS^tJ .$$

Here t is for transpose. The characteristic polynomial of such an S is reciprocal: if α is an eigenvalue, then α^{-1} also is. The Jordan structures for α and α^{-1} are the same.

Introduce an auxiliary scalar product $(\cdot, \cdot)$ on K, with the matrix I_{2n} in our Darboux basis. Then

$$\omega(a, b) = (a, \widetilde{J}b) ,$$

where $\widetilde{J}$ is the operator on K with the matrix J. Setting $q = \operatorname{Re} z$ and $p = \operatorname{Im} z$ makes K a complex Hermitian space, with the multiplication by $i = \sqrt{-1}$ being the application of $\widetilde{J}$. The Hermitian structure is

$$(a, b) + i\omega(a, b)\,.$$

From this,

$$Gl(n, \mathbf{C}) \bigcap O(2n) = Gl(n, \mathbf{C}) \bigcap Sp(2n) = O(2n) \bigcap Sp(2n) = U(n)\,.$$

Remark. The image of the unit sphere $S_1^{2n-1} : q^2 + p^2 = 1$ under a linear symplectomorphism can belong to a cylinder $q_1^2 + p_1^2 \le r$ only if $r \ge 1$.

The non-linear analog of this result is rather non-trivial: $S_1^{2n-1} \in T^*\mathbf{R}^n$ (in the standard Euclidean structure) cannot be symplectically embedded into the cylinder $\{q_1^2 + p_1^2 < 1\} \times T^*\mathbf{R}^{n-1}$. This is Gromov's theorem on symplectic camel.

Thus, for $n > 1$, symplectomorphisms form a thin subset in the set of diffeomorphisms preserving the volume ω^n.

The dimension k of a linear subspace $L^k \subset K$ and the rank r of the restriction of the bilinear form ω on it are the complete set of $Sp(2n)$-invariants of L.

Define the skew-orthogonal complement $L^\angle$ of L as

$$L^\angle = \{v \in K | \omega(v, u) = 0 \quad \forall u \in L\}\,.$$

So $\dim L^\angle = 2n - k$. The kernel subspace of the restriction of ω to L is $L \bigcap L^\angle$. Its dimension is $k - r$.

A subspace is called isotropic if $L \subset L^\angle$ (hence $\dim L \le n$).
Any line is isotropic.

A subspace is called co-isotropic if $L^\angle \subset L$ (hence $\dim L \ge n$).
Any hyperplane H is co-isotropic. The line $H^\angle$ is called the characteristic direction on H.

A subspace is called Lagrangian if $L^\angle = L$ (hence $\dim L = n$).

Lemma. *Each Lagrangian subspace $L \subset K$ has a regular projection to at least one of the 2^n co-ordinate Lagrangian planes (p_I, q_J), along the complementary Lagrangian plane (p_J, q_I). Here $I \bigcup J = \{1, \ldots, n\}$ and $I \bigcap J = \emptyset$.*

Proof. Let L_q be the intersection of L with the q-space and $\dim L_q = k$. Assume $k > 0$, otherwise L projects regularly onto the p-space. The plane L_q has a regular projection onto some q_I-plane (along q_J) with $|I| = k$. If L does not project regularly to the p_J-plane (along (q, p_I)) then L contains a vector $v \in (q, p_I)$ with a non-trivial p_I-component. Due to this non-triviality, the intersection of the skew-orthogonal complement $v^\angle$ with the q-space has a $(k-1)$-dimensional projection to q_I (along q_J) and so does not contain L_q. This contradicts to L being Lagrangian. □

A Lagrangian subspace L which projects regularly onto the q-plane is the graph of a self-adjoint operator S from the q-space to the p-space with its matrix symmetric in the Darboux basis.

Splitting $K = L_1 \bigoplus L_2$ with the summands Lagrangian is called a polarisation. Any two polarisations are symplectomorphic.

The Lagrangian Grassmanian $Gr_L(2n)$ is diffeomorphic to $U(n)/O(n)$. Its fundamental group is $\mathbf{Z}$.

The Grassmanian $Gr_k(2n)$ of isotropic k-spaces is isomorphic to $U(n)/(O(k) + U(n-k))$.

Even in a non-linear setting a symplectic structure has no local invariants (unlike a Riemannian structure) according to the classical

Darboux Theorem. *Any two symplectic manifolds of the same dimension are locally symplectomorphic.*

Proof. We use the homotopy method. Let ω_t, $t \in [0,1]$, be a family of germs of symplectic forms on a manifold coinciding at the distinguished point A. We are looking for a family $\{g_t\}$ of diffeomorphisms such that $g_t^*\omega_t = \omega_0$ for all t. Differentiate this by t:

$$\mathcal{L}_{v_t}\omega_t = -\gamma_t$$

where $\gamma_t = \partial\omega_t/\partial t$ is a known closed 2-form and $\mathcal{L}_{v_t}$ is the Lie derivative along the vector field to find. Since $\mathcal{L}_v = i_v d + d i_v$, we get

$$d i_{v_t}\omega_t = -\gamma_t \,.$$

Choose a 1-form α_t vanishing at A and such that $d\alpha_t = -\gamma_t$. Due to the non-degeneracy of ω_t, the equation $i_{v_t}\omega_t = \omega(\cdot, v_t) = \alpha_t$ has a unique solution v_t vanishing at A.qed

Weinstein's Theorem. *A submanifold of a symplectic manifold is defined, up to a symplectomorphism of its neighborhood, by the restriction of the symplectic form to the tangent vectors to the ambient manifold at the points of the submanifold.*

In a similar local setting, the inner geometry of a submanifold defines its outer geometry:

Givental's Theorem. *A germ of a submanifold in a symplectic manifold is defined, up to a symplectomorphism, by the restriction of the symplectic structure to the tangent bundle of the submanifold.*

Proof of Givental's Theorem. It is sufficient to prove that if the restrictions of two symplectic forms, ω_0 and ω_1, to the tangent bundle of a submanifold $G \subset M$ at point A coincide, then there exits a local diffeomorphism of M fixing G point-wise and sending one form to the other. We may assume that the forms coincide on $T_A M$.

We again use the homotopy method, aiming to find a family of diffeomorphism-germs g_t, $t \in [0,1]$, such that

$$g_t|_G = id_G\,, \quad g_0 = id_M\,, \qquad g_t^*(\omega_t) = \omega_0 \quad (*) \qquad \text{where} \quad \omega_t = \omega_0 + (\omega_1 - \omega_0)t\,.$$

Differentiating (∗) by t, we again get

$$\mathcal{L}_{v_t}(\omega_t) = d(i_{v_t}\omega_t) = \omega_0 - \omega_1$$

where v_t is the vector field of the flow g_t. Using the "relative Poincaré lemma", it is possible to find a 1-form α so that $d\alpha = \omega_0 - \omega_1$ and α vanishes on G. Then the required vector field v_t exists since ω_t is non-degenerate. □

The Darboux theorem is a particular case of Givental's theorem: take a point as a submanifold.

If at each point x of a submanifold L of a symplectic manifold M the subspace T_xL is Lagrangian in the symplectic space T_xM, then L is called Lagrangian.

Examples

1. In T^*N, the following are Lagrangian submanifolds: the zero section of the bundle, fibres of the bundle, graph of the differential of a function on N.

2. The graph of a symplectomorphism is a Lagrangian submanifold of the product space (it has regular projections onto the factors). An arbitrary Lagrangian submanifold of the product space defines a so-called Lagrangian relation.

3. Weinstein's theorem implies that a tubular neighborhood of a Lagrangian submanifold L in any symplectic space is symplectomorphic to a tubular neighborhood of the zero section in T^*N.

A fibration with Lagrangian fibres is called Lagrangian.

Locally all Lagrangian fibrations are symplectomorphic (the proof is similar to that of the Darboux theorem).

A cotangent bundle is a Lagrangian fibration.

Let $\psi : L \to T^*N$ be a Lagrangian embedding and $\rho : T^*N \to N$ the fibration. The product $\rho \circ \psi : L \to N$ is called a Lagrangian mapping. It critical values

$$\Sigma_L = \{q \in N | \exists p : (p,q) \in L,\ \operatorname{rank} d(\rho \circ \psi) < n\}$$

form the caustic of the Lagrangian mapping. The equivalence of Lagrangian mappings is that up to fibre-preserving symplectomorphisms of the ambient symplectic space. Caustics of equivalent Lagrangian mappings are diffeomorphic.

Hamiltonian vector fields

Given a real function $h : M \to \mathbf{R}$ on a symplectic manifold, define a Hamiltonian vector field v_h on M by the formula

$$\omega(\cdot, v_h) = dh\,.$$

This field is tangent to the level hypersurfaces $H_c = h^{-1}(c)$:

$$\forall a \in H_c \quad dh(T_aH_c) = 0 \qquad \Longrightarrow \qquad T_aH_c = v_h^{\angle}\,, \quad \text{but} \quad v_h \in v_h^{\angle}\,.$$

The directions of v_h on the level hypersurfaces H_c of h are the characteristic directions of the tangent spaces of the hypersurfaces.

Associating v_h to h, we obtain a Lie algebra structure on the space of functions:

$$[v_h, v_f] = v_{\{h,f\}} \qquad \text{where} \quad \{h, f\} = v_h(f)\,,$$

the latter being the Poisson bracket of the Hamiltonians h and f.

A Hamiltonian flow (even if h depends on time) consists of symplectomorphisms. Locally (or in $\mathbf{R}^{2n}$), any time-dependent family of symplectomorphisms that starts from the identity is a phase flow of a time-dependent Hamiltonian. However, for example, on a torus $\mathbf{R}^2/\mathbf{Z}^2$ (the quotient of the plane by an integer lattice) the family of constant velocity displacements are symplectomorphisms but they cannot be Hamiltonian since a Hamiltonian function on a torus must have critical points.

Given a time-dependent Hamiltonian $\widetilde{h} = \widetilde{h}(t, p, q)$, consider the extended space $M \times T^*\mathbf{R}$ with auxiliary co-ordinates (s, t) and the form $pdq - sdt$. An auxiliary (extended) Hamiltonian $\widehat{h} = -s + \widetilde{h}$ determines a flow in the extended space generated by the vector field

$$\dot{p} = -\frac{\partial \widehat{h}}{\partial q} \qquad\qquad \dot{q} = -\frac{\partial \widehat{h}}{\partial p}$$

$$\dot{t} = -\frac{\partial \widehat{h}}{\partial s} = 1 \qquad\qquad \dot{s} = \frac{\partial \widehat{h}}{\partial t}.$$

The restrictions of this flow to the t = const sections are essentially the flow mappings of $\widetilde{h}$.

The integral of the extended form over a closed chain in $M \times \{t_o\}$ is preserved by the $\hat{h}$-Hamiltonian flow. Hypersurfaces $-s + \widetilde{h}$ = const are invariant. When $\widetilde{h}$ is autonomous, the form pdq is also a relative integral invariant.

A (transversal) intersection of a Lagrangian submanifold $L \subset M$ with a Hamiltonian level set $H_c = h^{-1}(c)$ is an isotropic submanifold L_c. All Hamiltonian trajectories emanating from L_c form a Lagrangian submanifold $exp_H(L_c) \subset M$. The space Ξ_{H_c} of the Hamiltonian trajectories on H_c inherits, at least locally, an induced symplectic structure. The image of the projection of $exp_H(L_c)$ to Ξ_{H_c} is a Lagrangian submanifold there. This is a particular case of a symplectic reduction which will be discussed later.

Example. The set of all oriented straight lines in $\mathbf{R}^n_q$ is T^*S^{n-1} as a space of characteristics of the Hamiltonian $h = p^2$ on its level $p^2 = 1$ in $K = \mathbf{R}^{2n}$.

1.2. Contact geometry

An odd-dimensional manifold M^{2n+1} equipped with a maximally non-integrable distribution of hyperplanes (contact elements) in the tangent spaces of its points is called a contact manifold.

The maximal non-integrability means that if locally the distribution is determined by zeros of a 1-form α on M then $\alpha \wedge (d\alpha)^n \neq 0$ (cf. the Frobenius condition $\alpha \wedge d\alpha = 0$ of complete integrability).

Examples

1. A projectivised cotangent bundle PT^*N^{n+1} with the projectivisation of the Liouville form $\alpha = pdq$. This is also called the space of contact elements on N. The spherisation of PT^*N^{n+1} is a 2-fold covering of PT^*N^{n+1} and its points are co-oriented contact elements.

2. The space J^1N of 1-jets of functions on N^n. (Two functions have the same m-jet at a point x if their Taylor polynomials of degree k at x coincide). The space of all 1-jets at all points of N has local co-ordinates $q \in N$, $p = df(q)$ which are the partial derivatives of a function at q, and $z = f(q)$. The contact form is $pdq - dz$.

Contactomorphisms are diffeomorphisms preserving the distribution of contact elements.

Contact Darboux theorem. *All equidimensional contact manifolds are locally contactomorphic.*

An analog of Givental's theorem also holds.

Symplectisation

Let $\widetilde{M}^{2n+2}$ be the space of all linear forms vanishing on contact elements of M. The space $\widetilde{M}^{2n+2}$ is a "line" bundle over M (fibres do not contain the zero forms). Let

$$\widetilde{\pi} : \widetilde{M} \to M$$

be the projection. On $\widetilde{M}$, the symplectic structure (which is homogeneous of degree 1 with respect to fibres) is the differential of the canonical 1-form $\widetilde{\alpha}$ on $\widetilde{M}$ defined as

$$\widetilde{\alpha}(\xi) = p(\widetilde{\pi}_*\xi)\,, \qquad \xi \in T_p\widetilde{M}\,.$$

A contactomorphism F of M lifts to a symplectomorphism of $\widetilde{M}$:

$$\widetilde{F}(p) := (F^*_{F(x)})^{-1}p\,.$$

This commutes with the multiplication by constants in the fibres and preserves $\widetilde{\alpha}$. The symplectisation of contact vector fields (= infinitesimal contactomorphisms) yields Hamiltonian vector fields with homogeneous (of degree 1) Hamiltonian functions $h(rx) = rh(x)$.

Assume the contact structure on M is defined by zeros of a fixed 1-form β. Then M has a natural embedding $x \mapsto \beta_x$ into $\widetilde{M}$.

Using the local model $J^1\mathbf{R}^n$, $\beta = pdq - dz$, of a contact space we get the following formulas for components of the contact vector field with a homogeneous Hamiltonian function $K(x) = h(\beta_x)$ (notice that $K = \beta(X)$ where X is the corresponding contact vector field):

$$\dot{z} = pK_p - K, \quad \dot{p} = -K_q - pK_z, \quad \dot{q} = K_p,$$

where the subscripts mean the partial derivations.

Various homogeneous analogs of symplectic properties hold in contact geometry (the analogy is similar to that between affine and projective geometries).

In particular, a hypersurface (transversal to the contact distribution) in a contact space inherits a field of characteristics.

Contactisation

To an exact symplectic space M^{2n} associate $\widehat{M} = \mathbf{R} \times M$ with an extra co-ordinate z and take the 1-form $\alpha = \lambda - dz$. This gives a contact space.
Here the vector field $\chi = -\frac{\partial}{\partial z}$ satisfies $i_\chi \alpha = 1$ and $i_\chi d\alpha = 0$. Such a field is called a Reeb vector field. Its direction is uniquely defined by a contact structure. It is transversal to the contact distribution. Locally, projection along χ produces a symplectic manifold.

A Legendrian submanifold $\widehat{L}$ of M^{2n+1} is an n-dimensional integral submanifold of the contact distribution. This dimension is maximal possible for integral submanifolds.

Examples

1. To a Lagrangian $L \subset T^*M$ associate $\widehat{L} \subset J^1M$:

$$\widehat{L} = \{(z, p, q) \mid z = \int pdq, \ (p, q) \in L\}.$$

Here the integral is taken along a path on L joining a distinguished point on L with the point (p, q). Such an $\widehat{L}$ is Legendrian.

2. The set of all covectors annihilating tangent spaces to a given submanifold (or variety) $W_0 \subset N$ form a Legendrian submanifold (variety) in PT^*N.

3. If the intersection I of a Legendrian submanifold $\widehat{L}$ with a hypersurface Γ in a contact space is transversal, then I is transversal to the characteristic vector field on Γ. The set of characteristics emanating from I form a Legendrian submanifold.

A Legendrian fibration of a contact space is a fibration with Legendrian fibres. For example, $PT^*N \to N$ and $J^1N \to J^0N$ are Legendrian. Any two Legendrian fibrations of the same dimension are locally contactomorphic.

The projection of an embedded Legendrian submanifold $\widehat{L}$ to the base of a Legendrian fibration is called a Legendrian mapping. Its image is called the wave front of $\widehat{L}$.

Examples

1. Embed a Legendrian submanifold $\widehat{L}$ into J^1N. Its projection to J^0N, wave front $W(\widehat{L})$, is a graph of a multivalued action function $\int pdq + c$ (again we integrate along paths on the Lagrangian submanifold $L = \pi_1(\widehat{L})$, where $\pi_1 : J^1N \to T^*N$ is the projection dropping the z co-ordinate). If $q \in N$ is not in the caustic Σ_L of L, then over q the wave front $W(\widehat{L})$ is a collection of smooth sheets.

If at two distinct points $(p', q), (p'', q) \in L$ with a non-caustical value q, the values z of the action function are equal, then at (z, q) the wave front is a transversal intersection of graphs of two regular functions on N.

The images under the projection $(z,q) \mapsto q$ of the singular and transversal self-intersection loci of $W(\widehat{L})$ are respectively the caustic Σ_L and so-called Maxwell (conflict) set.

2. To a function $f = f(q)$, $q \in \mathbf{R}^n$, associate its Legendrian lifting $\widehat{L} = j^1(f)$ (also called the 1-jet extension of f) to $J^1\mathbf{R}^n$. Project $\widehat{L}$ along the fibres parallel to the q-space of another Legendrian fibration

$$\pi_1^\wedge(z,p,q) \mapsto (z - pq, p)$$

of the same contact structure $pdq - dz = -qdp - d(z - pq)$. The image $\pi_1^\wedge(\widehat{L})$ is called the Legendre transform of the function f. It has singularities if f is not convex.

This is an affine version of the projective duality (which is also related to Legendrian mappings). The space PT^*P^n (P^n is the projective space) is isomorphic to the projectivised cotangent bundle $PT^*P^{n\wedge}$ of the dual space $P^{n\wedge}$. Elements of both are pairs consisting of a point and a hyperplane, containing the point. The natural contact structures coincide. The set of all hyperplanes in P^n tangent to a submanifold $S \subset P^n$ is the front of the dual projection of the Legendrian lifting of S.

Wave front propagation

Fix a submanifold $W_0 \subset N$. It defines the (homogeneous) Lagrangian submanifold $L_0 \subset T^*N$ formed by all covectors annihilating tangent spaces to W_0.

Consider now a Hamiltonian function $h : T^*N \to \mathbf{R}$. Let I be the intersection of L_0 with a fixed level hypersurface $H = h^{-1}(c)$. Consider the Lagrangian submanifold $L = exp_H(I) \subset H$ which consists of all the characteristics emanating from I. It is invariant under the flow of H.

The intersections of the Legendrian lifting $\widehat{L}$ of L into J^1N ($z = \int pdq$) with co-ordinate hypersurfaces $z = const$ project to Legendrian submanifolds (varieties) $\widehat{L}_z \subset PT^*N$. In fact, the form pdq vanishes on each tangent vector to $\widehat{L}_z$. In general, the dimension of $\widehat{L}_z$ is $n-1$.

The wave front of $\widehat{L}$ in J^0N is called the big wave front. It is swept out by the family of fronts W_z of the $\widehat{L}_z$ shifted to the corresponding levels of the z-co-ordinate. Notice that, up to a constant, the value of z at a point over a point (p,q) is equal to $z = \int p\frac{\partial h}{\partial p}dt$ along a segment of the Hamiltonian trajectory going from the initial I to (p,q).

When h is homogeneous of degree k with respect to p in each fibre, then $z_t = kct$. Let $I_t \subset L$ be the image of I under the flow transformation g_t for time t. The projectivised I_t are Legendrian in PT^*N. The family of their fronts in N is $\{W_{kct}\}$. So the W_t are momentary wave fronts propagating from the initial W_0. Their singular loci sweep out the caustic Σ_L.

The case of a time-depending Hamiltonian $h = h(t,p,q)$ reduces to the above by considering the extended phase space $J^1(N \times \mathbf{R})$, $\alpha = pdq - rdt - dz$. The image of the initial Legendrian subvariety $\widehat{L}_0 \subset J^1(N \times \{0\})$ under g_t is a Legendrian $L_t \subset J^1(N \times \{t\})$.

When z can be written locally as a regular function in q, t it satisfies the Hamilton-Jacobi equation $-\frac{\partial z}{\partial t} + h(t, \frac{\partial z}{\partial q}, q) = 0$.

2. Generating families

2.1. Lagrangian case

Consider a co-isotropic submanifold $C^{n+k} \subset M^{2n}$. The skew-orthogonal complements $T_c^{\angle}C$, $c \in C$, of tangent spaces to C define an integrable distribution on C. Indeed, take two regular functions whose common zero level set contains C. At each point $c \in C$, the vectors of their Hamiltonian fields belong to $T_c^{\angle}C$. So the corresponding flows commute. Trajectories of all such fields emanating from $c \in C$ form a smooth submanifold I_c integral for the distribution.

By Givental's theorem, any co-isotropic submanifold is locally symplectomorphic to a co-ordinate subspace $p_I = 0$, $I = \{1, \dots, n-k\}$, in $K = \mathbf{R}^{2n}$. The fibres are the sets $q_J = \text{const}$.

Proposition. *Let L^n and C^{n+k} be respectively Lagrangian and co-isotropic submanifolds of a symplectic manifold M^{2n}. Assume L meets C transversally at a point a. Then the intersection $X_0 = L \bigcap C$ is transversal to the isotropic fibres I_c near a.*

The proof is immediate. If $T_a X_0$ contains a vector $v \in T_a I_c$, then v is skew-orthogonal to $T_a L$ and also to $T_a C$, that is to any vector in $T_a M$. Hence $v = 0$.

Isotropic fibres define the fibration $\xi : C \to B$ over a certain manifold B of dimension $2k$ (defined at least locally). We can say that B is the manifold of isotropic fibres.

It has a well-defined induced symplectic structure ω_B. Given any two vectors u, v tangent to B at a point b take their liftings, that is vectors $\widetilde{u}, \widetilde{v}$ tangent to C at some point of $\xi^{-1}(b)$ such that their projections to B are u and v. The value $\omega(\widetilde{u}, \widetilde{v})$ depends only on the vectors u, v. For any other choice of liftings the result will be the same. This value is taken for the value of the two-form ω_B on B.

Thus, the base B gets a symplectic structure which is called a symplectic reduction of the co-isotropic submanifold C.

Example. Consider a Lagrangian section L of the (trivial) Lagrangian fibration $T^*(\mathbf{R}^k \times \mathbf{R}^n)$. The submanifold L is the graph of the differential of a function $f = f(x, q)$, $x \in \mathbf{R}^k$, $q \in \mathbf{R}^n$. The dual co-ordinates y, p are given on L by $y = \frac{\partial f}{\partial x}$, $p = \frac{\partial f}{\partial q}$. Therefore, the intersection $\widetilde{L}$ of L with the co-isotropic subspace $y = 0$ is given by the equations $\frac{\partial f}{\partial x} = 0$. The intersection is transversal iff the rank of the matrix of the derivatives of these equations, with respect to x and q, is k. If so, the symplectic reduction of $\widetilde{L}$ is a Lagrangian submanifold L_r in $T^*\mathbf{R}^n$ (it may not be a section of $T^*\mathbf{R}^n \to \mathbf{R}^n$).

This example leads to the following definition of a generating function (the idea is due to Hörmander).

Definition. A generating family of the Lagrangian mapping of a submanifold $L \subset T^*N$ is a function $F : E \to \mathbf{R}$ defined on a vector bundle E over N such that

$$L = \left\{ \quad (p, q) \quad | \quad \exists x : \frac{\partial F(x, q)}{\partial x} = 0, \quad p = \frac{\partial F(x, q)}{\partial q} \quad \right\}.$$

Here $q \in N$, and x is in the fibre over q. We also assume that the following Morse condition is satisfied:

$$0 \text{ is a regular value of the mapping} \quad (x, q) \mapsto \frac{\partial F}{\partial x}.$$

The latter guarantees L being a smooth manifold.

Remark. The points of the intersection of L with the zero section of T^*N are in one-to-one correspondence with the critical points of the function F. In symplectic topology, when interested in such points, it is desirable to avoid a possibility of having no critical points at all (as it may happen on a non-compact manifold E).

Therefore, dealing with global generating families defining Lagrangian submanifolds globally, generating families with good behavior at infinity should be considered.

A generating family F is said to be quadratic at infinity (QI) if it coincides with a fibre-wise quadratic non-degenerate form $Q(x, q)$ outside a compact.

On the topological properties of such families and on their rôle in symplectic topology see the papers by C.Viterbo, for example [5].

Existence and uniqueness (up to a certain equivalence relation) of QI generating families for Lagrangian submanifolds which are Hamiltonian isotopic to the zero section in T^*N of a compact N was proved by Viterbo, Laundeback and Sikorav in the 80s:

Given any two QI generating families for L, there is a unique integer m and a real ℓ such that $H^k(F_b, F_a) = H^{k-m}(F_{b-\ell}, F_{a-\ell})$ for any pair of $a < b$. Here F_a is the inverse image under F of the ray $\{t \leq a\}$.

However, we shall need a local result which is older and easier.

Existence

Any germ L of a Lagrangian submanifold in $T^*\mathbf{R}^n$ has a regular projection to some (p_J, q_I) co-ordinate space. In this case there exists a function $f = f(p_J, q_I)$ (defined up to a constant) such that

$$L = \left\{ \quad (p, q) \quad | \quad q_J = -\frac{\partial f}{\partial p_J}, \quad p_I = \frac{\partial f}{\partial q_I} \quad \right\}.$$

Then the family $F_J = xq_J + f(x, q_I)$, $x \in \mathbf{R}^{|J|}$, is generating for L. If $|J|$ is minimal possible, then $\mathrm{Hess}_{xx} F_J = \mathrm{Hess}_{p_J p_J} f$ vanishes at the distinguished point.

Uniqueness

Two family-germs $F_i(x,q)$, $x \in \mathbf{R}^k$, $q \in \mathbf{R}^n$, $i = 1, 2$, at the origin are called $\mathcal{R}_0$-equivalent if there exists a diffeomorphism $\mathcal{T} : (x,q) \mapsto (X(x,q), q)$ (i.e., preserving the fibration $\mathbf{R}^k \times \mathbf{R}^n \to \mathbf{R}^n$) such that $F_2 = F_1 \circ \mathcal{T}$.

The family $\Phi(x,y,q) = F(x,q) \pm y_1^2 \pm \cdots \pm y_m^2$ is called a stabilisation of F.

Two family-germs are called stably $\mathcal{R}_0$-equivalent if they are $\mathcal{R}_0$-equivalent to appropriate stabilisations of the same family (in a lower number of variables).

Lemma. *Up to addition of a constant, any two generating families of the same germ L of a Lagrangian submanifold are stably $\mathcal{R}_0$-equivalent.*

Proof. Morse Lemma with parameters implies that any function-germ $F(x,q)$ (with zero value at the origin which is taken as the distinguished point) is stably $\mathcal{R}_0$-equivalent to $\widetilde{F}(y,q) \pm z^2$ where $x = (y,z)$ and the matrix $\mathrm{Hess}_{yy}\widetilde{F}|_0$ vanishes. Clearly $\widetilde{F}(y,q)$ is a generating family for L if we assume that $F(x,q)$ is.

Since the matrix $\partial^2\widetilde{F}/\partial y^2$ vanishes at the origin, the Morse condition for $\widetilde{F}$ implies that there exists a subset J of indices such that the minor $\partial^2\widetilde{F}/\partial y \partial q_J$ is not zero at the origin. Hence the mapping

$$\Theta : (y,q) \mapsto (p_J, q) = (\partial\widetilde{F}/\partial q_J, q)$$

is a local diffeomorphism. The family $G = \widetilde{F} \circ \Theta^{-1}$, $G = G(p_J, q)$, is also a generating family for L.

The variety $\partial\widetilde{F}/\partial y = 0$ in the domain of Θ is mapped to the Lagrangian submanifold L in the (p,q)-space by setting $p = \partial\widetilde{F}/\partial q$ and forgetting y. Therefore, the variety $X = \{\partial G/\partial p_J = 0\}$ in the (p_J, q)-space is the image of L under its (regular) projection $(p,q) \mapsto (p_J, q)$.

Compare now G and the standard generating family F_J defined above (with p_J in the role of x). We may assume their values at the origin coinciding. Then the difference $G - F_J$ has vanishing 1-jet along X. Since X is a regular submanifold, $G - F_J$ is in the square of the ideal $\mathcal{I}$ generated by the equations of X, that is by $\partial F_J/\partial p_J$.

The homotopy method applied to the family $A_t = F_J + t(G - F_J)$, $0 \le t \le 1$, shows that G and F_J are $\mathcal{R}_0$-equivalent. Indeed, it is clear that the homological equation

$$-\frac{\partial A_t}{\partial t} = F_J - G = \frac{\partial A_t}{\partial p_J}\dot{p}_J$$

has a smooth solution $\dot{p}_J$ since $F_J - G \in \mathcal{I}^2$ while the $\partial A_t/\partial p_J$ generate $\mathcal{I}$ for any fixed t. □

2.2. Legendrian case

Definition. A generating family of the Legendrian mapping $\pi|_L$ of a Legendrian submanifold $L \subset PT^*(N)$ is a function $F : E \to \mathbf{R}$ defined on a vector bundle E

over N such that

$$L = \left\{ (p,q) \mid \exists x : \quad F(x,q) = 0\,, \quad \frac{\partial F(x,q)}{\partial x} = 0\,, \quad p = \frac{\partial F(x,q)}{\partial q} \right\},$$

where $q \in N$ and x is in the fibre over q, provided that the following Morse condition is satisfied:

$$0 \text{ is a regular value of the mapping } (x,q) \mapsto \{F, \frac{\partial F}{\partial x}\}\,.$$

Definition. Two function family-germs $F_i(x,q)$, $i = 1,2$, are called V-equivalent if there exists a fibre-preserving diffeomorphism $\Theta : (x,q) \mapsto (X(x,q),q)$ and a function $\Psi(x,q)$ not vanishing at the distinguished point such that $F_2 \circ \Theta = \Psi F_1$.

Two function families are called stably V-equivalent if they are stabilisations of a pair of V-equivalent functions (may be in a lower number of variables x).

Theorem. *Any germ $\pi|_L$ of a Legendrian mapping has a generating family. All generating families of a fixed germ are stably V-equivalent.*

Proof. For an n-dimensional N, we use the local model $\pi_0 : J^1N' \to J^0N'$, $N' = \mathbf{R}^{n-1}$, for the Legendrian fibration.

Consider the projection $\pi_1 : J^1N' \to T^*N'$ restricted to L. Its image is a Lagrangian germ $L_0 \subset T^*N$. If $F(x,q)$ is a generating family for L_0, then $F(x,q) - z$ considered as a family of functions in x with parameters $(q,z) \in J^0N' = N$ is a generating family for L and vice versa. Now the theorem follows from the Lagrangian result and an obvious property: multiplication of a Legendrian generating family by a function-germ not vanishing at the distinguished point gives a generating family. After multiplication by an appropriate function Ψ, a generating family (satisfying the regularity condition) takes the form $F(x,q) - z$ where (q,z) are local co-ordinates in N. □

Remarks

A symplectomorphism φ preserving the bundle structure of the standard Lagrangian fibration $\pi : T^*\mathbf{R}^n \to \mathbf{R}^n$, $(q,p) \mapsto q$ has a very simple form

$$\varphi : (q,p) \mapsto \left(Q(q), DQ^{-1*}(q)(p + df(q))\right)\,,$$

where $DQ^{-1*}(q)$ is the dual of the derivative of the inverse mapping of the base of the fibration, $Q \circ \pi = \pi \circ \varphi$, and f is a function on the base.

To see this, it is sufficient to write in the co-ordinates the equation $\varphi_*\lambda - \lambda = df$.

The above formula shows that fibres of any Lagrangian fibration possess a well-defined affine structure.

Consequently, a contactomorphism ψ of the standard Legendrian fibration $PT^*\mathbf{R}^n \to \mathbf{R}^n$ acts by projective transformations in the fibres:

$$\psi : (q,p) \mapsto (Q(q), DQ^{-1*}(q)p)\,.$$

Hence, there is a well-defined projective structure on the fibres of any Legendrian fibration.

We also see that Lagrangian equivalences act on generating families as $\mathcal{R}$-equivalences $(x,q) \mapsto (X(x,q), Q(q))$ and additions of function in parameters q.

Legendrian equivalences act on Legendrian generating families just as $\mathcal{R}$-equivalences.

We see that the results of this section relate local singularities of caustics and wave fronts to those of discriminants and bifurcation diagrams of families of functions depending on parameters. In particular, this explains the famous results of Arnold and Thom on the classification of stable singularities of low-dimensional wave fronts by the discriminants of the Weyl groups.

The importance of the constructions introduced above for various applications is illustrated by the examples of the next section.

2.3. Examples of generating families

1. Consider a Hamiltonian $h : T^*\mathbf{R}^n \to \mathbf{R}$ which is homogeneous of degree k with respect to the impulses p: $h(\tau p, q) = \tau^k h(p,q)$, $\tau \in \mathbf{R}$.

An initial submanifold $W_0 \subset \mathbf{R}^n$ (initial wave front) defines an exact isotropic $I \subset H_c = h^{-1}(c)$. Assume I is a manifold transversal to v_h. Put $c = 1$.

The exact Lagrangian flow-invariant submanifold $L = exp_h(I)$ is a cylinder over I with local co-ordinates $\alpha \in I$ and time t from a real segment (on which the flow is defined).

Assume that in a domain $U \subset T^*R^n \times \mathbf{R}$ the restriction to L of the phase flow g_t of v_h is given by the mapping $(\alpha, t) \mapsto (Q(\alpha,t), P(\alpha,t))$ with $\frac{\partial P}{\partial \alpha, t} \neq 0$. Then the following holds.

Proposition.

a) *The family* $F = P(\alpha,t)(q - Q(\alpha,t)) + kt$ *of functions in* α, t *with parameters* $q \in \mathbf{R}^n$ *is a generating family of* L *in the domain* U.

b) *For any fixed* t, *the family* $\widetilde{F}_t = P(\alpha,t)(q - Q(\alpha,t))$ *is a Legendrian generating family of the momentary wave front* W_t.

The proof is an immediate verification of the Hörmander definition using the fact that value of the form pdq on each vector tangent to $g_t(I)$ vanishes and on the vector v_h it is equal to $p\frac{\partial h}{\partial p} = kh = k$.

2. Let $\varphi : T^*\mathbf{R}^n \to T^*\mathbf{R}^n$, $(q,p) \mapsto (Q,P)$ be a symplectomorphism close to the identity. Thus the system of equations $q' = Q(q,p)$ is solvable for q. Write its solution as $q = \widetilde{q}(q', p)$.

Assume the Lagrangian mapping of a Lagrangian submanifold L has a generating family $F(x,q)$. Then the following family G of functions in x, q, p with parameters q' is a generating family of $\varphi(L)$:

$$G(x,p,q;q') = F(x,\widetilde{q}) + p(\widetilde{q} - q) + S(p,q')\,.$$

Here $S(q',p)$ is the "generating function" in the sense of Hamiltonian mechanics of the canonical transformation φ, that is

$$dS = PdQ - pdq\,.$$

Notice that, if φ coincides with the identity mapping outside a compact, then G is a quadratic form at infinity with respect to the variables (q,p).

The expression $p(\widetilde{q}-q)+S(p,q')$ from the formula above is the generating family of the symplectomorphism φ.

3. Represent a symplectomorphism φ of $T^*\mathbf{R}^n$ into itself homotopic to the identity as a product of a sequence symplectomorphisms each of which is close to the identity. Iterating the previous construction, we obtain a generating family of $\varphi(L)$ as a sum of the initial generating family with the generating families of each of these transformations. The number of the variables becomes very large, $\dim(x)+2mn$, where m is the number of the iterations. Namely, consider a partition of the time interval $[0,T]$ into m small segments $[t_i,t_{i+1}]$, $i=0,\ldots,m-1$. Let $\varphi=\varphi_m\circ\varphi_{m-1}\circ\cdots\circ\varphi_1$ where $\varphi_i:(Q_i,P_i)\mapsto(Q_{i+1},P_{i+1})$ is the flow map on the interval $[t_i,t_{i+1}]$. Then the generating family is

$$G(x,Q,P,q)=F(x,Q_0)+\sum_{i=0}^{m-1}\left(P_i(U(_iQ_{i+1},P_i)-Q_i)+S_i(P_i,Q_{i+1})\right),$$

where:

- $Q=Q_0,\ldots,\ Q_{m-1},\ q=Q_m,\quad Q_i\in\mathbf{R}^n,\quad q\in\mathbf{R}^n,$
- S_i is a generating function of φ_i,
- $U_i(Q_{i+1},P_i)$ are the solutions of the system of equations $Q_{i+1}=Q_{i+1}(Q_i,P_1)$ defined by φ_i.

One can show that if φ is a flow map for time $t=1$ of a Hamiltonian function which is convex with respect to the impulses then the generating family G is also convex with respect to the P_i and these variables can be removed by the stabilisation procedure. This provides a generating family of $\varphi(L)$ depending just on x,Q,q which are usually taken from a compact domain. Therefore, the function attains minimal and maximal values on the fibre over point q, this property being important in applications.

References

[1] V.I. Arnold, *Singularities of caustics and wave fronts*, Kluwer Academic Publ., Dordrecht-Boston-London, 1990.

[2] V.I. Arnold, V.V. Goryunov, O.V. Lyashko and V.A. Vassiliev, *Singularities II. Classification and Applications*, Encyclopaedia of Mathematical Sciences, vol. 39, Dynamical Systems VIII, Springer Verlag, Berlin a.o., 1993.

[3] V.I. Arnold, S.M. Gusein-Zade and A.N. Varchenko, *Singularities of Differentiable maps. Vol. I*, Monographs in Mathematics **82**, Birkhäuser, Boston, 1985.

[4] D. McDuff and D. Salomon, *Introduction to symplectic topology*, Springer Verlag, 1995.

[5] C. Viterbo, *Generating functions, symplectic geometry and applications*, Proc. Intern. Congr. Math., Zurich, 1994.

Victor V. Goryunov
Department of Mathematical Sciences
University of Liverpool
Liverpool L69 3BX, UK
e-mail: goryunov@liv.ac.uk

V.M. Zakalyukin
Department of Mechanocs and Mathematics
Moscow State University
Leninskie gory, 1
119992 Moscow, Russia
e-mail: vzakal@liv.ac.uk

Real and Complex Singularities
Trends in Mathematics, 187–199

F-manifolds from Composed Functions

Ignacio de Gregorio and David Mond

Mathematics Subject Classification (2000). Primary 32S30, 58K60; Secondary 53D45.

1. Introduction

In this note we show how to endow the base of a versal deformation of a composite singularity with an F-manifold structure, as defined by Hertling and Manin in [9], and in particular with a pointwise, integrable multiplication on the tangent bundle. This is closely related to, but not a special case of, K. Saito's construction of a Frobenius manifold structure on the base-space of a versal deformation of a function with isolated critical point.

Let us clarify the notion of versality we are concerned with. Consider a function $f : (Y, y_0) \to \mathbf{C}$ and a map-germ $F : (X, x_0) \to (Y, y_0)$. Deformations of F give rise to some, but not all, of the possible deformations of the composite $f \circ F$. In this context, a deformation $\mathscr{F}$ of F is versal if, up to the usual notion of equivalence, it contains all the deformations of $f \circ F$ which can be achieved by deforming F. A precise definition is given below. For now, we point out that even when $f \circ F$ has non-isolated singularity, F may have a finite-dimensional versal deformation in the sense being considered. Exactly this is the case in Damon's theory of almost free divisors, [4].

The ideas of the previous paragraph are made precise by means of the action on the space of map-germs $(X, x_0) \to (Y, y_0)$ of the subgroup $\mathcal{K}_f$ of the contact group $\mathcal{K}$. We give the definition at the start of Section 2. The $\mathcal{K}_f$-equivalence of F_1 and F_2 equivalence implies (though it is not implied by) right-equivalence of the composites $f \circ F_1$ and $f \circ F_2$. A $\mathcal{K}_f$-miniversal deformation $\mathscr{F} : (X \times S, (x_0, 0)) \to (Y, y_0)$ of $F : (X, x_0) \to (Y, y_0)$ can be constructed by the usual procedures of singularity theory; loosely speaking, the tangent space $T_0 S$ is isomorphic to the quotient $T^1_{\mathcal{K}_f} F$ of the space $\theta(F)$ of infinitesimal deformations of F by the tangent space to the $\mathcal{K}_f$-orbit of F.

If $f:(\mathbf{C}^n,0)\to\mathbf{C}$ is a germ of function with isolated singularity, $T^1 f := \mathcal{O}_{\mathbf{C}^n}/J_f$ is naturally a ring, and it is this that Saito uses to define the multiplicative structure on the tangent bundle of the base of a versal deformation of f. It turns out (see Section 2) that $T^1_{\mathcal{K}_f}F$ is equal to

$$\frac{\theta(F)}{tF(\theta_X)+F^*(\mathrm{Der}(-\log f))}$$

and does not appear to carry a multiplicative structure. However, contraction with $F^*(df)$ defines an $\mathcal{O}_X$-linear epimorphism

$$\theta(F)=F^*(\theta_Y)\to F^*(J_f)$$

which passes to the quotient to give a natural $\mathcal{O}_X$-linear epimorphism

$$T^1_{\mathcal{K}_f}F\to\frac{F^*(J_f)}{J_{f\circ F}},$$

whose target does have a multiplicative structure. Once again, let $\mathscr{F}:(X\times S,(x_0,0))\to(Y,y_0)$ be a $\mathcal{K}_f$-versal deformation of F. Write $F_s(x)=\mathscr{F}(x,s)$. A multiplicative structure on TS is therefore defined at those points $s\in S$ such that for all $x\in\operatorname{supp}T^1_{\mathcal{K}_f}F_s$, the epimorphism

$$\left(T^1_{\mathcal{K}_f}F_s\right)_x\longrightarrow\left(\frac{F_s^*(J_f)}{J_{f\circ F_s}}\right)_x \tag{1}$$

is injective. From Proposition 3.3 (due to Jim Damon) it follows that under certain rather weak conditions on the divisor $E:=f^{-1}(0)$,

(∗) there is a proper analytic subset B of S, such that for $s\in S-B$

$$\operatorname{supp}T^1_{\mathcal{K}_f}F_s\cap V(F_s^*(J_f))=\emptyset.$$

If $s\in S-B$, the image of (1) is all of $\mathcal{O}_{X,x}/J_{f\circ F_s}$. Less obviously, if $s\in S-B$ the morphism of (1) is also injective (Lemma 4.3 of [6]). We will use (1) and the relative Kodaira-Spencer map of the deformation to endow $T(S-B)$ with a well-defined multiplication with unit. In Section 3 we prove a transversality lemma which allows us to show that for s in a certain non-empty open set in S, the critical points of $f\circ F_s$ off $F_s^{-1}(E)$ are generically non-degenerate and that the critical values are generically pairwise distinct.

To transfer the multiplicative structure to the tangent sheaf of the base, the relative Kodaira-Spencer map

$$\theta_S\to\pi_*(T^1_{\mathcal{K}_f/S}\mathscr{F})$$

must be an isomorphism, and in particular $T^1_{\mathcal{K}_f/S}\mathscr{F}$ must be free over the base. Section 4 recalls the arguments given in [5] and [6] to prove freeness in three cases: where $E:=f^{-1}(0)$ is a free divisor, and where $\dim X$ is equal to $m_0:=\dim Y-\dim E_{\mathrm{Sing}}$ or to m_0-1.

If $\dim X\geq m_0$, the generic fibres $D_s=F_s^{-1}(E)$ will contain singularities; they are only *partial* smoothings of $D:=F^{-1}(E)$. The most extreme case is where

E is a (singular) free divisor. In this case, D_s, like E, will be singular in codimension 1. Nevertheless, in all cases every fibre D_s has the homotopy type of a wedge of spheres of middle dimension. This is because we have "triviality at the boundary" (recall that under the assumption that $\operatorname{supp} T^1_{\mathcal{K}_f} = \{0\}$, F is transverse to E away from 0), and thus the vanishing homology of D_s is accounted for by the isolated critical points which move off D as s moves away from 0 (cf [11]). The number of spheres in this wedge for a generic parameter-value s is called by Damon the "singular Milnor number" of D. We will denote it by $\mu_E(F)$. In Section 3, we show, following the argument of [5] Section 5, that if

(i) $T^1_{\mathcal{K}_f/S}\mathscr{F}$ is free over $\mathcal{O}_S$, and
(ii) condition $(*)$ holds,

then $\mu_E(F) = \dim_{\mathbf{C}} T^1_{\mathcal{K}_f}F$. Hence, in these favorable circumstances, our F-manifold $S - B$ supports a locally trivial holomorphic fibration whose fibre D_s has homology concentrated in middle dimension, where its rank is equal to the dimension of S. We expect that using the Gauss-Manin connection on this fibration it will be possible to go some way further towards endowing S with a Frobenius structure, but here we do not attempt this.

The authors are grateful to the referee for a very careful reading of the manuscript and a number of helpful comments.

2. Background and notation

Throughout this paper, X and Y will denote the germs $(\mathbf{C}^m, 0)$ and $(\mathbf{C}^n, 0)$ respectively. We consider a fixed map $f : Y \to \mathbf{C}$, and classify map-germs $X \to Y$ as follows: $F_1 : X \to Y$ and $F_2 : X \to Y$ are $\mathcal{K}_f$-equivalent if there exists a germ of diffeomorphism $\Phi : X \times Y \to X \times Y$ such that

(i) Φ covers a diffeomorphism $\phi : X \to X$ (i.e., there is a germ of diffeomorphism $\phi : X \to X$ such that $\pi_X \circ \Phi = \phi \circ \pi_X$)
(ii) Φ preserves the level sets of f; more precisely, $f \circ \pi_Y \circ \Phi = f \circ \pi_Y$, and
(iii) $\Phi(\operatorname{graph}(F_1)) = \operatorname{graph}(F_2)$.

Observe that $\mathcal{K}_f$ contains the group $\mathscr{R}$ of right-equivalences. One calculates that the extended tangent space to the group action on a germ $F : X \to Y$ is

$$T\mathcal{K}_f F = tF(\theta_X) + F^*(\operatorname{Der}(-\log f))$$

where $\operatorname{Der}(-\log f)$ is the $\mathcal{O}_Y$-module of germs of vector-fields tangent to all the level sets of f. We denote the quotient $\theta(F)/T\mathcal{K}_f F$ by $T^1_{\mathcal{K}_f}F$. It is easy to show that

$$F_1 \sim_{\mathcal{K}_f} F_2 \quad \Rightarrow f \circ F_1 \sim_{\mathscr{R}} f \circ F_2,$$

but the converse does not always hold.

The group $\mathcal{K}_f$ is geometric, in the sense of Damon [1], and so the usual properties hold; in particular, if $T^1_{\mathcal{K}_f}F$ has finite length then a deformation $\mathscr{F}$:

$X \times U \to Y$ of F is versal if the initial velocities $\partial\mathscr{F}/\partial s_i|_{s=0}$ generate $T^1_{\mathcal{K}_f}F$ over $\mathbf{C}$, and miniversal if they form a basis.

Closely related to $\mathcal{K}_f$ is the group $\mathcal{K}_E$, introduced by Damon in [2], in which part 2 of the definition above is weakened to the requirement that Φ preserve only the level set $X \times E$ of $f \circ \pi_Y$. It is an immediate consequence of Nakayama's lemma that

$$\operatorname{supp} T^1_{\mathcal{K}_E}F = \{x \in X : F \not\pitchfork E \text{ at } x\},$$

where transversality to E is understood to mean transversality to the distribution $\operatorname{Der}(-\log E)$.

Damon showed in [3] that if

$$\begin{array}{ccc} Y_0 & \xrightarrow{H} & Y_1 \\ \uparrow & & \uparrow F \\ X_0 & \xrightarrow{h} & X_1 \end{array}$$

is a fibre square in which H is a right-left stable map-germ with discriminant (or image, if $\dim Y_0 < \dim Y_1$) equal to E, and $F \pitchfork H$, then

$$T^1_{\mathcal{K}_E}F \simeq T^1_{h:X_0 \to X_1} := \frac{\theta(h)}{th(\theta_{X_0}) + \omega h(\theta_{X_1})}.$$

Our F-manifold structure is therefore closely related to the theory of right-left equivalence of map-germs, and of right-left versal unfoldings.

3. Transversality

To prove a number of properties of $\mathcal{K}_f$-versal deformations, we will use a local transversality lemma. Recall that $X^{(r)}$ is the subset of the r-fold cartesian product X^r consisting of r-tuples of pairwise distinct points, that ${}_rJ^k(X,Y)$ is the restriction of $(J^k(X,Y))^r$ to $X^{(r)}$, and that

$${}_rj^k_x\mathscr{F} : X^{(r)} \times S \to {}_rJ^k(X,Y),$$

the *relative r-fold multi-jet extension map*, is defined by

$${}_rj^k_x\mathscr{F}(x_1, \dots, x_r, s) = \big(j^kF_s(x_1), \dots, j^kF_s(x_r)\big).$$

Lemma 3.1. *Let $W \subset {}_rJ^k(X,Y)$ be a $\mathcal{K}_f$ invariant submanifold. If $\mathscr{F} : X \times S \to Y$ is a $\mathcal{K}_f$-versal deformation of a germ $F : X \to Y$, then ${}_rj^k_x\mathscr{F} \pitchfork W$.*

Proof. It is possible to find a deformation $\widetilde{\mathscr{F}} : X \times S \times U \to Y$ of $\mathscr{F}$ such that ${}_rj^k_x\widetilde{\mathscr{F}} : X \times S \times U \to J^k(X,Y)$ is transverse to W. For example, identifying X and Y with $\mathbf{C}^m$ and $\mathbf{C}^n$ respectively, we take as U the space of polynomial maps $p : \mathbf{C}^m \to \mathbf{C}^n$ with each component of degree $\leq N$, and define

$$\widetilde{\mathscr{F}}(x, s, p) = \mathscr{F}(x, s) + p(x).$$

If N is sufficiently large then ${}_rj^k_x\widetilde{\mathscr{F}}$ is a submersion, and in particular transverse to W.

As $\mathscr{F}$ is $\mathcal{K}_f$-versal, so is $\widetilde{\mathscr{F}}$. Versality of $\mathscr{F}$ implies that $\widetilde{\mathscr{F}}$ is $\mathcal{K}_{f\mathrm{un}}$-equivalent to the deformation $i^*\mathscr{F}$ induced from $\mathscr{F}$ by some map of base-spaces $i : S\times U \to S$. Versality of $\widetilde{\mathscr{F}}$ implies that i is a submersion, and in particular locally surjective. As W is $\mathcal{K}_f$-invariant, the transversality of $j^k_x\widetilde{\mathscr{F}}$ to W implies that $j^k_x i^*(\mathscr{F}) : X \times S \times U \to J^k(X,Y)$ is also transverse to W. But $j^k_x i^*(\mathscr{F}) = (j^k_x\mathscr{F})\circ(\mathrm{id}_X \times i)$. As $\mathrm{id}_X \times i$ is surjective, it follows that $j^k_x\mathscr{F}$ is transverse to W. □

In what follows, when we consider a map $F : X \to Y$, we write $D = V(f\circ F)$. If $\mathscr{F} : X\times S \to Y$ is a deformation of F, we write $F_s(x) := \mathscr{F}(x,s)$, $\mathscr{D} := \mathscr{F}^{-1}(E)$ and $D_s = F_s^{-1}(E)$.

From our transversality lemma we derive first a statement about the behavior of perturbations F_s of F off the zero set D_s of $f\circ F_s$. The reason we want this is that typically, D_s will have non-isolated singularities.

Here is a simple example. Define $f : \mathbf{C}^4 \to \mathbf{C}$ by $f(y_1,y_2,y_3,y_4) = y_1y_2y_3y_4$, let $E = f^{-1}(0)$ and let $F : \mathbf{C}^3 \to \mathbf{C}^4$ be given by $F(x_1,x_2,x_3) = (x_1,x_2,x_3,x_1 + x_2 + x_3)$. $\mathrm{Der}(-\log E)$ is well known to be generated by the vector fields $y_i\partial/\partial y_i$ for $i = 1,\dots,4$, and $\mathrm{Der}(-\log f)$ consists of all linear combinations $\sum a_iy_i\partial/\partial y_i$ where $\sum_i a_i = 0$. Thus $T\mathcal{K}_fF$ is generated over $\mathcal{O}_X$ by

$$\frac{\partial}{\partial y_1} + \frac{\partial}{\partial y_4},\ \frac{\partial}{\partial y_2} + \frac{\partial}{\partial y_4},\ \frac{\partial}{\partial y_3} + \frac{\partial}{\partial y_4}$$

and by

$$x_1\frac{\partial}{\partial y_1} - x_2\frac{\partial}{\partial y_2},\ x_1\frac{\partial}{\partial y_1} - x_3\frac{\partial}{\partial y_3},\ x_1\frac{\partial}{\partial y_1} - (x_1+x_2+x_3)\frac{\partial}{\partial y_4}.$$

The quotient $T^1_{\mathcal{K}_f}F$ has length 1, and is generated by the class of $\partial/\partial y_4$. The two drawings below show the real part of $D = F^{-1}(E)$ and $D_s = F_s^{-1}(E)$, where F_s is the deformation $F_s = F + s\partial/\partial y_4$, for $s < 0$. Both surfaces have non-isolated singularities; the defining equation $x_1x_2x_3(x_1+x_2+x_3+s) = 0$ of the second also has an isolated singularity at $(-s/4,-s/4,-s/4)$ with Milnor number 1, inside the chamber which has opened up as s moves away from zero.

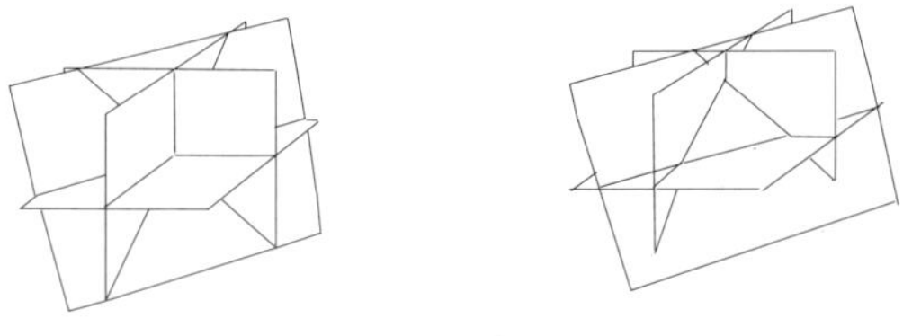

Proposition 3.2. *If $\mathscr{F} : X \times S \to Y$ is a $\mathcal{K}_f$-miniversal deformation of F, then*

(i) *the variety $\sum^{rel}_{f\circ\mathscr{F}}$ defined by the ideal $J^{rel}_{f\circ\mathscr{F}} := (\{\partial(f\circ\mathscr{F})/\partial x_i : i = 1,\dots,m\})$ is non-singular off $\mathscr{D}$.*

(ii) *for $s \in S - B_1$, each critical point of $f\circ F_s$ off D_s is non-degenerate, and*

(iii) *for $s \in S - B_2$,*

the values of $f\circ F_s$ at these critical points are all distinct.

Proof. Apply 3.1 taking as W the submanifold of $J^1(X,Y)$ consisting of 1-jets (x,y,A) with the property that $y \notin E$ and the image of the linear map A lies in $\ker d_y f$. For each $y \notin E$, $\ker d_y f$ has dimension $n-1$, and so the space of admissible matrices A has dimension $m(n-1)$. Thus the codimension of W in $J^1(X,Y)$ is m. Clearly $\sum^{\mathrm{rel}}_{f\circ\mathscr{F}-\mathscr{D}} = (j^1_x f \circ \mathscr{F})^{-1}(W)$ and is therefore smooth by 3.1. This proves (i).

Let $W_1 \subset J^2(X,Y)$ be the set consisting of jets $j^2H(x)$ such that $H(x) \notin E$, $d_x(f{\circ}H) = 0$ and the Hessian determinant of $f{\circ}H$ vanishes at x. Although not a manifold, W_1 is an analytic set and can be stratified. Its open stratum has codimension $m+1$. The set B_1 is the closure of $\pi_S\big((j^2\mathscr{F})^{-1}(W_1)\big)$.

The projection $\pi : \sum^{\mathrm{rel}}_{f\circ\mathscr{F}} - \mathscr{D} \to S$ is finite. If it were not, then for some $s \in S$, $f{\circ}F_s$ would have a non-isolated singularity off D_s. But the length of $T^1_{\mathcal{K}_f}F_s$ is upper semi-continuous, and for $s=0$ it is finite. Thus, B_1 is a hypersurface.

To ensure that the critical values of the critical points off the zero level are all distinct, let us consider the submanifold $W_2 \subset {}_2J^1(X,Y)$ consisting of jets $(x_1,y_1,A_1,x_2,y_2,A_2)$ such that $f(y_1) = f(y_2) \neq 0$, and $d_{y_i}f{\circ}A_i = 0$ for $i=1,2$. As $\mathcal{K}_f$ leaves the level sets of f unchanged, W_2 is indeed $\mathcal{K}_f$-invariant. The codimension of W_2 in ${}_2J^1(X,Y)$ is $2m+1$, so transversality of ${}_2j^1_x\mathscr{F}$ to W_2 means that the set $(x_1,x_2,s) \in X^{(2)} \times S$ such that x_1, x_2 are critical points of $f{\circ}F_s$ not in D_s and with equal critical values, is empty or has dimension $\dim S - 1$. In particular, the closure B_2 of its projection to S is a hypersurface (or empty), and if s is not in B_2 then the values of $f{\circ}F_s$ at its critical points off D_s are pairwise distinct. □

A divisor E is *holonomic* at x if the logarithmic partition of E is locally finite (and thus a stratification) in some neighborhood U of E. Holonomicity is an analytic condition, and thus the set of points where it fails is an analytic subset of E. We say that E is *holonomic in codimension* k if this subset has codimension at least $k+1$.

In similar vein, E is *Euler-homogeneous* at x if there is a germ of vector field χ at x such that

(i) $\chi(x) = 0$, and
(ii) $\chi \cdot f = f$ (where f is a reduced equation for E).

Note that this property is independent of the choice of defining equation f. We say that E is *strongly Euler-homogeneous in codimension* k if there is a Whitney stratification of E such that E is Euler-homogeneous at every point of each stratum of codimension $\leq k$.

Proposition 3.3. (*J.N. Damon,* [4]) *Suppose that $E = V(f)$ is holonomic and strongly Euler-homogeneous in codimension m. Then there is a proper analytic subset Δ of the semi-universal base-space S of $\mathscr{F}$ such that if $s \notin \Delta$,*

$$\operatorname{supp}\big(T^1_{\mathcal{K}_f}F_s\big) \cap D_s = \emptyset.$$

Proof. Let $\mathscr{S} = \{E_\alpha\}$ be a Whitney stratification of E, in which each stratum of codimension $\leq m$ is logarithmic. Any $\mathcal{K}_f$-versal deformation of F is logarithmically transverse to E, and thus is transverse to $\mathscr{S}$. By Sard's Theorem, the set Δ_α of critical values in S of the projection $\mathscr{F}^{-1}(E_\alpha) \to S$ has measure zero, and so also does $\Delta := \bigcup_\alpha \Delta_\alpha$. If $s \notin \Delta$ then F_s is transverse to each stratum E_α of $\mathscr{S}$. Since E is holonomic in codimension m, this means that F_s meets only holonomic strata of E, and so in fact F_s is logarithmically transverse to E itself.

For $s \in S - \Delta$, we have

$$d_x F_s(T_x X) + T^{\log}_{F_s(x)} E = T_{F_s(x)} Y \tag{2}$$

for all $x \in F_s^{-1}(E)$. Let be a germ of Euler vector field, vanishing at $F_s(x)$, such that $\chi \cdot f = f$. Then $\mathrm{Der}(-\log E)_{F_s(x)} = \langle \chi \rangle_{F_s(x)} + \mathrm{Der}(-\log f)_{F_s(x)}$, where $\langle \chi \rangle$ is the $\mathcal{O}_Y$ module generated by χ. Since χ vanishes at $F_s(x)$, from (2) we obtain

$$d_x F_s(T_x X) + \mathrm{Der}(-\log f)(F_s(x)) = T_{F_s(x)} Y \tag{3}$$

and now Nakayama's Lemma gives $(T\mathcal{K}_f F_s)_x = \theta(F_s)_x$. □

Corollary 3.4. *With the condition of Proposition* 3.3, *suppose that* $\mathscr{F} : X \times S \to Y$ *is a* $\mathcal{K}_f$*-versal deformation of* $F : X \to Y$. *If also* $T^1_{\mathcal{K}_f/S}\mathscr{F}$ *is free over* S, *then the generic fibre* $D_s := F_s^{-1}(E)$ *has the homotopy type of a wedge of* $\tau = \dim S$ *spheres of dimension* $m - 1$.

Proof. By a theorem of Siersma, ([11]), D_s has the homotopy type of a wedge of $(m-1)$-spheres, whose number is equal to the sum of the Milnor numbers of the critical points of $f \circ F_s$ which move off D_s as s moves off 0. By the hypothesis of freeness, we have

$$\tau := \dim_{\mathbf{C}} T^1_{\mathcal{K}_f} F = \sum_i \dim_{\mathbf{C}} (T^1_{\mathcal{K}_f} F_s)_{x_i}$$

where the x_i are the points of the support of $T^1_{\mathcal{K}_f} F_s$. These fall into two subsets: those in D_s and those outside it. If $s \in S - \Delta$, the first subset is empty.

If $x_i \notin D_s$ then the morphism (1)

$$F^*(df) : (T^1_{\mathcal{K}_f} F_s)_{x_i} \to \mathcal{O}_{X,x_i}/J_{f \circ F_s}$$

is an isomorphism. It follows that if $s \notin \Delta$, all points of $\mathrm{supp}\, T^1_{\mathcal{K}_f} F_s$ are outside D_s and moreover

$$\tau = \sum_i \dim_{\mathbf{C}} \mathcal{O}_{X.x_i}/J_{f \circ F_s}. \qquad \square$$

Let

$$B = \{s \in S : \mathrm{supp}\big(T^1_{\mathcal{K}_f} F_s\big) \cap V\big(\mathscr{F}^*(J_f)\big) \neq \emptyset\}.$$

Since $V\big(F_s^*(J_f)\big) \subset D_s$, (3.3) assures us that if $V(f)$ is holonomic and strongly Euler homogeneous in codimension m, then B is a proper analytic subset of S.

4. When is the relative T^1 free over the base?

Suppose that $\mathcal{O}_Y/J_f$ is Cohen-Macaulay, and let m_0 be the codimension of $V(J_f)$ in Y. Let $\mathscr{F} : X \times S \to Y$ be a deformation of a germ $F : X \to Y$ for which $\operatorname{supp} T^1_{\mathcal{K}_f} F = \{0\}$. There are three cases where we can show that the relative module $T^1_{\mathcal{K}_f/S}\mathscr{F}$ is free over S. These are

(i) where f is a defining equation for a free divisor E, such that $f \in J_f$,
(ii) where $\dim X = m_0$, and
(iii) where $\dim X = m_0 - 1$.

Proofs of all three are straightforward, and may be found, for example, in [5], [6], though for completeness we sketch them here.

1. *Free divisors:* $\operatorname{Der}(-\log f)$ is a direct summand of the free module $\operatorname{Der}(-\log E)$, with complementary summand generated by a vector field χ such that $\chi \cdot f = f$, and so is free on $n-1$ generators. Thus, the presentation

$$\theta_{X\times S/S} \oplus \mathscr{F}^*(\operatorname{Der}(-\log f)) \to \theta(\mathscr{F}) \to T^1_{\mathcal{K}_f/S}\mathscr{F} \to 0$$

can be read as

$$\mathcal{O}^m_{X\times S} \oplus \mathcal{O}^{n-1}_{X\times S} \to \mathcal{O}^n_{X\times S} \to T^1_{\mathcal{K}_f/S}\mathscr{F} \to 0. \tag{4}$$

Since $\operatorname{supp} T^1_{\mathcal{K}_f/S}\mathscr{F}$ is finite over S, $\dim T^1_{\mathcal{K}_f/S}\mathscr{F} \leq \dim S$; on the other hand, from (4) it follows that the codimension of $\operatorname{supp} T^1_{\mathcal{K}_f/S}\mathscr{F}$ is no greater than dim X. Thus $\dim T^1_{\mathcal{K}_f/S}\mathscr{F} = \dim S$, and now from (4) it follows, by the theorem of Eagon-Northcott, that $T^1_{\mathcal{K}_f/S}\mathscr{F}$ has depth equal to dim S. Since it is finite over $\mathcal{O}_S$, it is a free $\mathcal{O}_S$-module.

2. *The cases $m = m_0$ and $m = m_0 - 1$*

Lemma 4.1. *If $m \leq m_0$ and $T^1_{\mathcal{K}_f}F$ has finite length then $f\circ F$ has an isolated singularity.*

Proof. First, because $\operatorname{supp} T^1_{\mathcal{K}_f} F = \{0\}$, the restriction of F to $X - \{0\}$ is transverse to every level set of f. We have $\dim X \leq \operatorname{codim} V(J_f) \leq \operatorname{codim} E_\alpha$ for each stratum E_α of any Whitney stratification of E contained in $V(J_f)$, and so $F^{-1}(E_\alpha)$ must consist of isolated points. Thus the germ of $F^{-1}(V(J_f))$ consists at most of $\{0\}$. At every point $x \notin F^{-1}(V(J_f))$, the transversality of F to the level set of f through $F(x)$ means that x is not a critical point of $f\circ F$. □

The multiplicity, μ, of the critical point of $f\circ F$ is preserved in any deformation. When $m = m_0$, the exact sequence of Corollary 1.3 of [6] reduces to

$$0 \to T^1_{\mathcal{K}_f}F \to \mathcal{O}_X/J_{f\circ F} \to \mathcal{O}_X/F^*(J_f) \to 0. \tag{5}$$

The lengths of the second and third non-trivial terms in this short exact sequence are conserved (the latter because $\mathcal{O}_Y/J_f$ is Cohen-Macaulay), and hence so is the length of the first. This implies that $T^1_{\mathcal{K}_f/S}\mathscr{F}$ is free over $\mathcal{O}_S$.

When $m = m_0 - 1$, the exact sequence acquires an extra term, and becomes

$$0 \to \mathrm{Tor}_1^{\mathcal{O}_Y}(\mathcal{O}_Y/J_f, \mathcal{O}_X) \to T^1_{\mathcal{K}_f}F \to \mathcal{O}_X/J_{f\circ F} \to \mathcal{O}_X/F^*(J_f) \to 0 \qquad (6)$$

An easy argument ([6] Lemma 4.3(i)) shows that the lengths of the modules $\mathrm{Tor}_1^{\mathcal{O}_Y}(\mathcal{O}_Y/J_f, \mathcal{O}_X)$ and $\mathcal{O}_X/F^*(J_f)$ are equal, so that the length of $T^1_{\mathcal{K}_f}F$ is equal to μ. As μ is conserved, so is the length of $T^1_{\mathcal{K}_f}F$, and so once again $T^1_{\mathcal{K}_f/S}\mathscr{F}$ is free over $\mathcal{O}_S$.

5. Multiplication on the tangent bundle of the base

Let $\mathscr{F}$ be a $\mathcal{K}_f$-miniversal deformation of some germ $F\colon X \to Y$ for which $T^1_{\mathcal{K}_f}F$ has finite length. Suppose that E is strongly Euler-homogeneous and holonomic in codimension m, and that $T^1_{\mathcal{K}_f/S}\mathscr{F}$ is free over S. By Prop. 3.3, there is a proper analytic subset B of the base space S, such that for $s \in S - B$, $\mathrm{supp}\, T^1_{\mathcal{K}_f}F_s$ does not meet $V\big(F_s^*(J_f)\big)$. For such s,

$$(T^1_{\mathcal{K}_f}F_s)_x \simeq \mathcal{O}_{X,x}/J_{f\circ F_s}$$

for each $x \in \mathrm{supp}\, T^1_{\mathcal{K}_f}F_s$, and indeed

$$\pi_*(T^1_{\mathcal{K}_f/S}\mathscr{F}) \simeq \pi_*(\mathcal{O}_{(X\times S)-\mathscr{D}}/J^{\mathrm{rel}}_{f\circ\mathscr{F}}). \qquad (7)$$

Because $T^1_{\mathcal{K}_f/S}\mathscr{F}$ is free over S, the Kodaira-Spencer map gives an isomorphism of free sheaves

$$\theta_S \simeq \pi_*(T^1_{\mathcal{K}_f/S}\mathscr{F})$$

on all of S. Composing this with the isomorphism (7) we get an isomorphism

$$\theta_{S-B} \simeq \pi_*(\mathcal{O}_{(X\times S)-\mathscr{D}}/J^{\mathrm{rel}}_{f\circ\mathscr{F}})\big|_{S-B} \qquad (8)$$

and it is this that we use to define a multiplication on the tangent sheaf, just as in the case of deformations of isolated hypersurface singularities.

Hertling and Manin introduced the notion of F-manifold in [9]. A complex manifold with an associative and commutative multiplication $\star$ on the tangent bundle is called an F-manifold if:

(i) (*unity*) there exists a *global* vector field e such that $e \star u = u$ for any $u \in \theta_M$ and,

(ii) (*integrability*) $\mathrm{Lie}_{u\star v}(\star) = u \star \mathrm{Lie}_v(\star) + \mathrm{Lie}_u(\star) \star v$ for any $u, v \in \theta_M$.

The main consequence of this definition is the integrability of multiplicative subbundles of TM, namely, if in a neighborhood U of a point $p \in M$ we can decompose TU as a sum $A \oplus B$ of multiplicatively closed subbundles with unity, then A and B are integrable.

An *Euler vector field* E for M is defined by the condition

$$\mathrm{Lie}_E(\star) = \star$$

Theorem 5.1. *The complement $S - B$ with the multiplication induced from* (8) *is an F-manifold with Euler vector field E_S coming from the class of $f \circ \mathscr{F}$ in $\mathcal{O}_{(X\times S)-\mathscr{D}}/J^{rel}_{f\circ\mathscr{F}}$ via the isomorphism*

$$\theta_{S-B} \simeq \pi_*\big(\mathcal{O}_{(X\times S)-\mathscr{D}}/J^{rel}_{f\circ\mathscr{F}}\big).$$

Proof. It is enough to show that the integrability condition holds in an open and dense subset of $S - B$. According to Prop. 3.2, there exists a proper analytic subvariety B_1 such that for $s \in S - B_1$, the composite $f \circ F_s$ has only non-degenerate critical points. In a neighborhood $U \subset S - B$ of such a point, the integrability condition is equivalent to the image L of the map

$$\operatorname{supp} T^1_{K_f}\mathscr{F} \ni (x,s) \mapsto d_{(x,s)}(f\circ\mathscr{F}) \in T^*_s S \tag{9}$$

being a Lagrangian subvariety of T^*S (see [8], Th. 3.2). If α denotes the canonical 1-form on T^*S and $p\colon T^*S \to S$ the projection, it is easy to check that the diagram

$$\begin{array}{ccc} \pi_*(T^1_{K_f/S}\mathscr{F}) & \longrightarrow & p_*\mathcal{O}_L \\ & \nwarrow \quad \nearrow & \\ & \theta_S & \end{array}$$

is commutative. The homomorphism on the right-hand side is given by evaluation, so that it can also be expressed as $u \mapsto \alpha(\tilde{u})$ where $\tilde{u}$ is a lift of $u \in \theta_S$ to θ_{T^*S}. Hence $\alpha|_L$ is the relative differential of $(f \circ \mathscr{F})$ when thought of as a map on L via the identification (9). It follows that $\alpha|_L$ is exact and hence closed, so that L is Lagrangian.

The statement about the Euler vector field is an easy calculation that we leave to the reader (see [8], Th. 3.3.). □

Remark 5.2. It follows that the critical values of $f\circ\mathscr{F}$ are local coordinates around a generic point in the base. For a point s where $\operatorname{supp} T^1_{\mathcal{K}_f} F_s$ consists of m different points, the algebra T_sS decomposes in 1-dimensional subalgebras with unity. Hence there exist coordinates $(u_1,\ldots,u_m)$ such that $(\partial/\partial_i)\star(\partial/\partial_j) = \delta_{ij}\partial/\partial_i$. These special coordinates are known as *canonical coordinates.* Writing the Euler vector field E in these coordinates and using the fact that $d(f\circ\mathcal{F}) = \alpha|_L$, we see that the canonical coordinates coincide, up to a constant, with the critical values of $f\circ\mathscr{F}$.

6. Morphisms of F-manifolds

In the cases where $f\circ F$ has an isolated singularity, we can compare its $\mathcal{R}$-miniversal deformation with the $\mathcal{K}_f$-versal deformation $\mathscr{F}$ of F. Suppose that $f\circ\mathscr{F}$, thought of as a deformation of $f\circ F$, is (up to $\mathcal{R}_{\text{e-un}}$-equivalence) induced from some other, say $G : X \times W \to \mathbf{C}$. Then we have a fiber square

$$\begin{array}{ccccc} X \times S & & \xrightarrow{\Phi} & & X \times W \\ \Big\downarrow & \searrow^{f\circ\mathscr{F}} & & \swarrow^{G} & \Big\downarrow \\ \pi_S & & \mathbf{C} & & \pi_W \\ \Big\downarrow & & & & \Big\downarrow \\ S & & \xrightarrow[i]{} & & W \end{array} \tag{10}$$

where Φ is the $\mathcal{R}_{\text{e-un}}$-equivalence, of the form

$$\Phi(x, s) = (\varphi(x, s), i(s)) = (\varphi_s(x), i(s)),$$

and i is the inducing map from the base-space of $f \circ \mathscr{F}$ to the base space of G.

Lemma 6.1. *There is a commutative diagram*

$$\begin{array}{ccc} (\pi_S)_*(T^1_{\mathcal{K}_f/S}\mathscr{F}) & \longrightarrow & (\pi_S)_*(\mathcal{O}_{X\times S}/J^{rel}_{f\circ\mathscr{F}}) \\ \uparrow & & \uparrow \\ \theta_S & \xrightarrow{ti} & \theta(i) \end{array} \tag{11}$$

The vertical arrow on the left-hand side is the Kodaira-Spencer map of $\mathscr{F}$ as a $\mathcal{K}_f$-deformation whereas the one on the right-hand side is the pull-back (by a morphism explained below) of that of G as an $\mathcal{R}$-deformation of $f \circ F$. The right-hand vertical morphism is an isomorphism when G is miniversal.

Proof. The morphism of the top row is just $\mathscr{F}^*(df)$, defined on a section $\xi = \sum_j \xi_j \partial/\partial y_j$ of $\theta(\mathscr{F})$ by

$$\sum_{i=1}^{n} \xi_j \partial/\partial y_j \mapsto \sum_j \big((\partial f/\partial y_j) \circ \mathscr{F}\big)\xi_j$$

and passing to the quotient as before. The vertical morphism of the right is defined on the generators $\partial/\partial w_j$ by

$$(\partial/\partial w_j) \mapsto \big(\partial G/\partial w_j\big) \circ \Phi$$

and extended by $\mathcal{O}_S$-linearity. It is straightforward to check that the diagram commutes.

Recall that $T^1_{\mathcal{K}_f/S}\mathscr{F}$ and $\mathcal{O}_{X\times S}/J^{\text{rel}}_{f\circ\mathscr{F}}$ are both finite over S, so that in particular

$$\pi_*\big(\mathcal{O}_{X\times S}/J^{\text{rel}}_{f\circ\mathscr{F}}\big)_s \simeq \bigoplus_x \mathcal{O}_{X\times S,(x,s)}/J^{\text{rel}}_{f\circ\mathscr{F}}.$$

Reducing the right-hand vertical morphism modulo the maximal ideal $m_{S,s}$, we obtain the reduced Kodaira Spencer map of G at $i(s)$, or rather, its translation by the isomorphism φ_s^* (composition with φ_s), as shown in the diagram

$$\begin{array}{ccc} \bigoplus_x (\mathcal{O}_{X,x}/J_{f\circ F_s}) & & \\ & \nwarrow & \\ \uparrow & & \bigoplus_x (\mathcal{O}_{X,\varphi_s(x)}/J_{g_{i(s)}}) \\ & \nearrow & \\ T_{i(s)}W & & \end{array} \tag{12}$$

in which the northeast arrow is the reduced Kodaira-Spencer map of G at s and the northwest arrow is φ_s^*. Miniversality of G implies that the northeast arrow in (12) is an isomorphism. Hence, so is the right-hand vertical map in (11), since its source and target are both free modules of rank $\mu(f \circ F)$ and its reduction modulo $m_{S,s}$ is an isomorphism. □

Proposition 6.2. *Assume that G in (10) is an $\mathcal{R}_e$-miniversal deformation of the isolated singularity $f \circ F$, and that $\mathscr{F}$ is a $\mathcal{K}_f$-miniversal deformation of F. Then*

(i) *if $m = m_0$, i is a local immersion into the discriminant Δ_G of G,*
(ii) *if $m = m_0 - 1$, the critical locus $\mathcal{C}$ of i is $\pi_S(V(\mathscr{F}^*J_f))$ and $i\colon S - \mathcal{C} \to T - \Delta$ is an unramified covering.*

Proof. In each of the two cases, since G is a $\mathcal{R}_e$-miniversal deformation, the vertical arrows of (11) are both isomorphisms. The exact sequences (6) and (5) show that the support of the cokernel of ti is exactly the projection to S of supp $\mathcal{O}_{X\times S}/\mathscr{F}^*J_f$. As this last module is Cohen-Macaulay and supported inside $\mathscr{D} = V(f \circ \mathscr{F})$ we see that $D_s = V(f \circ F_s)$ is singular. Hence the set of values where i is not submersive is contained in Δ_G.

To conclude, the exact sequence (6) says that i is a local immersion in the case $m = m_0$, whereas for $m = m_0 - 1$, (5) says the critical locus $\mathcal{C}$ is the projection by π_S of $\operatorname{supp} \operatorname{Tor}_1^{\mathcal{O}_Y}(\mathcal{O}_Y/J_f, \mathcal{O}_{X\times S})$ and hence equal to $\pi_S(V(\mathscr{F}^*J_f))$. □

Remark 6.3. Let $G : X \times T \to \mathbf{C}$ be a miniversal deformation of the composite function $f\circ F$. The Kodaira-Spencer map $\rho_G\colon \theta_T \to \mathcal{O}_{X\times T}/J_G^{\mathrm{rel}}$ defines Saito's F-manifold structure on T (see, e.g., [8] Chapter 5), and hence on $\theta(i) = \theta_T \otimes \mathcal{O}_S$. The Euler vector field E_T of this F-manifold is given by the class of G in the relative Jacobian algebra. The mapping i respects the multiplication and Euler vector field in the sense that

$$\begin{aligned} ti(u \star v) &= ti(u) \star ti(v) \\ ti(E_S) &= E_T \circ i \end{aligned} \tag{13}$$

Thus, the restriction of i to $S - B$ is a morphism of F-manifolds.

Note that in the case $m = m_0$ the multiplication in TS is defined even at the points of B, since now $\operatorname{Tor}_1^{\mathcal{O}_Y}(\mathcal{O}_Y/J_f, \mathcal{O}_X)$ vanishes and $(T^1_{\mathcal{K}_f}F_s)_x$ can be identified with $\left(F_s^*(J_f)/J_{f\circ F}\right)_x$ for each $x \in \operatorname{supp} T^1_{\mathcal{K}_f}F_s$. Nevertheless, since the multiplication at a point $s \in B$ lacks a unit, the unit vector field is not defined at s, and we cannot refer to all of S as an F-manifold. Similarly, the Euler field on $S - B$, which corresponds to the class of $f \circ \mathscr{F}$ in $\pi_*\big(\mathcal{O}_{X\times S}/J_{f\circ\mathscr{F}}\big)$ under the identifications

$$\theta_{S-B} \simeq \pi_*\big(T^1_{\mathcal{K}_f/S}\mathscr{F}\big) \simeq \pi_*\big(\mathcal{O}_{(X\times S)-\mathscr{D}}/J_{f\circ\mathscr{F}}\big),$$

is no longer defined at points of B where the class of $f\circ\mathscr{F}$ does not lie in the sheaf $\pi_*\big(\mathscr{F}^*(J_f)/J^{\mathrm{rel}}_{f\circ\mathscr{F}}\big)$.

Remark 6.4. The statement (2) in 6.2 implies a conjecture in [7]. Let $F\colon \mathbf{C}^2 \to \mathrm{Sym}_n$ be a family of $n\times n$-symmetric matrices and $f\colon \mathrm{Sym}_n \to \mathbf{C}$ the determinant. Then the subvariety $\sum \subset S$ corresponding to values of the parameter space for which F_s intersects the set of matrices of corank at least 2 is $(\pi_S)_* V(\mathscr{F}^* J_f)$ and hence coincides with $\mathcal{C}$.

References

[1] J.N. Damon, *The unfolding and determinacy theorems for subgroups of $\mathcal{A}$ and $\mathcal{K}$*, Memoirs Amer. Math. Soc. 306, 1984.

[2] J.N. Damon, *Deformations of sections of singularities and Gorenstein surface singularities,* Amer. J. Math. 109 (1987) 695–722.

[3] J.N.Damon, *$\mathcal{A}$-equivalence and the equivalence of sections of images and discriminants*, in *Singularity Theory and its Applications, Warwick* 1989, *Part* 1, D. Mond and J. Montaldi Eds., Springer Lecture Notes in Math 1462, 1991.

[4] J.N. Damon, *Higher Multiplicities and almost free divisors and complete intersections*, Memoirs Amer. Math. Soc. 589, 1996.

[5] J.N. Damon and D. Mond, *$\mathcal{A}$-codimension and the vanishing topology of discriminants*, Invent. Math. **106** (1991), 217–242.

[6] V.V. Goryunov and D. Mond, *Tjurina and Milnor number of matrix singularities*, math.AG/0307025 at arXiv.org, 2003, *to appear in* Proc. London Math. Soc.

[7] V.V. Goryunov and V.M. Zakalyukin, *Simple symmetric matrix singularities and the subgroups of Weyl groups A_μ, D_μ, E_μ*, Mosc. Math. J. 3 (2003), 507–530.

[8] C. Hertling, *Frobenius manifolds and moduli spaces for singularities*, Cambridge Tracts in Mathematics 151, 2002.

[9] C. Hertling and Y. I. Manin, *Weak Frobenius manifolds*, Internat. Math. Res. Notices 1999, no. 6, 277–286.

[10] J.N. Mather, *Stability of C^∞ mappings VI: The nice dimensions*, Proceedings of Liverpool Singularities-Symposium, I (1969/70), 207–253. Lecture Notes in Math., Vol. 192, 1971.

[11] D. Siersma, *Vanishing cycles and special fibres*, in *Singularity Theory and its Applications, Warwick* 1989, *Part* 1, D. Mond and J. Montaldi Eds., Springer Lecture Notes in Math 1462, 1991.

Ignacio de Gregorio and David Mond
Mathematics Institute, University of Warwick
Coventry CV4 7AL, England
e-mail: ignacio@maths.warwick.ac.uk
e-mail: mond@maths.warwick.ac.uk

Real and Complex Singularities
Trends in Mathematics, 201–208

On Equisingularity of Families of Maps $(\mathbb{C}^n, 0) \to (\mathbb{C}^{n+1}, 0)$

Kevin Houston

Abstract. A classical theorem of Briançon, Speder and Teissier states that a family of isolated hypersurface singularities is Whitney equisingular if, and only if, the μ^*-sequence for a hypersurface is constant in the family. This paper shows that the constancy of relative polar multiplicities and the Euler characteristic of the Milnor fibres of certain families of non-isolated singularities is equivalent to the Whitney equisingularity of a family of corank 1 maps from n-space to $n+1$-space. The number of invariants needed is $4n-2$, which greatly improves previous general estimates.

Mathematics Subject Classification (2000). Primary 32S15; Secondary 32S30, 32S60.

Keywords. Whitney stratification, equisingularity.

1. Introduction

Given a family of maps it is useful to have conditions that imply that the family is, in some sense, trivial. Suppose that we have a family of complex hypersurfaces such that each member has an isolated singularity. The family is called *Whitney equisingular* if the singular set of the variety formed by the whole family is a stratum in a Whitney stratification. This implies, for example, that, for each pair of hypersurfaces, there is a homeomorphism between the ambient spaces that takes one hypersurface to the other.

An overall aim of the theory is to find invariants of the elements of the family, the constancy of which implies, or is equivalent to, this Whitney equisingularity. In the isolated hypersurface case the constancy of the μ^*-sequence of Teissier (see for example [13]) is equivalent to the Whitney equisingularity of the family.

This work was initially carried out when the author visited ICMC at USP in São Carlos, Brazil. He is grateful to Maria Ruas for the hospitality during his stay and expresses his thanks to EPSRC for funding (Grant reference GR/S48639/01). Thanks are also due to the referee who made a number of useful comments.

One can consider what happens for maps, rather than varieties, i.e., when is a family of complex analytic maps $F : \mathbb{C}^n \times \mathbb{C} \to \mathbb{C}^p \times \mathbb{C}$ equisingular? There is again a notion of Whitney equisingularity. This time the parameter spaces $\{0\} \times \mathbb{C}$ in $\mathbb{C}^n \times \mathbb{C}$ and $\{0\} \times \mathbb{C}$ in $\mathbb{C}^p \times \mathbb{C}$ have to be strata in a stratification of the map.

In a series of papers Gaffney, (and also in conjunction with others such as Gassler and Massey) has produced some sterling work in answering this question for families where the members have an isolated instability at the origin. In [1] he introduced some new invariants, and proved that the constancy of these invariants within a family is equivalent to Whitney equisingularity.

However, the invariants are very difficult to manipulate, even in low dimensional cases, as the number of invariants is quite large and the method of description varies greatly from one invariant to another. For $\mathbb{C}^2$ to $\mathbb{C}^3$ we need 10 invariants; for $\mathbb{C}^3$ to $\mathbb{C}^4$ we need 20. At present there is no formula in the literature that allows one to calculate the number required.

Nonetheless, because of relationships between them, it is possible to reduce considerably the number of invariants needed in each case. For example, for a family of corank 1 maps from $\mathbb{C}^2$ to $\mathbb{C}^3$ the constancy of only one invariant is required. It should be noted that there was a significant amount of investigation done in [1] to show that this really is the only invariant needed. The $\mathbb{C}^3$ to $\mathbb{C}^4$ case is tackled in [8] where the 20 invariants are reduced to only 8. (The definition of 'reducing' is somewhat vague; one could reduce to one invariant merely by adding together all these upper semi-continuous invariants. The heuristic requirement is that the invariants should be calculable and that they should not be decomposable into other ones.)

The main result (Theorem 3.3) is that we can use relative polar multiplicities and the Euler characteristic of hypersurfaces to produce a Whitney equisingularity result in the case of $p = n+1$, i.e., the image of F is a hypersurface, and where the stable singularities of F have corank 1, (i.e., the differential of F at these points is, at worst, corank 1). So, in particular, the theorem holds when $n \leq 5$.

We reduce the number of invariants to $4n-2$, which is a considerable saving, when n is large (which here means bigger than 3). This saving is achieved, not through using Gaffney's work in [1], but his subsequent work with Gassler, [3], and Massey, [2].

2. Notation and basic definitions

In this section we give the definitions related to equisingularity for the sets and the complex analytic maps that concern us, and we reproduce the definitions of two sequences from [3], which in the main theorem will be used to control equisingularity.

Standard definitions from Singularity Theory, such as finite $\mathcal{A}$-determinacy, can be found in [16]. A differentiable map is called *corank 1* if its differential has corank at most 1 at all points.

Often we shall need to move from a germ and choose a representative, or a smaller neighborhood, etc. Since this is entirely standard and is obvious when it occurs, no explicit mention shall be made of the details as they will be distracting to the exposition.

Definition 2.1. *Let X be complex analytic set and Y a subset of X. We say that X is* Whitney equisingular along *Y if Y is a stratum of some Whitney stratification of X.*

This has been the subject of considerable investigation, see [2] for a survey. Gaffney has studied the notion for the more general case of maps, see [1]. We recall Thom's condition A_f before stating the definition of Whitney equisingularity.

Definition 2.2. *Let $f : X \to Y$ be a complex analytic map. Two strata A and B of a Whitney stratification of X are said to satisfy the* Thom A_f condition *with respect to f at a point $p \in B$ if the differential df has constant rank on A and for any sequence of points $p_i \in A$ such that p_i converges to p and $\ker d_{p_i}(f|_A)$ converges to some T (in the appropriate Grassmannian), then $\ker d_p(f|_B) \subseteq T$. We say f satisfies the* Thom A_f *condition if all pairs of strata satisfy the condition.*

Example 2.3. *Let $f : X \to Y$ be a finite complex analytic map such that X and Y are Whitney stratified so that strata map to strata by local diffeomorphisms. Then, f satisfies the Thom A_f condition as the kernels are all $\{0\}$.*

Now, the main definition is given.

Definition 2.4. *Let $F : (\mathbb{C}^n \times \mathbb{C}, 0 \times 0) \to (\mathbb{C}^p \times \mathbb{C}, 0 \times 0)$ be a family of maps $F(x, t) = (f_t(x), t)$ such that each $f_t : (\mathbb{C}^n, 0) \to (\mathbb{C}^p, 0)$ has an isolated instability at the origin, (i.e., each f_t is finitely $\mathcal{A}$-determined).*

We say that F is Whitney equisingular *if $\mathbb{C}^n \times \mathbb{C}$ and $\mathbb{C}^p \times \mathbb{C}$ can be Whitney stratified so that*

(i) *F satisfies Thom's A_F condition, and*
(ii) *the sets $S = \{0\} \times \mathbb{C} \subseteq \mathbb{C}^n \times \mathbb{C}$, and $T = \{0\} \times \mathbb{C} \subseteq \mathbb{C}^p \times \mathbb{C}$ are strata. (That is the 'parameter axes' are strata.)*

Remark 2.5. *This means, by the Thom-Mather Second Isotopy Lemma, that the members of the family are topologically equivalent.*

Let $f : (\mathbb{C}^N, 0) \to (\mathbb{C}, 0)$ be a complex analytic function, and denote the Jacobian ideal by $J(f)$:

$$J(f) = \left(\frac{\partial f}{\partial z_1}, \ldots, \frac{\partial f}{\partial z_N} \right)$$

for coordinates $z_1, \ldots, z_N$ in $\mathbb{C}^N$.

Definition 2.6. *The* blowup of $\mathbb{C}^N$ along the Jacobian ideal, *denoted $Bl_{J(f)}\mathbb{C}^N$, is the closure in $\mathbb{C}^N \times \mathbb{P}^{N-1}$ of the graph of the map*

$$\mathbb{C}^N \backslash V(J(f)) \to \mathbb{P}^{N-1}, \qquad x \mapsto \left(\frac{\partial f}{\partial z_1}(x) : \cdots : \frac{\partial f}{\partial z_N}(x) \right),$$

where $V(J(f))$ is zero-set of $J(f)$.

A hyperplane h in $\mathbb{P}^{N-1}$ can be pulled back by the natural projection $p : \mathbb{C}^N \times \mathbb{P}^{N-1} \to \mathbb{P}^{N-1}$ to a Cartier divisor, H, on $Bl_{J(f)}\mathbb{C}^N$, (provided $Bl_{J(f)}\mathbb{C}^N$ is not contained in the product of $\mathbb{C}^N$ and h). We call this a *hyperplane on* $Bl_{J(f)}\mathbb{C}^N$.

Let $b : \mathbb{C}^N \times \mathbb{P}^{N-1} \to \mathbb{C}^N$ be the other natural projection. For suitably generic hyperplanes $h_1, \dots, h_k$ in $\mathbb{P}^{N-1}$, the multiplicity at the origin of $b(H_1 \cap \dots \cap H_k \cap Bl_{J(f)}\mathbb{C}^N)$ is an analytic invariant of $V(J(f))$, see [3].

Definition 2.7. *The* kth relative polar multiplicity of f *is the multiplicity of the variety* $b(H_1 \cap \dots \cap H_k \cap Bl_{J(f)}\mathbb{C}^N)$ *at the origin. It is denoted by* $m_k(f)$.

Remark 2.8. *Full details of this construction and proofs of the various assertions can be found in* [3] *where the authors also show that the situation can be generalized to ideals other than the Jacobian.*

We can now define another sequence of invariants; again these have a topological nature.

Definition 2.9. ([2] p. 238) *Let* $f : (\mathbb{C}^{n+1}, 0) \to (\mathbb{C}, 0)$ *be a complex analytic function and* $L_i \subseteq \mathbb{C}^{n+1}$ *be a generic* i*-dimensional linear subspace. Denote the Euler characteristic of the Milnor Fibre of* $f|L_i$ *by* $\chi^i(f)$.

From this we can define a sequence

$$\chi^*(f) := \left(\chi^{n+1}(f), \dots, \chi^2(f)\right).$$

In the case of an isolated singularity, this (effectively) reduces to the standard μ^*-sequence in Equisingularity Theory.

Remark 2.10. *It transpires that the number* $\chi^1(f)$ *is not needed in the theory in* [3] *and so is omitted.*

Example 2.11. *If* f *defines the Swallowtail singularity, (i.e., the image of the stable map* $(x, y, z) \mapsto (x, y, z^4 + xz^2 + yz)$*), then* $\chi^3(f) = 1$*, (see, for example,* [11] *page 54), and* $\chi^2(f) = 6$*. The latter can be calculated using a program such as* `Singular`.

In general, it is not known how to calculate the homology of the Milnor Fibre of a non-isolated singularity. In some cases it is possible to calculate the Euler characteristic in practice, for example, using Massey's theorem that it is equal to the alternating sum of the Lê numbers, see [11].

3. Main theorems

Let $X \subseteq \mathbb{C}^N \times \mathbb{C}$ be a family of hypersurface germs defined by $H : \mathbb{C}^N \times \mathbb{C} \to \mathbb{C} \times \mathbb{C}$, and where $H(x, t) = (h_t(x), t)$ and Y is the parameter stratum $\{0\} \times \mathbb{C} \subseteq \mathbb{C}^N \times \mathbb{C}$.

We can now tease out the important elements of the proof of Theorem 6.6 of [2] to prove the following.

Lemma 3.1. *Suppose that $X\backslash Y$ is Whitney stratified such that, at each point of a stratum in $X\backslash Y$, X is locally analytically a product of a normal slice and the stratum. Suppose further that the complex link of every stratum in $X\backslash Y$, and not in the non-singular part of X, is not contractible.*

Then, the family is Whitney equisingular along Y if, and only if, the sequence

$$\left(m_{N-1}(h_t), \ldots, m_1(h_t), \chi^N(h_t), \ldots, \chi^2(h_t)\right)$$

is independent of $t \in Y$.

Proof. Suppose that the sequence is constant. Theorem 6.5 of [2] states that the sequence being constant in the family implies that the non-singular part of X is Whitney regular along Y.

We can now deal with the strata in the singular part of X. Suppose that R is a stratum of X of dimension r. Take the normal slice to R at the point p, i.e., the set $M \cap X$ where M is a manifold transverse to R with $M \cap R = \{p\}$. Since X has a product structure we can assume that (M,p) is $(\mathbb{C}^{N-r},p)$. Then, $M \cap X$ will be a hypersurface defined locally at p by the germ $g : (\mathbb{C}^{N-r},p) \to (\mathbb{C},0)$ say.

By definition, the complex link of the stratum R is the complex link of $M \cap X$ at the point p. This complex link is homotopically equivalent to a wedge of spheres (since the space is a hypersurface, see [4] p187), the number of which is the multiplicity of the relative polar curve of g, see Massey [10] page 365. Since, by assumption, this number is positive, the polar curve is non-empty. The example on page 235 of [2] shows that this implies that the origin of $\mathbb{C}^{N-r}$ is the image of a component of the exceptional divisor of $Bl_{J(g)}\mathbb{C}^{N-r}$. Since X has an analytic product structure along R this means that the closure of R is the image of a component of the exceptional divisor of $Bl_{J(H)}(\mathbb{C}^N)$. Thus, by the assumptions of the statement of the lemma and by using Theorem 6.5 of [2], we conclude that R is Whitney regular along Y.

The converse is just Theorem 6.3 of [3]. □

It seems likely that requiring that the complex links are non-contractible is necessary. This is because the topology of functions is intimately connected with complex links. (In [14], Tibăr shows that the Milnor fibre of a function with an isolated singularity on a complex analytic set is homotopically equivalent to a bouquet of suspensions of the complex links of the strata of the set.)

There are not many general results on the non-contractibility of complex links, see Section 4 of [15] for examples of hypersurfaces with a stratum that has a contractible complex link and for a theorem that states that complete intersections with a singular locus of dimension less than 2 have non-contractible complex links.

We can use the above lemma to prove our main theorem. First we need a definition:

Definition 3.2. *Let $p : A \to B$ be a continuous map. Then the* double point space of p in the source *is the set*

$$\text{closure}\,\{a \in A \mid \textit{there exists } a' \in A \textit{ such that } p(a) = p(a'),\ a \neq a'\}.$$

For a finitely $\mathcal{A}$-determined map-germ $f : (\mathbb{C}^n, 0) \to (\mathbb{C}^{n+1}, 0)$ the double point space is a hypersurface in $\mathbb{C}^n$.

Theorem 3.3. *Let $F : (\mathbb{C}^n \times \mathbb{C}, 0 \times 0) \to (\mathbb{C}^{n+1} \times \mathbb{C}, 0 \times 0)$ be a 1-parameter family of finitely $\mathcal{A}$-determined map-germs whose locus of instability is $\{0\} \times \mathbb{C} \subseteq \mathbb{C}^{n+1} \times \mathbb{C}$.*

Suppose that the stable singularity types appearing in F are corank 1 (e.g., $n \leq 5$ or each f_t is corank 1).

Then, F is Whitney equisingular if, and only if,

$$(m_n(h_t), \ldots, m_1(h_t), \chi^{n+1}(h_t), \ldots, \chi^2(h_t)) \text{ and}$$
$$(m_{n-1}(g_t), \ldots, m_1(g_t), \chi^n(g_t), \ldots, \chi^2(g_t))$$

are independent of $t \in Y$, where h_t is the function defining the image of f_t, and g_t is the function defining the double point set in the source of f_t.

Proof. Since the stable types are corank 1 and the map F is stable outside the instability locus of the members, we can stratify the source and target by stable type to get a Whitney stratification outside S and T (the parameter axes), such that the double point set in the source is a union of strata. (The stratification is finite as there are only finitely many right-left equivalence classes of stable germs of corank 1.) Furthermore, this stratification by stable types means that locally the spaces have a product structure – the one arising from unfolding minimal stable maps. Also, since F is finite, the strata map to strata by a local diffeomorphism. Thus, outside S and T the map is Thom A_F, see Example 2.3.

Theorem 7.3 of [6] states that the complex link of a stratum of the image of a corank 1 stable map is homotopically equivalent to a single sphere (except for the 'top' stratum which is the non-singular part of the image and hence has an empty complex link). Thus, we can apply Lemma 3.1 to the family of hypersurfaces giving the image of F to show that the image of F is Whitney equisingular along T.

Now, the double point set of F is also a family of hypersurfaces. Furthermore, outside S, it is also the image of a stable corank 1 map, see Proposition 3.5.1 of [5]. Thus, again applying Lemma 3.1, the double point set is Whitney equisingular along S.

Since S maps to T and this map is a local diffeomorphism we see that F is Thom A_F. Since source and target are Whitney stratified we conclude that F is Whitney equisingular. □

Thus we can reduce the number of invariants required to $4n - 2$ invariants. This is a considerable saving. For example, for $n = 2$ we get 6 invariants compared to Gaffney's original 10, and for $n = 3$ we get 10 rather than the original 20. Note however, that in the former case Gaffney reduced to 1 invariant and in the latter Pérez reduced the number of invariants to 8. It is in the cases where $n > 3$ that the theorem comes into its own. So far no-one has attempted to tackle the large task of enumerating precisely Gaffney's invariants for $n = 4$ or the even greater task of reducing through utilizing relationships between them.

Whilst at the meeting in Luminy Marcelo Saia informed me that, in [9], Jorge Pérez and Saia show how the number of Gaffney's original invariants can be cut, more or less, in half for corank 1 maps. This is rather suggestive as, in a similar vein, Corollary 8.8 of [1] states that, for maps in the theorem with $n = 2$, the map F is Whitney equisingular if its *image* is Whitney equisingular along the parameter axis. Combining this observation with the result in [9] we can conjecture that the same is true for more general n. If this were the case, then it would imply that the theorem above could be improved further as we could drop the assumption concerning the sequences associated with the double point set, i.e., we would require only the $2n$ invariants controlling the Whitney equisingularity of the image and could discard those in the source.

References

[1] T. Gaffney, Polar multiplicities and equisingularity of map germs, *Topology*, 32 (1993), 185–223.

[2] T. Gaffney and D.B. Massey, Trends in Equisingularity, in *Singularity Theory*, eds Bill Bruce and David Mond, London Math. Soc. Lecture Notes 263, Cambridge University Press 1999, 207–248.

[3] T. Gaffney and R. Gassler, Segre numbers and hypersurface singularities, *J. Algebraic Geometry*, 8, (1999), 695–736.

[4] M. Goresky and R. Macpherson, *Stratified Morse Theory*, Springer Verlag, Berlin, 1988.

[5] V.V. Goryunov, Semi-simplicial resolutions and homology of images and discriminants of mappings, *Proc. London Math. Soc.* 70 (1995), 363–385.

[6] K. Houston, On the topology of augmentations and concatenations of singularities, *Manuscripta Mathematica* 117, (2005), 383–405.

[7] V.H. Jorge Pérez, Polar multiplicities and equisingularity of map germs from $\mathbb{C}^3$ to $\mathbb{C}^3$, *Houston Journal of Mathematics* 29, (2003), 901–924.

[8] V.H. Jorge Pérez, Polar multiplicities and Equisingularity of map germs from $\mathbb{C}^3$ to $\mathbb{C}^4$, in *Real and Complex Singularities*, eds. David Mond and Marcelo José Saia, Dekker Lecture Notes in Pure and Applied Mathematics Vol 232, Marcel Dekker 2003, 207–226.

[9] V.H. Jorge Pérez and Marcelo Saia, Euler obstruction, Polar multiplicities and equisingularity of map germs in $\mathcal{O}(n, p)$, $n < p$, *Notas do ICMC*, no 157, 2002.

[10] D. Massey, Numerical invariants of perverse sheaves, *Duke Mathematical Journal*, 73(2), (1994), 307–369.

[11] D. Massey, *Lê Cycles and Hypersurface Singularities*, Springer Lecture Notes in Mathematics 1615, (1995).

[12] D. Mond, Vanishing cycles for analytic maps, in *Singularity Theory and its Applications*, SLNM 1462, D. Mond, J. Montaldi (Eds.), Springer Verlag Berlin, 1991, pp. 221–234.

[13] B. Teissier, Multiplicités polaires, section planes, et conditions de Whitney, in *Algebraic Geometry, Proc. La Rábida*, 1981, eds J.M. Aroca, R. Buchweitz, M. Giusti, and M. Merle, Springer Lecture Notes In Math 961 (1982), 314–491.

[14] M. Tibăr, Bouquet decomposition of the Milnor fiber, *Topology* 35, (1996), 227–241.

[15] M. Tibăr, Limits of tangents and minimality of complex links, *Topology* 42, (2003), 629–639.

[16] C.T.C. Wall, Finite determinacy of smooth map-germs, *Bull. Lond. Math. Soc.*, 13 (1981), 481–539.

Kevin Houston
School of Mathematics
University of Leeds
Leeds, LS2 9JT, U.K.
e-mail: `e-mail: k.houston@leeds.ac.uk`

Real and Complex Singularities
Trends in Mathematics, 209–217

Projected Wallpaper Patterns

Isabel S. Labouriau and Eliana M. Pinho

Abstract. Consider a periodic function f of two variables with symmetry Γ and let $\mathcal{L} \subset \Gamma$ be the subgroup of translations. The Fourier expansion of a periodic function is a sum over $\mathcal{L}^*$, the dual of the set $\mathcal{L}$ of all the periods of f. After projecting f, some of its original symmetry remains. We describe the symmetries of the projected function, starting from Γ and from the structure of $\mathcal{L}^*$.

Mathematics Subject Classification (2000). Primary 52C20; Secondary 20H15, 82D25, 35B10, 58E09, 51M20.

Keywords. Tilings in two dimensions, crystallographic groups, periodic solutions, group-invariant bifurcations.

1. Introduction and preliminaries

Patterns with spatial periodicity are observed in many physical systems such as crystals, convection experiments, chemical reactions, Faraday wave experiments (see Rucklidge et al. [10]), or magnetic perturbation of a liquid crystal (Chillingworth and Golubitsky [3]). Moreover, one method used in the study of bifurcation [6] of problems equivariant under the Euclidean group $\mathbf{E}(2)$ is to look for periodic solutions – see [2, 4, 5]. The patterns observed in reaction-diffusion experiments in thin layers of gel are usually explained by two-dimensional models. However, some observed periodic solutions, like black-eye patterns, are not expected in two-dimensions. Gomes [7] suggests that black-eye patterns are the projection into the plane of a three-dimensional repetitive solution.

We may ask in general how a projection transforms repetitive patterns. This paper presents a first step in answering this question.

Both authors had financial support from Fundação para a Ciência e a Tecnologia (FCT), Portugal, through programs POCTI and POSI of Quadro Comunitário de Apoio III (2000–2006) with national and EU (FEDER) funding. E. M. Pinho was partly supported by the grant SFRH/BD/13334/2003 of FCT and by UBI-Universidade da Beira Interior, Portugal.

If $f : \mathbf{R}^2 \longrightarrow \mathbf{R}$ has two noncolinear periods then its symmetry group is a plane crystalographic group, $\Gamma \leq \mathbf{E}(2)$, and its level sets form a periodic pattern. We start with a pattern in $\mathbf{R}^2$ and project it into $\mathbf{R}$. The new pattern, the level sets of a function in $\mathbf{R}$, may be periodic or invariant under reflections. We relate the existence of these symmetries to properties of Γ and of $\mathcal{L}^*$, the dual of the set $\mathcal{L}$ of all the periods of f. The set $\mathcal{L}^*$ arises naturally in the Fourier expansion of f and the symmetries in Γ impose restrictions on Fourier coefficients.

The result is formulated in a form that simplifies the generalization to dimension n instead of 2. At the end of the paper we indicate the main technical complications appearing in higher dimension.

We write elements of $\mathbf{E}(2) = \mathbf{R}^2 \dot{+} \mathbf{O}(2)$ in the form (v_δ, δ), with $v_\delta \in \mathbf{R}^2$ representing a translation and $\delta \in \mathbf{O}(2)$. They act in $f : \mathbf{R}^2 \longrightarrow \mathbf{R}$ with the scalar action (see [8]):

$$(v_\delta, \delta) \cdot f(x) = f((v_\delta, \delta)^{-1}) \cdot x) = f(\delta^{-1}x - \delta^{-1}v_\delta).$$

We assume that Γ is a plane crystalographic group – see [1, 11] for general results and definitions. Denote by $\mathcal{L}$ the subgroup of the translations in Γ, a module over the integers, also called a lattice. If f is Γ-invariant, then in particular elements of $\mathcal{L}$ are periods of f. A pattern and the lattice $\mathcal{L}$ may not have the same symmetries: see Figure 1.

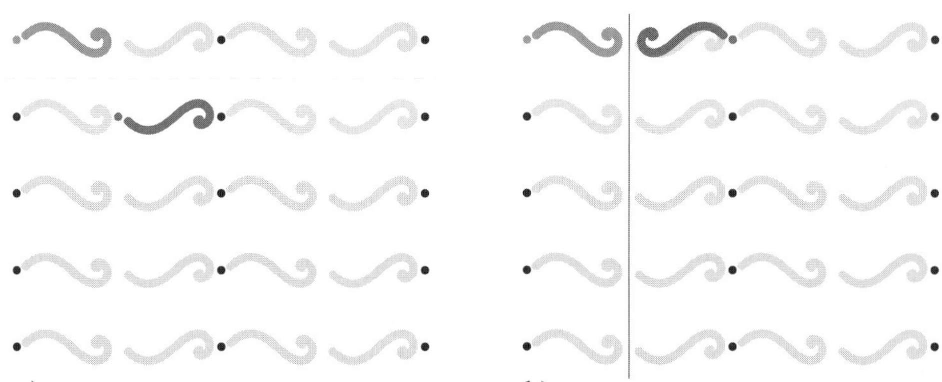
a) b)

FIGURE 1. a) The lattice (black dots) is not invariant under the glide reflection transforming the grey motif into the darker one. However this is a symmetry of the lighter pattern. b) The lighter pattern is not invariant under the reflection on the black line, although this is a symmetry of the lattice (black dots).

2. Symmetries and projection

Let X_Γ be a vector space of Γ-invariant functions $f : \mathbf{R}^2 \longrightarrow \mathbf{R}$, having unique formal Fourier expansions of the form:

$$f(x,y) = \sum_{k\in\mathcal{L}^*} \omega_k(x,y)C(k),$$

where $\mathcal{L}^*$ is the dual lattice and $\omega_k(x,y) = \mathrm{e}^{2\pi i<k,(x,y)>}$.

The elements of $\mathcal{L}^*$ are those $k \in \mathbf{R}^2$ such that $< k,l >\in \mathbf{Z}$ for all $l \in \mathcal{L}$, where $< k,l >$ is the usual inner product in $\mathbf{R}^2$.

Given $y_0 > 0$, define the projection of a function $f \in X_\Gamma$ to be the function

$$\Pi_{y_0}(f)(x) = \int_0^{y_0} f(x,y)dy \qquad x,y \in \mathbf{R}.$$

We assume that in X_Γ we have,

$$\Pi_{y_0}(f)(x) = \sum_{k\in\mathcal{L}^*} \int_0^{y_0} \omega_k(x,y)C(k)dy$$

and that X_Γ contains, for all $k \in \mathcal{L}^*$, the real and imaginary parts of $I_k(x,y) = \sum_{\delta\in\mathbf{J}} \omega_{\delta k}(-v_\delta)\omega_{\delta k}(x,y)$, where $\mathbf{J} \sim \Gamma/\mathcal{L}$ is the largest subgroup of $\mathbf{O}(2)$ that leaves $\mathcal{L}$ invariant. Notice that these are the simplest Γ-invariant functions.

The first step in obtaining the symmetries of the projected functions is to relate the (v_α,α)-invariance to restrictions on Γ and on $\mathcal{L}^*$. This is the main result in this paper – Proposition 2.1 below – where we characterize situations giving rise to (v_α,α)-invariance of the projected function. Here (v_α,α) is either a reflection $(\alpha = -1)$ or a translation $(\alpha = 1)$ so we are asking for conditions ensuring that projected functions are even or periodic, respectively.

For $\alpha \in \{1,-1\}$, let $\alpha_+ \in \{I,-\sigma\}$ and $\alpha_- = \sigma\alpha_+ \in \{\sigma,-I\}$, where

$$\alpha_+ = \begin{pmatrix} \alpha & 0 \\ 0 & 1 \end{pmatrix} \qquad \text{and} \qquad \sigma = \begin{pmatrix} 1 & 0 \\ 0 & -1 \end{pmatrix}.$$

Thus, an element $(v_+,\alpha_+) \in \Gamma$ is either a (glide) reflection on a vertical line when $\alpha = -1$, or a translation when $\alpha = 1$. Similarly $(v_-,\alpha_-) \in \Gamma$ is either a rotation of π $(\alpha = -1)$ or a reflection on a horizontal line when $\alpha = 1$ (and therefore $\alpha_- = \sigma$). Note that $\alpha_\pm = \alpha_\pm^{-1}$ and $\sigma = \sigma^{-1}$.

Proposition 2.1. *All functions in* $\Pi_{y_0}(X_\Gamma)$ *are invariant under the action of* $(v_\alpha,\alpha) \in \mathbf{R}\dot{+}\mathbf{O}(1)$ *if and only if one of the following conditions holds:*

A. $(v_+,\alpha_+) \in \Gamma$ *and for each* $k \in \mathcal{L}^*$,
either $\langle k,(0,y_0)\rangle \in \mathbf{Z}-\{0\}$ *or* $\langle k, v_+ - (v_\alpha,0)\rangle \in \mathbf{Z}$,

B. $(v_-,\alpha_-) \in \Gamma$ *and for each* $k \in \mathcal{L}^*$,
either $\langle k,(0,y_0)\rangle \in \mathbf{Z}-\{0\}$ *or* $\langle k, v_- - (v_\alpha,y_0)\rangle \in \mathbf{Z}$,

C. (v_σ,σ), $(v_+,\alpha_+) \in \Gamma$ *and, for each* $k \in \mathcal{L}^*$, *one of the conditions* C1, C2 *or* C3 *below holds:*

C1. $\langle k, (0, y_0)\rangle \in \mathbf{Z} - \{0\}$,
C2. $\langle k, v_+ - (v_\alpha, 0)\rangle \in \mathbf{Z}$,
C3. $\langle k, v_\sigma - (0, y_0)\rangle + \frac{1}{2} \in \mathbf{Z}$.

Parity of projected functions arises through the presence of either a (glide) reflection or a rotation by π in Γ, plus conditions relating its translation part to the projection width and to $\mathcal{L}^*$. Similarly, periods of the projected functions arise either from reflections in Γ, or from periods of the original function, plus extra conditions on these and on y_0 in relation to $\mathcal{L}^*$.

A more concise formulation of this result is possible using the subsets of $\mathcal{L}^*$ defined below. Let $\mathcal{M}_+^*$ and $\mathcal{M}_-^*$ be the modules

$$\mathcal{M}_+^* = \{k \in \mathcal{L}^* : \langle k, v_+ - (v_\alpha, 0)\rangle \in \mathbf{Z}\} \text{ and}$$
$$\mathcal{M}_-^* = \{k \in \mathcal{L}^* : \langle k, v_- - (v_\alpha, y_0)\rangle \in \mathbf{Z}\},$$

and let

$$\mathcal{N}_{y_0}^* = \{k \in \mathcal{L}^* : \langle k, (0, y_0)\rangle \in \mathbf{Z} - \{0\}\},$$
$$\mathcal{N}_\sigma^* = \{k \in \mathcal{L}^* : \langle k, v_\sigma - (0, y_0)\rangle + 1/2 \in \mathbf{Z}\}.$$

The last two sets are not modules. The smallest modules generated by each of them are, respectively, $\overline{\mathcal{N}_{y_0}^*} = \mathcal{N}_{y_0}^* \cup \mathcal{M}_{y_0}^*$ and $\overline{\mathcal{N}_\sigma^*} = \mathcal{N}_\sigma^* \cup \mathcal{M}_\sigma^*$, where all the unions are disjoint and $\mathcal{M}_{y_0}^*$ and $\mathcal{M}_\sigma^*$ are the modules

$$\mathcal{M}_{y_0}^* = \{k \in \mathcal{L}^* : \langle k, (0, y_0)\rangle = 0\} \text{ and}$$
$$\mathcal{M}_\sigma^* = \{k \in \mathcal{L}^* : \langle k, v_\sigma - (0, y_0)\rangle \in \mathbf{Z}\}.$$

Properties of $\mathcal{N}_\sigma^*$: Let $m_1, m_2 \in \mathbf{Z}$. If $g_1, g_2 \in \mathcal{N}_\sigma^*$ then

$$m_1 g_1 + m_2 g_2 \in \begin{cases} \mathcal{M}_\sigma^* & \text{if} \quad m_1 + m_2 \text{ even} \\ \mathcal{N}_\sigma^* & \text{if} \quad m_1 + m_2 \text{ odd} \end{cases}. \tag{1}$$

Proposition 2.1 can therefore be written the following way:

Proposition 2.2. *All functions in $\Pi_{y_0}(X_\Gamma)$ are invariant under the action of $(v_\alpha, \alpha) \in \mathbf{R}\dot{+}\mathbf{O}(1)$ if and only if one of the following conditions holds:*

A. $(v_+, \alpha_+) \in \Gamma$ *and* $\mathcal{L}^* = \mathcal{N}_{y_0}^* \cup \mathcal{M}_+^*$,
B. $(v_-, \alpha_-) \in \Gamma$ *and* $\mathcal{L}^* = \mathcal{N}_{y_0}^* \cup \mathcal{M}_-^*$,
C. (v_σ, σ), $(v_+, \alpha_+) \in \Gamma$ *and* $\mathcal{L}^* = \mathcal{N}_{y_0}^* \cup \mathcal{M}_+^* \cup \mathcal{N}_\sigma^*$.

For $D(k_1) = \sum_{k_2:(k_1,k_2)\in\mathcal{L}^*} C(k_1, k_2) \int_0^{y_0} \omega_{k_2}(y)dy$, the projection of $f \in X_\Gamma$ may be written, with $\mathcal{L}_1^* = \{k_1 : (k_1, k_2) \in \mathcal{L}^*\}$, as

$$\Pi_{y_0}(f)(x) = \sum_{k_1 \in \mathcal{L}_1^*} \omega_{k_1}(x) D(k_1).$$

Thus $\Pi_{y_0}(f)$ is (v_α, α)-invariant if and only if

$$\sum_{k_1 \in \mathcal{L}_1^*} \omega_{k_1}(x) D(k_1) = \sum_{k_1 \in \mathcal{L}_1^*} \omega_{k_1}(\alpha\, x)\omega_{k_1}(-\alpha\, v_\alpha) D(k_1), \tag{2}$$

or, equivalently, $D(k_1) = \omega_{k_1}(-v_\alpha) D(\alpha\, k_1)$, for all $k_1 \in \mathcal{L}_1^*$.

In the next section we show that each condition of Proposition 2.1 leads to the restrictions on the coefficients $D(k_1)$ above. Reciprocally, when those restrictions are imposed on the projection of $I_k(x, y)$, for all $k \in \mathcal{L}^*$, this implies the conditions of Proposition 2.1.

3. Proof of Proposition 2.2

Let $f \in X_\Gamma$ and $(v_\alpha, \alpha) \in \mathbf{R}\dot{+}\mathbf{O}(1)$. If $\Pi_{y_0}(f)$ is (v_α, α)-invariant then $\Pi_{y_0}(f)(x) = \Pi_{y_0}(f)(\alpha\, x - \alpha\, v_\alpha)$, which is equivalent to (2). The right-hand side of (2) equals $\sum_{k_1 \in \mathcal{L}_1^*} \omega_{\alpha k_1}(x)\omega_{\alpha k_1}(v_\alpha)D(k_1)$. Since $\alpha(\mathcal{L}_1^*) = \mathcal{L}_1^*$ and Fourier expansions are unique, then for each $k_1 \in \mathcal{L}_1^*$, we have:

$$D(k_1) - \omega_{k_1}(-v_\alpha)D(\alpha\, k_1) = 0. \tag{3}$$

Proof – sufficiency. The difference in (3) may be written as

$$\sum_{k_2:(k_1,k_2)\in\mathcal{L}^*} C(k_1, k_2)G(k_1, k_2)\int_0^{y_0} \omega_{k_2}(y)dy. \tag{4}$$

In each case we compute $G(k_1, k_2)$ and use the conditions on $\mathcal{L}^*$.

Suppose $\alpha_+ \in \mathbf{J}$. Then all the Fourier coefficients of any $f \in X_\Gamma$ satisfy $C(k) = \omega_k(-v_+)C(\alpha_+\, k)$ and $G(k_1, k_2) = 1 - \omega_k(v_+ - (v_\alpha, 0))$. Thus $G(k_1, k_2) = 0$ if $\langle k, v_+ - (v_\alpha, 0)\rangle \in \mathbf{Z}$.

If $(v_-, \alpha_-) \in \Gamma$ then $G(k_1, k_2) = 1 - \omega_k(v_- - (v_\alpha, y_0))$, since

$$\int_0^{y_0} \omega_{-k_2}(y)dy = \omega_{k_2}(-y_0)\int_0^{y_0} \omega_{k_2}(y)dy. \tag{5}$$

Then $G(k_1, k_2) = 0$ if $\langle k, v_- - (v_\alpha, y_0)\rangle \in \mathbf{Z}$.

When both (v_+, α_+) and (v_-, α_-) lie in Γ then

$$G(k_1, k_2) = 1 + \omega_k(v_\sigma)\omega_{k_2}(-y_0) - \omega_{k_1}(-v_\alpha)\left(\omega_k(v_+) + \omega_k(v_-)\omega_{k_2}(-y_0)\right).$$

Using $\omega_k(v_-) = \omega_k(v_\sigma)\omega_k(\sigma v_+)$ and $\omega_k(\sigma v_+ - v_+) = 1$ we get

$$G(k_1, k_2) = \left(1 - \omega_k(v_+ - (v_\alpha, 0))\right)\left(1 + \omega_k(v_\sigma - (0, y_0))\right).$$

If either $1 - \omega_k(v_+ - (v_\alpha, 0)) = 0$ or $1 + \omega_k(v_\sigma - (0, y_0)) = 0$ then $G(k_1, k_2) = 0$.

It follows from the conditions on $\mathcal{L}^*$ that for each $k \in \mathcal{L}^*$ either $\int_0^{y_0} \omega_{k_2}(y)dy = 0$ or $G(k_1, k_2) = 0$ and thus (3) holds for all $k \in \mathcal{L}^*$. □

Proof – necessity. For $D'(\delta, k) = \omega_{\delta k}(-v_\delta)\int_0^{y_0} \omega_{\delta k|_2}(y)dy$, the projections of I_k, with $k \in \mathcal{L}^*$, are

$$\Pi_{y_0}(I_k)(x) = \sum_{\tilde{k}_1 \in \mathbf{J}k|_1} \omega_{\tilde{k}_1}(x) \sum_{\tilde{k}_2:(\tilde{k}_1,\tilde{k}_2)\in\mathbf{J}k} D'(\delta, \tilde{k}),$$

where $\delta k|_j$ denotes the jth coordinate of δk. If $\Pi_{y_0}(I_k)$ is (v_α, α)-invariant then, by (3),

$$\sum_{\delta\in\mathbf{J}^I(k)} D'(\delta, k) - \omega_{k_1}(-v_\alpha)\sum_{\delta\in\mathbf{J}^\alpha(k)} D'(\delta, k) = 0,$$

where $\mathrm{J}^I(k) = \{\delta \in \mathbf{J} : \delta k|_1 = k_1\}$ and $\mathrm{J}^\alpha(k) = \{\delta \in \mathbf{J} : \delta k|_1 = \alpha\, k_1\}$. Let $\mathbf{J}^I = \{I, \sigma\} \cap \mathbf{J}$ and $\mathbf{J}^\alpha = \{\alpha_+, \alpha_-\} \cap \mathbf{J}$. We list some properties of $\mathrm{J}^I(k)$ and $\mathrm{J}^\alpha(k)$ in Lemma 3.1 below. Then, in Lemma 3.2, we describe the set

$$\mathcal{O}^* = \left\{k \in \mathcal{L}^* : \mathrm{J}^I(k) = \mathbf{J}^I \ \wedge \ \mathrm{J}^\alpha(k) = \mathbf{J}^\alpha\right\}.$$

A geometrical characterization of the complement of $\mathcal{O}^*$ in $\mathcal{L}^*$ is given in Lemma 3.3 and in Lemma 3.4 we reformulate the cases of Lemma 3.2 in terms of $\mathcal{L}^*$ instead of $\mathcal{O}^*$, completing the proof. □

Lemma 3.1. *For $k \in \mathcal{L}^*$, the sets $\mathrm{J}^I(k)$ and $\mathrm{J}^\alpha(k)$ satisfy:*

1. $\mathrm{J}^I(k) = \{\delta \in \mathbf{J} : \ \delta k = k \vee \delta k = \sigma k\}$.
2. $\mathrm{J}^\alpha(k) = \{\delta \in \mathbf{J} : \ \delta k = \alpha_+ k \vee \delta k = \alpha_- k\}$.
3. $\mathbf{J}^I \subset \mathrm{J}^I(k)$, $\mathbf{J}^\alpha \subset \mathrm{J}^\alpha(k)$ *and* $\mathrm{J}^I(0,0) = \mathrm{J}^\alpha(0,0) = \mathbf{J}$.
4. *Let $k = (k_1, k_2) \neq (0,0)$. If $\delta \in \mathrm{J}^I(k) - \mathbf{J}^I$ then $\delta k = (k_1, -|\delta| k_2)$ and if $\delta \in \mathrm{J}^\alpha(k) - \mathbf{J}^\alpha$ then $\delta k = \alpha(k_1, -|\delta| k_2)$, where $|.|$ is the determinant.*

Proof. Properties 1 and 2 follow by orthogonality of $\mathbf{J}$ and Property 3 is immediate from this and the definitions.

For Property 4, let $\delta \in \mathrm{J}^I(k) - \mathbf{J}^I$ and $k \neq (0,0)$. If $\delta k = k$ then $|\delta| = -1$, since an element of $\mathbf{O}(2)$ with determinant 1, other than the identity, does not fix any point besides the origin. Similarly if $\delta k = \sigma k$ then $|\sigma\delta| = -1$ and $|\delta| = 1$. Now suppose $\delta \in \mathrm{J}^\alpha(k) - \mathbf{J}^\alpha$ and $k \neq (0,0)$. Thus, either $\alpha_+\delta = k$ or $\alpha_+\delta = \sigma k$. As $\alpha_+\delta \in \mathrm{J}^I(k) - \mathbf{J}^I$, we may apply the previous result to $\alpha_+\delta$, and the property follows. □

Lemma 3.2. *Suppose that $\sum_{\delta \in \mathrm{J}^I(k)} D'(\delta, k) = \omega_{k_1}(-v_\alpha) \sum_{\delta \in \mathrm{J}^\alpha(k)} D'(\delta, k)$ for all $k = (k_1, k_2) \in \mathcal{L}^*$. Then one of the following cases holds:*

1. $\mathbf{J}^I = \{I\}$, $\mathbf{J}^\alpha = \emptyset$ *and* $\mathcal{O}^* \subset \mathcal{N}^*_{y_0}$,
2. $\mathbf{J}^I = \{I, \sigma\}$, $\mathbf{J}^\alpha = \emptyset$ *and* $\mathcal{O}^* \subset \left(\mathcal{N}^*_{y_0} \cup \mathcal{N}^*_\sigma\right)$,
3. $\mathbf{J}^I = \{I\}$, $\mathbf{J}^\alpha = \{\alpha_+\}$ *and* $\mathcal{O}^* \subset \left(\mathcal{N}^*_{y_0} \cup \mathcal{M}^*_+\right)$,
4. $\mathbf{J}^I = \{I\}$, $\mathbf{J}^\alpha = \{\alpha_-\}$ *and* $\mathcal{O}^* \subset \left(\mathcal{N}^*_{y_0} \cup \mathcal{M}^*_-\right)$,
5. $\mathbf{J}^I = \{I, \sigma\}$, $\mathbf{J}^\alpha = \{\alpha_+, \alpha_-\}$ *and* $\mathcal{O}^* \subset \left(\mathcal{N}^*_{y_0} \cup \mathcal{M}^*_+ \cup \mathcal{N}^*_\sigma\right)$.

Proof. If $\mathbf{J}^\alpha = \emptyset$ and $k \in \mathcal{O}^*$ then by hypothesis $\sum_{\delta \in \mathbf{J}^I} D'(\delta, k) = 0$. By (5), if $\sigma \in \mathbf{J}$ then $(1 + \omega_k(v_\sigma - (0, y_0))) \int_0^{y_0} \omega_{k_2}(y) dy = 0$ and $\int_0^{y_0} \omega_{k_2}(y) dy = 0$ if $\sigma \notin \mathbf{J}$. Cases 1 and 2 follow because $\int_0^{y_0} \omega_{k_2}(y) dy = 0$ implies $k \in \mathcal{N}^*_{y_0}$ and $1 + \omega_k(v_\sigma - (0, y_0)) = 0$ implies $k \in \mathcal{N}^*_\sigma$.

In case 3 we have $(1 - \omega_{k_1}(-v_\alpha)\omega_k(v_+)) \int_0^{y_0} \omega_{k_2}(y) dy = 0$ and the result follows because $1 - \omega_{k_1}(-v_\alpha)\omega_k(v_+) = 0$ implies $k \in \mathcal{M}^*_+$.

In case 4, $(1 - \omega_{k_1}(-v_\alpha)\omega_k(v_-)\omega_{k_2}(-y_0)) \int_0^{y_0} \omega_{k_2}(y) dy = 0$ and either $k \in \mathcal{N}^*_{y_0}$ or $1 - \omega_{k_1}(-v_\alpha)\omega_k(v_-)\omega_{k_2}(-y_0) = 0$, which implies $k \in \mathcal{M}^*_-$.

The hypothesis in case 5 yields $G(k_1, k_2) \int_0^{y_0} \omega_{k_2}(y) dy = 0$, where

$$G(k_1, k_2) = 1 + \omega_k(v_\sigma)\omega_{k_2}(-y_0) - \omega_{k_1}(-v_\alpha)\left(\omega_k(v_+) + \omega_k(v_-)\omega_{k_2}(-y_0)\right),$$

as in the proof of sufficiency in Proposition 2.1. Therefore, either $k \in \mathcal{N}^*_{y_0}$ or $G(k_1, k_2) = 0$. In the second case the result follows because either

$$(1 - \omega_k(v_+ - (v_\alpha, 0))) = 0 \quad \text{or} \quad (1 + \omega_k(v_\sigma - (0, y_0))) = 0. \qquad \square$$

Let $\mathcal{P}^* = \{k \in \mathcal{L}^* : \mathrm{J}^I(k) \neq \mathbf{J}^I \vee \mathrm{J}^\alpha(k) \neq \mathbf{J}^\alpha\}$ be the complement of $\mathcal{O}^*$ in $\mathcal{L}^*$.

Lemma 3.3. *$\mathcal{P}^*$ lies in a finite union of lines through the origin.*

Proof. $\mathcal{P}^*$ may be written as a finite union of submodules

$$\mathcal{P}^* = \bigcup_{\delta \in \mathbf{J} - \mathbf{J}^I} \mathcal{M}^*_{\delta, I} \cup \bigcup_{\delta \in \mathbf{J} - \mathbf{J}^\alpha} \mathcal{M}^*_{\delta, \alpha}$$

for $\mathcal{M}^*_{\delta,\xi} = \{k \in \mathcal{L}^* : \delta k = \xi(k_1, -|\delta| k_2)\}$ and $\xi = I, \alpha$. If δ is a rotation then for $k \in \mathcal{M}^*_{\delta,\xi}$ we have $\delta k = \pm(k_1, -k_2)$, *i.e.*, k lies on the line fixed by $\pm\sigma\delta$. Therefore $\mathcal{M}^*_{\delta,\xi}$ is the intersection of those lines with $\mathcal{L}^*$. Similarly, if δ is a reflection then $\mathcal{M}^*_{\delta,\xi}$ is the intersection of $\mathcal{L}^*$ with a line fixed either by δ or by $-\delta$. $\square$

Lemma 3.4. *If $\sum_{\delta \in \mathrm{J}^I(k)} D'(\delta, k) = \omega_{k_1}(-v_\alpha) \sum_{\delta \in \mathrm{J}^\alpha(k)} D'(\delta, k)$ for all $k = (k_1, k_2)$ in $\mathcal{L}^*$, then one of the following cases holds:*

A. $\mathbf{J}^\alpha = \{\alpha_+\}$ *and* $\mathcal{L}^* = \mathcal{N}^*_{y_0} \cup \mathcal{M}^*_+$,
B. $\mathbf{J}^\alpha = \{\alpha_-\}$ *and* $\mathcal{L}^* = \mathcal{N}^*_{y_0} \cup \mathcal{M}^*_-$,
C. $\mathbf{J}^\alpha = \{\alpha_+, \alpha_-\}$ *and* $\mathcal{L}^* = \mathcal{N}^*_{y_0} \cup \mathcal{M}^*_+ \cup \mathcal{N}^*_\sigma$.

Proof. Let $k \in \mathcal{L}^* - \{(0,0)\}$ and observe that

$$\left(\mathcal{M}^*_{y_0} \cap \mathcal{P}^*\right) - \{(0,0)\} = \emptyset. \tag{6}$$

Let $g = (1/n)k \in \mathcal{L}^*$, $n \in \mathbf{Z}$, have minimal norm and choose $h \in \mathcal{L}^*$ such that $\mathcal{L}^* = \{g, h\}_{\mathbf{Z}}$. Let $\mathcal{Q}^*_k = \{k + mh : m \in \mathbf{Z}\}$. Since $\mathcal{Q}^*_k$ is contained in a line in $\mathbf{R}^2$ that does not go through the origin, by Lemma 3.3, the set $\mathcal{Q}^*_k \cap \mathcal{P}^*$ is finite.

For $k \in \mathcal{L}^* - \{(0,0)\}$ there are three possibilities for $\mathcal{Q}^*_k \cap \overline{\mathcal{N}^*_{y_0}}$: it is either the empty set, or a set with only a point, or an infinite set of equally spaced points. This happens because $\overline{\mathcal{N}^*_{y_0}}$ is a module and if $k + m_1 h \neq k + m_2 h \in \mathcal{Q}^*_k \cap \overline{\mathcal{N}^*_{y_0}}$, then $(m_2 - m_1)h \in \overline{\mathcal{N}^*_{y_0}}$ and $\{k + m_1 h + m(m_2 - m_1)h : m \in \mathbf{Z}\}$ is a subset of $\left(\mathcal{Q}^*_k \cap \overline{\mathcal{N}^*_{y_0}}\right)$. A characteristic period, τ_{y_0}, is given by the smallest difference between two elements of $\mathcal{Q}^*_k \cap \overline{\mathcal{N}^*_{y_0}}$.

The same three possibilities hold for $\mathcal{Q}^*_k \cap \mathcal{N}^*_\sigma$. Although $\mathcal{N}^*_\sigma$ is not a module, the smallest difference between two elements of $\mathcal{Q}^*_k \cap \mathcal{N}^*_\sigma$ defines a period $\tau_\sigma \in \mathcal{M}^*_\sigma$, by (1). Thus, whenever $\mathcal{Q}^*_k \cap \mathcal{N}^*_\sigma$ has more than one element, if $k + m_1 h \in \mathcal{N}^*_\sigma$ then $\{k + m_1 h + m\tau_\sigma : m \in \mathbf{Z}\} = \mathcal{Q}^*_k \cap \mathcal{N}^*_\sigma$.

Repeating the construction for $\mathcal{Q}^*_k \cap \mathcal{M}^*_+$ and $\mathcal{Q}^*_k \cap \mathcal{M}^*_-$ we may define characteristic periods τ_+ and τ_-, respectively, when these sets have more than one element.

We complete the proof following the cases of Lemma 3.2.

Case 1) From $\mathcal{L}^* = \mathcal{N}^*_{y_0} \cup \mathcal{P}^*$, we get $\mathcal{M}^*_{y_0} \subset \mathcal{P}^*$ and, by (6), $\mathcal{M}^*_{y_0} = \{(0,0)\}$. Moreover, $\mathcal{Q}^*_k \cap \mathcal{N}^*_{y_0}$ must be infinite because $\mathcal{Q}^*_k \cap \mathcal{P}^*$ is finite. Thus, the period τ_{y_0} exists and $\mathcal{Q}^*_k - \overline{\mathcal{N}^*_{y_0}}$ is either empty or infinite. From $\left(\mathcal{Q}^*_k - \overline{\mathcal{N}^*_{y_0}}\right) \subset (\mathcal{Q}^*_k \cap \mathcal{P}^*)$ it follows that $\mathcal{L}^* = \overline{\mathcal{N}^*_{y_0}}$. Since $\sigma \in \mathbf{J}$, then $\mathcal{M}^*_{y_0} \neq \{(0,0)\}$ and so case 1) cannot occur.

Case 2) Here $\mathcal{L}^* = \mathcal{N}^*_{y_0} \cup \mathcal{N}^*_{\sigma} \cup \mathcal{P}^*$ which implies $\mathcal{M}^*_{y_0} \subset (\mathcal{N}^*_{\sigma} \cup \{(0,0)\})$, by (6). Moreover, $\mathcal{M}^*_{y_0} \neq \{(0,0)\}$ since $\sigma \in \mathbf{J}$. Suppose $\tilde{k} \in \mathcal{M}^*_{y_0}$ and $\tilde{k} \neq (0,0)$, then, $\tilde{k} \in \mathcal{N}^*_{\sigma}$ and $2\tilde{k} \in \mathcal{M}^*_{y_0}$. However, $2\tilde{k} \notin \mathcal{N}^*_{\sigma}$, by (1), and so case 2) is also impossible.

Case 3) We follow the arguments of case 1) using the least common multiple of the existing periods, τ_{y_0} or τ_+, instead of τ_{y_0}. Therefore $k \in \left(\mathcal{N}^*_{y_0} \cup \mathcal{M}^*_+\right)$ and condition A follows because $(0,0) \in \mathcal{M}^*_+$.

Case 4) This is like case 3) with $\mathcal{M}^*_-$ and τ_- instead of $\mathcal{M}^*_+$ and τ_+, yielding condition B.

Case 5) Here $\mathcal{Q}^*_k - \left(\mathcal{N}^*_{y_0} \cup \mathcal{M}^*_+ \cup \mathcal{N}^*_{\sigma}\right) = \emptyset$ because at least one of the periods τ_{y_0}, τ_+ or τ_σ exists and condition C follows. □

The extension of this result to functions $f : \mathbf{R}^n \longrightarrow \mathbf{R}$, with $n > 2$, is in preparation [9]. Condition C of the Proposition has a more complicated formulation because $\sigma v_+ - v_+ \in \mathcal{L}$ is not always true. Moreover, $\sigma(\mathcal{L}^*_1) = \mathcal{L}^*_1$ may also fail, and thus condition (3) does not always hold. In higher dimension Lemma 3.3 no longer holds and this in turn changes the proof of Lemma 3.4. However, the overall structure of the proof remains.

References

[1] Armstrong, M.A., *Groups and Symmetry*, Springer-Verlag, 1988.

[2] Bosch Vivancos, I., Chossat, P. and Melbourne, I., *New planforms in systems of partial differential equations with Euclidean symmetry*, Arch. Rat. Mech. Anal. **131** (1995) 199–224.

[3] Chillingworth, D. and Golubitsky, M., *Symmetry and pattern formation for a planar layer of nematic liquid crystal*, J. Math. Phys., Vol. 44, No. 9 (2003) 4201–4219.

[4] Dionne, B. and Golubitsky, M., *Planforms in two and three dimensions*, Z. Angew. Math. Phys. **43** (1992) 36–62.

[5] Dionne, B., Silber, M. and Skeldon, A.C. *Stability results for steady, spatially periodic planforms*, Nonlinearity **10** (1997) 321–353.

[6] Golubitsky, M. and Stewart, I., *The Symmetry Perspective*, Prog. Math. **200**, Birkhäuser Verlag, 2002

[7] Gomes, M.G.M., *Black-eye patterns: A representation of three-dimensional symmetries in thin domains*, Phys. Rev. E **60** (1999) 3741–3747.

[8] Melbourne, I., *Steady-state bifurcation with Euclidean symmetry*, Trans. Amer. Math. Soc. **351** (1999) 1575–1603.

[9] Pinho, E.M., *Symmetries of Projected Symmetric Patterns*, Thesis, University of Porto – in preparation.

[10] Rucklidge, A.M., Silber, M. and Fineberg, J., *Secondary instabilities of hexagons: a bifurcation analysis of experimentally observed Faraday wave patterns*, in Bifurcation, Symmetry and Patterns (Buescu, J., Castro, S., Dias, A.P. and Labouriau, I. eds.), Trends in Mathematics, Birkhäuser Verlag (2003) 101–114.

[11] Senechal, M., *Quasicrystals and Geometry*, Cambridge University Press, 1995.

Isabel S. Labouriau
Centro de Matemática da Universidade do Porto
Rua do Campo Alegre, 687
4169-007 Porto, Portugal
e-mail: islabour@fc.up.pt

Eliana M. Pinho
Departamento de Matemática Aplicada
Faculdade de Ciências da Universidade do Porto
Rua do Campo Alegre, 687
4169-007 Porto, Portugal
e-mail: eopinho@fc.up.pt

Real and Complex Singularities
Trends in Mathematics, 219–228

Modular Lines for Singularities of the T-series

Bernd Martin

Abstract. Unimodular functions have a μ-constant line in their miniversal unfoldings. Their miniversal deformations on the other hand contain a non-trivial τ-constant stratum only for the three cases of elliptic singularities. In computer experiments we found six sub-series of the T-series, which have a modular line in the their miniversal deformations. The singular locus of the family restricted to such a line splits into an elliptic singularity and another one of A_k-type, such that the deformation is τ-constant along the modular line. Each modular line can be patched together with the modular line of the associated elliptic singularity, completing it at infinity. All computations are based on the author's algorithm for computing modular spaces as flatness stratum of the relative cotangent cohomology inside a deformation.

Mathematics Subject Classification (2000). Primary 14B07; Secondary 32S30.

Keywords. Unimodal Singularities, Deformations of singularities, Modularity.

1. Introduction

The notion of a modular space has been introduced for complete complex varieties and for analytic polyhedron by Palamodov, and in a formal context by Laudal, cf. [Pal78], [Pal93], [Lau79]. One possible approach to constructing a kind of moduli of isolated singularities is restricting the miniversal family to subgerms that have a universal property for all families induced from it.

The author has developed an algorithm for calculating non-trivial examples of modular deformations for complete intersection singularities and for space curve singularities by connecting the modular property with flatness of the relative Tjurina module, [Mar02], [Mar03]. Further extensions of that concept are described in [HM04]. The flatness stratum may be computed by an obstruction calculus for lifting flatness of relative Artinian modules. Hence we can compute the jets of the flatness stratum. In general a power series representation of the defining ideal of the stratum is unavoidable as output of the algorithm. But an algebraic defining ideal of the flatness stratum is seen for small examples, as in the cases discussed below. The implementation and all computations are done in SINGULAR, [GPS02]. This has been described in detail elsewhere. Here we restrict ourselves to applications of this algorithm.

We present a new perspective on the classification of complex unimodular hypersurface singularities with respect to analytic $\mathcal{K}$-equivalence of germs. In particular we can compute series of compactifications of the moduli line of the three elliptic hypersurface singularities at infinity to a global modular deformation over a projective line. Hence, the compactification of a modular family is not unique. We find various splittings of the singularity under a modular deformation into a multi-germ. They have inspired the idea for a general combinatorial pattern of the modular strata of all T-series singularities. The existence of splitting lines can be checked afterwards by hand. Other examples give rice to the formulation of several hypotheses that are unproven so far. For completeness we also mention experimental results on the modular strata of the exceptional unimodular functions obtained by computer experiments.

2. Modular deformations

First, recall the definition and some basic properties of modularity, cf. for instance [Pal93], [Mar03], [HM04].

Definition 2.1. *Let $F : X \to S$ be a deformation of a complex germ X_0. A subgerm $M \subseteq S$ of the base germ is called* modular *if the following universal property holds: If $\varphi : T \to M$ and $\psi : T \to S$ are morphisms such that the induced deformations $\varphi^*(F_{|M})$ and $\psi^*(F)$ over T are isomorphic, then $\varphi = \psi$.*

Note that a unique maximal modular germ exists with respect to any deformation. The maximal modular germs in different miniversal deformations of X_0 are isomorphic with a unique isomorphism of germs, by definition. Hence we can speak of the *modular space* of the singularity X_0. It is sufficient to check the modular property for deformations over Artinian germs T. A representation of a deformation $\mathcal{X} \to \mathcal{M}$ is called (globally) modular if any subgerm of the deformation is modular.

The above definition coincides with Laudal's notion of a *pro-representing substratum* of a deformation functor. The basic characterizations of modularity in terms of the cotangent cohomology were already discussed by Palamodov and by Laudal in different contexts.

Proposition 2.2. *Given a miniversal deformation $F : X \to S$ of an isolated singularity X_0, the following conditions are equivalent for a subgerm of the base space $M \subseteq S$:*

i) *M is* modular.
ii) *M is* infinitesimally modular, *i.e., injectivity of the relative Kodaira-Spencer map $T^0(S, \mathcal{O}_M) \longrightarrow T^1(X/S, \mathcal{O}_S)_{|M}$ holds.*

iii) *M has the* lifting property *of vector fields of the special fiber:* $T^0(X/S, \mathcal{O}_S)_{|M} \longrightarrow T^0(X_0, \mathbb{C})_{|M}$ *is surjective.*

The tangent space of the modular space is identified with those tangent directions in $T(S) \cong T^1(X_0)$ on which the Lie bracket $[-,-] : T^0 \times T^1 \to T^1$ of the tangent cohomology in degree $(0,1)$ vanishes.

The motivation for this notion is the following observation: Take a small representation $\mathcal{M}$ of the modular germ $M(X_0)$. Isomorphisms of the fibers over $\mathcal{M}$ induce a discrete equivalence relation on $\mathcal{M}$. If there would be a 'moduli space' of singularities its germ at X_0 must coincide with the quotient of M by the induced equivalence relation. Hence, up to a finite covering the modular germ M should coincide with the germ of the moduli space. If a separated moduli space were to exist a modular family would have to have the following property:

A modular deformation over a punctured disc can be induced by at most one modular deformation over the complete disc.

Below we give several counter examples to that property.

The third characterization implies that the support of the modular space is the τ-constant stratum. The modular space of a quasihomogeneous isolated complete intersection singularity is reduced, [Ale85]. Hence it is the smooth τ-constant stratum corresponding to deformations on the Newton face. The ADE-singularities (i.e., the simple ones) are all quasihomogeneous. Their zero-graded part of the Tjurina algebra $T^1(X_0)_0$ vanishes and there are no deformations on the Newton face, hence the modular spaces are simple points.

In general, we have to expect a complicated non-reduced structure on the modular space. While the singular locus of a μ-constant family is always irreducible a splitting singular locus over a τ-constant family was first found in [Mar02] by computing non-trivial examples of modular spaces. More examples are given below.

Our algorithm for computing the modular spaces is based on the *flatness criterion*, here formulated for our situation.

Proposition 2.3.
Let X_0 be an isolated complete intersection singularity with miniversal deformation $F : X \to S$. Then the modular space coincides with the flatness stratum of the relative Tjurina module $T^1(X/S)$ as $\mathcal{O}_S$-module.

3. Unimodular singularities

Recall the classification of the unimodular hypersurface singularities. We find 14 exceptional singularities and the singularities of the T-series, [AGZV85]. We may restrict the discussion to three variables, because the co-rank of the Hessian of these singularities is either 2 or 3. It is easy to check that any exceptional one-parameter unimodular unfolding is written as $f_0(x)+\lambda h(x)$, where $f_0(x)$ is quasihomogeneous and $h(x)$ the associated Hesse monomial, i.e., the class of the determinant of the Hesse matrix of f_0 modulo the Jacobian ideal $J(f_0)$. Such a family splits into two

$\mathcal{K}$-classes: A quasihomogeneous one ($\lambda = 0$) and a semi-quasihomogeneous one of *Hesse*-type ($\lambda \neq 0$).

The singularities of the T-series are defined by the equations

$$T_{p,q,r} : \; x^p + y^q + z^r + \lambda xyz, \quad \frac{1}{p} + \frac{1}{q} + \frac{1}{r} \leq 1.$$

For $\frac{1}{p} + \frac{1}{q} + \frac{1}{r} < 1$, $\lambda \neq 0$, the singularity $T_{p,q,r}$ is called *hyperbolic* and its $\mathcal{K}$-class is independent of λ. The Newton boundary has three maximal faces and the singularity is neither quasihomogeneous nor semi quasihomogeneous. We have $\tau(T_{p,q,r}) = \mu(T_{p,q,r}) - 1 = p + q + r - 2$.

In exactly three cases we have $\frac{1}{p} + \frac{1}{q} + \frac{1}{r} = 1$ and the singularities are quasihomogeneous. They are called the *parabolic* singularities P_8, X_9 and J_{10} in Arnold's notation or *elliptic* hypersurface singularities $\tilde{E}_6$, $\tilde{E}_7$ and $\tilde{E}_8$, in K. Saito's paper, [Sai74]:

$$\begin{aligned} &\tilde{E}_6 = P_8 = T_{3,3,3} : \; x^3 + y^3 + z^3 + \lambda xyz, \; \lambda^3 \neq -3^3, \; \tau = \mu = 8; \\ &\tilde{E}_7 = X_9 = T_{4,4,2} : \; x^4 + y^4 + z^2 + \lambda xyz, \; \lambda^4 \neq 2^6, \; \tau = \mu = 9; \\ &\tilde{E}_8 = J_{10} = T_{6,3,2} : \; x^6 + y^3 + z^2 + \lambda xyz, \; \lambda^6 \neq 2^4 3^3, \; \tau = \mu = 10. \end{aligned}$$

It is well known that the $\mathcal{K}$-class is not determined by λ. The $\mathcal{K}$-equivalence induces a discrete equivalence relation on the λ-line with the indicated gaps. Its quotient is an affine line parametrized by the classical j-invariant of elliptic curves, i.e., a real sphere with one gap. Other normal forms of the elliptic singularities exists, where the connection to elliptic curves and the j-invariant is better seen, for instance:

$$\begin{aligned} &\tilde{E}_6 : \; x(x - y)(x - \nu y) - yz^2, \\ &\tilde{E}_7 : \; xy(x - y)(x - \nu y) - z^2, \\ &\tilde{E}_8 : \; x(x - y^2)(x - \nu y^2) - z^2, \end{aligned}$$

and $\nu \in \mathbb{C} - \{0, 1\}$, $j = \frac{4}{27} \frac{(\nu^2 - \nu + 1)^3}{\nu^2(\nu - 1)^2}$, see [Sai74].

Note, that the families $T_{4,4,2}(\lambda)$ and $T_{6,3,2}(\lambda)$ are not contained in a miniversal family. They form a double covering of the τ-constant line in a miniversal deformation, which can be realized by the substitution $z \mapsto z - (1/2)\lambda xy$:

$$x^4 + y^4 + z^2 + \lambda xyz \;\mapsto\; x^4 + y^4 + z^2 - \frac{1}{4}\lambda^2 x^2 y^2,$$

$$x^6 + y^3 + z^2 + \lambda xyz \mapsto \; x^6 + y^3 + z^2 - \frac{1}{4}\lambda^2 x^2 y^2.$$

In all three cases the $\mathcal{K}$-equivalence relations on the ν-lines are induced by an action of the permutation group S_3, while on the λ-lines the relations are induced by more complicated actions of finite groups. The tetrahedron group A_4 acts on the λ-line of the Hesse normal form of cubics $T_{3,3,3}$. The action is described with all details in the book of Brieskorn and Knörrer, cf. [BK81]. The obvious action by the third (resp. 4th and 6th root of unity) does not give the full equivalence relation.

4. Modular spaces of hyperbolic singularities

Hyperbolic singularities are the first type, that has a Newton boundary with more than one maximal face. We want to find the modular spaces, expecting either a fat point or a germ of a curve. A miniversal deformation is given by the family

$$F(x,y,z,t,u,v) := f + t_0 + \sum_{i=0}^{p-1} t_i x^i + \sum_{j=0}^{q-1} u_j x^j + \sum_{k=0}^{r-1} v_k x^k.$$

By a direct computation of the Lie bracket we obtain the dimension of the tangent space of the modular spaces equal to the co-rank of the singularity.

Proposition 4.1. *For a hyperbolic singularity $T_{p,q,r}$, $\frac{1}{p}+\frac{1}{q}+\frac{1}{r}<1$, the tangent directions of the modular space in miniversal family F given above are corresponding to the monomials $\{x^{p-1}, y^{q-1}, z^{r-1}\}$ if $p \geq q \geq r > 2$, or $\{x^{p-1}, y^{q-1}\}$ if $p \geq q \geq r = 2$ respectively.*

Proof. We only discuss the case $r > 2$. The annihilator of f in the Milnor algebra is the maximal ideal, $\mathrm{Ann}(f) = (x,y,z) \mod (J(f))$, because $\tau = \mu - 1$. Consequently $T^0(f)$ is generated as module by three derivations δ_x, δ_y, δ_z. A derivation δ_x with $\delta_x(f) = xf$ is given by $\delta_x := (\frac{1}{p}x^2 + \varepsilon y^{q-2}z^{r-2})\frac{\partial}{\partial x} + (\frac{1}{q}xy - \varepsilon z^{r-1})\frac{\partial}{\partial y} + (\frac{1}{r}xz + \varepsilon yz)\frac{\partial}{\partial x}$, with $\varepsilon := (1 - \frac{1}{p} - \frac{1}{q} - \frac{1}{r})(1 + x^{p-3}y^{q-3}z^{r-3})^{-1}$ a unit. Interchanging the variables cyclically we find the other derivations. The Lie bracket $[\delta, t]$, $\delta \in T^0(f)$ and $t \in T^1(f)$, is defined by the class of $(\delta(t) - h_\delta t)$ in $T^1(f) \cong \mathbb{C}\{x,y,z\}/(f, J(f))$, where $\delta(f) = h_\delta f$. We get $[\delta_x, x^{p-1}] = (\frac{1}{p} - 1)x^p \mod (f, J(f))$ and $[\delta_x, t] = 0$ for all $t \neq x^{p-1}$ from a monomial basis of $T^1(f)$. □

Careful inspection identifies six hyperbolic singularities with a τ-constant line in the miniversal deformation. They are characterized by the condition that exactly one of the parameters (p,q,r) differs by one from one of the elliptic types:

Proposition 4.2. *The following hyperbolic singularities are adjacent to an elliptic singularity with the same Tjurina number:*

$$\begin{array}{llll} T_{4,3,3} & & & \Longrightarrow \tilde{E}_6, \\ T_{4,4,3}, & T_{5,4,2} & & \Longrightarrow \tilde{E}_7, \\ T_{6,3,3}, & T_{6,4,2}, & T_{7,3,2} & \Longrightarrow \tilde{E}_8. \end{array}$$

They have a one-parameter τ-constant family in their miniversal deformations.

For instance the family $f_t := x^4 + y^3 + z^3 + xyz + tx^3$ in t is a τ-constant deformation, hence the base line-germ is contained in the modular space of $T_{4,3,3}$. The generic fiber singularity is $\tilde{E}_6$. Moreover, this modular deformation fits into the λ-line of $\tilde{E}_6(\lambda)$ at infinity: $f_t \sim_{\mathcal{K}} \tilde{E}_6(t^{(-1/3)})$ for $t \neq 0$. More exactly, the inducing morphism is a threefold covering from the t-line, $t \neq 0$, to the λ-line, $\lambda \neq 0$. The same holds in the other cases in a similar way. At least for $\tilde{E}_7$ and $\tilde{E}_8$ there are, hence, different ways to compactify the j-line at infinity.

The t-line is the reduced τ-constant stratum over the zero-section of the miniversal family, but it cannot be the whole modular space, because of the embedding dimension is greater than one. Indeed we find by computer in four cases as modular space the t-line with an embedded fat point. For the remaining two cases the Hilbert function of the local ring of the modular space stabilizes with value 2 for $T_{6,4,2}$ and with value 3 for $T_{6,3,3}$, causing two or three branches in the modular space. Clearly we have two copies of a τ-constant line by symmetry in y, z for $T_{6,3,3}$. The other two components turn out to be splitting lines, see below. We list the equations of modular spaces and their primary decomposition for three of the indicated cases as example:

The ideal of the modular space of $T_{4,3,3}$ is given by

$$(t_0,t_1,t_2,7u_1 + 6t_3v_2,v_1 + 6v_2^2,t_3u_2 - 7v_2^2,7u_2^2 - t_3v_2,u_2v_2,t_3^2v_2,t_3v_2^2,v_2^3).$$

It has two primary components: The line defined by the ideal

$$(t_0,t_1,t_2,u_1,u_2,v_1,v_2)$$

and a fat point of multiplicity 6 with equations

$$(t_0,t_1,t_2,7u_1 + 6t_3v_2,v_1 + 6v2^2,t_3^2,t_3u_2 - 7v_2^2,7u_2^2 - t_3v_2,u_2v_2,t_3v_2^2,v_2^3).$$

The ideal of the modular space of $T_{4,4,3}$ has the following form

$$(v_1,t_0,u_3v_2-14t_1,u_2v_2,u_1v_2,t_3v_2-14u_1,t_2v_2,t_1v_2,3u_3^2-16u_2,u_2u_3+12u_1, u_1u_3, t_3u_3,t_2u_3,t_1u_3,u_2^2,u_1u_2,t_3u_2,t_2u_2,t_1u_2,u_1^2,t_3u_1,t_2u_1,t_1u_1,3t_3^2-16t_2,- t_2t_3 + 12t_1,t_1t_3,t_2^2,t_1t_2,t_1^2).$$

It has two primary components: The line defined by ideal

$$(v_1,\ u_3,\ u_2,\ u_1,\ t_3,\ t_2,\ t_1,\ t_0)$$

and a fat point of multiplicity 8 with equations

$$(v_1,\ t_0,\ v_2^2,\ u_3v_2 - 14t_1,\ u_2v_2,\ u_1v_2,\ t_3v_2 - 14u_1,\ t_2v_2,\ t_1v_2,\ 3u_3^2 - 16u_2, u_2u_3 + 12u_1,\ u_1u_3,\ t_3u_3,\ t_2u_3,\ t_1u_3,\ u_2^2,\ u_1u_2,\ t_3u_2,\ t_2u_2,\ t_1u_2,\ u_1^2,\ t_3u_1, t_2u_1,\ t_1u_1,\ 3t_3^2 - 16t_2,\ t_2t_3 + 12t_1,\ t_1t_3,\ t_2^2,\ t_1t_2,\ t_1^2).$$

The modular space of $T_{6,4,2}$ is given by the ideal

$$(u_2,u_1,t_3,t_2,t_1,t_0,v_1^2,u_3v_1,t_5v_1,t_4v_1,t_5u_3 - 20v_1,t_4u_3,t_5^2 - 4t_4).$$

It has three primary components: The expected line defined by the ideal

$$(v_1, u_2, u_1,t_5,t_4,t_3,t_2,t_1,t_0)\ ,$$

another line (a splitting line, see below) defined by

$$(v_1,u_3,u_2,u_1,t_3,t_2,t_1,t_0,t_5^2 - 4t_4),$$

and a fat point of multiplicity 4 with equations

$$(u_2,u_1,t_4,t_3,t_2,t_1,t_0,v_1^2,u_3v_1,t_5v_1,u_3^2,t_5u_3 - 20v_1,t_5^2).$$

For all other hyperbolic singularities the Tjurina number at the origin drops outside the special fiber $(t, u, v) \neq \mathbf{0}$. Hence, one should expect a zero-dimensional modular space. Using the computer we obtain the jet of the modular space up to high order for all small values of the parameters (p, q, r). In many cases the output was a fat point as expected, but in various cases the Hilbert function stabilized at a value of 1, 2 or 3. This would imply one or more curve components in the modular space.

A careful analysis of the computable examples yields a combinatorial pattern of modular curves. Any of the six exceptional cases from above is the first member of a sub-series: These series are characterized by the property that two indices coincide with the indices of an elliptic singularity:

$T_{k,3,3}$, $k > l = 4$, associated to $\tilde{E}_6$,
$T_{k,4,2}$, $k > l = 5$, associated to $\tilde{E}_7$,
$T_{k,4,4}$, $k > l = 3$, associated to $\tilde{E}_7$,
$T_{k,3,2}$, $k > l = 7$, associated to $\tilde{E}_8$,
$T_{k,6,2}$, $k > l = 4$, associated to $\tilde{E}_8$,
$T_{k,6,3}$, $k > l = 3$, associated to $\tilde{E}_8$.

We can determine the curve components as smooth branches, such that the critical locus of the induced deformation over any branch is reducible and has constancy of the global Tjurina number of the singular multi-germ of the fibers.

Proposition 4.3. *Let f_0 be one singularity of one of the above series, $d := k - l$, then*

i) *the ideal*

$$(t_0, \ldots t_{l-1}, g_2, \ldots, g_d, u, v), \quad g_i := \binom{d}{i} t_{k-1}^i - d^i t_{k-i},$$

defines a modular line of the singularity f_0,

ii) *the corresponding modular family of multi-germs $f_a = f_0 + x^l(x+a)^d$, $a \in \mathbb{C}$, is τ-constant, and*

iii) *the modular family has a splitting singular locus if $d > 1$: the singularity at the origin of the special fiber $\{f_0 = 0\}$ splits for $a \neq 0$ into the associated elliptic singularity at the origin of the general fiber $\{f_a = 0\}$ and an A_{d-1}-singularity at the point $(-a, 0, 0)$.*

Proof. The last statement is checked by inspection. The equation $f_a = 0$ defines obviously at zero the associated elliptic singularity. Substituting $x := x - a$ we obtain $f_a(x - a, y, z) = x^d(x - a)^l + y^q + z^r + xyz - ayz$. This function has an A_{d-1}-point at the origin. Moreover the critical point $(-a, 0, 0)$ of f_a moves to the origin with $a \to 0$.

Consider an affine family of hypersurfaces $V(f_a(\mathbf{x})) \subset \mathbb{A}^n \times \mathbb{A}^1$, $a \in \mathbb{A}^1$. Its global Tjurina number, i.e., the sum of Tjurina numbers of all singularities in the fiber is semi-continuous: $\tau_a = \sum_{p \in Sing(f_a)} \tau(f_a, p) \leq \tau_0$. Consider only those components of the relative singular locus $V(f_a, \partial f_a / \partial \mathbf{x})$ that contain the origin $(\mathbf{0}, 0) \in A^n \times \mathbb{A}^1$. We add only the Tjurina numbers of singular point of these components. This restricted Tjurina number is a global Tjurina number of a multi-germ formed by the associated singularities of the fiber. It is still semi-continuous. The germ at the origin of the special fiber deforms into this multi-germ. If the restricted Tjurina number of the multi-germ is equal to $\tau(f_0, 0)$, it has to be a τ-*constant family* of multi-germs.

The difference of $\tau(f_0, 0)$ and the Tjurina number of the associated elliptic singularity is $(d-1)$. By semi-continuity any other component besides that of the A_{d-1}-points of the relative singular locus of the affine family cannot contain the origin. Hence the singularity of the special fiber splits exactly into two singularities, this splitting is τ-constant and the family is modular. □

Note that only the leading singularity of each sub-series does not have a splitting locus. On the other hand five special cases exist which have more than one modular line: three cases with two modular lines and two cases with three modular lines. They correspond either to overcrossings of the series or to extra symmetries with respect to variables:

Corollary 4.4.

i) $T_{6,3,3}$ *has three modular lines, one line of splitting singular type* $\tilde{E}_6 + A_2$ *and two lines of singular type* $\tilde{E}_8$.
ii) $T_{4,4,4}$ *has three modular lines all of splitting singular type* $\tilde{E}_7 + A_1$.
iii) $T_{6,4,2}$ *has two modular lines, one of splitting singular type* $\tilde{E}_7 + A_1$ *and one line of singular type* $\tilde{E}_8$.
iv) $T_{6,6,2}$ *has two modular lines, both of splitting singular type* $\tilde{E}_8 + A_2$.
v) $T_{6,6,3}$ *has two modular lines, both of splitting singular type* $\tilde{E}_8 + A_3$.

We cannot proof that we have found all modular spaces of the T-series with a one-dimensional support. But all computed examples underline this hypothesis:

The above modular lines are all occurring modular curves. The modular lines are primary components of the modular spaces.

The modular spaces of these singularities has another embedded fat point at the origin in all known examples with the exception of the very symmetric case $T_{4,4,4}$. All other modular spaces are expected to be zero-dimensional.

5. Modular fat points of exceptional unimodular functions

For completeness we give the result for the 14 exceptional unimodular singularities. We already pointed out that the modular space for the quasihomogeneous $\mathcal{K}$-class of an exceptional singularity is a simple point. The semi-quasihomogeneous ones are of Hesse type and have a trivial τ-constant stratum, too. This holds for nearly all singularities of Hesse type, apart from few exceptional cases, where a monomial of type $h(x)/x_i$ lies on the quasi-homogeneous face. Our list does not contain any such exceptions. Hence their modular spaces are fat points. They are all computable. The results are listed below. We get the following surprising coincidence of Hilbert functions of the Milnor algebra $Q(f) = \mathbb{C}\{x\}/J(f)$ and the local ring of the modular space:

Proposition 5.1. *The 14 exceptional unimodular semi-quasihomogeneous functions $f = f_0 + h$ have a fat point modular space $M(f)$ of multiplicity $\tau + 1$. It holds*

$$hilb(\mathcal{O}_{M(f)}) = hilb(Q(f)) = hilb(Q(f_0)).$$

We shall discuss one example:

$$S_{11}: \quad f_0 = x^4 + y^2z + xz^2, \quad h = x^3z, \quad \tau(f) = 10.$$

A miniversal deformation is given by

$$F = f + t_1 + t_2x + t_3y + t_4z + t_5x^2 + t_6xy + t_7xz + t_8x^3 + t_9x^2y + t_{10}x^2z.$$

We compute the reduced standard basis of the ideal of the fat point modular space. It is generated by 12 polynomials:

$$\begin{aligned}
p_1 &= 4t_1 + t_{10}^3t_8,\\
p_2 &= 132300\, t_2 - 44100\, t_{10}^2t_8 + 40729\, t_{10}^3t_8,\\
p_3 &= 11t_3 - 3t_{10}^2t_9,\\
p_4 &= 19404000\, t_4 - 3880800\, t_{10}^3 + 12012000\, t_{10}^2t_8 - 5681813\, t_{10}^3t_8,\\
p_5 &= 465696000\, t_5 + 232848000\, t_{10}t_8 - 279417600\, t_{10}^3 + 744282000\, t_{10}^2t_8\\
&\quad - 386202211\, t_{10}^3t_8,\\
p_6 &= 1617\, t_6 + 693\, t_{10}t_9 - 359\, t_{10}^2t_9,\\
p_7 &= 1397088000\, t_7 + 465696000\, t_{10}^2 - 1047816000\, t_{10}t_8 + 312404400\, t_{10}^3\\
&\quad - 811849500\, t_{10}^2t_8 + 409182569\, t_{10}^3t_8,\\
p_8 &= 7276500\, t_9^2 + 3234000\, t_{10}t_8 - 1455300\, t_{10}^3 + 4398625\, t_{10}^2t_8 - 2053928\, t_{10}^3t_8,\\
p_9 &= 231t_9t_8 - 52t_{10}^2t_9,\\
p_{10} &= 1293600\, t_8^2 - 1293600\, t_{10}^3 + 3742200\, t_{10}^2t_8 - 1893181\, t_{10}^3t_8,\\
p_{11} &= 1155\, t_{10}^4 - 4121\, t_{10}^3t_8,\\
p_{12} &= t_{10}^3t_9.
\end{aligned}$$

The first 7 polynomials are of the form $t_i + g_i(t_8, t_9, t_{10})$, $i = 1, \ldots, 7$. Hence the local ring $\mathcal{O}_M$ of the fat point is isomorphic to $\mathbb{C}\{t_8, t_9, t_{10}\}/(p_8, \ldots, p_{12})$. Its Hilbert sequence $(1, 3, 3, 3, 1)$ coincides with that of the Milnor algebra. It is not clear if $\mathcal{O}_M$ and $Q(f)$ are isomorphic.

6. More questions than answers

The computer based investigation of the unimodular function open a number of new questions and hypotheses. Any splitting line of a hyperbolic singularity can be considered as a compactification of the j-line of the associated elliptic singularity that induces a (globally) modular family. Hence the gap of the j-line can be closed by a singularity of arbitrarily high Tjurina number and possibly with greater corank $3 > 2$. One may ask:

- Are there other singularities compactifying the j-lines (of higher co-rank or non-hypersurfaces)?
- Are there other types of the splitting singular point than A_k?
- Are there examples with more than one splitting point?

Hyperbolic singularities, which do not belong to one of the above sub-series are expected to have a fat point modular space. As all examples fulfil the statement (5.1), one may hypothesise:

> *Any hyperbolic singularity f_0 not belonging to one of the six sub-series fulfill the Hilbert function equality:* $\mathrm{hilb}(\mathcal{O}_{M(f_0)}) = \mathrm{hilb}(Q(f_0))$.

These singularities are not of Hesse type. The only similarity of these singularities is the difference of Milnor number and Tjurina number $\mu - \tau = 1$. Whilst the involved computations are highly non-trivial, there is some hope to show that it is more than a strange coincidence.

References

[Ale85] A.G. Alexandrov, *Cohomology of a quasihomogeneous complete intersection*, Izv. Akad. Nauk SSSR Ser. Mat. **49**(3), (1985), 467–510.

[AGZV85] V.I. Arnol'd, S.M. Gusein-Zade, A.N. Varchenko, it Singularities of differentiable maps Vol. I, Brikhäuser, (1985).

[BK81] E. Brieskorn, H. Knörrer, *Ebene algebraische Kurven*, Birkhäuser, (1981).

[Lau79] O.A. Laudal, *Formal moduli of algebraic structures,* Lecture Notes in Math. **754**, Springer-Verlag, Berlin-New York, (1979).

[HM04] T. Hirsch, B. Martin, *Deformations with sections: Cotangent cohomology, flatness and modular subspaces*, Contem. math. and its appl., **15**, (2004), 49–70.

[Mar02] B. Martin, *Algorithmic computation of flattenings and of modular deformations*, J. Symbolic Computation **34** (2002), no. 3, 199–212.

[Mar03] B. Martin, *Modular deformation and Space Curve Singularities*, Rev. Mat. Iberoamericana **19** (2003), no. 2, 613–621.

[Pal78] V.P. Palamodov, *Moduli and versal deformations of complex spaces*, in: Variétés analytiques compactes, LNM **683** Springer-Verlag, Berlin-New York, (1978), 74–115.

[Pal93] V.P. Palamodov, *Deformation of analytic polyhedrons*, J. Algebraic Geometry **2**, (1993), 263–294.

[Sai74] K. Saito, *Einfach-elliptische Singularitäten*, Invent. Math. **23**, (1974), 289–325.

[GPS02] G.-M. Greuel, G. Pfister, H. Schnemann, SINGULAR 2.0, University of Kaiserslautern, 2002, `http://www.singular.uni-kl.de`.

Bernd Martin
BTU Cottbus
Inst. Mathematik
PF 101324
D-03046 Cottbus, Germany
e-mail: `martin@math.tu-cottbus.de`

Real and Complex Singularities
Trends in Mathematics, 229–246

Do Moduli of Goursat Distributions Appear on the Level of Nilpotent Approximations?

Piotr Mormul

Abstract. It is known that Goursat distributions (subbundles in the tangent bundles having the tower of consecutive Lie squares growing in ranks very slowly, always by one) possess, from corank 8 onwards, numerical moduli of the local classification, in both C^∞ and real analytic categories. (Whereas up to corank 7 that classification is discrete, as shown in a series of papers, the last in that series being [13].)
A natural question, first asked by A.Agrachev in 2000, is whether the moduli of Goursat distributions descend to the level of nilpotent approximations: whether they are stiff enough to survive the passing to the nilpotent level. In the present work we show that it is not the case for the first modulus appearing in corank 8 (and the only one known to-date in that corank).

Mathematics Subject Classification (2000). Primary 58A30; Secondary 58K50, 17B66, 17B30.

Keywords. Nilpotent approximation, Goursat distribution, local classification, continuous invariant, modulus.

1. Introduction

We want to use throughout the present work several notions related to the nilpotent approximations of geometric distributions. Those approximations have been, over the past 30 years, rather painfully finding their way to the mathematical usage, in different contexts and under different names. We just say that they can be viewed as a far reaching generalization of the notion of *linearization* of a single vector field. The linearization of a vector field v at a point, although simplifying geometry a big deal, retains some basic local properties of v. Likewise, the nilpotent (or graded) approximations simplify enormously the geometry of distributions without losing the most essential (mainly nonholonomic) traits of them.

Research supported by Polish KBN grant 2 P03A 010 22.

In a coordinate language, first steps bringing those objects to existence were made in [9] and decisive ones in [4], [2], [8], with substantial later simplifications proposed in [6]. Local coordinates in which nilpotent approximations can be viewed (something like night glasses in nonholonomic geometry) have a separate history of their own. They are useful in sub-Riemannian geometry, too, from the basic (nonholonomic) Ball-Box Theorem onwards – see [6] and the entire book *Sub-Riemannian Geometry* containing that contribution. We note that there exists also a purely sub-Riemannian version of nilpotent approximations, introduced (and used) in [3]. On the coordinate-free side there exists a keystone (if not published) short text [1] and recent contribution [5], itself a forerunner to a bigger work.

The adjective 'nilpotent' is related to the fact that, whatever local generators of a given distribution D, the simplified (or: trimmed) generators of the approximation around a point p generate a nilpotent Lie algebra, over the reals, of precisely known nilpotency order stemming from the geometry of D in the vicinity of p – equal to the nonholonomy degree of D at p. (In [5] there is given a much deeper explanation and interpretation of nilpotent approximations making reference to the concept of *nonholonomic tangent spaces.*) In the present paper, we will consequently use the abbreviation NA for 'nilpotent approximation'.

Until recently, not many examples of NAs have been effectively computed in dimensions exceeding four (in [2] examples are just illustrating definitions and are mainly in dimension 3, in [19] – in dimension 5, in [14] – in dimension 6). The analysis of invariant parameters (moduli) in families of distributions and their hypothetical surviving after descending to the simpler level of NAs has – to our knowledge – not yet been investigated.

An ideal environment for these directions of research seems to be the world of Goursat distributions. On the one side, they are sufficiently tight and possessing clear polynomial presentations as to allow for efficient computations. For instance, they are free of *functional* moduli. On the other, are abundant with *numeric* moduli, found not earlier than in the end of XX century. In fact, the very question of the descending of moduli of Goursat objects to the nilpotent level was asked by Agrachev in the year 2000 and has since sparked an entire line of work, like [14] or [17], leading also to the present text.
It is known that the moduli of Goursat objects show up not earlier than in corank 8. As a matter of fact, in corank 8, to-date only one 1-parameter family $\mathcal{F}$ of pairwise nonequivalent (and also unimodal in Arnold's sense) germs has been found; in certain appropriate local coordinates these are the objects (5.1).

The aim of the paper is to show, in our Theorem 5.1, that, despite strong opposite expectations, that modulus of Goursat distributions does *not* descend to the nilpotent approximations. The message of the paper is that all NAs of the objects in the family $\mathcal{F}$ can be computed to the very end and turn out to be pairwise equivalent, hence moduliless.

The techniques used in the paper can be divided into two categories. In the (more elementary) beginning they consist of our shortcuts to a procedure proposed

by Bellaïche in [6]. Those shortcuts were developed recently in [14] and [17] for the computation of NAs of Goursat germs in dimensions not exceeding 8. When applied to the objects treated in [17], they quickly lead to the very output of the (long) algorithm of [6]. The objects addressed in the present paper live in dimension 10 but have some affinity to those handled in [17] (are their two steps' prolongations, or: grandchildren). Because of that computing the NAs for them *can* spring from those previously done computations.
In the (more advanced) continuation the handling of the members of $\mathcal{F}$ goes well beyond the Bellaïche procedure. In that advanced part we have not even attempted to apply his algorithm to $\mathcal{F}$.[1] Instead, we exploit to the limit the particular Goursat character of our distributions and produce needed adapted coordinates in a concise way. That main body of the paper is included in Section 5, with all turning points (but not all underlying computations) presented.

We note here that our procedures work thanks to a series of surprising properties of the members of $\mathcal{F}$ which enormously simplify computations. In Section 5.5, we nickname those mysterious properties 'the flatness of Goursat distributions'.

2. Nilpotent approximation of a distribution at a point

For any distribution D of rank d on an n-dimensional, smooth or real analytic, manifold M (i.e., a rank-d subbundle in the tangent bundle TM) its *small flag* is the nested sequence

$$V_1 \subset V_2 \subset V_3 \subset V_4 \subset \cdots$$

of modules (or: presheaves of modules) of vector fields, of the same category as M, tangent to M: $V_1 = D$, $V_{j+1} = V_j + [D, V_j]$ for $j = 1, 2, \ldots$ The *small growth vector* at $p \in M$ is the sequence (n_j) of linear dimensions at p of the modules V_j: $n_j = \dim V_j(p)$. Naturally, $n_1 = d$ independently of p.

D is *completely nonholonomic* when at every point of M its small growth vector attains (sooner or later) the highest value $n = \dim M$. We truncate that vector after the first appearance of n in it. The length d_{NH} of the truncated vector is called the *nonholonomy degree* of D at p.

In the theory that we only sketch here (cf. [9], [2], [8], [6]; this list is not complete) important are the *weights* w_i related to the small flag at a point: $w_1 = \cdots = w_d = 1$, $w_{d+1} = \cdots = w_{n_2} = 2$ (no value 2 among them when $n_2 = n_1$ $(= d)$, and generally

$$w_{n_j+1} = \cdots = w_{n_{j+1}} = j + 1$$

(no value $j + 1$ among the w's when $n_j = n_{j+1}$) for $j = 1, 2, \ldots$

Definition 2.1. For a completely nonholonomic distribution D on M, coordinates $z_1, z_2, \ldots, z_n$ around $p \in M$ (centered at p) are *linearly adapted* at p when $D(p) = (\partial_1, \ldots, \partial_d)$, $V_2(p) = (\partial_1, \ldots, \partial_d, \ldots, \partial_{n_2})$, and so on until $V_{d_{\mathrm{NH}}}(p) =$

[1] When applied with naked force, it quickly becomes cumbersome. There will be, in our opinion, no Bellaïche algorithm' software for many years to come.

$(\partial_1, \dots, \partial_n) = T_pM$. Here and in the sequel we skip writing 'span' before a set of vector fields' generators.
For such linearly adapted coordinates one defines (as in [6]) their *weights* $w(z_i) = w_i,\ i = 1, \dots, n$.

On the other hand, given a completely nonholonomic D, every smooth function f on M has, at any point $p \in M$, its *nonholonomic order* $\text{nord}(f)$ with respect to D (for simplicity of notation, we skip writing its dependence on p). By definition, it is the minimal order of a nonholonomic derivative of f that is non-zero at p.[2]

It follows directly from the above definitions that, for linearly adapted coordinates, their nonholonomic orders **do not exceed** their weights.

Definition 2.2. Linearly adapted coordinates $z_1, \dots, z_n$ are *adapted* when $\text{nord}(z_i)$ *equals* $w(z_i)$ for $i = 1, \dots, n$.

It is rather laborious to show, but adapted coordinates always exist, see in this respect [2], [8], [6], [19]. (This last and recent reference summarizes on Bellaïche's variables in a way perhaps more readable than [6] itself.) Moreover, adapted variables are by far not unique; there remains plenty of freedom behind the requirement being imposed on linearly adapted coordinates that the nonholonomic orders be maximal possible. On the other hand, by certain omissions, there occasionally appear in the literature (e.g., Theorem 1.5 in [7]) too simplistic formulae for upgrading linearly adapted variables to adapted ones.

In adapted coordinates it is reasonable to attach quasihomogeneous weights also to monomial vector fields (this definition goes back to the work [18] in the theory of differential operators; for geometric distributions see in this respect [2], p. 215). Namely,

$$w(z_{i_1} \cdots z_{i_k} \partial_j) = w(z_{i_1}) + \cdots + w(z_{i_k}) - w(z_j)\,. \tag{2.1}$$

The gist of the concept of adaptedness resides in the following

Proposition 2.3. *Every smooth vector field X with values in D has in its Taylor expansion, in arbitrary coordinates adapted for the relevant germ of D, only terms of weights not smaller than -1 that can be grouped in homogeneous summands* $X = X^{(-1)} + X^{(0)} + X^{(1)} + \cdots$

(superscripts mean the weights defined by (2.1)). We denote by $\widehat{X}$ the lowest ('nilpotent') summand $X^{(-1)}$. That is, $\widehat{X} = X^{(-1)}$.

When a distribution D has around p local generators (vector fields) $X_1, \dots, X_d$, then

Definition 2.4. The distribution $\widehat{D} = \big(\widehat{X_1}, \dots, \widehat{X_d}\big)$, defined on M locally around p, is called the *nilpotent approximation* of D at p.

It is proved in Proposition 5.20 in [6] that this object $\widehat{D}$ is well defined, independently of the adapted coordinates being used. Its basic property is

---

[2] $+\infty$ is not excluded.

Proposition 2.5. *At the reference point p, the small flag of $\widehat{D}$ coincides with that of D at p. Hence $\widehat{D}$ has at p the same small growth vector as D (and hence the same nonholonomy degree d_{NH}, too).*

This property is crucial. It shows that, in the occurrence, much simpler objects retain some basic geometric characteristics of the initial objects. One can deeply trim a distribution germ without losing essential information! This opportunity can only support one's hope for the survival of moduli in nilpotent approximations.

Attention. There is, however, one warning pointing in the opposite direction: unlike the small growth vector at the reference point, the *big growth vector* of a distribution D at a point (the sequence of linear dimensions at a point of the *big flag* of D – the tower of modules of vector fields – consecutive Lie squares $D \subset [D, D] \subset [[D, D], [D, D]] \subset \cdots$) is **not** preserved under passing to the approximation. See in this respect Corollary 4.3 later on, and also pp. 258–59 in [14]. One time the proof of Theorem 5.1 finished, Corollary 5.3 shows that poor performance of the big flag also for the members of the family $\mathcal{F}$.

3. Kumpera-Ruiz normal forms for Goursat distributions

In the sequel we deal uniquely with Goursat distributions – a rather restricted class of objects for which preliminary (local) polynomial normal forms of [11] exist with real parameters only, and no functional moduli.

A distribution $D \subset TM$ is Goursat when it is rank-2 and the big growth vector of D is, at every point $p \in M$, just $[2, 3, 4, \ldots, n-1, n]$, where $n = \dim M \geq 4$. The number $n - 2 \geq 2$ is called the *length* of the [big] flag of D.

(Sometimes the assumption $\operatorname{rk} D = 2$ is being dropped in this definition, like, for instance, in [12], [15], [16]. Both variants locally lead to the same theory, because there always splits off an integrable corank-2 subdistribution in D. In fact, that splitting object is the Cauchy-characteristic subdistribution of D.)

There exists a very basic partition of Goursat germs into disjoint *geometric classes* encoded by words of length $n - 2$ over the alphabet G, S, T, with two first letters always G and such that never a T goes directly after a G. Their construction has its roots in the pioneering work [10] of Jean, in which the author used a trigonometric, not polynomial, presentation of Goursat objects. That construction, with some natural subsequent improvements, has been reproduced in detail in Section 1.1 of [16].

In dimension 4 there is but one class GG, in dimension 5 – only GGG and GGS, in dimension 6 – GGGG, GGSG, GGST, GGSS, GGGS.

The union of all geometric classes ('quarks') of fixed length with letters S in *fixed* positions in the codes is called, after [12], a *Kumpera-Ruiz class* (a 'particle') of Goursat germs of that corank. For instance, in dimension 6 the two geometric classes GGSG and GGST build up one KR class $**S*$. In dimension 7 the geometric classes GGSGG, GGSTG, and GGSTT build $**S**$, etc.

What are the mentioned polynomial (local) presentations of Goursat objects? The essence of the contribution [11], given in the notation of vector fields and taking into account posterior works, is as follows. One constructs first a (not unique, depending on a number of real parameters) rank-2 distribution on $(\mathbb{R}^n(x^1, \dots, x^n), 0)$ departing from the code of a geometric class $\mathcal{C}$.

When the code starts with precisely s letters G, one puts $\overset{1}{Y} = \partial_1$, $\overset{2}{Y} = \overset{1}{Y} + x^3\partial_2, \dots, \overset{s+1}{Y} = \overset{s}{Y} + x^{s+2}\partial_{s+1}$. When $s < n-2$, then the $(s+1)$st letter in $\mathcal{C}$ is S. More generally, if the mth letter in $\mathcal{C}$ is S, and $\overset{m}{Y}$ is already defined, then

$$\overset{m+1}{Y} = x^{m+2}\overset{m}{Y} + \partial_{m+1}\,.$$

But there can also be T's or G's after an S. If the mth letter in $\mathcal{C}$ is not S, and $\overset{m}{Y}$ is already defined, then

$$\overset{m+1}{Y} = \overset{m}{Y} + \left(c^{m+2} + x^{m+2}\right)\partial_{m+1}\,,$$

where a real constant c^{m+2} is not absolutely free but

- equal to 0 when the mth letter in $\mathcal{C}$ is T,
- not equal to 0 when the mth letter is G going directly after a string ST…T (or after a short string S).

Now, on putting $\mathbf{X} = \partial_n$ and $\mathbf{Y} = \overset{n-1}{Y}$, and understanding $(\mathbf{X}, \mathbf{Y})$ as the germ at $0 \in \mathbb{R}^n$, we have

Theorem 3.1 ([11]). *Any Goursat germ D on a manifold of dimension n, sitting in a geometric class $\mathcal{C}$, can be put (in certain local coordinates) in a form $D = (\mathbf{X}, \mathbf{Y})$, with certain constants in the writing of the field $\mathbf{Y}$ corresponding to* G*'s past the first* S *in $\mathcal{C}$.*

This will be the main tool for us. In the next Section it will be applied to one particular geometric class in length 6 whose NAs have been computed in [17]. We mean the class $\mathcal{C}$ = GGSGSG. It is the (double) prolongation by the sequence SG of the very class GGSG in length 4 in which there was discovered, [14], the phenomenon of the loss of strong nilpotency among nilpotentizable (that is, in our terminology, weakly nilpotent) distributions. Note also that one more sequence SG of prolongations makes from $\mathcal{C}$ the family $\mathcal{F}$ = GGSGSGSG – the object of the present work.

In order to show how quickly the complexity of the nilpotent matter grows, let us mention that the NAs for GGSG were computed in *hours* and just waited for an interpretation. Later the NAs for $\mathcal{C}$ were computed in *weeks*, then still needed a substantial simplification, and eventually suggested that the loss of strong nilpotency was likely to be frequent in Goursat world. Whereas the computation of NAs for GGSGSGSG was scattered over an interval of *years* (Section 5 shows the essence of those developments) and the answer it brought was unexpected.

Open Question. In the paper's end, in Remark 5.4 we list an infinite series of multi-modal Goursat distributions for which Agrachev's question is widely open. And right here point to one geometric class GGGSGSGGG that also conceals, [15], a modulus of the local classification. This class, although living in the underlying dimension 11, not 10, is simpler than $\mathcal{F}$. For it is of codimension 2, not 3 like $\mathcal{F}$ (codimension of a geometric class, cf. [16], is the number of *not* G letters in its code). This notwithstanding, we do not know if its modulus descends to the NA's level.

4. Nilpotent approximation in the class GGSGSG

In order to close up the subject, we apply Theorem 3.1 to Goursat distributions living on 8-dimensional manifolds, around points where their germs sit in the geometric class GGSGSG. For a given such germ D, in certain coordinates x^1, $x^2, \ldots, x^8$ in $\mathbb{R}^8$, one has

$$D = (\mathbf{X},\ \mathbf{Y}) = \left(\partial_8,\ x^5x^7\partial_1 + x^3x^5x^7\partial_2 + x^4x^5x^7\partial_3 + x^7\partial_4 + X^6x^7\partial_5 + \partial_6 + X^8\partial_7\right), \tag{4.1}$$

the germ at $0 \in \mathbb{R}^8$, where, for simplicity, the constant in $X^6 = 1 + x^6$ is already normalized to 1, and the constant c in $X^8 = c + x^8$ is not zero.[3] The small growth vector of D at the reference point can be either computed directly or found in the literature. It is, and independently of c, $[2,\ 3,\ 4,\ 5_2,\ 6_2,\ 7_4,\ 8]$, where the subscripts in this context mean (here and in the sequel) the numbers of repetitions of a given integer. Therefore, the weights w_i $(i = 1, \ldots, 8)$ are

$$1,\ 1,\ 2,\ 3,\ 4,\ 6,\ 8,\ 12. \tag{4.2}$$

Due to nonzero constants present in (4.1), the original variables $x^1, \ldots, x^8$ are not linearly adapted. Watching the small flag of the distribution (4.1) at 0 and improving the x variables accordingly, the coordinates

$$x^8,\ x^6,\ x^7 - cx^6,\ x^4,\ x^5 - x^4,\ x^1,\ x^3,\ x^2 \tag{4.3}$$

already are linearly adapted and so get their respective weights (4.2). One of objectives in [17] was to improve further the functions (4.3), keeping their linear parts at 0, to certain adapted coordinates $z_1, z_2, \ldots, z_8$. To that end, we firstly applied the recursive procedure from [6]. Later we found a much shorter way, consequently exploiting the Goursat character of the objects. Arriving in both cases at exactly the same adapted variables upgrading (4.3). Namely,

$$z_1 = x^8, \qquad z_2 = x^6, \qquad z_3 = x^7 - cx^6, \qquad z_4 = x^4 - \frac{c}{2}\left(x^6\right)^2, \tag{4.4}$$

[3] this c could also be normalized to 1 but we refrain from doing so in order to better follow the further corrections and improvements of the coordinates. In the pseudo-normal form (5.1) for GGSGSGSG in Section 5, a similar constant c will already be crucial, not superfluous. That is to say, that Kumpera-Ruiz form will be just normal.

$$z_5 = x^5 - x^4 - \frac{c}{3}\left(x^6\right)^3, \quad z_6 = x^1 - \frac{c}{2}x^4\left(x^6\right)^2 + \frac{c^2}{8}\left(x^6\right)^4 - \frac{c^2}{15}\left(x^6\right)^5, \tag{4.5}$$

$$z_7 = x^3 - \frac{c^2}{4}x^4\left(x^6\right)^4 + \frac{c^3}{12}\left(x^6\right)^6 - \frac{c^3}{42}\left(x^6\right)^7, \tag{4.6}$$

$$z_8 = x^2 - \frac{c^4}{48}x^4\left(x^6\right)^8 + \frac{c^5}{120}\left(x^6\right)^{10} - \frac{13\,c^5}{5544}\left(x^6\right)^{11}.$$

Remark 4.1. It is also important to note that z_1 through z_5 in (4.4) and (4.5) are but replicas, with only an upward shift in indices, of the relevant formulae, obtained by the Bellaïche algorithm, for the NAs of the objects in the class GGSG, see p. 258 in [14]. (The departure objects sitting in GGSG have to be written in Kumpera-Ruiz coordinates with the additive constant next to x^6 not as 1, but as $c \neq 0$. In [14] that constant is normalized to 1.)

Now one can write D in these z coordinates, cf. [17], then extract $\widehat{D}$. In the outcome, $\widehat{\mathbf{X}} = \mathbf{X} = \partial_1$, and

$$\widehat{\mathbf{Y}} = \partial_2 + z_1\partial_3 + z_3\partial_4 + z_2z_3\partial_5 + \Big(z_3z_4 + \frac{c}{3}z_2^3z_3 + cz_2z_5\Big)\partial_6 + \Big(cz_2z_4^2 + cz_2^2z_3z_4$$
$$+ \frac{c^2}{2}z_2^3z_5 + \frac{c^2}{3}z_2^4z_4 + \frac{c^2}{6}z_2^5z_3\Big)\partial_7 + \Big(\frac{c^3}{4}z_2^5z_4^2 + \frac{c^3}{6}z_2^6z_3z_4 + \frac{c^2}{2}z_2^3z_7$$
$$+ \frac{c^4}{24}z_2^7z_5 + \frac{3\,c^4}{28}z_2^8z_4 + \frac{13\,c^4}{504}z_2^9z_3 + \frac{c^5}{126}z_2^{11}\Big)\,\partial_8\,.$$

It is an explicitly given vector field, although its expression is illegible. It turns out that among sets of adapted coordinates there are good and better ones, and certain are as good as night glasses. In fact,

Proposition 4.2 ([17]). *The nilpotent approximations of Goursat germs in the geometric class* GGSGSG *can be written down using only two adapted coordinates* z_1, z_2.

In the proof, p. 1620 in [17], one passes to newer, still adapted coordinates $z_1, z_2, \dots, z_{10}$ in which $\widehat{\mathbf{X}} = \partial_1$ and

$$\widehat{\mathbf{Y}} = \partial_2 + z_1\partial_3 - z_1z_2\partial_4 - \frac{1}{2}z_1z_2^2\partial_5 + \frac{c}{24}z_1z_2^4\partial_6 + \frac{c^2}{240}z_1z_2^6\partial_7 - \frac{c^4}{34560}z_1z_2^{10}\partial_8\,. \tag{4.7}$$

There is still the constant c in (4.7) which can clearly be eliminated by a rescaling of variables. Reiterating, that constant has been redundant in the input objects (4.1) and it is redundant in the output.

Corollary 4.3. *The nilpotent approximations of germs in* GGSGSG *have the big growth vector at the reference points equal to the small one, hence equal to* $[2,\,3,\,4,\,5_2,\,6_2,\,7_4,\,8]$.

(Compare Proposition 2.5 above and the proof of Proposition 3 in [17].) This is to be compared with the fact that the initial germs are Goursat, hence have the big growth vector $[2,3,4,5,6,7,8]$.

5. Nilpotent approximation in the geometric class GGSGSGSG

The aim of the present work is to prove the following

Theorem 5.1. *The modulus of the local classification residing in the geometric class* GGSGSGSG *disappears on the level of nilpotent approximations. That is, the nilpotent approximations of the members of this class are all equivalent.*

Our proof will firstly focus on finding certain adapted coordinates, in order to get hold of the NAs of distributions in GGSGSGSG. We will start with our shortcut to Bellaïche (by extending and extrapolating Remark 4.1) but later diverge radically from [6]: the last three, most involved adapted coordinates will be sought in a new, Goursat-motivated way. The obtained NAs will at first be illegible. Special efforts will be needed, like those underlying Proposition 4.2, to make them legible. A key role will be played by resonances among coefficients of quasi-homogeneous polynomials, similar to those holding true for GGSGSG, which have made possible the radical simplifications of [17], recalled in the previous chapter.

Proof. To get started, let us write the Kumpera-Ruiz pseudo-normal forms that in the occurrence are exact local models,

$$D = (\mathbf{X},\, \mathbf{Y}) = \big(\partial_{10},\; x^9x^7x^5\partial_1 + x^9x^7x^5x^3\partial_2 + x^9x^7x^5x^4\partial_3 + x^9x^7\partial_4 \\ + x^9x^7X^6\partial_5 + x^9\partial_6 + x^9X^8\partial_7 + \partial_8 + X^{10}\partial_9\big)\,, \qquad (5.1)$$

where $X^6 = 1+x^6$, $X^8 = 1+x^8$, $X^{10} = c+x^{10}$ with the value $c \neq 0$ being univocally tied to D (different values parametrizing different and pairwise nonequivalent D's). It is a matter of some computations (with the underlying papers [10], [16] at hand) that the small growth vector of D at the reference point $0 \in \mathbb{R}^{10}$ is, regardless of the value of $c \neq 0$, $[2,\, 3,\, 4,\, 5_2,\, 6_2,\, 7_4,\, 8_4,\, 9_8,\, 10]$, and hence the weights $w_1, w_2, \ldots, w_{10}$ are

$$1,\; 1,\; 2,\; 3,\; 4,\; 6,\; 8,\; 12,\; 16,\; 24. \qquad (5.2)$$

Clearly, the KR variables used in (5.1) are not linearly adapted. After some Lie bracket manipulations over the distribution (5.1), one quickly improves them to linearly adapted coordinates

$$x^{10},\;\; x^8,\;\; x^9 - cx^8,\;\; x^6,\;\; x^7 - x^6,\;\; x^4,\;\; x^5 - x^4,\;\; x^1,\;\; x^3,\;\; x^2\,. \qquad (5.3)$$

Then, an upgrading to adapted coordinates happens to be waiting for us, as regards the first seven variables in this list. We mean that the affinity observed between the NAs in GGSG and GGSGSG (Remark 4.1) extends onto the pair GGSGSG and GGSGSGSG, too. The formulas for adapted $z_1, z_2, \ldots, z_7$ are, modulo a shift in indices, due replicas of (4.4), (4.5), (4.6) (hence, formally, come from Bellaïche, too). In fact,

$$z_1 = x^{10}, \quad z_2 = x^8, \quad z_3 = x^9 - cx^8, \quad z_4 = x^6 - \frac{c}{2}\big(x^8\big)^2, \quad z_5 = x^7 - x^6 - \frac{c}{3}\big(x^8\big)^3,$$

$$z_6 = x^4 - \frac{c}{2}x^6\left(x^8\right)^2 + \frac{c^2}{8}\left(x^8\right)^4 - \frac{c^2}{15}\left(x^8\right)^5,$$
$$z_7 = x^5 - x^4 - \frac{c^2}{4}x^6\left(x^8\right)^4 + \frac{c^3}{12}\left(x^8\right)^6 - \frac{c^3}{42}\left(x^8\right)^7.$$

The justification repeats word for word the one given in [17] for GGSGSG. We know, therefore, that the functions $z_1, z_2, \ldots, z_7$ are already of nonholonomic orders $w_1, w_2, \ldots, w_7$ (see (5.2)) at 0, respectively.

How a check would look like for, say, z_7? Clearly, $\mathbf{X}z_7 = 0$ identically and the function $\mathbf{Y}z_7$ satisfies $\operatorname{nord}(\mathbf{Y}z_7) \geq 7$, implying $\operatorname{nord}(z_7) \geq w_7$ (whereas the $\leq$ inequality holds for any linearly adapted coordinate). Indeed,

$$\mathbf{Y}z_7 = X^6x^7x^9 - x^7x^9 - \frac{c^2}{4}x^9\left(x^8\right)^4 - c^2x^6\left(x^8\right)^3 + \frac{c^3}{2}\left(x^8\right)^5 - \frac{c^3}{6}\left(x^8\right)^6 \tag{5.4}$$

and $X^6x^7x^9 - x^7x^9 = x^6x^7x^9$. To check orders, it suffices to express the RHS of (5.4) in the z variables, $x^6 = z_4 + \frac{c}{2}z_2^2$, $x^7 = z_5 + z_4 + \frac{c}{2}z_2^2 + \frac{c}{3}z_2^3$, $x^8 = z_2$, $x^9 = z_3 + cz_2$ and open all brackets, finding only terms of nonholonomic orders ≥ 7 at 0.

Note that, in parallel, the first seven components of the generator $\mathbf{Y}$ are got expressed in the z variables,

$$\begin{aligned}\mathbf{Y} = {} & \partial_2 + z_1\partial_3 + z_3\partial_4 + z_2z_3\partial_5 \\ & + \Big(\big(\text{a term of order } > 5\big) + z_3z_4 + \frac{c}{3}z_2^3z_3 + cz_2z_5\Big)\partial_6 \\ & + \Big(\big(\text{terms of orders } > 7\big) \\ & \qquad + cz_2z_4^2 + cz_2^2z_3z_4 + \frac{c^2}{2}z_2^3z_5 + \frac{c^2}{3}z_2^4z_4 + \frac{c^2}{6}z_2^5z_3\Big)\partial_7 + \cdots,\end{aligned}$$

these seven components being formally identical with those in [17] (compare (6), (7), (9) there).

5.1. Upgrading of x^1

Starting from handling x^1, we put forward – solely in the Goursat environment – shorter ways of improving functions than via the Bellaïche algorithm. In fact, in [6] the terms correcting a given linearly adapted coordinate, say y_q, were sought in the pool of monomials $y_1^{\alpha_1}y_2^{\alpha_2}\cdots y_l^{\alpha_l}$ defined by the condition

$$2 \leq \alpha_1w_1 + \alpha_2w_2 + \cdots + \alpha_lw_l < w_q\,.$$

Concerning x^1, for our correcting terms there will happen an equality $= w_q$, and there will even be inequalities $> w_q$ for the ones correcting x^3 and x^2.

The variable x^1 is of nonholonomic order 8 instead of 12 ($= w_8$), because $\mathbf{X}x^1 = 0$ and $\mathbf{Y}x^1 = x^9x^7x^5$ has order $1 + 2 + 4 = 7$. Its upgrading to an adapted z_8 is short, if surprising. In the beginning, a try $z_8 = x^1 - \frac{1}{2}\left(x^4\right)^2 + \cdots$ prompts by itself, for

$$\mathbf{Y}\left(x^1 - \frac{1}{2}\left(x^4\right)^2 + \cdots\right) = x^9x^7x^5 - x^9x^7x^4 + \cdots = x^9x^7(x^5 - x^4) + \cdots,$$

and $x^5 - x^4$ is among the functions (5.3). Now

$$x^9x^7(x^5 - x^4) = x^9x^7\left(z_7 + \frac{c^2}{4}z_2^4x^6 - \frac{c^3}{12}z_2^6 + \frac{c^3}{42}z_2^7\right) \tag{5.5}$$

has order 9 and its to-be-removed terms are (in the z variables that are still under construction) z_2^9, z_2^{10}, $z_2^8z_3$, $z_2^7z_4$. Hence further improvements are necessary. A natural idea is to get rid in (5.5) of the entire term $x^9x^7z_2^6$ (of order 9). We seek it via $z_8 = x^1 - \frac{1}{2}\left(x^4\right)^2 + *\,x^4\left(x^8\right)^6 + \cdots$ and expand

$$\begin{aligned}\mathbf{Y}z_8 &= x^9x^7\left(x^5 - x^4 + *\left(x^8\right)^6\right) + 6*x^4\left(x^8\right)^5 + \cdots\\ &= x^9x^7\left(z_7 + \underline{\frac{c^2}{4}\left(z_4 + \frac{c}{2}z_2^2\right)z_2^4} + \underline{\left(* - \frac{c^3}{12}\right)z_2^6} + \frac{c^3}{42}z_2^7\right) + 6*x^4\left(x^8\right)^5 + \cdots.\end{aligned}$$

There does to have $\frac{c^3}{8} + * - \frac{c^3}{12} = 0$, or $* = -\frac{c^3}{24}$. With this correction,

$$\begin{aligned}\mathbf{Y}\Big(x^1 - \frac{1}{2}\left(x^4\right)^2 - \frac{c^3}{24}x^4\left(x^8\right)^6 + \cdots\Big) &= x^9x^7\left(z_7 + \frac{c^2}{4}z_2^4z_4 + \frac{c^3}{42}z_2^7\right) +\\ &\quad - \frac{c^3}{4}z_2^5\left(z_6 + \frac{c}{2}z_2^2\left(z_4 + \frac{c}{2}z_2^2\right) - \frac{c^2}{8}z_2^4 + \frac{c^2}{15}z_2^5\right) + \cdots\\ &= \left(z_3 + cz_2\right)\left(z_5 + z_4 + \frac{c}{2}z_2^2 + \frac{c}{3}z_2^3\right)\left(z_7 + \frac{c^2}{4}z_2^4z_4 + \frac{c^3}{42}z_2^7\right) +\\ &\quad - \frac{c^3}{4}z_2^5z_6 - \frac{c^4}{8}z_2^7z_4 - \frac{c^5}{32}z_2^9 - \frac{c^5}{60}z_2^{10} + \cdots\end{aligned}$$

After opening brackets, the only terms here of orders smaller than 11 are z_2^9 and z_2^{10}. One eliminates them instantly by taking, eventually,

$$z_8 = x^1 - \frac{1}{2}\left(x^4\right)^2 - \frac{c^3}{24}x^4\left(x^8\right)^6 + \frac{c^5}{320}\left(x^8\right)^{10} + \frac{c^5}{2310}\left(x^8\right)^{11}. \tag{5.6}$$

Indeed,

$$\begin{aligned}\mathbf{Y}z_8 = \big(\text{terms of orders} > 11\big) &+ \frac{c^2}{2}z_2^3z_7 - \frac{c^3}{4}z_2^5z_6 + \frac{c^3}{4}z_2^5z_4^2 + \frac{c^3}{8}z_2^6z_3z_4\\ &+ \frac{3\,c^4}{28}z_2^8z_4 + \frac{c^4}{84}z_2^9z_3 + \frac{c^5}{126}z_2^{11}.\end{aligned}$$

All terms explicited here are of order 11, while this function z_8 keeps being linearly adapted, hence of nonholonomic order not exceeding its weight 12. These arguments together imply that $\mathrm{nord}(z_8) = 12$.

5.2. Upgrading of x^3

We improve the function x^3 of order 12 ($\mathbf{Y}x^3 = x^9x^7x^5 \cdot x^4$ is of order $7+4=11$) to an adapted z_9 of order 16. The first correcting term also prompts itself, $z_9 = x^3 - \frac{1}{3}\left(x^4\right)^3 + \cdots$, and causes

$$
\begin{aligned}
\mathbf{Y}\left(x^3 - \frac{1}{3}\left(x^4\right)^3 + \cdots\right) &= x^9x^7x^5x^4 - x^9x^7\left(x^4\right)^2 + \cdots \\
&= x^9x^7x^4\left(x^5 - x^4\right) + \cdots \\
&= x^9x^7x^4\left(z_7 + \frac{c^2}{4}z_2^4z_4 + \frac{c^3}{24}z_2^6 + \frac{c^3}{42}z_2^7\right) + \cdots . \qquad (5.7)
\end{aligned}
$$

After opening brackets and writing in the z variables, the unwanted terms in (5.7) are, quite like in the previous section, z_2^{13}, z_2^{14}, $z_2^{12}z_3$, $z_2^{11}z_4$. Our idea was to eliminate in (5.7) the entire term $x^9x^7x^4z_2^6$ (of order 13) by means of $z_9 = x^3 - \frac{1}{3}\left(x^4\right)^3 + *\left(x^4\right)^2\left(x^8\right)^6 + \cdots$. Then

$$
\mathbf{Y}z_9 = x^9x^7x^4\left(z_7 + \frac{c^2}{4}z_2^4z_4 + \left(\frac{c^3}{24} + 2*\right)z_2^6 + \frac{c^3}{42}z_2^7\right) + 6*\left(x^4\right)^2z_2^5 + \cdots ,
$$

implying $* = -\frac{c^3}{48}$. So for $z_9 = x^3 - \frac{1}{3}\left(x^4\right)^3 - \frac{c^3}{48}\left(x^4\right)^2\left(x^8\right)^6 + \cdots$ one has

$$
\begin{aligned}
\mathbf{Y}z_9 &= x^9x^7x^4\left(z_7 + \frac{c^2}{4}z_2^4z_4 + \frac{c^3}{42}z_2^7\right) - \frac{c^3}{8}z_2^5\left(x^4\right)^2 + \cdots \\
&= \left(z_3 + cz_2\right)\left(z_5 + z_4 + \frac{c}{2}z_2^2 + \frac{c}{3}z_2^3\right)\left(z_6 + \frac{c}{2}z_2^2z_4 + \frac{c^2}{8}z_2^4 + \frac{c^2}{15}z_2^5\right) \\
&\quad \times \left(z_7 + \frac{c^2}{4}z_2^4z_4 + \frac{c^3}{42}z_2^7\right) \\
&\quad - \frac{c^3}{8}z_2^5\left(z_6 + \frac{c}{2}z_2^2z_4 + \frac{c^2}{8}z_2^4 + \frac{c^2}{15}z_2^5\right)^2 + \cdots . \qquad (5.8)
\end{aligned}
$$

After opening all brackets in (5.8), there show up only terms of orders not smaller than 15 and two benign terms z_2^{13}, z_2^{14}. Consequently, we take

$$
z_9 = x^3 - \frac{1}{3}\left(x^4\right)^3 - \frac{c^3}{48}\left(x^4\right)^2\left(x^8\right)^6 + \frac{c^7}{7168}\left(x^8\right)^{14} + \frac{c^7}{25200}\left(x^8\right)^{15} \qquad (5.9)
$$

and obtain

$$
\begin{aligned}
\mathbf{Y}z_9 = \left(\text{terms of orders } > 15\right) &+ \frac{c^4}{16}z_2^7z_7 - \frac{c^5}{32}z_2^9z_6 + \frac{c^5}{16}z_2^9z_4^2 + \frac{c^5}{64}z_2^{10}z_3z_4 \\
&+ \frac{13\,c^6}{672}z_2^{12}z_4 + \frac{c^6}{672}z_2^{13}z_3 + \frac{31\,c^7}{25200}z_2^{15}
\end{aligned}
$$

in which all evidenced summands are of orders ≥ 15, implying $\operatorname{nord}(z_9) = 16$.

5.3. Refining x^2 to z_{10}

This is the most involved part; naked x^2 has at zero order 20 instead of 24 (for $\mathrm{nord}\big(\mathbf{Y}x^2\big) = \mathrm{nord}\big(x^9x^7x^5 \cdot x^3\big) = 7 + 12 = 19$). Having sought $z_{10} = x^2 + \cdots$, or else

$$\mathbf{Y}z_{10} \;=\; x^9x^7\left(z_7 + x^4 + \cdots\right)\left(z_9 + \frac{1}{3}\left(x^4\right)^3 + \cdots\right) + \cdots ,$$

we wanted to have in $\mathbf{Y}z_{10}$ *no* order-19 term $x^9x^7x^4\frac{1}{3}\left(x^4\right)^3$. Hence, tentatively, took $z_{10} = x^2 - \frac{1}{15}\left(x^4\right)^5 + \cdots$. And, needless to say, this improvement has proved insufficient.

In fact, the nonholonomic derivative $\mathbf{Y}\left(x^2 - \frac{1}{15}\left(x^4\right)^5\right) = x^9x^7x^5x^3 - \frac{1}{3}x^9x^7\left(x^4\right)^4$, when expressed in the (adapted!) variables $z_1, \ldots, z_9$ (a fairy long procedure), has the following terms of orders smaller than 23:

$$\frac{13\,c^{11}}{516096}z_2^{21} + \frac{1387\,c^{11}}{19353600}z_2^{22} + \frac{13\,c^{10}}{516096}z_2^{20}z_3 + \frac{65\,c^{10}}{129024}z_2^{19}z_4\,. \tag{5.10}$$

On the grounds of formulas (5.6) and (5.9) one guesses that, in z_{10}, there should also be used a term $*\left(x^4\right)^4\left(x^8\right)^6$. This time, however, unlike for z_8 and z_9 in the previous sections, there is no quick guess concerning the value of $*$. The way out is to express the derivative

$$\mathbf{Y}\left(\left(x^4\right)^4\left(x^8\right)^6\right) \;=\; 4\,x^9x^7\left(x^4\right)^3\left(x^8\right)^6 + 6\left(x^4\right)^4\left(x^8\right)^5$$

likewise in the coordinates $z_1, \ldots, z_9$ and then notice that the terms of orders smaller than 23 are precisely

$$\frac{11\,c^8}{2048}z_2^{21} + \frac{23\,c^8}{1920}z_2^{22} + \frac{c^7}{256}z_2^{20}z_3 + \frac{5\,c^7}{64}z_2^{19}z_4\,. \tag{5.11}$$

In both expressions (5.10) and (5.11) the coefficients at $z_2^{20}z_3$ and $z_2^{19}z_4$ are in the same proportion 1:20. In fact, the second pair of them times $\frac{13\,c^3}{2016}$ equals the first pair. This observation prompts the value $* = -\frac{13\,c^3}{2016}$. Then, for

$$z_{10} \;=\; x^2 - \frac{1}{15}\left(x^4\right)^5 - \frac{13\,c^3}{2016}\left(x^4\right)^4\left(x^8\right)^6 + \cdots$$

one has in $\mathbf{Y}z_{10}$, as the low orders' terms, only (possibly) z_2^{21} and z_2^{22}. After handling the coefficients at z_2^{21} and z_2^{22} in (5.10)–(5.11), the eventual formula reads

$$z_{10} \;=\; x^2 - \frac{1}{15}\left(x^4\right)^5 - \frac{13\,c^3}{2016}\left(x^4\right)^4\left(x^8\right)^6 + \frac{13\,c^{11}}{30277632}\left(x^8\right)^{22} + \frac{c^{11}}{4121600}\left(x^8\right)^{23}$$

The derivative of this z_{10} along $\mathbf{Y}$ is $\mathbf{Y}z_{10} =$

$$\text{(terms of orders } > 23) + \frac{c^4}{16}z_2^7 z_9 + \frac{c^8}{3072}z_2^{15}z_7 - \frac{c^9}{6144}z_2^{17}z_6 + \frac{17\,c^9}{21504}z_2^{17}z_4^2 + \frac{13\,c^9}{86016}z_2^{18}z_3z_4 + \frac{229\,c^{10}}{1075200}z_2^{20}z_4 + \frac{47\,c^{10}}{3225600}z_2^{21}z_3 + \left(\frac{c^{12}}{258048} + \frac{293\,c^{11}}{24192000}\right)z_2^{23},$$

with all displayed terms of order 23. Hence z_{10} is of order 24 as needed.

Remark 5.2. The common 'resonant' proportion of coefficients 1:20 is crucial for the above upgrading. Realizing this, one can better interpret the surprising opportunities met in upgrading of x^1 and x^3. Indeed, similar mechanisms have been in action for the pairs of functions $x^1 - \frac{1}{2}(x^4)^2$, $x^4(x^8)^6$ and $x^3 - \frac{1}{3}(x^4)^3$, $(x^4)^2(x^8)^6$. In the $\mathbf{Y}$-derivatives of the former pair, the terms $z_2^8 z_3$ and $z_2^7 z_4$ appear in the same proportion 1:8. And in the $\mathbf{Y}$-derivatives of the latter pair, the terms $z_2^{12}z_3$ and $z_2^{11}z_4$ appear in the same proportion 1:12.

At this moment the entire nilpotent approximation is got hold of, albeit the modulus c still shows up in many places. In order to proceed, we write the approximation down as $(\widehat{\mathbf{X}}, \widehat{\mathbf{Y}})$, where $\widehat{\mathbf{X}} = \partial_1$, and

$$\begin{aligned}\widehat{\mathbf{Y}} = {} & \partial_2 + z_1\partial_3 + z_3\partial_4 + z_2z_3\partial_5 + \left(z_3z_4 + \frac{c}{3}z_2^3z_3 + cz_2z_5\right)\partial_6 + \Big(cz_2z_4^2 + cz_2^2z_3z_4 \\ & + \frac{c^2}{2}z_2^3z_5 + \frac{c^2}{3}z_2^4z_4 + \frac{c^2}{6}z_2^5z_3\Big)\partial_7 \quad + \quad \Big(\frac{c^2}{2}z_2^3z_7 - \frac{c^3}{4}z_2^5z_6 + \frac{c^3}{4}z_2^5z_4^2 \\ & + \frac{c^3}{8}z_2^6z_3z_4 + \frac{3\,c^4}{28}z_2^8z_4 + \frac{c^4}{84}z_2^9z_3 + \frac{c^5}{126}z_2^{11}\Big)\,\partial_8 + \Big(\frac{c^4}{16}z_2^7z_7 - \frac{c^5}{32}z_2^9z_6 \\ & + \frac{c^5}{16}z_2^9z_4^2 + \frac{c^5}{64}z_2^{10}z_3z_4 + \frac{13\,c^6}{672}z_2^{12}z_4 + \frac{c^6}{672}z_2^{13}z_3 + \frac{31\,c^7}{25200}z_2^{15}\Big)\,\partial_9 \\ & + \Big(\frac{c^4}{16}z_2^7z_9 + \frac{c^8}{3072}z_2^{15}z_7 - \frac{c^9}{6144}z_2^{17}z_6 + \frac{17\,c^9}{21504}z_2^{17}z_4^2 + \frac{13\,c^9}{86016}z_2^{18}z_3z_4 \\ & + \frac{229\,c^{10}}{1075200}z_2^{20}z_4 + \frac{47\,c^{10}}{3225600}z_2^{21}z_3 + \left(\frac{c^{12}}{258048} + \frac{293\,c^{11}}{24192000}\right)z_2^{23}\Big)\,\partial_{10}\,.\end{aligned} \tag{5.12}$$

5.4. Improving z_{10}, z_9, z_8, z_7, z_6

The issue is whether (5.12) can be simplified. • In the first round we eliminate there the variable z_9, taking as a new 10th coordinate $z_{10} - \frac{c^4}{128}z_2^8z_9$. This expression is labelled z_{10} again (until the end of paper we will use, for compactness, the same letters for newer and newer coordinates). • In the second round we get rid of the variables z_7, z_6. This is obtained by passing to the new 8th, 9th, and 10th variables

$$z_8 + \frac{c^3}{24}z_2^6z_6 - \frac{c^2}{8}z_2^4z_7\,, \qquad z_9 + \frac{c^5}{320}z_2^{10}z_6 - \frac{c^4}{128}z_2^8z_7\,,$$

$$z_{10} + \frac{c^8}{98304}z_2^{16}z_7 - \frac{c^9}{221184}z_2^{18}z_6\,,$$

respectively. • In the third round the variable z_5 is left out by passing to the, new again, 6th, 7th, 8th, 9th, and 10th variables

$$z_6 - \frac{c}{2} z_2^2 z_5\,, \qquad z_7 - \frac{c^2}{8} z_2^4 z_5\,, \qquad z_8 + \frac{c^4}{384} z_2^8 z_5\,,$$

$$z_9 + \frac{c^6}{15360} z_2^{12} z_5\,, \qquad z_{10} - \frac{c^{10}}{35389440} z_2^{20} z_5\,,$$

respectively. At this point the expression (5.12) gets reduced to

$$\begin{aligned}
\widehat{\mathbf{Y}} &= \partial_2 + z_1\partial_3 + z_3\partial_4 + z_2z_3\partial_5 + \Big(z_3z_4 - \frac{c}{6} z_2^3 z_3\Big)\partial_6 \\
&+ \Big(\underline{c z_2 z_4^2 + c z_2^2 z_3 z_4} + \frac{c^2}{3} z_2^4 z_4 + \frac{c^2}{24} z_2^5 z_3\Big)\partial_7 \\
&+ \left(\underline{\frac{c^3}{8} z_2^5 z_4^2 + \frac{c^3}{24} z_2^6 z_3 z_4} + \frac{11\,c^4}{168} z_2^8 z_4 + \frac{61\,c^4}{8064} z_2^9 z_3 + \frac{c^5}{126} z_2^{11}\right)\partial_8 \\
&+ \left(\underline{\frac{7c^5}{128} z_2^9 z_4^2 + \frac{7c^5}{640} z_2^{10} z_3 z_4} + \frac{15\,c^6}{896} z_2^{12} z_4 + \frac{139\,c^6}{107520} z_2^{13} z_3 + \frac{31\,c^7}{25200} z_2^{15}\right)\partial_9 \\
&+ \left(\underline{\frac{215\,c^9}{688128} z_2^{17} z_4^2 + \frac{215\,c^9}{6193152} z_2^{18} z_3 z_4} + \frac{481\,c^{10}}{7372800} z_2^{20} z_4 + \frac{11539\,c^{10}}{3715891200} z_2^{21} z_3 \right. \\
&\qquad \left. + \Big(\frac{c^{12}}{258048} + \frac{121\,c^{11}}{48384000}\Big) z_2^{23}\right)\partial_{10}\,. \qquad (5.13)
\end{aligned}$$

5.5. Flatness of the nilpotent approximations in question

It is clear that • passing to the new 6th variable $z_6 - \frac{1}{2} z_4^2$ eliminates the term z_3z_4 in the ∂_6-component of (5.13). Then a natural wish is to eliminate the terms with z_3z_4 *and* with z_4^2 in the 7th, 8th, 9th, and 10th components in (5.13) – the terms underlined there.

The coefficients with which these terms appear – coefficients emerging from a long line of preceding simplifications! – are flat, or: perfectly bound together. Because of that it is possible to kill them in pairs, by just • taking new 7th, 8th, 9th, and 10th coordinates

$$z_7 - \frac{c}{2} z_2^2 z_4^2\,, \qquad z_8 - \frac{c^3}{48} z_2^6 z_4^2\,, \qquad z_9 - \frac{7c^5}{1280} z_2^{10} z_4^2\,, \qquad z_{10} - \frac{215\,c^9}{12386304} z_2^{18} z_4^2$$

(in the last ∂_{10}-component, because $18 \cdot 688128 = 12386304 = 2 \cdot 6193152$).[4]

[4] We want to underline that *without* that kind of flatness the nilpotent approximations in question would have had much different properties, and the modulus have been likely to persist after taking the approximation. That possibility can be turned into a rigorous construction in a somewhat *perturbed* Goursat world that we intend to present elsewhere. Thus producing uni-, bi-, etc. -modal families of pairwise non-equivalent, strongly nilpotent in the sense of [14], distribution germs.

After these simplifications,

$$\widehat{\mathbf{Y}} = \partial_2 + z_1\partial_3 + z_3\partial_4 + z_2z_3\partial_5 - \frac{c}{6}z_2^3z_3\partial_6 + \Big(\frac{c^2}{3}z_2^4z_4 + \frac{c^2}{24}z_2^5z_3\Big)\partial_7$$
$$+ \left(\frac{11\,c^4}{168}z_2^8z_4 + \frac{61\,c^4}{8064}z_2^9z_3 + \frac{c^5}{126}z_2^{11}\right)\partial_8 + \left(\frac{15\,c^6}{896}z_2^{12}z_4 + \frac{139\,c^6}{107520}z_2^{13}z_3\right.$$
$$\left. + \frac{31\,c^7}{25200}z_2^{15}\right)\partial_9 + \left(\frac{481\,c^{10}}{7372800}z_2^{20}z_4 + \frac{11539\,c^{10}}{3715891200}z_2^{21}z_3\right.$$
$$\left. + \Big(\frac{c^{12}}{258048} + \frac{121\,c^{11}}{48384000}\Big)z_2^{23}\right)\partial_{10}\,.$$

Now the way is open towards the eventual simplification – getting rid of z_4 and z_3. In fact (after tracing the coefficients down with care), the following – new again – 4th through 10th coordinates are appropriate:

$$z_4 - z_2z_3\,, \qquad z_5 - \frac{1}{2}z_2^2z_3\,, \qquad z_6 + \frac{c}{24}z_2^4z_3\,, \qquad z_7 - \frac{c^2}{15}z_2^5z_4 + \frac{c^2}{240}z_2^6z_3\,,$$
$$z_8 - \frac{11\,c^4}{1512}z_2^9z_4 - \frac{c^4}{34560}z_2^{10}z_3 - \frac{c^5}{1512}z_2^{12},$$
$$z_9 - \frac{15\,c^6}{11648}z_2^{13}z_4 - \frac{c^6}{2795520}z_2^{14}z_3 - \frac{31\,c^7}{403200}z_2^{16},$$
$$z_{10} - \frac{481\,c^{10}}{154828800}z_2^{21}z_4 + \frac{c^{10}}{16349921280}z_2^{22}z_3 - \Big(\frac{c^{12}}{6193152} + \frac{121\,c^{11}}{1161216000}\Big)z_2^{24}\,.$$

In the previous coordinates z_1, z_2, z_3 and the above-described new ones, the generators of the nilpotent approximations in question are $\widehat{\mathbf{X}} = \partial_1$ and

$$\widehat{\mathbf{Y}} = \partial_2 + z_1\partial_3 - z_1z_2\partial_4 - \frac{1}{2}z_1z_2^2\partial_5 + \frac{c}{24}z_1z_2^4\partial_6 + \frac{c^2}{240}z_1z_2^6\partial_7 \qquad (5.14)$$
$$- \frac{c^4}{34560}z_1z_2^{10}\partial_8 - \frac{c^6}{2795520}z_1z_2^{14}\partial_9 + \frac{c^{10}}{16349921280}z_1z_2^{22}\partial_{10}\,.$$

This strikingly resembles the expressions (4.7) for the nilpotent approximation within the class GGSGSG, remembering of course that the meanings of c in Sections 4 and 5 are different. There c was just the highest constant in the departure KR form (reducible in itself and kept just for better readability of the algebraic side of computations), here c is a modulus of the local classification. In fact, we were sticking in Section 4 to a redundant constant *for* the purpose of this comparison. The algebra in Goursat world is so rigid that also in (4.7) the first six components are nothing but those of the nilpotent approximation in the class GGSG, when the additive constant standing next to x^6, in the relevant KR form, is kept as $c \neq 0$, and not normalized to 1 as in (4.1).

Endly, it is a matter of simple rescaling (if cardinal because getting rid of the modulus c) to reduce the description (5.14) to the form $\widehat{\mathbf{X}} = \partial_1$ and

$$\widehat{\mathbf{Y}} = \partial_2 + z_1\partial_3 + z_1z_2\partial_4 + z_1z_2^2\partial_5 + z_1z_2^4\partial_6 + z_1z_2^6\partial_7 + z_1z_2^{10}\partial_8 + z_1z_2^{14}\partial_9 + z_1z_2^{22}\partial_{10}\,.$$

At this moment, at long last, Theorem 5.1 is proved. □

Corollary 5.3. *The nilpotent approximations of germs in* GGSGSGSG *have the big growth vector at the reference points equal to the small one, hence equal to* $[2,\ 3,\ 4,\ 5_2,\ 6_2,\ 7_4,\ 8_4,\ 9_8,\ 10]$.

This information should be compared with the fact that the initial germs have the big growth vector $[2, 3, 4, 5, 6, 7, 8, 9, 10]$.

Remark 5.4. The answer to Agrachev' question is not known already for the bimodal geometric class GGSGSGS$\underline{\text{G}}$S$\underline{\text{G}}$ living in dimension 12, having moduli in flag' members denoted by the underlined letters. Nor is it known for the trimodal class GGSGSGS$\underline{\text{G}}$S$\underline{\text{G}}$S$\underline{\text{G}}$ in dimension 14, with independent moduli related to the $\underline{\text{G}}$'s, and so on onwards, with arbitrarily long strings of S$\underline{\text{G}}$ in the codes. This infinite series of geometric classes of quickly growing modalities (see Remark 4 in [13] which covers all these classes) is very representative for Goursat, offering the highest known to-date ratio modality : length in Goursat world.

All we have at present is an organized plan, shortcutting and replacing repetitions of our procedures from Section 5, for computing the NAs in these classes. The answer to question, however, depends on the *realization* of that plan. At present it is impossible to predict the outcome of this project.

References

[1] A.A. Agrachev, *Canonical Nilpotentization of Control Systems.* Preprint, 1997 (3 pages).

[2] ——, R.V. Gamkrelidze, A.V. Sarychev, *Local invariants of smooth control systems.* Acta Appl. Math. **14** (1989), 191–237.

[3] ——, J-P. Gauthier, *On subanalyticity of Carnot–Carathéodory distances.* Ann. Inst. H. Poincaré – AN **18** (2001), 359–382.

[4] ——, A.V. Sarychev, *Filtrations of a Lie algebra of vector fields and nilpotent approximation of control systems.* Dokl. Akad. Nauk SSSR, **295** (1987); English transl. in Soviet Math. Dokl. **36** (1988), 104–108.

[5] ——, A. Marigo, *Nonholonomic tangent spaces: intrinsic construction and rigid dimensions.* Electronic Res. Announc. Amer. Math. Soc. **9** (2003), 111–120.

[6] A. Bellaïche, *The tangent space in sub-Riemannian geometry.* In: A. Bellaïche, J.-J. Risler (Eds.), Sub-Riemannian Geometry. Birkhäuser, 1996, 1–78.

[7] —— et al., *Geometry of Nonholonomic Systems.* In: J.-P. Laumond (Ed.), Robot Motion Planning and Control, Lecture Notes in Control and Inform. Sci. **229**, Springer, 1998, 54–91.

[8] R.M. Bianchini, G. Stefani, *Graded approximations and controllability along a trajectory.* SIAM J. Control Optimiz. **28** (1990), 903–924.

[9] H. Hermes, *Nilpotent approximations of control systems and distributions.* SIAM J. Control Optimiz. **24** (1986), 731–736.

[10] F. Jean, *The car with N trailers: characterisation of the singular configurations.* ESAIM: Control, Optimisation and Calculus of Variations (http://www.edpsciences.org/cocv/) **1** (1996), 241–266.

[11] A. Kumpera, C. Ruiz, *Sur l'équivalence locale des systèmes de Pfaff en drapeau.* In: F. Gherardelli (Ed.), Monge–Ampère Equations and Related Topics. Ist. Alta Math. F. Severi, Rome 1982, 201–248.

[12] R. Montgomery, M. Zhitomirskii, *Geometric approach to Goursat flags.* Ann. Inst. H. Poincaré – AN **18** (2001), 459–493.

[13] P. Mormul, *Local classification of rank-2 distributions satisfying the Goursat condition in dimension* 9. In: P. Orro et F. Pelletier (Eds.), Singularités et Géométrie sous-riemannienne, Collection Travaux en cours **62**, Hermann, 2000, 89–119.

[14] ——, *Goursat distributions not strongly nilpotent in dimensions not exceeding seven.* In: A. Zinober, D. Owens (Eds.), Nonlinear and Adaptive Control, Lecture Notes in Control and Inform. Sci. **281**, Springer, 2003, 249–261.

[15] ——, *Real moduli in local classification of Goursat flags.* Hokkaido Math. J. **34** (2005), 1–35.

[16] ——, *Geometric classes of Goursat flags and the arithmetics of their encoding by small growth vectors.* Central European J. Math. **2** (2004), 859–883.

[17] ——, *Kumpera–Ruiz algebras in Goursat flags are optimal in small lengths.* J. Math. Sciences **126** (2005), 1614–1629.

[18] L.P. Rothschild, E.M. Stein, *Hypoelliptic differential operators and nilpotent groups.* Acta Math. **137** (1976), 247–320.

[19] M. Vendittelli, G. Oriolo, F. Jean, J.-P. Laumond, *Nonhomogeneous nilpotent approximations for nonholonomic systems with singularities.* IEEE Trans. Automat. Control **49** (2004), 261–266.

Piotr Mormul
Institute of Mathematics
Warsaw University
Banach st. 2
PL-02-097 Warsaw, Poland
e-mail: mormul@mimuw.edu.pl

Real and Complex Singularities
Trends in Mathematics, 247–262

Calculation of Mixed Hodge Structures, Gauss-Manin Connections and Picard-Fuchs Equations

Hossein Movasati

Abstract. In this article we introduce algorithms which compute iterations of Gauss-Manin connections, Picard-Fuchs equations of Abelian integrals and mixed Hodge structure of affine varieties of dimension n in terms of differential forms. In the case $n = 1$ such computations have many applications in differential equations and counting their limit cycles. For $n > 3$, these computations give us an explicit definition of Hodge cycles.

Mathematics Subject Classification (2000). Primary 14C30; Secondary 32S65.

Keywords. Mixed Hodge structure of affine varieties, Gauss-Manin connection.

1. Introduction

The theory of abelian integrals which arises in polynomial differential equations of the type $\dot{x} = P(x, y)$, $\dot{y} = Q(x, y)$ is one of the most fruitful areas which needs a special attention from algebraic geometry and in particular singularity theory. The reader is referred to the articles [6], [10] and [3] for a history and applications of such abelian integrals in differential equations. The book [1] and its references contains the theory of such integrals in the local case. In this article we deal with computational aspects of such integrals. All polynomial objects which we use are defined over $\mathbb{C}$.

Let us be given a polynomial f in $n+1$ variables $x_1, x_2, \ldots, x_{n+1}$, a polynomial differential n-form ω and a continuous family of n-dimensional oriented cycles $\delta_t \subset L_t := f^{-1}(t)$. The protagonist of this article is the integral $\int_{\delta_t} \omega$, called the abelian integral. Computations related to these integrals become easier when we put a certain kind of tameness condition on f (see §2). For such a tame polynomial we can write $\int_{\delta_t} \omega$ as:

$$\sum_{\beta \in I} p_\beta(t) \int_{\delta_t} \eta_\beta, \tag{1.1}$$

where $\eta_\beta, \beta \in I$ is a class of differential n-forms constructed from a basis of the Milnor vector space of f and p_β's are polynomials in t (see §5 for the algorithm which produces p_β's). The Gauss-Manin connection $\nabla\omega$ has the following basic property

$$\frac{\partial}{\partial t}\int_{\delta_t}\omega = \int_{\delta_t}.\nabla\omega. \tag{1.2}$$

The above term can be written in the form (1.1) with p_β's rational functions in t with poles in the critical values of f (see §6 for the algorithm which produces p_β's). The nth cohomology of a smooth fiber L_t is canonically isomorphic to $\Omega^n_{L_t}/d\Omega^{n-1}_{L_t}$, where $\Omega^i_{L_t}$ is the restriction of polynomial differential i-forms to L_t, and carries two natural filtrations called the weight and the Hodge filtrations (a mixed Hodge structure consists of these filtrations and a real structure satisfying certain axioms). These filtrations are generalizations of classical notions of differential forms of the first, second and third type for Riemann surfaces in higher dimensional varieties. The reader who is not interested in the case $n > 1$ is invited to follow the article with $n = 1$ and with the usual notions of differential forms of the first, second and third type. How to calculate these filtrations by means of differential forms is the main theorem of [9] and related algorithms are explained in §7. Last but not least, our protagonist satisfies a Picard-Fuchs equation $\sum_{i=0}^{k} p_i(t)\frac{\partial^i}{\partial t^i} = 0$, where p_i's are polynomials in t. The algorithm which produces p_i's is explained in §8. The theory of abelian integrals can be studied even in the case $n = 0$, i.e., f is a polynomial in one variable. Since some open problems, for instance infinitesimal Hilbert Problem (see [6]), can be also stated in this case, we have included §9. All the algorithms explained in this article are implemented in a library of SINGULAR. This together with some examples are explained in §10. Applications of our computations in differential equations and particularly in direction of the article [3] is a matter of future work.

2. Tame polynomials and Brieskorn modules

We start with a definition.

Definition 2.1. A polynomial $f \in \mathbb{C}[x]$ is called (weighted) tame if there exist natural numbers $\alpha_1, \alpha_2, \dots, \alpha_{n+1} \in \mathbb{N}$ such that $Sing(g) = \{0\}$, where $g = f_d$ is the last homogeneous piece of f in the graded algebra $\mathbb{C}[x]$, $deg(x_i) = \alpha_i$.

The multiplicative group $\mathbb{C}^*$ acts on $\mathbb{C}^{n+1}$ in the following way:

$$\lambda^* : (x_1, x_2, \dots, x_{n+1}) \to (\lambda^{\alpha_1}x_1, \lambda^{\alpha_2}x_2, \dots, \lambda^{\alpha_{n+1}}x_{n+1}),\ \lambda \in \mathbb{C}^*.$$

The polynomial (resp. the polynomial form) ω in $\mathbb{C}^{n+1}$ is (weighted) homogeneous of degree $d \in \mathbb{N}$ if $\lambda^*(\omega) = \lambda^d\omega,\ \lambda \in \mathbb{C}^*$. Fix a homogeneous polynomial g of degree d and with an isolated singularity at $0 \in \mathbb{C}^{n+1}$. Let A_g be the affine space of all tame polynomials $f = f_0 + f_1 + \cdots + f_{d-1} + g$. The space A_g is parameterized by the coefficients of $f_i, i = 0, 1, \dots, d-1$. The multiplicative group $\mathbb{C}^*$ acts on A_g by

$$\lambda \bullet f = \frac{f \circ \lambda^*}{\lambda^d} = \lambda^{-d} f_0 + \lambda^{-d+1} f_1 + \cdots + \lambda^{-1} f_d + g.$$

The action of $\lambda \in \mathbb{C}^*$ takes $\lambda \bullet f = 0$ biholomorphically to $f = 0$.

Let $f \in \mathsf{A}_g$. We choose a basis $x^I := \{x^\beta \mid \beta \in I\}$ of monomials for the Milnor $\mathbb{C}$-vector space

$$V := \mathbb{C}[x]/\text{jacob}(g).$$

Define

$$w_i := \frac{\alpha_i}{d},\ 1 \le i \le n+1,\ \eta := (\sum_{i=1}^{n+1} (-1)^{i-1} w_i x_i \widehat{dx_i}),\ L_t := f^{-1}(t), t \in \mathbb{C}, \quad (2.1)$$

$$A_\beta := \sum_{i=1}^{n+1} (\beta_i + 1) w_i,\ \eta_\beta := x^\beta \eta,\ \omega_\beta = x^\beta dx,\ (\beta \in I),$$

where $\widehat{dx_i} = dx_1 \wedge \cdots \wedge dx_{i-1} \wedge dx_{i+1} \wedge \cdots \wedge dx_{n+1}$. Note that $A_\beta = \frac{\deg(x^{\beta+1})}{d}$. It turns out that x^I is also a basis of $V_f := \mathbb{C}[x]/\text{jacob}(f)$ and so f and g have the same Milnor numbers (see the conclusion after Lemma 4 of [9]). We denote it by μ . We denote by P the set of critical points of f and by $C := f(P)$ the set of critical values of f. We will also use P for a polynomial in $\mathbb{C}[x]$. This will not make any confusion.

Let $\Omega^i, i = 1, 2, \dots, n+1$ (resp. $\Omega^i_j,\ j \in \mathbb{N} \cup \{0\}$) be the set of polynomial differential i-forms (resp. homogeneous degree j polynomial differential i-forms) in $\mathbb{C}^{n+1}$. The Milnor vector space of f can be rewritten in the form $V := \frac{\Omega^{n+1}}{df \wedge \Omega^n}$. The Brieskorn modules

$$H' = H'_f := \frac{\Omega^n}{df \wedge \Omega^{n-1} + d\Omega^{n-1}},\quad H'' = H''_f = \frac{\Omega^{n+1}}{df \wedge d\Omega^{n-1}}$$

of f are $\mathbb{C}[t]$-modules in a natural way: $t.[\omega] = [f\omega],\ [\omega] \in H'$ resp. $\in H''$. They are defined in the case $n > 0$. The case $n = 0$ is treated separately in §9.

3. Mixed Hodge structures

In this section we assume that the reader is familiar with the notion of mixed Hodge structure in the cohomologies of an affine variety (see [7, 2]).

Definition 3.1. Let H be one of H' or H''. If $H = H''$ then by restriction of ω on $L_c,\ c \in \mathbb{C} \backslash C$ we mean the residue of $\frac{\omega}{f-c}$ in L_c and by $\int_\delta \omega,\ \delta \in H_n(L_c, \mathbb{Z})$ we mean $\int_\delta \text{residue}(\frac{\omega}{f-c})$. It is natural to define the Hodge and weight filtrations of H as follows: $W_m H,\ m \in \mathbb{Z}$ (resp. $F^k H,\ k \in \mathbb{Z}$) consists of elements $\omega \in H$ such that the restriction of ω on all $L_c,\ c \in \mathbb{C} \backslash C$ belongs to $W_m H^n(L_c, \mathbb{C})$ (resp. $F^k H^n(L_c, \mathbb{C})$).

Each piece of the mixed Hodge structure of H is a $\mathbb{C}[t]$-module. In the same way we define the mixed Hodge structure of the localization of H over multiplicative subgroups of $\mathbb{C}[t]$. In the case $n=1$ our definition can be simplified as follows: We have the filtrations $\{0\}=W_0\subset W_1\subset W_2=H$ and $0=F^2\subset F^1\subset F^0=H$, where

$$W_1=\{\omega\in H \mid \omega \text{ restricted to a regular fiber has not residue at infinity }\},$$

$$F^1=\{\omega\in H \mid \omega \text{ restricted to a regular fiber has poles of order} \geq 1 \text{ at infinity}\}.$$

In particular, $W_1\cap F^1$ is the set of all $\omega\in H$ such that ω restricted to a regular compactified fiber is of the first kind. For the notion of compactification of $\mathbb{C}^2$ and infinity see [3] and [8]. The projection of $F^\bullet$ in $\mathrm{Gr}_m^W H := W_m/W_{m-1}$ gives us the filtration $\bar{F}^\bullet$ in $\mathrm{Gr}_m^W H$ and we define $\mathrm{Gr}_F^k\mathrm{Gr}_m^W H=\bar{F}^k/\bar{F}^{k+1}$.

Definition 3.2. Suppose that H is a free $\mathbb{C}[t]$-module. The set $B=\cup_{m,k\in\mathbb{Z}}B_m^k\subset H$ is a basis of H compatible with the mixed Hodge structure if B_m^k form a basis of $\mathrm{Gr}_F^k\mathrm{Gr}_m^W H$.

For a $\mathbb{C}[t]$-module M and a set $C\subset\mathbb{C}$, we denote by M_C the localization of M on the multiplicative subset of $\mathbb{C}[t]$ generated by $\{t-c\mid c\in C\}$. The following theorem gives a basis of a localization of H which is compatible with the mixed Hodge structure. It is proved in [9]. Our aim in this article is to explain the algorithms which lead to the calculation of such a basis.

Theorem 3.3. *Let $b\in\mathbb{C}\backslash C$ be a regular value of $f\in\mathbb{C}[x]$. If f is a (weighted) tame polynomial then $\mathrm{Gr}_m H'=0$ for $m\neq n,n+1$ and there exist a map $\beta\in I\to d_\beta\in\mathbb{N}\cup\{0\}$ and $C\subset\tilde{C}\subset\mathbb{C}$ such that $b\notin\tilde{C}$ and*

$$\nabla^k\eta_\beta,\ \beta\in I,\ A_\beta=k \tag{3.1}$$

form a basis of $\mathrm{Gr}_F^{n+1-k}\mathrm{Gr}_{n+1}^W H'_{\tilde{C}}$ and the forms

$$\nabla^k\eta_\beta, A_\beta+\frac{1}{d}\leq k\leq A_\beta+\frac{d_\beta}{d} \tag{3.2}$$

form a basis of $\mathrm{Gr}_F^{n+1-k}\mathrm{Gr}_n^W H'_{\tilde{C}}$. The same is true for $H''_{\tilde{C}}$ replacing $\nabla^k\eta_\beta$ with $\nabla^{k-1}\omega_\beta$.

In the above theorem $\nabla: H\to H_C$ is the Gauss-Manin connection associated to f (see §6).

4. Quasi-homogeneous singularities

Let $f=g$ be a weighted homogeneous polynomial with an isolated singularity at origin. It is well known that H' (resp. H'') is freely generated by $\eta_\beta,\ \beta\in I$ (resp. $\omega_\beta,\beta\in I$). In this section we explain the algorithm which writes every element in H' (resp. H'') of g as a $\mathbb{C}[t]$-linear combination of η_β's (resp. ω_β's). Recall that

$$dg\wedge d(P\widehat{dx_i,dx_j})=(-1)^{i+j+\epsilon_{i,j}}(\frac{\partial g}{\partial x_j}\frac{\partial P}{\partial x_i}-\frac{\partial g}{\partial x_i}\frac{\partial P}{\partial x_j})dx,$$

where $\epsilon_{i,j} = 0$ if $i < j$ and $= 1$ if $i > j$ and $\widehat{dx_i, dx_j}$ is dx without dx_i and dx_j (we have not changed the order of $dx_1, dx_2, \ldots$ in dx).

Proposition 4.1. *For a monomial $P = x^\beta$ we have*

$$\frac{\partial g}{\partial x_i} P dx = \frac{d}{d \cdot A_\beta - \alpha_i} \frac{\partial P}{\partial x_i} g dx + dg \wedge d(\sum_{j \neq i} \frac{(-1)^{i+j+1+\epsilon_{i,j}} \alpha_j}{d \cdot A_\beta - \alpha_i} x_j P \widehat{dx_i, dx_j}). \quad (4.1)$$

Proof. The proof is a straightforward calculation.

$$\sum_{j \neq i} \frac{(-1)^{i+j+1+\epsilon_{i,j}} \alpha_j}{d \cdot A_\beta - \alpha_i} dg \wedge d(x_j P \widehat{dx_i, dx_j})$$

$$= \frac{-1}{d \cdot A_\beta - \alpha_i} \sum_{j \neq i} (\alpha_j \frac{\partial g}{\partial x_j} \frac{\partial (x_j P)}{\partial x_i} - \alpha_j \frac{\partial g}{\partial x_i} \frac{\partial (x_j P)}{\partial x_j}) dx$$

$$= \frac{-1}{d \cdot A_\beta - \alpha_i} ((d \cdot g - \alpha_i x_i \frac{\partial g}{\partial x_i}) \frac{\partial P}{\partial x_i} - P \frac{\partial g}{\partial x_i} \sum_{j \neq i} \alpha_j (\beta_j + 1)) dx$$

$$= \frac{-1}{d \cdot A_\beta - \alpha_i} (d \cdot g \frac{\partial P}{\partial x_i} - \alpha_i \beta_i P \frac{\partial g}{\partial x_i} - P \frac{\partial g}{\partial x_i} \sum_{j \neq i} \alpha_j (\beta_j + 1)) dx.$$

In the above equalities ds means the differential of s and $d \cdot s$ means the multiplication of d, the degree of g, with s. □

We use the above Proposition to write every $Pdx \in \Omega^{n+1}$ in the form

$$Pdx = \sum_{\beta \in I} p_\beta(g) \omega_\beta + dg \wedge d\xi, \quad (4.2)$$

$$p_\beta \in \mathbb{C}[t],\ \xi \in \Omega^{n-1}, \deg(p_\beta(g)\omega_\beta, dg \wedge d\xi) \leq \deg(Pdx).$$

- Input: The homogeneous polynomial g and $P \in \mathbb{C}[x]$ representing $[Pdx] \in H''$. Output: $p_\beta, \beta \in I$ and ξ satisfying (4.2). We write

$$Pdx = \sum_{\beta \in I} c_\beta x^\beta . dx + dg \wedge \eta,\ \deg(dg \wedge \eta) \leq \deg(Pdx). \quad (4.3)$$

  Then we apply (4.1) to each monomial component $\tilde{P} \frac{\partial g}{\partial x_i}$ of $dg \wedge \eta$ and then we write each $\frac{\partial \tilde{P}}{\partial x_i} dx$ in the form (4.3). The degree of the components which make Pdx not to be of the form (4.2) always decreases and finally we get the desired form.

To find a similar algorithm for H' we note that if $\eta \in \Omega^n$ is written in the form

$$\eta = \sum_{\beta \in I} p_\beta(g) \eta_\beta + dg \wedge \xi + d\xi_1,\ p_\beta \in \mathbb{C}[t], \xi, \xi_1 \in \Omega^{n-1}, \quad (4.4)$$

where each piece in the right-hand side of the above equality has degree less than $\deg(\eta)$ then

$$d\eta = \sum_{\beta \in I} (p_\beta(g) A_\beta + p'_\beta(g) g) \omega_\beta - dg \wedge d\xi \quad (4.5)$$

and the inverse of the map $\mathbb{C}[t] \to \mathbb{C}[t]$, $p \mapsto A_\beta.p + p'.t$ is given by $\sum_{i=0}^k a_i t^i \mapsto \sum_{i=1}^k \frac{a_i}{A_\beta+i} t^i$.

Since in the case of a quasi-homogeneous singularity $f = g$ we have $\nabla(\omega_\beta) = \frac{A_\beta - 1}{t}\omega_\beta$ and $\nabla(\eta_\beta) = \frac{A_\beta}{t}\eta_\beta$ (see §6), Theorem 3.3 in this case reduces to:

Theorem 4.2. (*Steenbrink,* [11]) *For a weighted homogeneous polynomial g, the set*

$$B = \cup_{k=1}^n B_{n+1}^k \cup \cup_{k=0}^n B_n^k$$

with

$$B_{n+1}^k = \{\eta_\beta \mid A_\beta = n-k+1\}, B_n^k = \{\eta_\beta \mid n-k < A_\beta < n-k+1\},$$

is a basis of H' compatible with the mixed Hodge structure. The same is true for H'' replacing η_β with ω_β.

5. A basis of H' and H''

Proposition 5.1. *For every tame polynomial $f \in \mathsf{A}_g$ the forms ω_β, $\beta \in I$ (resp. η_β, $\beta \in I$) form a basis of the Brieskorn module H'' (resp. H') of f. More precisely, every $\omega \in \Omega^{n+1}$ (resp. $\omega \in \Omega^n$) can be written*

$$\omega = \sum_{\beta \in I} p_\beta(f)\omega_\beta + df \wedge d\xi,\ p_\beta \in \mathbb{C}[t],\ \xi \in \Omega^{n-1},\ \deg(p_\beta) \le \frac{\deg(\omega)}{d} - A_\beta \quad (5.1)$$

$$\Big(\textit{resp.}\quad \omega = \sum_{\beta \in I} p_\beta(f)\eta_\beta + df \wedge \xi + d\xi_1,\ p_\beta \in \mathbb{C}[t],\ \xi \in \Omega^{n-1},\ \deg(p_\beta) \le \frac{\deg(\omega)}{d} - A_\beta\Big). \quad (5.2)$$

This proposition is proved in [9] Proposition 1. The proof also gives us the following algorithm to find all the unknown data in the above equalities.

- Input: The tame polynomial f and $P \in \mathbb{C}[x]$ representing $[Pdx] \in H''$. Output: $p_\beta, \beta \in I$ and ξ satisfying (5.1).

  We use the algorithm of §4 and write an element $\omega \in \Omega^{n+1}, \deg(\omega) = m$ in the form:

$$\omega = \sum_{\beta \in I} p_\beta(g)\omega_\beta + dg \wedge d\psi,\ p_\beta \in \mathbb{C}[t],\ \psi \in \Omega^{n-1},\ \deg(p_\beta(g)\omega_\beta), \deg(dg \wedge d\psi) \le m$$

  This is possible because g is homogeneous. We have

$$\omega = \sum_{\beta \in I} p_\beta(f)\omega_\beta + df \wedge d\psi + \omega',\ \omega' = \sum_{\beta \in I}(p_\beta(g) - p_\beta(f))\omega_\beta + d(g-f) \wedge d\psi.$$

  The degree of ω' is strictly less than m and so we repeat what we have done at the beginning and finally we write ω as a $\mathbb{C}[t]$-linear combination of ω_β's.

The algorithm for H' is similar. The statement about degrees is the direct consequence of the proof and (4.2).

6. Gauss-Manin connection

Let $S(t) \in \mathbb{C}[t]$ such that

$$S(f)dx = df \wedge \eta_f,\ \eta_f = \sum_{i=1}^{n+1} (-1)^{i-1} p_i \widehat{dx_i} \in \Omega^{n-1}.$$

For instance one can take $S(t) := det(A_f - t.I)$, where A_f is the multiplication by f linear map from $V_f := \mathbb{C}[x]/\text{jacob}(f)$ to itself. The Gauss-Manin connection $\nabla = \nabla_{\frac{\partial}{\partial t}}$ associated to the fibration $f = t,\ t \in \mathbb{C}$ on H'' turns out to be the map

$$\nabla : H'' \to H''_C, \nabla([Pdx]) = \frac{[(Q_P - P.S'(f))dx]}{S},\ P \in \mathbb{C}[x],$$

where

$$Q_P = \sum_{i=1}^{n+1} \left(\frac{\partial P}{\partial x_i} p_i + P \frac{\partial p_i}{\partial x_i} \right) \tag{6.1}$$

satisfying the Leibniz rule, where for a set $\tilde{C} \subset \mathbb{C}$ by $H''_{\tilde{C}}$ we mean the localization of H'' on the multiplicative subgroup of H'' generated by $t - c,\ c \in \tilde{C}$. Using the Leibniz rule one can extend ∇ to a function from H''_C to itself and so the iteration $\nabla^k = \nabla \circ \nabla \cdots \nabla$ k times, makes sense. It is given by

$$\nabla^k = \frac{\nabla_{k-1} \circ \nabla_{k-2} \circ \cdots \circ \nabla_0}{S(t)^k}, \tag{6.2}$$

where

$$\nabla_k : H'' \to H'',\ \nabla_k([Pdx]) = [(Q_P - (k+1)S'(t)P)dx].$$

To calculate $\nabla : H' \to H'_C$ we use the fact that

$$\nabla^k \omega = \frac{\nabla^{k-1} d\omega}{df},\ \omega \in H',$$

where $d : H' \to H''$ is taking differential and is well defined. The main property of ∇ is (1.2). Usually the iteration of the Gauss-Manin connection produces polynomial forms with huge number of monomials. But fortunately our Brieskorn module H'' (resp. H') has already the canonical basis $\omega_\beta,\ \beta \in I$ (resp. $\eta_\beta,\ \beta \in I$) and after writing ∇ the obtained coefficients are much more easier to read. In H'' one can write

$$S(t)\nabla(\omega_\beta) = \sum_{\beta' \in I} p_{\beta,\beta'} \omega_{\beta'},\ p_{\beta,\beta'} \in \mathbb{C}[t],\ \deg(p_{\beta,\beta'}) \leq \deg(S) - 1 + A_\beta - A_{\beta'}. \tag{6.3}$$

The bound on degrees can be obtained as follows:

$$S(f)\omega_\beta = df \wedge \eta,\ \Rightarrow d \cdot \deg(S) + d \cdot A_\beta = d + \deg(\eta).$$

$$\deg(p_{\beta,\beta'}) \leq \frac{\deg(d\eta)}{d} - A_{\beta'} = \deg(S) - 1 + A_\beta - A_{\beta'}.$$

The Gauss-Manin connection ∇ has two nice properties:

1. Griffiths transversality theorem: For all $i = 1, 2, \ldots, n+1$ we have
$$S(t)\nabla(F^i) \subset F^{i-1}.$$
2. Residue killer: For all $\omega \in H$ there exists a $k \in \mathbb{N}$ such that $\nabla^k \omega \in W_n$

For the first one see [5]. The second one for $n = 1$ is proved in Lemma 2.3 of [8]. The proof for $n > 1$ is similar and uses the fact that the residue as a function in t for a cycle around infinity is a polynomial in t.

7. The numbers $d_\beta, \ \beta \in I$

Let f be a tame polynomial with the last homogeneous part g, F be its homogenization and
$$V = \mathbb{C}[x, x_0] / < \frac{\partial F}{\partial x_i} \mid i = 1, 2, \ldots, n+1 > .$$
We consider V as a $\mathbb{C}[x_0]$-module and it is easy to show that V is freely generated by $x^I := \{x^\beta, \beta \in I\}$. Let
$$A_F : V \to V, \ A_F(G) = \frac{\partial F}{\partial x_0} G, \ G \in V.$$

Proposition 7.1. *The matrix of A_F in the basis x^I is of the form $d \cdot [x_0^{K_{\beta,\beta'}} c_{\beta,\beta'}]$, where $K_{\beta,\beta'} := d - 1 + \deg(x^\beta) - \deg(x^{\beta'})$ and $A_f := [c_{\beta,\beta'}]$ is the multiplication by f in the Milnor vector space of f. In particular, if $A_{\beta'} - A_\beta \geq 1$ then $c_{\beta,\beta'} = 0$ and*
$$\det(A_F - t.x_0^{d-1} I) = \det(A_f - t.I) x_0^{(d-1)\mu}.$$

Proof. Since the polynomial F is weighted homogeneous, we have $\sum_{i=0}^{n+1} \alpha_i x_i \frac{\partial F}{\partial x_i} = d \cdot F$ and so $x_0 \frac{\partial F}{\partial x_0} = d.F$ in V (note that $\alpha_0 = 1$ by definition). Let
$$F.x^\beta = \sum_{\beta' \in I} x^{\beta'} c_{\beta,\beta'}(x_0) + \sum_{i=1}^{n+1} \frac{\partial F}{\partial x_i} q_i, \ c_{\beta,\beta'}(x_0) \in \mathbb{C}[x_0], \ q_i \in \mathbb{C}[x_0, x]. \quad (7.1)$$
Since the left-hand side is homogeneous of degree $d + \deg(x^\beta)$ we can assume that the pieces of the write hand side are also homogeneous of the same degree. This can be done by taking an arbitrary equation (7.1) and subtracting the unnecessary parts. □

Let $\tilde{C}$ be a finite subset of $\mathbb{C}$ and $\mathbb{C}[t]_{\tilde{C}}$ be the localization of $\mathbb{C}[t]$ on its multiplicative subgroup generated by $t - c, \ c \in \tilde{C}$ and $F_t = F - t.x_0^d$. From now on we work with $\mathbb{C}[t]_{\tilde{C}}[x_0, x]$ instead of $\mathbb{C}[x_0, x]$ and redefine V using $\mathbb{C}[t]_{\tilde{C}}[x_0, x]$. Let
$$V_{\tilde{C}} = \mathbb{C}[t]_{\tilde{C}}[x_0, x] / \left\langle \frac{\partial F_t}{\partial x_0}, \frac{\partial F}{\partial x_i}, \mid i = 1, 2, \ldots, n+1 \right\rangle .$$

It is useful to reformulate $V_{\tilde{C}}$ in the following way: Let $R := \mathbb{C}[t]_{\tilde{C}}[x_0]$ be the set of polynomials in x_0 with coefficients in $\mathbb{C}[t]_{\tilde{C}}$ and $A_t = A_F - t.d.x_0^{d-1}I$. We have

$$V_{\tilde{C}} = V/\left\langle \frac{\partial F_t}{\partial x_0} q \mid q \in V \right\rangle = R^\mu / A_t.R^\mu.$$

Here R^μ is the set of $\mu \times 1$ matrices with entries in R. We consider the statement:

$*(\tilde{C})$: There is a function $\beta \in I \to d_\beta \in \mathbb{N} \cup \{0\}$ such that the $\mathbb{C}[t]_{\tilde{C}}$-module $V_{\tilde{C}}$ is freely generated by

$$\{x_0^{\beta_0} x^\beta, 0 \leq \beta_0 \leq d_\beta - 1, \beta \in I\}. \tag{7.2}$$

To prove the statement $*(\tilde{C})$ we may introduce a kind of Gaussian elimination in A_t and simplify it. For this reason we introduce the operation $GE(\beta_1, \beta_2, \beta_3)$. For $\beta \in I$ let $(A_t)_\beta$ be the β-th row of A_t.

- Input: A_t, $\beta_1, \beta_2, \beta_3 \in I$ with $A_{\beta_1} \leq A_{\beta_2}$. Output: a matrix A'_t and a finite subset B of $\mathbb{C}$.

  We replace $(A_t)_{\beta_2}$ with

  $$-\frac{(A_t)_{\beta_2,\beta_3}}{(A_t)_{\beta_1,\beta_2}} * (A_t)_{\beta_1} + (A_t)_{\beta_2}$$

  and we set $B = zero(c(t))$, where $(A_t)_{\beta_1,\beta_2} = c(t).x_0^{K_{\beta_1,\beta_2}}$. Since for all $\beta_4 \in I$ we have

  $$K_{\beta_2,\beta_3} + K_{\beta_1,\beta_4} = K_{\beta_1,\beta_3} + K_{\beta_2,\beta_4}.$$

  The obtained matrix A'_t is of the form $[x_0^{K_{\beta,\beta'}} c'_{\beta,\beta'}]$ and $c'_{\beta_2,\beta_3} = 0$. If the matrix B_t is obtained from A_t by applying the above operation and $B \subset \tilde{C}$ then $A_t.R^\mu = B_t R^\mu$.

  We give an example of algorithm which calculates d_β's for for some finite set $\tilde{C} \subset \mathbb{C}$:

- Input: A_t. Output: $d_\beta, \beta \in I$ and a finite set $\tilde{C} \subset \mathbb{C}$.

  We identify I with $\{1, 2, \ldots, \mu\}$ and assume that

  $$\beta_1 \leq \beta_2 \Rightarrow A_{\beta_1} \geq A_{\beta_2}.$$

  The algorithm has μ steps indexed by $\beta = \mu, \mu - 1, \ldots, 1$. We define the set $\tilde{C}$ to be empty. In $\beta = \mu$ we have $A(\beta) = A_t$. In the step β we find the first β_1 such that $A(\beta)_{\beta,\beta_1} \neq 0$ and put $d_{\beta_1} = d - 1 + \deg(x^\beta) - \deg(x^{\beta_1})$. For $\beta_2 = \beta - 1, \ldots, 1$ we make $GE(\beta, \beta_2, \beta_1)$ and define $\tilde{C} = \tilde{C} \cup \cup_{\beta_2=1}^{\beta-1} B_{\beta_2}$, where B_{β_2} is obtained during $GE(\beta, \beta_2, \beta_1)$. The numbers d_β's obtained in this way prove the statement $*(\tilde{C})$.

The advantage of this algorithm is that in many cases it gives $\tilde{C} = C$. We do not have a proof for $*(C)$. One can also fix a value $c \in \mathbb{C} \backslash C$ and apply the above algorithm for A_c. In this case we do not care about $\tilde{C}$ during the algorithm. The obtained d_β's make the statement $*(\tilde{C})$ true for some $\tilde{C} \subset \mathbb{C}$ with $c \notin \tilde{C}$. We prove the following weak statements:

Proposition 7.2. *There is a function $\beta \in I \to d_\beta \in \mathbb{N} \cup \{0\}$ such that the $\mathbb{C}[t]_C$-module V' is generated by $\{x_0^{\beta_0} x^\beta, 0 \le \beta_0 \le d_\beta - 1, \beta \in I\}$.*

Proof. We have

$$V' = R^\mu / A_t R^\mu \overset{b}{\cong} A_t^{-1} R^\mu / R^\mu = \frac{A_t^{\mathrm{adj}} R^\mu}{x_0^{\mu(d-1)}} / R^\mu.$$

The isomorphism b in the middle is obtained by acting A_t^{-1} from left on R^μ and adj makes the adjoint of a matrix. Now for $\beta \in I$ let d_β be the pole order of βth arrow of $\frac{A_t^{\mathrm{adj}}}{x_0^{\mu(d-1)}}$. The numbers d_β are the desired numbers. It is easy to see that $\{x_0^{\beta_0} x^\beta, 0 \le \beta_0 \le d_\beta, \beta \in I\}$ generates V'. □

Proposition 7.3. *There is a subset $\tilde{C} \subset \mathbb{C}$ such that the statement $*(\tilde{C})$ is true with $d_\beta = d - 1, \beta \in I$.*

Proof. We identify I with $\{1, 2, \dots, \mu\}$ and assume that

$$\beta_1 \le \beta_2 \Rightarrow A_{\beta_1} \ge A_{\beta_2}.$$

By various use of operation GE on A_t we make all the entries of $(A_t)_{\beta,\mu} = 0,\ \beta \in I \backslash \{\mu\}$. We repeat this for $(A_t)_{\beta,\mu-1} = 0, \beta \in I \backslash \{\mu, \mu - 1\}$ and after μ-times we get a lower triangular matrix. We always divide on a polynomial on t with leading coefficient one and so division by zero does not occur. □

Proposition 7.4. *Let $*(\tilde{C})$ be valid with $d_\beta, \beta \in I$. Then*

$$A_\beta < n + 1,\ d_\beta < d(n + 2 - A_\beta),\ \sum_{\beta \in I} d_\beta = \mu(d - 1).$$

Proof. The first one is already in Steenbrink's Theorem 4.2. The second inequality is obtained by applying the first inequality associated to $F - cx_0^d$ for some $c \in \mathbb{C} \backslash \tilde{C}$:

$$A_{(d_\beta - 1, \beta)} = A_\beta + \frac{d_\beta - 1 + 1}{d} < n + 2.$$

The Milnor number of $F - cx_0^d$ is $\sum_{\beta \in I} d_\beta$ and equals to the Milnor number of $g - cx_0^d$ which is $\mu(d - 1)$. □

Suppose that $*(\tilde{C})$ is valid with $d_\beta,\ \beta \in I$. Define

$$I_{n+1}^k = \{\beta \in I \mid A_\beta = n + 1 - k\},\ I_n^k = \{\beta \in I \mid A_\beta + \frac{1}{d} \le n + 1 - k \le A_\beta + \frac{d_\beta}{d}\}.$$

We can restate Theorem 3.3 in the following way: For a tame polynomial f, the set

$$B = \cup_{k=1}^n B_{n+1}^k \cup \cup_{k=0}^n B_n^k$$

with

$$B_{n+1}^k = \{\nabla^{n-k} \omega_\beta \mid \beta \in I_{n+1}^k\}, B_n^k = \{\nabla^{n-k} \omega_\beta \mid \beta \in I_n^k\},$$

is a basis of $H''_{\tilde{C}}$ compatible with the mixed Hodge structure. The same is true for $H'_{\tilde{C}}$ replacing $\nabla^{n-k} \omega_\beta$ with $\nabla^{n+1-k} \eta_\beta$. Unfortunately, this theorem gives us a basis of a localization H compatible with the mixed Hodge structure. In §10 we have computed such bases for the Brieskorn module itself.

To handle easier the pieces of the mixed Hodge structure of $H_{\tilde{C}}$ we make the following table.

0	1		2		…	n		$n+1$
I^n_n	I^n_{n+1}	I^{n-1}_n	I^{n-1}_{n+1}	I^{n-2}_n	…	I^1_n	I^1_{n+1}	I^0_n

In the case $n = 1$ we have the table

0	1	2
I^1_1	I^1_2	I^0_1

$$I^1_1 = \left\{\beta \in I \mid A_\beta + \frac{1}{d} \leq 1 \leq A_\beta + \frac{d_\beta}{d}\right\},\ I^0_1 = \left\{\beta \in I \mid A_\beta + \frac{1}{d} \leq 2 \leq A_\beta + \frac{d_\beta}{d}\right\},$$

$$I^1_2 = \{\beta \in I \mid A_\beta = 1\}.$$

The forms ω_β, $\beta \in I^1_1$ form a basis of $F^1 \cap W_1$ and the forms $\omega_\beta, \beta \in I^2_1$ form a basis of H''/W_1. Now to obtain a basis of $W_1/(F^1 \cap W_1)$ we must modify $\nabla\omega_\beta$, $\beta \in I^0_1$.

8. Picard-Fuchs equations

It is a well-known fact that for a polynomial $f \in \mathbb{C}[x]$ and $\omega \in H$ the integral $I(t) := \int_{\delta_t} \omega$ satisfies

$$\left(\sum_{i=0}^{k} p_i(t) \frac{\partial^i}{\partial t^i}\right) I_t = 0,\ p_i(t) \in \mathbb{C}[t] \tag{8.1}$$

called Picard-Fuchs equation, where $\delta_t \in H_n(L_t, \mathbb{Z})$ is a continuous family of topological cycles. When f is tame, it is possible to calculate p_i' as follows:

We write

$$\nabla^i(\omega) = \sum_{\beta \in I} p_{i,\beta}\omega_\beta$$

and define the $k \times \mu$ matrix $A = [p_{i,\beta}]$, where i runs through $1, 2, \ldots, k$ and $\beta \in I$. Let k be the smallest number such that the the rows of A_{k-1} are $\mathbb{C}(t)$-linear independent. Now, the rows of A_k are $\mathbb{C}(t)$-linear dependent and this gives us (after multiplication by a suitable element of $\mathbb{C}[t]$)

$$\sum_{i=0}^{k} p_i(t)\nabla^i(\omega) = 0,\ p_i(t) \in \mathbb{C}[t].$$

Using the formula (1.2) and integrating the above equality, we get the equation (8.1).

9. Polynomials in one variable, $n = 0$

The theory developed in §2 does not work for the case $n = 0$. For a polynomial of degree d in one variable $\dim(H^0(L_t, \mathbb{C})) = d$ but $\mu = d - 1$. However, if we use the following definition of homology and cohomology for a discrete topological space M,

$$H_0(M, \mathbb{Z}) = \Big\{ m = \sum_i a_i m_i \mid a_i \in \mathbb{Z},\ m_i \in M \mid \deg(m) = \sum_i a_i = 0 \Big\},$$

$$H^0(M, \mathbb{C}) = \{ f : H_0(M, \mathbb{Z}) \to \mathbb{C} \text{ linear} \} / \{ f \mid \text{ f is constant on } M \},$$

then

$$H' = \mathbb{C}[x]/\mathbb{C}[f],\ H'' = \mathbb{C}[x]dx/f'\mathbb{C}[f]dx,\ I = \{1, x, x^2, \dots, x^{d-2}\}, \mu = d - 1.$$

In this case

$$\int_\delta \omega = \sum_i a_i \omega(p_i), \quad \text{where } \delta = \sum_i a_i p_i,\ a_i \in \mathbb{Z},\ p_i \in f^{-1}(t),\ \omega \in H'.$$

If, for instance, $f' = 0$ has d distinct roots then every vanishing cycle in L_t is a difference of two points of L_t. The set $B = \{x, x^2, \dots, x^{d-1}\}$ form a basis of H' and its ∇ which is $\{dx, xdx, \dots x^{d-2}dx\}$ (up to multiplication by some constants) form a basis of H''. The first fact is easy to see. We write $f = a_d x^d + f_0$ and for a polynomial $p(x) \in \mathbb{C}[x]$ whenever we find some x^d we replace it with $\frac{f - f_0}{a_d}$ and at the end we get $p(x) = p_0(f) + \sum_{i=1}^{d-1} p_i(f) x^i$ or equivalently $p = \sum_{i=1}^{d-1} p_i(t) x^i$ in H'. There is no $\mathbb{C}[t]$-linear relation between the elements of B because B restricted to each regular fiber is of dimension d. We write

$$\begin{aligned} p(x)dx &= \sum_{i=0}^{d-2} q_i(f) x^i dx + q_{d-1}(f) x^{d-1} dx \\ &= \left(\sum_{i=0}^{d-2} q_i(f) x^i dx - \frac{q_{d-1}(f) f_0'}{d.a_d} dx \right) + \frac{q_{d-1}(f) f'}{d.a_d} dx \end{aligned}$$

and this proves the statement for H''.

Proposition 4.1 can be stated in the case $n = 0$ as follows: The only case in which $dA_\beta - \alpha_i = 0$ is when $n = 0$ and $P = 1$. In the case $n = 0$ for $P \neq 1$ we have

$$\frac{\partial g}{\partial x_i}.Pdx = \frac{d}{d.A_\beta - \alpha_i} \frac{\partial P}{\partial x_i} g dx$$

and if $P = 1$ then $\frac{\partial g}{\partial x_i}.Pdx$ is zero in H''. The argument in (4.4) and (4.5) can be done also in the case $n = 0$. In this case if

$$\eta = \sum_{\beta \in I} p_\beta(g) \eta_\beta + p(g),\ p, p_\beta \in \mathbb{C}[t], \tag{9.1}$$

where each piece in the right-hand side of the above equality has degree less than $\deg(\eta)$ then

$$d\eta = \sum_{\beta \in I} (p_\beta(g)A_\beta + p'_\beta(g)g)\omega_\beta + p'(g)dg. \tag{9.2}$$

In the case $n = 0$, we have only the set $I_0^0 = \{A_\beta + \frac{1}{d} \leq 1 \leq A_\beta + \frac{d_\beta}{d}\}$ and this is equal to I. We have $d_\beta < d.(n+2-A_\beta) = 2d - \beta - 1 =$ and $A_\beta = \frac{\beta+1}{d}$. We conclude that

$$d \leq d_\beta + \beta + 1 < 2d.$$

Now the infinitesimal Hilbert problem (see [6] Problem 7) can be stated in the case $n = 0$. Can one give an effective solution to this problem in this case? The positive answer to this question may give light into the the problem in the case $n = 1$. It is worked out in [4].

10. Examples

All the algorithms explained in this article are implemented in a library of SINGULAR. It can be downloaded from the authors homepage. The procedure `okbase` makes a permutation on the output of `kbase` and gives us the set x^I with $\deg(x^\beta)$ decreasing. The algorithms in §4 after (4.2) are implemented in the procedures `linear1, linear2`. The procedures `linear` and `linearp` are for the algorithms in §5. Based on the observations in §9, these procedures work also for the case $n = 0$. The procedure `nabla` uses the formulas (6.1) and (6.2) and computes ∇ and its iterations. The procedure `nablamat` calculates the matrix $\frac{1}{S(t)}[p_{\beta,\beta'}]$ in (6.3). The calculation of the polynomial S in §6 is implemented in the procedure `S`. Using Proposition 7.1, the procedure `muldF` calculates A_F. The algorithm for d_β's is implemented in the procedure `dbeta`. The procedure `changebase` calculates the matrix of the basis of the Brieskorn module $H''_{\tilde{C}}$ obtained in Theorem 3.3 in the canonical basis $\omega_\beta, \beta \in I$. The procedure `Imk` gives us $x^\beta, \beta \in I_m^k,\ m = n, n+1, k = 0, 1, \ldots n$ with the order $I_n^n, I_n^{n-1}, \ldots, I_n^0, I_{n+1}^n, I_{n+1}^{n-1}, \ldots, I_{n+1}^1$. The procedure `PFeq` calculates p_i's in (8.1).

Theorem 3.3 does not give a basis of the Brieskorn module compatible with the mixed Hodge structure. In the following examples we obtain such bases for some examples of f by modifying the one given in §3.3 (we do not have a general method for every f).

For all the examples below we download the author's library and matrix.lib.

10.1. Examples, $n = 0$

Example. $f = x^5 - 5x$, $P = \{\epsilon^i \mid i = 0, 1, 2, 3\}$, $C = \{-4\epsilon^i \mid i = 0, 1, 2, 3\}$, where $\epsilon = e^{\frac{2\pi i}{d-1}}$ is the d-th root of unity.

```
> ring r0=(0,t),x, dp;
> int d=5; poly f=x^d-d*x;
> poly Sf=S(f); Sf;
(t4-256)
> list l=nablamat(f,Sf);
> l[1]; print(l[2]);
1/(5t4-1280)
(-t3), 128, (-48t),(16t2),
(4t2), (-2t3),192, (-64t),
(-16t),(8t2), (-3t3),256,
64, (-32t),(12t2),(-4t3)
//This is the matrix of nabla in the canonical
//basis x^3,x^2,x^1,1.
>PFeq(f,1);
_[1,1]=6144
_[1,2]=(35625t)
_[1,3]=(33375t2)
```

```
_[1,4]=(8750t3)
_[1,5]=(625t4-160000)
```

The residues of $\frac{dx}{f-t}$ at its poles satisfy the Picard-Fuchs equation

$$6144 + 35625t\frac{\partial}{\partial t} + 33375t^2\frac{\partial^2}{\partial t^2} + 8750t^3\frac{\partial^3}{\partial t^3} + (625t^4 - 160000)\frac{\partial^4}{\partial t^4} = 0.$$

10.2. Examples, $n = 1$

For the examples below we define

```
ring r1=(0,t), (x,y), dp;
```

Example. $f = xy(x + y - 1)$.

```
> poly f= x2y+xy2-xy ;
> poly g=lasthomo(f); g;
x2y+xy2
> okbase(std(jacob(g)));
_[1]=y2
_[2]=y
_[3]=x
_[4]=1
> print(muldF(f-par(1)));
(-3t+1/18)*x2,-1/18*x3, 0, 0,
1/6*x, (-3t-1/6)*x2,0, 0,
1/6*x, -1/6*x2, (-3t)*x2,0,
1/2, -1/2*x, 0, (-3t)*x2
> poly Sf=S(f); Sf;
(t4+1/27t3)
//We can take Sf=t*(t+1/27);
> list l1=nablamat(f,Sf);
> l1[1]; " "; print(l1[2]);
1/(54t2+2t)

(18t+1),(-18t-1),0,(-2t),
1, -1, 0,(-6t),
1, -1, 0,(-6t),
3, -3, 0,(-18t)
//--------------
> dbeta(f,par(1));
0,2,2,4
> Imk(f,par(1));
[1]:
   [1]:
      [1]:
         1
   [2]:
      [1]:
         1
[2]:
   [1]:
      [1]:
         x
      [2]:
         y
> list l3=changebase(f,Sf,par(1));
> print(l3[1]); " "; print(l3[2]); det(l3[2]);
1,3/(54t2+2t),1,1

0,0, 0,1,
1,-1,0,(-6t),
0,0, 1,0,
0,1, 0,0
1
//--------------
> dbeta(f);
2,2,2,2
> Imk(f);
[1]:
   [1]:
      [1]:
         1
   [2]:
      [1]:
         y2
[2]:
   [1]:
      [1]:
         x
      [2]:
         y
> list l2=changebase(f,Sf);
> print(l2[1]); " "; print(l2[2]); det(l2[2]);
1,1/(54t2+2t),1,1

0, 0, 0,1,
(18t+1),(-18t-1),0,(-2t),
0, 0, 1,0,
0, 1, 0,0
(18t+1)
//The obtained basis does not work for the
//fiber c=-1/18.
//--------------
> PFeq(f,1, Sf);
_[1,1]=6
_[1,2]=(54t+1)
_[1,3]=(27t2+t)
_[1,4]=0
_[1,5]=0
```

We get the following basis of H'' compatible with mixed Hodge structure.

$f = xy(x + y - 1)$

$\mathrm{Gr}_F^1\mathrm{Gr}_1^W H''$	$[1]$
$\mathrm{Gr}_F^0\mathrm{Gr}_1^W H''$	$[y^2] - [y] - 6t[1]$
$\mathrm{Gr}_F^1\mathrm{Gr}_2^W H''$	$[x], [y]$

The integrals $I = \int_{\delta_t} \frac{dx\wedge dy}{f-t}$ satisfy the Picard-Fuchs equation

$$6 + (54t + 1)\frac{\partial I}{\partial t} + (27t^2 + t)\frac{\partial^2 I}{\partial t^2} = 0$$

Example. $f = 2(x^3 + y^3) - 3(x^2 + y^2)$, $P = \{(0,0), (0,1), (1,0), (1,1)\}$, $C = \{0, -1, -1, -2\}$,

```
> poly f= 2*x3+2*y3-3*x2-3*y2 ;
> poly g=lasthomo(f); g;
2*x3+2*y3
> okbase(std(jacob(g)));
_[1]=xy
_[2]=y
_[3]=x
_[4]=1
>S(f);
(t4+4t3+5t2+2t)
//We can put
>poly Sf=t*(t+1)*(t+2);
> list l2=changebase(f,Sf);
> print(l2[1]); " "; print(l2[2]); det(l2[2]);
1,-1/(6t+12),1,1

0, 0,0,1,
-2,1,1,0,
0, 0,1,0,
0, 1,0,0
-2
```

$f = 2(x^3 + y^3) - 3(x^2 + y^2)$

$\mathrm{Gr}_F^1\mathrm{Gr}_1^W H''$	$[1]$
$\mathrm{Gr}_F^0\mathrm{Gr}_1^W H''$	$[2xy - x - y]$
$\mathrm{Gr}_F^1\mathrm{Gr}_2^W H''$	$[x], [y]$

Example. $f = x^4 + y^4 - x$.

```
> poly f= x4+y4-x ;
> poly g=lasthomo(f);
> okbase(std(jacob(g)));
_[1]=x2y2
_[2]=xy2
_[3]=x2y
_[4]=y2
_[5]=xy
_[6]=x2
_[7]=y
_[8]=x
_[9]=1
> poly Sf=S(f); Sf;
(t9+81/256t6+2187/65536t3+19683/16777216)
//We can take
>Sf=t^3+27/256;
> dbeta(f,par(1));
2,2,2,5,2,2,5,2,5
> Imk(f,par(1));
[1]:
   [1]:
      [1]:
         1
      [2]:
         x
      [3]:
         y
   [2]:
      [1]:
         y
      [2]:
         y2
      [3]:
         x2y2
   [2]:
      [1]:
         [1]:
            x2
         [2]:
            xy
         [3]:
            y2
> list l3=changebase(f,Sf,par(1));
> print(l3[1]); " "; print(l3[2]); det(l3[2]);
1,1,1,4/(256t3+27),24/(256t3+27),1/(256t3+27),1,1,1
0, 0, 0,0,0,0,0, 0,1,
0, 0, 0,0,0,0,0, 1,0,
0, 0, 0,0,0,0,1, 0,0,
0, 0, 9,0,0,0,(-16t2),0,0,
3, (-2t),0,0,0,0,0, 0,0,
(128t2),9, 0,0,0,0,0, 0,0,
0, 0, 0,0,0,1,0, 0,0,
0, 0, 0,0,1,0,0, 0,0,
0, 0, 0,1,0,0,0, 0,0
(2304t3+243) // 9*256*Sf;
> matrix A=l3[2];
> A[6,1..ncols(A)]=
((-128*t2)/3)*submat(A,5,1..ncols(A))+
submat(A,6,1..ncols(A));
> A[5,1..ncols(A)]=
2*t*submat(A,6,1..ncols(A))+submat(A,5,1..ncols(A));
print(A);
0,0,0,0,0,0,0, 0,1,
0,0,0,0,0,0,0, 1,0,
0,0,0,0,0,0,1, 0,0,
0,0,9,0,0,0,(-16t2),0,0,
1,0,0,0,0,0,0, 0,0,
0,1,0,0,0,0,0, 0,0,
0,0,0,0,0,1,0, 0,0,
0,0,0,0,1,0,0, 0,0,
0,0,0,1,0,0,0, 0,0
```

We obtain the following table

$f = x^4 + y^4 - x$

$\mathrm{Gr}_F^1 \mathrm{Gr}_1^W H''$	$[1], [x], [y]$
$\mathrm{Gr}_F^0 \mathrm{Gr}_1^W H''$	$9[x^2y] - 16t^2[y], [x^2y^2], [xy^2]$
$\mathrm{Gr}_F^1 \mathrm{Gr}_2^W H''$	$[x^2], [xy], [y^2]$

We make the following remark

```
> reduce(9*x2*y-16*(f^2)*y, std(jacob(f)));
0
```

References

[1] V.I. Arnold, S.M. Gusein-Zade, A.N. Varchenko, *Singularities of differentiable maps. Vol. II.* Monodromy and asymptotics of integrals. Monographs in Mathematics, **83** Birkhäuser Boston, Inc., Boston, MA, 1988.

[2] P. Deligne, *Théorie de Hodge. II.* Inst. Hautes Études Sci. Publ. Math. **40** (1971), 5–57.

[3] L. Gavrilov, *The infinitesimal* 16*th Hilbert problem in the quadratic case.* Invent. Math. **143**(3) (2001), 449–497.

[4] L. Gavrilov, H. Movasati. *The infinitesimal* 16*th Hilbert problem in dimension zero.* Preprint 2005 (math.CA/0507061).

[5] P.A. Griffiths. *Infinitesimal variations of Hodge structure. III. Determinantal varieties and the infinitesimal invariant of normal functions.* Compositio Math. **50**(2-3) (1983), 267–324.

[6] Yu. Ilyashenko. *Centennial history of Hilbert's* 16*th problem.* Bull. Amer. Math. Soc. (N.S.), **39**(3) (2002), 301–354.

[7] V.S. Kulikov, P.F. Kurchanov. *Complex algebraic varieties: periods of integrals and Hodge structures.* Algebraic geometry, III, 1–217, 263–270, Encyclopaedia Math. Sci., 36, Springer, Berlin, 1998.

[8] H. Movasati. *Center conditions: rigidity of logarithmic differential equations.* J. Differential Equations, **197**(1) (2004), 197–217.

[9] H. Movasati. *Mixed Hodge structure of affine varieties.* To appear in Ann. Ins. Four. (2005).

[10] H. Movasati. *Abelian integrals in holomorphic foliations.* Revista Matemática Iberoamericana, **20**(1) (2004), 183–204.

[11] J. Steenbrink. *Intersection form for quasi-homogeneous singularities.* Compositio Math. **34**(2) (1977) 211–223.

Hossein Movasati
IMPA – Instituto de Matematica Pura e Aplicada
Estrada Dona Castorina 110
Jardim Botanico 22460-320
Rio de Janeiro – RJ Brazil
www.impa.br/∼hossein/
e-mail: `hossein@impa.br`

Real and Complex Singularities
Trends in Mathematics, 263–287

Whitney Equisingularity, Euler Obstruction and Invariants of Map Germs from $\mathbb{C}^n$ to $\mathbb{C}^3$, $n > 3$

Victor H. Jorge Pérez, Eliris C. Rizziolli and Marcelo J. Saia

Abstract. We study how to minimize the number of invariants that is sufficient for the Whitney equisingularity of a one parameter deformation of any finitely determined holomorphic germ $f : (\mathbb{C}^n, 0) \to (\mathbb{C}^3, 0)$, with $n > 3$. Gaffney showed in [3] that the invariants for the Whitney equisingularity are the 0-stable invariants and the polar multiplicities of the stable types of the germ. First we describe all stable types which appear in these dimensions. Then we find relationships between the polar multiplicities of the stable types in the singular set and also in the discriminant. When $n > 3$, for any germ f there is an hypersurface in $\mathbb{C}^n$, which is of special interest, the closure of the inverse image of the discriminant by f, which possibly is with non isolated singularities. For this hypersurface we apply results of Gaffney and Gassler [6], and Gaffney and Massey [7], to show how the Lê numbers control the polar invariants of the strata in this hypersurface. Gaffney shows that the number of invariants needed is $4n + 10$. In the corank one case we reduce this number to $2n + 2$. The polar multiplicities are also an interesting tool to compute the local Euler obstruction of a singular variety, see [12]. Here we apply this result to obtain explicit algebraic formulae to compute the local Euler obstruction of the stable types which appear in the singular set and also for the stable types which appear in the discriminant, of corank one map germs from $\mathbb{C}^n$ to $\mathbb{C}^3$ with $n \geq 3$.

Mathematics Subject Classification (2000). 58C27.

Keywords. Whitney equisingularity, Euler obstruction, Polar multiplicities, stable invariants.

The first named author was partially supported by CAPES - Grant 3131/04-3.
The third named author is partially supported by CNPq - Grant 300880/03-6.

1. Introduction

Gaffney describes in [3] the following problem: "Given a 1-parameter family of map germs $F\colon \mathbb{C} \times \mathbb{C}^n, (0,0) \to \mathbb{C} \times \mathbb{C}^p, (0,0)$, find analytic invariants whose constancy in the family implies the family is Whitney equisingular." He shows that for the class of finitely determined map germs of discrete stable type, the Whitney equisingularity of such a family is guaranteed by the invariance of the zero stable types and the polar multiplicities associated to all stable types.

The number of invariants depends on the dimensions (n,p) and it can be very big according to n and p are big. Then a natural question arises: "For a fixed pair of dimensions (n,p), what is the minimum number of invariants in Gaffney's theorem that are necessary to guarantee the Whitney equisingularity of the family?"

In the case of corank one map germs, Vohra in [8] used Gaffney's approach to study map germs from n-space ($n \geq 3$), to the plane. Recently the case $n < p$ was described by Jorge Perez and Saia and the case $n = p$ by Levcovitz, Jorge Perez and Saia.

In this paper we also consider germs of corank one and investigate the case $(n,3)$ with $n > 3$. We reduce the number of invariants needed by finding relations among them and using the fact that they are upper semi-continuous. To obtain these relations we apply a result of Lê-Greuel to all strata which are related with complete intersections with isolated singularity, ICIS for short. When $n \geq p$, there are some stable types which appear in the source which, possibly are with non isolated singularities, for these sets we apply results of Lê and Teissier to show how the Lê numbers control the invariants of these strata.

Another invariant that is associated to the polar varieties is their local Euler obstruction. Here we apply results of Gonzales-Sprinberg [9] and Lê and Teissier (see [12]), to obtain explicit algebraic formulae for the Euler obstruction of the stable types of map germs from $\mathbb{C}^n$ to $\mathbb{C}^3$.

2. Notation and preliminaries

We follow Gaffney in [3] and denote by $\mathcal{O}(n,p)$ the set of origin preserving germs of holomorphic mappings from $\mathbb{C}^n$ to $\mathbb{C}^p$, $\mathcal{O}_e(n,p)$ denotes the set of germs at the origin but not necessarily origin preserving.

For a germ $f \in \mathcal{O}_e(n,p)$, We denote the singular set of f by $S(f)$. It consists of all points where the rank of the derivative of f is less than $min(n,p)$, $J(f)$ denotes the ideal generated by the set of $p \times p$ minors of the derivative of f. The critical set $\Sigma(f)$ of f is the set of points $x \in \mathbb{C}^n$ such that $J(f)(x) = 0$. The discriminant $\Delta(f)$ of f is the image of $\Sigma(f)$ by f. The determinant of the derivative of a germ f in $\mathcal{O}_e(n,n)$ is denoted by $J[f]$.

Our interest is in $\mathcal{A}$-finitely determined map-germs, $\mathcal{A}$ denotes the usual Mather group of germs of holomorphic diffeomorphisms in the source and in the target.

Let $F : (\mathbb{C}^s \times \mathbb{C}^n, (0,0)) \to (\mathbb{C}^s \times \mathbb{C}^p, (0,0))$ be a versal unfolding of such a map germ f.

Definition 2.1. A stable type $\mathcal{Q}$ appears in F if for any representative $F = (id, f_u(x))$ of F, there exists a point $(u,y) \in \mathbb{C}^s \times \mathbb{C}^p$ such that the germ $f_u : (\mathbb{C}^n, S) \to (\mathbb{C}^p, y)$ is a stable germ of type $\mathcal{Q}$, $S = f^{-1}(y) \cap \Sigma(f_u)$. The points (u,y) and (u,x) with $x \in S$ are called points of stable type $\mathcal{Q}$ in the target and in the source, respectively.

Definition 2.2. If $f \in \mathcal{O}(n,p)$ is stable, denote the set of points in $\mathbb{C}^p$ of type $\mathcal{Q}$ by $\mathcal{Q}(f)$ and $\mathcal{Q}_S(f) = f^{-1}(\mathcal{Q}(f)) - \mathcal{Q}_\Sigma(f)$, where $\mathcal{Q}_\Sigma(f) = f^{-1}(\mathcal{Q}(f)) \cap \Sigma(f)$.

If f is finitely determined, denote

$$\overline{\mathcal{Q}(f)} = (\{0\} \times \mathbb{C}^p) \cap \overline{\mathcal{Q}(F)}$$

and

$$\overline{\mathcal{Q}_S(f)} = (\{0\} \times \mathbb{C}^n) \cap \overline{\mathcal{Q}_S(F)},$$
$$\overline{\mathcal{Q}_\Sigma(f)} = (\{0\} \times \mathbb{C}^n) \cap \overline{\mathcal{Q}_\Sigma(F)},$$

the bar means the closure of this set. We say that a stable $\mathcal{Q}$ is a zero-dimensional stable type for the pair (n,p) if $\mathcal{Q}(f)$ has dimension 0, where f is a representative of the stable type $\mathcal{Q}$.

We observe that the set $\overline{\mathcal{Q}(F)} = \cap F(j^{(p+1)} F^{-1}(\overline{\mathcal{A}z_i}))$ is closed and analytic, where z_i is the $p+1$ jet of the stable type $\mathcal{Q}$ and $\mathcal{A}z_i$ is the $\mathcal{A}$-orbit of z_i.

A finitely determined germ f has discrete stable type if there exist a versal unfolding F of f in which appears only a finite number of stable types. If (n,p) is in the nice range of dimensions or in this boundary, then any finitely determined germ f has a discrete stable type.

Suppose that $\mathcal{Q}(F) = \{p_1, \ldots, p_r\}$ is the set of points of zero-dimensional type, where F is a versal unfolding of f. The 0-stable invariant of type $\mathcal{Q}$ of f, denoted by $m(f;\mathcal{Q})$ is the multiplicity of the ideal $m_s\mathcal{O}_{\overline{\mathcal{Q}(F)},(0,0)}$ in $\mathcal{O}_{\overline{\mathcal{Q}(F)},(0,0)}$, where m_s denotes the ideal generated by the coordinates of the space of parameters $\mathbb{C}^s$.

Let $F : (\mathbb{C} \times \mathbb{C}^n, (0,0)) \to (\mathbb{C} \times \mathbb{C}^p, (0,0))$, $F = (t, \overline{f}(t,x))$, be a 1-parameter unfolding of a finitely determined germ f, such that $\overline{f}(t,-)$ preserves the origin for all t. Let $T := \mathbb{C} \times \{0\}$. F is a *good unfolding* of f if there exist neighborhoods U, W of the origin in $\mathbb{C} \times \mathbb{C}^n$ and $\mathbb{C} \times \mathbb{C}^p$ respectively such that $F^{-1}(W) = U$, F maps $U \cap \Sigma(F) - T$ to $W - T$ and if $(t_0, y_0) \in W - T$, then the germ $f_{t_0} : \mathbb{C}^n, S \to \mathbb{C}^p, y_0$ is stable, where $S = F^{-1}(t_0, y_0) \cap \Sigma(F)$.

A good unfolding is *excellent* if all the 0-stable invariants are constant in the unfolding and f is of discrete type. In the equidimensional case $n = p$, it is also assumed that the degree of f, $\delta(f) = \dim_{\mathbb{C}} \frac{\mathcal{O}_n}{f^*(m_n)\mathcal{O}_n}$, is constant in the unfolding.

An unfolding F of f is *Whitney equisingular* along the parameter space T if there exists a regular stratification of the source and the target, with T a stratum of the source and the target and these stratifications are Whitney equisingular along T, i.e., satisfy the Whitney conditions **a** and **b** and Thom's A_F condition.

It is shown in [3] that if f has discrete stable type, and F is a versal unfolding which only a finite number of stable types, then there exists a regular stratification of the source and the target given by the stable types of F, but the unfolding F is not Whitney equisingular, since T is not a stratum of the source and the target.

One of the questions of main interest is to show when an excellent unfolding $F : (\mathbb{C} \times \mathbb{C}^n, (0,0)) \to (\mathbb{C} \times \mathbb{C}^p, (0,0))$ of a finitely determined germ $f \in \mathcal{O}(n,p)$ is Whitney equisingular. Using the polar invariants, i.e., the polar multiplicities of the polar varieties of the stable types (defined by Teissier in [20]) and Thom's isotopy lemmas, Gaffney showed the following principal result.

Theorem 2.3. ([3], p. 207) *Suppose that $F : (\mathbb{C} \times \mathbb{C}^n, (0,0)) \to (\mathbb{C} \times \mathbb{C}^p, (0,0))$ is an excellent unfolding of a finitely determined germ $f \in \mathcal{O}(n,p)$. Also suppose that the polar invariants of all stable types defined in:*

1. *the discriminant $\Delta(f_t) = f_t(\Sigma(f_t))$,*
2. *the singular set $\Sigma(f_t)$ and also in the set*
3. *$X(f_t) = \overline{(f_t^{-1}(\Delta(f_t)) - \Sigma(f_t))}$,*

are constant at the origin for all t. Then the unfolding is Whitney equisingular.

The theorem remains valid if we replace "an excellent unfolding" in the hypothesis by "a 1-parameter unfolding which, when stratified by stable types and by the parameter axis T, has only the parameter axis T as 1-dimensional stratum at the origin" ([8]).

Here we reduce the number of invariants needed by finding relations among them. From the fact that they are upper semi-continuous the relations will allow us to reduce the number of invariants required in Gaffney's theorem.

We remark that in the case of corank one map germs, the stable types which appear in the set $\Sigma(f_t)$ are ICIS and the stable types which appear in $\Delta(f_t)$ are also related to ICIS which are in $\mathbb{C}^n$. For these sets we shall apply the following results.

Theorem 2.4. (Lê-Greuel, [11], page 47) *Let X_1 be an ICIS, with a singularity at $0 \in \mathbb{C}^n$. Let X be an ICIS defined in X_1 by $f_k = 0$, and let $f_1, \dots, f_{k-1}$ be the generators of the ideal that defines X_1 at 0 in $\mathbb{C}^n$. Then*

$$\mu(X_1, 0) + \mu(X, 0) = \dim_{\mathbb{C}} \frac{\mathcal{O}_n}{(f_1, \dots, f_{k-1}, J(f_1, \dots, f_k))}.$$

In the case of a zero-dimensional ICIS we can use the following simpler formula.

Let $f : \mathbb{C}^k, 0 \to \mathbb{C}^k, 0$ be a germ such that $X = f^{-1}(0)$ is an ICIS. Then $\mu(X, 0) = \delta(f) - 1$ (see [13] p. 78). Another elementary result that we use here is the following. Let $f : \mathbb{C}^n, 0 \to \mathbb{C}^n, 0$ be a finitely determined germ. Then $f : \Sigma(f) \subset \mathbb{C}^n, 0 \to \Delta(f) \subset \mathbb{C}^n, 0$ is bimeromorphic (see [4] p. 154, or [13]).

When $n \geq 3$, the stable types which appear in $X(F) = \overline{(F^{-1}(\Delta(F)) - \Sigma(F))}$ possibly are not ICIS. In this case, we use the associate Lê numbers to control all invariants needed to show the Whitney equissingularity of these stable types along T. We apply the results of Lê and Teissier given in [12] and of Gaffney and Gassler in [6].

3. The stable types in $\mathcal{O}(n,3)$

According to the theorem of Gaffney, the constancy of the polar invariants of all stable types defined in $\Delta(f_t)$, $\Sigma(f_t)$ and $X(f_t)$ is the condition for the Whitney equisingularity. The first step in the strategy to minimize the number of invariants is to describe all stable types which appear in these sets, as $(n,3)$ is in the range of the nice dimensions of Mather, any finitely determined map germ $f \in \mathcal{O}(n,3)$ is of discrete type, hence the stratification has a finite number of strata. In general, for any pair of dimensions (n,p) the description of the stable types can be done in terms of subschemes of multiple points of a germ f, as we can see in [18] for the case $n = p$ or in [19] for the case $n < p$.

Here we use the Thom-Boardman stratification of the singular set to describe the stable types which appear in $\Sigma(f)$, then we show the stable types in the discriminant $\Delta(f)$ of f and finally the stable types of $X(f) = \overline{(f^{-1}(\Delta(f)) - \Sigma(f))}$. For any Boardman symbol $i = (i_1, \ldots, i_r)$, we denote by $\Sigma^i(f)$ the set of points in $\Sigma(f)$ of type i.

Stratification of source

In the source there are two sets which are stratified, one of them is the singular set $\Sigma(f)$ whose stratification is done by the smooth parts of the following sets:

1. The 2-dimensional set $\Sigma^{n-2}(f) = \Sigma(f)$;
2. The 1-dimensional set $\Sigma^{n-2,1}(f)$;
3. The 1-dimensional set of double points $D^2_1(f|\Sigma(f))$, which we describe below.

The construction of the set of multiple points for any finitely determined map germ from $(\mathbb{C}^n, 0)$ to $(\mathbb{C}^p, 0)$, with $n \geq p$, is described in details by Goryunov in [10]. Here we resume this construction for the case $(n,3)$.

First, consider the set of double points of the restriction of f to $\Sigma(f)$, denoted by $D^2(f|\Sigma(f))$, which is a subset of $\mathbb{C}^{2n}$:
$D^2(f|\Sigma(f)) = \{(p_1, p_2) \in \mathbb{C}^{2n}, p_1 \neq p_2,$ with $p_1, p_2 \in \Sigma(f)$, and $f(p_1) = f(p_2)\}$. Then denote by $D^2_1(f|\Sigma(f))$ the projection of $D^2(f|\Sigma(f))$ to $\mathbb{C}^n$. We remember that the set $D^2_1(f|\Sigma(f))$ is part of the singular set of f.

If f is of corank 1, write $f(x_1, x_2, z_3, \ldots, z_n) = (x_1, x_2, g(x_1, x_2, z_3, \ldots, z_n))$, therefore we can describe $D^2(f|\Sigma(f))$ as the set $(x_1, x_2, z_3, \ldots, z_n, z^1_3, \ldots, z^1_n)$ such that $g(x_1, x_2, z_3, \ldots, z_n) = g(x_1, x_2, z^1_3, \ldots, z^1_n)$, which is in fact a subset of $\mathbb{C}^n \times \mathbb{C}^{n-2}$ and $D^2_1(f|\Sigma(f))$ is the projection of $D^2(f|\Sigma(f))$ to $\mathbb{C}^n$ with coordinates $(x_1, x_2, z_3, \ldots, z_n)$.

The other set to be stratified in the source is $X(f) = \overline{f^{-1}(f(\Sigma(f))) - \Sigma(f)}$, which has its interior in the regular set of f. This stratification is obtained by the inverse image of f of the stable types in the target and also the inverse image of the multiple points. It is formed by the smooth parts of the following sets.

1. The $(n-1)$-dimensional set $X(f) = \overline{(f^{-1}(f(\Sigma(f)) - \Sigma(f))}$;
2. the $(n-2)$-dimensional set $X_1(f) = \overline{(f^{-1}(f(\Sigma^{n-2,1}(f))) - \Sigma(f) \cap \Sigma^{n-2,1}(f))}$;
3. the $(n-2)$-dimensional set

$$X_2(f) = \overline{(f^{-1}(f(D^2_1(f|\Sigma(f)))) - \Sigma(f) \cap D^2_1(f|\Sigma(f)))};$$

4. the $(n-3)$-dimensional set $f^{-1}(0)$.

We remark that the set $f^{-1}(0)$ is an ICIS, while the other, possibly are not.

Stratification of the target

In the target the stratification is done in the discriminant of f, $\Delta(f) = f(\Sigma(f))$. It is formed by the smooth parts of the following sets.

1. The discriminant $\Delta(f) = f(\Sigma(f))$, which is 2-dimensional;
2. The 1-dimensional set $f(\Sigma^{(n-2,1)}(f))$;
3. The image of the double points of f, which is 1-dimensional and denoted by $f(D^2_1(f|\Sigma(f)))$.

Example: Let $F(x, y, z, w) = (x, y, g(x, y, z, w))$ with $g(x, y, z, w) = z^5 + xz^2 + yz + w^2$. Here $\Sigma(f) = \{(x, y, z, w) : g_z = 0 = g_w\} = \{5z^4 + 2xz + y = 0, \text{ and } w = 0\}$ is a surface in $\mathbb{C}^4$, the set $\Delta(f) = f(\Sigma(f)) \subset \mathbb{C}^3$ is two-dimensional and a parametrization for $\Delta(f)$ is given by $(x, z) \to (x, -5z^4 - 2xz, -4z^5 - xz^2)$. Therefore, the set $X(f)$ is an hypersurface in $\mathbb{C}^4$, with

$$X(f) = \left\{(x, y, z, w) \colon y = -4z^4 - 2xz\right\}.$$

We remark that, in this example the set $f^{-1}(0)$ is a curve in $X(f)$ with equations $x = 0, y = 0, z^5 + w^2 = 0$.

To a k-dimensional variety are associated $k + 1$ polar invariants. Since $\Sigma(f)$ and $\Delta(f)$ are of dimension 2, the sets $D^2_1(f|\Sigma(f))$, $\Sigma^{n-2,1}(f)$, $f(\Sigma^{n-2,1}(f))$ and $f(D^2_1(f|\Sigma(f)))$ are of dimension 1, there are 14 polar invariants defined on these sets. We also have $3n - 2$ polar multiplicities of the sets $X(f)$, $X_1(f)$, $X_2(f)$ and $n - 2$ polar multiplicities of the set $f^{-1}(0)$. Therefore to apply Theorem 2.3. to germs in $\mathcal{O}(n, 3)$ we need the constancy of $4n + 10$ invariants. In the following sections we show how they are related.

4. Polar invariants of the stable types in the discriminant

To show how the stable types are related in the discriminant we also follow the method developed by Gaffney in the cases $n = p = 2$ and $n = 2$, $p = 3$. The main idea is to compute the polar multiplicities associated to the stable types. The fact that $\Sigma(f)$, $\Sigma^{n-2,1}(f)$ and $D^2_1(f|\Sigma(f))$ are ICIS is strongly used here. From this we can apply Theorem 2.4 and also the results shown in the final part of Section 2.

The first relation is for the polar multiplicities of the discriminant, as it is 2-dimensional, there are 3 polar multiplicities, that we describe here:

We follow the definition done by Teissier in [20]. To compute $m_1(\Delta(f))$ it is needed to choose a generic projection $p_1 : \mathbb{C}^3 \to \mathbb{C}^2$ such that the degree of $p_{1|\Delta(f)}$ is the the multiplicity of $\Delta(f)$ at 0 and also the polar variety $P_1(\Delta(f))$ is $\overline{\Sigma(p_{1|\Delta^o(f)})}$; denote by $m_1(\Delta(f))$ its multiplicity.

To compute $m_2(\Delta(f))$, choose another linear generic projection $p_2 : \mathbb{C}^2 \to \mathbb{C}$ such that the degree of $(p_2 \circ p_1)_{|P_1(\Delta(f))}$ is $m_1(\Delta(f))$ and we also require $(p_2 \circ p_1)$ to be a generic projection which gives $m_2(\Delta(f))$.

To obtain the multiplicity $m_0(\Delta(f))$, consider the following diagram:

$$\begin{array}{ccc} \Sigma(f) \subset \mathbb{C}^n & \xrightarrow{f} & \Delta(f) \subset \mathbb{C}^3 \\ & \searrow^{p_1 \circ f} \quad \swarrow_{p_1} & \\ & (\mathbb{C}^2, 0) & \end{array}$$

From the choice of p_1, $m_0(\Delta(f)) = \deg(p_{1|\Delta(f)})$.

Next we give a relation between the polar multiplicities of $\Delta(f)$ in terms of the Milnor number of the singular set.

Theorem 4.1. *Let $f \in \mathcal{O}(n,3), n > 3$ be a finitely determined map germ. Then:*

$$m_2(\Delta(f)) - m_1(\Delta(f)) + m_0(\Delta(f)) = \mu(\Sigma(f)) + 1. \tag{I}$$

Proof. We have the following diagram:

$$\Sigma(f) \subset \mathbb{C}^n \xrightarrow{f} \Delta(f) \subset \mathbb{C}^3 \xrightarrow{p_1} \mathbb{C}^2 \xrightarrow{p_2} \mathbb{C}.$$

Now call $X_2 = V(p_2 \circ p_1 \circ f, J(f))$ and $X_1 = V(p_1 \circ f, J(f))$. As X_1 and X_2 are ICIS and subsets of $V(J(f)) = \Sigma(f)$, we apply Theorem 2.4 to obtain:

$$\mu(X_2) + \mu(X_1) = \dim_{\mathbb{C}} \frac{\mathcal{O}_n}{(p_2 \circ p_1 \circ f, J(f), J[p_1 \circ f, J(f)])}. \tag{1}$$

We remember that $J(f)$ denotes the ideal generated by the $p \times p$ minors of the derivative of f and $J[f]$ denotes the determinant of the derivative of $f \in \mathcal{O}_e(n,n)$.

Since $\Sigma(f)$ is also an ICIS we apply again Theorem 2.4 to $\Sigma(f) = V(J(f))$ and X_2 to get $\mu(\Sigma(f)) + \mu(X_2) = \dim_{\mathbb{C}} \frac{\mathcal{O}_n}{(J(f), J(p_2 \circ p_1 \circ f, J(f)))}$.

Then

$$\mu(X_2) = \dim_{\mathbb{C}} \frac{\mathcal{O}_n}{(J(f), J(p_2 \circ p_1 \circ f, J(f)))} - \mu(\Sigma(f)) \tag{2}$$

and

$$\dim_{\mathbb{C}} \frac{\mathcal{O}_n}{(J(f), J(p_2 \circ p_1 \circ f, J(f)))} - \mu(\Sigma(f)) + \mu(X_1)$$
$$= \dim_{\mathbb{C}} \frac{\mathcal{O}_n}{(p_2 \circ p_1 \circ f, J(f), J[p_1 \circ f, J(f)])}. \tag{3}$$

But, X_1 is 0-dimensional then

$$\mu(X_1) = \deg(p_1 \circ f, J(f)) - 1 \tag{4}$$

and

$$\dim_{\mathbb{C}} \frac{\mathcal{O}_n}{(J(f), J(p_2 \circ p_1 \circ f, J(f)))} - \mu(\Sigma(f)) + \deg(p_1 \circ f, J(f)) - 1$$
$$= \dim_{\mathbb{C}} \frac{\mathcal{O}_n}{(p_2 \circ p_1 \circ f, J(f), J[p_1 \circ f, J(f)])}. \tag{5}$$

From the equation (5) we shall obtain the equation (I) which gives the relationship between the polar multiplicities of the discriminant. In fact these multiplicities are implicitly described in the equation (5), as we shall see now.

Since $f : \Sigma(f) \to \Delta(f)$ is finite and bimeromorphic, $\deg(p_1|\Delta(f)) = \deg(p_1 \circ f|\Sigma(f)) = \deg(p_1 \circ f, J(f))$. Therefore

$$\deg(p_1 \circ f, J(f)) = m_0(\Delta(f)). \tag{i}$$

Now we find $m_1(\Delta(f))$, let $V' = V(J(f), J[p_1 \circ f, J(f)])$ and call $g : \mathbb{C}^3 \to \mathbb{C}$ the defining equation of $\Delta(f)$, that is $g^{-1}(0) = \Delta(f)$. As $f_{|V'}$ is finite and bimeromorphic we can also obtain $P_1(\Delta(f)) = \overline{V(J[g, p_1], J(f))}$.

From this definition of $P_1(\Delta(f))$, since $V' = V(J(f), J[p_1 \circ f, J(f)])$ we have that the projection of $p_2 \circ p_1 \circ f(V')$ in $\mathbb{C}$ gives the same image than the projection of $p_2(P_1(\Delta(f)))$ in $\mathbb{C}$.

$$\begin{array}{ccc} V' \subset \Sigma(f) & \xrightarrow{f} & \Delta(f) \supset P_1(\Delta(f)) \\ & \searrow^{p_2 \circ p_1 \circ f} \qquad \swarrow_{p_2} & \\ & \mathbb{C} & \end{array}$$

Then $\deg(p_2 \circ p_1 \circ f_{|V'}) = \deg(p_{2|P_1(\Delta(f))})$, from the choice of p_2, we have $\deg(p_{2|P_1(\Delta(f))}) = m_1(\Delta(f))$ and $m_1(\Delta(f)) = \deg(p_2 \circ p_1 \circ f_{|V'})$.

Since V' is I.C.I.S, the ring $\mathcal{O}_{V'}$ is Cohen-Macaulay, then

$$m_1(\Delta(f)) = \deg(p_2 \circ p_1 \circ f_{|V'}) = \dim_{\mathbb{C}} \frac{\mathcal{O}_n}{(p_2 \circ p_1 \circ f, J(f), J[p_1 \circ f, J(f)])}. \tag{ii}$$

Now we shall show that

$$m_2(\Delta(f)) = \dim_{\mathbb{C}} \frac{\mathcal{O}_n}{(J(f), J(p_2 \circ p_1 \circ f, J(f)))}.$$

Since this multiplicity involves the stable types, choose an s parameters versal unfolding F of f, to get

$$\begin{array}{rcl} F : \Sigma(F) \subset \mathbb{C}^s \times \mathbb{C}^n & \longrightarrow & \Delta(F) \subset \mathbb{C}^s \times \mathbb{C}^3 \\ (x, u) & \longmapsto & (u, f_u(x)). \end{array}$$

From the fact that p_2 is generic and linear, we have

$$\Sigma(((\pi_s, p_2 \circ p_1) \circ F)|\Sigma(F)) = V(J[F], J((\pi_s, p_2 \circ p_1) \circ F, J[F])) = V \subset \mathbb{C}^n \times \mathbb{C}^s.$$

We remark that $m_2(\Delta(f))$ is controlled by the degree of the projection $(\pi_s, p_2 \circ p_1)_{|V}$, or in other words, by the length $e_J(f)$ of the maximal ideal m_s in $\mathcal{O}_s$. Then $e_J(f) = \dim_{\mathbb{C}} \dfrac{\mathcal{O}_n}{(J(f), J(p_2 \circ p_1 \circ f, J(f)))}$.

The possible components of V are the closure of the sets: $F^{-1}(P_2(\Delta(F),\pi_s))$, $F^{-1}(A_3)$, $F^{-1}(A_{(1,2)})$ and $F^{-1}(A_{(1,1,1)})$, therefore we need to count the contribution for the degree of the projection $(\pi_s, p_2\circ p_1)$ restrict to each one of these components.

To do this we choose a generic parameter u and neighborhoods $U_2 \subset \mathbb{C}^s \times \mathbb{C}^n, U_1 \subset \mathbb{C}^s$ such that for all point in U_1 there exist $e_J(f)$ pré-images in $V\cap U_2$, counting its multiplicities.

Then we obtain

$$\begin{aligned} e_J(f) &= \sum_{x\in S} \dim_{\mathbb{C}} \frac{\mathcal{O}_{s+n,x}}{(m_s, J(f), J((\pi_s, p_2\circ p_1)\circ f, J(f)))} \\ &= \sum_{x\in S} \dim_{\mathbb{C}} \frac{\mathcal{O}_{n,x}}{(J(f_u), J(p_2\circ p_1\circ f_u, J(f_u)))}, \end{aligned}$$

where $S = {\pi_s}^{-1}(0)\cap V$.

Since the parameter u is generic we can consider that f_u is stable. Then to count the contribution of these components we need to use the normal forms of the stable types which appear in the dimensions $(n,3)$.

We present here an explicit description of the normal forms of all corank one stable map germs in $\mathcal{O}(n,3), n>3$. First we describe the mono germs, we remember that they are suspensions of the stable mono map germs which appear in $\mathcal{O}(3,3)$.

Stable Germs

1. Submersion Σ^0, (A_0): $f(x_1,x_2,\ldots,x_n) = (x_1,x_2,x_n)$
2. Fold Σ^{n-2}, (A_1): $f(x_1,x_2,\ldots,x_n) = (x_1,x_2,\pm x_3^2 \pm\cdots\pm x_{n-1}^2 \pm x_n^2)$
3. Cuspidal edge $\Sigma^{n-2,1}$, (A_2): $f(x_1,x_2,\ldots,x_n) = (x_1,x_2,\pm x_3^2 \pm\cdots\pm x_{n-1}^2 \pm x_n^3 + x_1x_n)$
4. Swallowtail $\Sigma^{n-2,1,1}$, (A_3): $f(x_1,x_2,\ldots,x_n) = (x_1,x_2,\pm x_3^2 \pm\cdots\pm x_{n-1}^2 \pm x_n^4 \pm x_1x_n \pm x_2x_n^2)$.

To describe the stable multigerms we consider the normal crossing between the stable mono germs.

Stable multigerms

1. Double points, $(A_{(1,1)})$:
$\{(x_1,x_2,\pm x_3^2\pm\cdots\pm x_{n-1}^2\pm x_n^2)(x_1,\pm x_2^2\pm x_4^2\pm\cdots\pm x_{n-1}^2\pm x_n^2,x_3)\}$.
2. Plane with a cuspidal edge, $(A_{(1,2)})$:
$\{(x_1,\pm x_2^2\pm x_4^2\pm\cdots\pm x_{n-1}^2\pm x_n^2,x_3);$
$(x_1,x_2,\pm x_3^2\pm\cdots\pm x_{n-1}^2\pm x_n^3+x_1x_n)\}$.
3. Triple points, $(A_{(1,1,1)})$:
$\{(x_1,x_2,\pm x_3^2\pm\cdots\pm x_{n-1}^2\pm x_n^2);\ (x_1,\pm x_2^2\pm x_4^2\pm\cdots\pm x_{n-1}^2\pm x_n^2,x_3);$
$(\pm x_1^2\pm x_4^2\pm\cdots\pm x_{n-1}^2\pm x_n^2,x_2,x_3)\}$.

Now we return to count **the contribution of the stable types:** we remark that these points can or not appear in V, depending of the type of the singularity.

- Singularity of type A_3, here

  $$(J(f_u), J(p_2 \circ p_1 \circ f_u, J(f_u))) = \left(x_3, \dots, x_{n-1}, 4{x_n}^3 + 2x_2x_n + x_1, \dots, \mathbf{1}\right)$$

  Hence

  $$\dim_{\mathbb{C}} \frac{\mathcal{O}_{n,x}}{(J(f_u), J(p_2 \circ p_1 \circ f_u, J(f_u)))} = \dim_{\mathbb{C}} \frac{\mathcal{O}_{n,x}}{(x_3, \dots, x_{n-1}, 4{x_n}^3 + 2x_2x_n + x_1, \dots, \mathbf{1})} = 0,$$

  and we obtain that there is no contribution of this stable type.

- Singularity of type $A_{(1,2)}$, here the ideal $J(f_u)$ is generated by the generators of the Jacobian ideals of the germs which define f_u, then

  $$J(f_u) = \left(x_2, x_3, \dots, x_{n-1}, x_n, 3{x_n}^2 + x_1\right) = (x_1, x_2, x_3, \dots, x_{n-1}, x_n)\,.$$

  Hence

  $$(J(f_u), J(p_2 \circ p_1 \circ f_u, J(f_u))) = (x_1, x_2, x_3, \dots, x_{n-1}, x_n, \mathbf{1})$$

  and

  $$\dim_{\mathbb{C}} \frac{\mathcal{O}_{n,x}}{(J(f_u), J(p_2 \circ p_1 \circ f_u, J(f_u)))} = \dim_{\mathbb{C}} \frac{\mathcal{O}_{n,x}}{(x_1, x_2, x_3, \dots, x_{n-1}, x_n, \mathbf{1})} = 0.$$

  And we obtain again that there is no contribution of the stable type $A_{(1,2)}$.

- Singularity of type $A_{(1,1,1)}$. Here $J(f_u) = (x_1, x_2, x_3, \dots, x_{n-1}, x_n)$ and

  $$(J(f_u), J(p_2 \circ p_1 \circ f_u, J(f_u))) = (x_1, x_2, x_3, \dots, x_{n-1}, x_n, \mathbf{1})\,.$$

  Then

  $$\dim_{\mathbb{C}} \frac{\mathcal{O}_{n,x}}{(J(f_u), J(p_2 \circ p_1 \circ f_u, J(f_u)))} = \dim_{\mathbb{C}} \frac{\mathcal{O}_{n,x}}{(x_1, x_2, x_3, \dots, x_{n-1}, x_n, \mathbf{1})} = 0.$$

  Again we obtain that there is no contribution of the stable type $A_{(1,1,1)}$.

Therefore we conclude that the components of V are in the closure of $F^{-1}(P_2(\Delta(F), \pi_s))$, then

$$e_J(f) = \sum_{x \in S} \dim_{\mathbb{C}} \frac{\mathcal{O}_{n,x}}{(J(f_u), J(p_2 \circ p_1 \circ f_u, J(f_u)))} = \deg(((\pi_s, p_2 \circ p_1) \circ F)|V).$$

Since $F_{|V}$ is finite and bimeromorphic we get

$$\deg(((\pi_s, p_2 \circ p_1) \circ F)|V) = \deg((\pi_s, p_2)|P_2(\Delta(F), \pi_s)) = m_2(\Delta(f)).$$

On the other side,

$$m_2(\Delta(f)) = e_J(f) = \dim_{\mathbb{C}} \frac{\mathcal{O}_n}{(J(f), J(p_2 \circ p_1 \circ f, J(f)))}. \qquad \text{(iii)}$$

Now we use (i), (ii) and (iii) in the equation (5) to obtain (I):

$$m_2(\Delta(f)) - m_1(\Delta(f)) + m_0(\Delta(f)) = \mu(\Sigma(f)) + 1. \qquad \square$$

Next we give the relation for the polar multiplicities of $f(\Sigma^{n-2,1}(f))$, as it is 1-dimensional, there are 2 polar multiplicities and the relation between them is given in terms of the Milnor number of the set $\Sigma^{n-2,1}(f)$ and also of the number of singularities of type A_3.

Remark 4.2. Defining ideal of $\Sigma^{n-2,1}(f)$ is $J_{(n-2,1)}(f) = I_n(d(f, I_3(d(f))))$, where $d(h)$ denotes the Jacobian matrix of a map germ h and $I_s(M)$ denotes the ideal generated by the s minors of some matrix M.

Then, for any corank one map germ $f(x_1, \dots, x_n) = (x_1, x_2, g(x_1, \dots, x_n))$ we obtain $J_{(n-2,1)}(f) = (g_{x_3}, g_{x_4}, \dots, g_{x_n}, M)$, where g_{x_i} denotes the partial derivative of g in the variable x_i, M is the determinant $\begin{vmatrix} g_{x_3^2} & \cdots & g_{x_3x_n} \\ g_{x_4x_3} & \cdots & g_{x_4x_n} \\ \vdots & \ddots & \vdots \\ g_{x_nx_3} & \cdots & g_{x_n^2} \end{vmatrix}$ and $g_{x_ix_j}$ denotes the partial derivative of g_{x_i} in the variable x_j.

Theorem 4.3. *For any corank one map germ f from C^n to $\mathbb{C}^3$:*

$$m_0(f(\Sigma^{n-2,1}(f))) - m_1(f(\Sigma^{n-2,1}(f))) = \sharp A_3 - \mu(\Sigma^{n-2,1}(f)) + 1. \qquad \text{(II)}$$

Proof. Since f is finitely determined, $\Sigma^{n-2,1}(f)$ has reduced structure, from the fact that f is of corank one $\Sigma^{n-2,1}(f) = V(J_{(n-2,1)}(f))$ is an ICIS, then to get the equation (II) we apply again Theorem 2.4. Choose a generic linear projection $p : \mathbb{C}^3 \to \mathbb{C}$ such that $X := \Sigma^{n-2,1}(f) \cap (p \circ f^{-1}(0))$ is an ICIS and

$$m_0(f(\Sigma^{n-2,1}(f))) = \deg(p|(f(\Sigma^{n-2,1}(f)))) = V(J_{(n-2,1)}(f), p \circ f).$$

We apply Theorem 2.5. for the sets $\Sigma^{n-2,1}(f)$ and X to obtain

$$\mu(\Sigma^{n-2,1}(f)) + \mu(X) = \dim_{\mathbb{C}} \frac{\mathcal{O}_n}{(J_{(n-2,1)}(f), J[J_{(n-2,1)}(f), p \circ f])}. \qquad (1)$$

From this equation we obtain the equation (II).

First we remark that X is an 0-dimensional ICIS, then

$$\mu(X) = \dim_{\mathbb{C}} \frac{\mathcal{O}_n}{(J_{(n-2,1)}(f), p \circ f)} - 1 = \deg((p \circ f)|\Sigma^{n-2,1}(f)).$$

Since $f|\Sigma^{n-2,1}(f)$ is bimeromorphic and finite,

$$\deg((p \circ f)|\Sigma^{n-2,1}(f)) = \deg(p|f(\Sigma^{n-2,1}(f))).$$

On the other side, from the choice of p, we obtain $\deg(p|f(\Sigma^{n-2,1}(f))) = m_0(f(\Sigma^{n-2,1}(f)))$.

Therefore $\deg((p\circ f)|\Sigma^{n-2,1}(f)) = m_0(f(\Sigma^{n-2,1}(f)))$ and

$$\mu(X) = \dim_{\mathbb{C}} \frac{\mathcal{O}_n}{\big(J_{(n-2,1)}(f), p\circ f\big)} - 1 = m_0(f(\Sigma^{n-2,1}(f))) - 1.$$

Then we have

$$m_0(f(\Sigma^{n-2,1}(f))) = \dim_{\mathbb{C}} \frac{\mathcal{O}_n}{\big(J_{(n-2,1)}(f), p\circ f\big)}. \qquad \text{(i)}$$

Our next step is to work with $m_1(f(\Sigma^{n-2,1}(f)))$ to get the equation (II).

Since $\dim(f(\Sigma^{n-2,1}(f))) = 1$ we need to consider all stable types which appear here, then we need to count the contribution of each one of the 0-stable types in $m_1(f(\Sigma^{n-2,1}(f)))$.

To obtain this, we consider a s-parameters versal unfolding F of f

$$\begin{aligned} F : \Sigma^{n-2,1}(F) \subset \mathbb{C}^s \times \mathbb{C}^n &\longrightarrow F(\Sigma^{n-2,1}(F)) \subset \mathbb{C}^s \times \mathbb{C}^3 \\ (u,x) &\longmapsto (u, f_u(x)). \end{aligned}$$

From the linearity of the generic projection p, $\Sigma(((\pi_s,p)\circ F)|\Sigma^{(n-2,1)}(F)) = V(J_{(n-2,1)}(F), J[(\pi_s,p)\circ F, J_{(n-2,1)}(F)]) = V \subset \mathbb{C}^s \times \mathbb{C}^n$ and we conclude that the $m_1(f(\Sigma^{n-2,1}(f)))$ is controlled by the degree of the projection π_s restrict to V, that is, by the length $e_J(f)$ of the maximal ideal m_s in the source $\mathcal{O}_s$. Then

$$e_J(f) = \dim_{\mathbb{C}} \frac{\mathcal{O}_n}{\big(J_{(n-2,1)}(f), J[p\circ f, J_{(n-2,1)}(f)]\big)}.$$

Since the possible components of V are the closure of the sets $F^{-1}(A_3)$, $F^{-1}(A_{(1,2)})$, $F^{-1}(A_{(1,1,1)})$ and $F^{-1}(P_1(F(\Sigma^{(n-2,1)}(F)),\pi_s))$, we need to count the contribution for the degree of the projection (π_s,p) restrict to each one of these components.

To do this we choose a generic parameter u and neighborhoods $U_2 \subset \mathbb{C}^s \times \mathbb{C}^n, U_1 \subset \mathbb{C}^s$ such that for each point in U_1 there exist $e_J(f)$ pre-images in $V\cap U_2$, counting its multiplicities.

Therefore, for $S = {\pi_s}^{-1}(0)\cap V$:

$$\begin{aligned} e_J(f) &= \sum_{x\in S} \dim_{\mathbb{C}} \frac{\mathcal{O}_{s+n,x}}{\big(m_s, J_{(n-2,1)}(F), J[(\pi_s,p)\circ F, J_{(n-2,1)}(F)]\big)} \\ &= \sum_{x\in S} \dim_{\mathbb{C}} \frac{\mathcal{O}_{n,x}}{\big(J_{(n-2,1)}(f_u), J[p\circ f_u, J_{(n-2,1)}(f_u)]\big)}. \end{aligned}$$

From the genericity of the parameter u we suppose that f_u is stable and to count the contribution of the components we use the normal forms.

Contribution of the stable types: For the type A_3, we have $J_{(n-2,1)}(f_u) = (x_3, x_4, \dots, x_{n-1}, 4{x_n}^3 + 2x_2x_n + x_1, 12{x_n}^2 + 2x_2\big)$ and

$$\big(J_{(n-2,1)}(f_u), J[p\circ f_u, J_{(n-2,1)}(f_u)]\big) = \big(x_3,\dots,x_{n-1}, 4{x_n}^3 + 2x_2x_n + x_1,$$
$$12{x_n}^2 + 2x_2, J(x_2, x_3, \dots, x_{n-1}, 4{x_n}^3 + 2x_2x_n + x_1, 12{x_n}^2 + 2x_2)\big).$$

But $J(x_2, x_3, \ldots, x_{n-1}, 4{x_n}^3 + 2x_2x_n + x_1, 12{x_n}^2 + 2x_2) = 24x_n$ and $\big(J_{(n-2,1)}(f_u), J[x_2, J_{(n-2,1)}(f_u)]\big) = \big(x_3, x_4, \ldots, x_{n-1}, 4{x_n}^3 + 2x_2x_n + x_1, 12{x_n}^2 + 2x_2, 24x_n\big)$.

Therefore

$$\dim_{\mathbb{C}} \frac{\mathcal{O}_{n,x}}{\big(J_{(n-2,1)}(f_u), J[x_2, J_{(n-2,1)}(f_u)]\big)} = 1$$

and we conclude that, here the contribution of the stable type A_3 is 1.

For the type $A_{(1,2)}$. Since f_u is a multi germ, the ideal $J_{(n-2,1)}(f_u)$ is generated by the generators of the iterated Jacobian ideals $J_{(n-2,1)}(f_1)$ and $J_{(n-2,1)}(f_2)$ of the germs which define f_u. Consider $f_1(x_1, \ldots, x_n) = (x_1, \pm x_2^2 \pm x_4^2 \pm \cdots \pm x_{n-1}^2 \pm x_n^2, x_3)$, and $f_2(x_1, \ldots, x_n) = (x_1, x_2, \pm x_3^2 \pm \cdots \pm x_{n-1}^2 \pm x_n^3 + x_1x_n)$.

Here $J_{(n-2,1)}(f_1) = (x_2, x_4, \ldots, x_{n-1}, x_n, 1)$ and $J_{(n-2,1)}(f_2) = (x_2, \ldots, x_n, \mathbf{1})$ then

$$\dim_{\mathbb{C}} \frac{\mathcal{O}_{n,x}}{\big(J_{(n-2,1)}(f_u), J(p \circ f_u, J_{(n-2,1)}(f_u))\big)} = \dim_{\mathbb{C}} \frac{\mathcal{O}_{n,x}}{(x_2, \ldots, x_n, \mathbf{1})} = 0$$

and there is no contribution of the stable type $A_{(1,2)}$.

Now we consider the type $A_{(1,1,1)}$, let $f_1(x_1, \ldots, x_n) = (x_1, x_2, \pm x_3^2 \pm \cdots \pm x_{n-1}^2 \pm x_n^2)$, $f_2(x_1, \ldots, x_n) = (x_1, \pm x_2^2 \pm x_4^2 \pm \cdots \pm x_{n-1}^2 \pm x_n^2, x_3)$ and $f_3(x_1, \ldots, x_n) = (\pm x_1^2 \pm x_4^2 \pm \cdots \pm x_{n-1}^2 \pm x_n^2, x_2, x_3)$.

Here use the results done for the ideal $J_{(n-2,1)}(f_1)$ in the case $A_{(1,2)}$ to conclude also that there is no contribution of the stable type $A_{(1,1,1)}$.

Therefore

$$e_J(f) = \sum_{x \in S} \dim_{\mathbb{C}} \frac{\mathcal{O}_{n,x}}{\big(J_{(n-2,1)}(f_u), J[p \circ f_u, J_{(n-2,1)}(f_u)]\big)} = \deg(((\pi_s, p) \circ F)|V).$$

Since $F|V$ is finite and bimeromorphic, we get

$$\deg(((\pi_s, p) \circ F)|V) = \deg((\pi_s, p)|P_1(F(\Sigma^{(n-2,1)}(F), \pi_s) = m_1(f(\Sigma^{(n-2,1)}(f)) + \sharp A_3$$

On the other side

$$e_J(f) = \dim_{\mathbb{C}} \frac{\mathcal{O}_n}{\big(J_{(n-2,1)}(f), J[p \circ f, J_{(n-2,1)}(f)]\big)}.$$

Then

$$m_1(f(\Sigma^{(n-2,1)}(f)) + \sharp A_3 = \dim_{\mathbb{C}} \frac{\mathcal{O}_n}{\big(J_{(n-2,1)}(f), J[p \circ f, J_{(n-2,1)}(f)]\big)}. \qquad \text{(ii)}$$

Using (i) and (ii) in the equation (1) we obtain the equation (II):

$$m_0(f(\Sigma^{n-2,1}(f))) - m_1(f(\Sigma^{n-2,1}(f))) = \sharp A_3 - \mu(\Sigma^{n-2,1}(f)) + 1. \qquad \square$$

Corollary 4.4. *From these results we obtain the following equality:*

$$m_1(\Delta(f)) = m_0(f(\Sigma^{n-2,1}(f))) \tag{III}$$

Proof. Since $f \in \mathcal{O}(n,3), n > 3$ is finitely determined and of corank one, we have $f(x_1, x_2, \ldots, x_n) = (x_1, x_2, g(x_1, x_2, \ldots, x_n))$ and $J_{(n-2,1)}(f) = (J(f), M)$, where M is the determinant $\begin{vmatrix} g_{x_3^2} & \cdots & g_{x_3x_n} \\ g_{x_4x_3} & \cdots & g_{x_4x_n} \\ \vdots & \ddots & \vdots \\ g_{x_nx_3} & \cdots & g_{x_n^2} \end{vmatrix}$.

In the proof of Theorems 4.1 e 4.3 we see that

$$m_1(\Delta(f)) = \dim_{\mathbb{C}} \frac{\mathcal{O}_n}{(p_2 \circ p_1 \circ f, J(f), J[p_1 \circ f, J(f)])}$$

and

$$\begin{aligned} m_0(f(\Sigma^{n-2,1}(f))) &= \dim_{\mathbb{C}} \frac{\mathcal{O}_n}{(J_{(n-2,1)}(f), p \circ f)} \\ &= \dim_{\mathbb{C}} \frac{\mathcal{O}_n}{(J_{(n-2,1)}(f), p \circ f)}. \end{aligned}$$

To obtain the equality $m_1(\Delta(f)) = m_0(f(\Sigma^{n-2,1}(f)))$, or

$$\dim_{\mathbb{C}} \frac{\mathcal{O}_n}{(p_2 \circ p_1 \circ f, J(f), J[p_1 \circ f, J(f)])} = \dim_{\mathbb{C}} \frac{\mathcal{O}_n}{(p_1 \circ f, J(f), M)}$$

we need to show that $J[p_1 \circ f, J(f)] = \langle M \rangle$.

Since f is of corank one, we choose $p_1(y_1, y_2, y_3) = (a_1y_1 + a_2y_2 + a_3y_3, b_1y_1 + b_2y_2 + b_3y_3)$, then $(p_1 \circ f)(x_1, x_2, \ldots, x_n) = (a_1x_1 + a_2x_2 + a_3g(x_1, \ldots, x_n), b_1x_1 + b_2x_2 + b_3g(x_1, \ldots, x_n))$. Then as p_1 is a generic projection we can choose $a_1 \neq 0$ and $b_2 \neq 0$ to get the result. □

Remark 4.5. Since all these invariants are upper semi continuous, from Theorem 4.1 we see that if $m_1(\Delta(f_t))$ and $\mu(\Sigma(f_t))$ are constants in the family, we obtain that $m_2(\Delta(f_t))$ and $m_0(\Delta(f_t))$ are also constants in the family.

Now we show how the polar multiplicities of $f(D_1^2(f|\Sigma(f)))$ are related. For this set there are two polar multiplicities and the relation between them is given in terms of the Milnor number of the set $f(D_1^2(f|\Sigma(f)))$ and also of the number of singularities of type A_3.

Theorem 4.6. *Let $f(\mathbb{C}^n, 0) \to (\mathbb{C}^3, 0)$ be a finitely determined map germ of corank one. Then:*

$$\begin{aligned} 2m_0(f(D_1^2(f|\Sigma(f)))) - 2m_1(f(D_1^2(f|\Sigma(f)))) + \mu(D_1^2(f|\Sigma(f))) \\ = 3\sharp A_{(1,2)} + 3\sharp A_3 + 6\sharp A_{(1,1,1)} + 1. \end{aligned} \tag{IV}$$

Proof. Call $\mathcal{I} = f^*(m_3)\mathcal{O}_{D_1^2(f|\Sigma(f))}$, from 8.1 of ([3], p. 209) and ([20], p. 294), we choose a linear generic projection $p : \mathbb{C}^3 \to \mathbb{C}$ in such a way that

$$e(\mathcal{I}) = \deg(p \circ (f|_{(D_1^2(f|\Sigma(f)))}) \text{ and } \deg(p|_{(f((D_1^2(f|\Sigma(f)))))}) = m_0(f(D_1^2(f|\Sigma(f)))).$$

Since $f|_{(D_1^2(f|\Sigma(f)))} - \{0\}$ is a two sheets recovering of $f(D_1^2(f|\Sigma(f))) - \{0\}$ we know that $e(\mathcal{I}) = 2\deg(p|_{(f((D_1^2(f|\Sigma(f)))))}) = 2m_0(f(D_1^2(f|\Sigma(f))))$.

Therefore

$$2m_0(f(D_1^2(f|\Sigma(f)))) = \deg(p \circ (f|_{(D_1^2(f|\Sigma(f)))}) = \deg(p \circ f, I_1^2(f|\Sigma(f)))$$

where $I_1^2(f|\Sigma(f))$ is the defining ideal of $(D_1^2(f|\Sigma(f)))$.

Call $X_2 = V(I_1^2(f|\Sigma(f))) = D_1^2(f|\Sigma(f))$ and $X_1 = V(I_1^2(f|\Sigma(f)), p \circ f) = D_1^2(f|\Sigma(f)) \cap (p \circ f)^{-1}(0)$.

As $D_1^2(f|\Sigma(f))$ is an ICIS, X_2 and X_1 are also ICIS, then applying Theorem 2.4 to X_2 and X_1,

$$\mu(X_2) + \mu(X_1) = \dim_{\mathbb{C}} \frac{\mathcal{O}_n}{(I_1^2(f|\Sigma(f)), J[I_1^2(f|\Sigma(f)), p \circ f])}. \tag{1}$$

Since X_1 is 0-dimensional, we obtain

$$\mu(X_1) = \deg(p \circ f, I_1^2(f|\Sigma(f))) - 1 = 2m_0(f(D_1^2(f|\Sigma(f)))) - 1. \tag{2}$$

Then

$$\begin{aligned} &\mu(D_1^2(f|\Sigma(f))) + 2m_0(f(D_1^2(f|\Sigma(f)))) - 1 \\ &= \dim_{\mathbb{C}} \frac{\mathcal{O}_n}{(I_1^2(f|\Sigma(f)), J[I_1^2(f|\Sigma(f)), p \circ f])}. \end{aligned} \tag{3}$$

To finish we need to show that

$$\begin{aligned} &\dim_{\mathbb{C}} \frac{\mathcal{O}_n}{(I_1^2(f|\Sigma(f)), J[I_1^2(f|\Sigma(f)), p \circ f])} \\ &= 2m_1(f(D_1^2(f|\Sigma(f)))) + 3\sharp A_{(1,2)} + 3\sharp A_3 + 6\sharp A_{(1,1,1)}. \end{aligned}$$

This equality appears when we study $m_1(f(D_1^2(f|\Sigma(f))))$, we remember that this is the multiplicity where the 0-stable types appear.

Choose a s-parameters versal unfolding F of f,

$$F : D_1^2(F|\Sigma(F)) \subset \mathbb{C}^s \times \mathbb{C}^n \longrightarrow F(D_1^2(F|\Sigma(F))) \subset \mathbb{C}^s \times \mathbb{C}^3, \quad F(x,u) = (u, f_u(x)).$$

From the linearity of the generic projection p we have

$$\Sigma(((\pi_s, p) \circ F)|D_1^2(F|\Sigma(F))) = V(I_1^2(F|\Sigma(F)), J((\pi_s, p) \circ F, I_1^2(F|\Sigma(F)))) = V.$$

Since the multiplicity $m_1(f(D_1^2(f|\Sigma(f))))$ is controlled by the degree of the projection $\pi_s : \mathbb{C}^n \times \mathbb{C}^s \to \mathbb{C}^s$ restrict to V, or in other words, by the length of $e_J(f)$ of the maximal ideal m_s in $\mathcal{O}_s$. Then

$$e_J(f) = \dim_{\mathbb{C}} \frac{\mathcal{O}_n}{(I_1^2(f|\Sigma(f)), J[I_1^2(f|\Sigma(f)), p \circ f])}.$$

The possible components of V are the closure of the sets:

$F^{-1}(P_1(F(D^2_1(F|\Sigma(F))),\pi_s))$, $F^{-1}(A_3)$, $F^{-1}(A_{(1,2)})$ and $F^{-1}(A_{(1,1,1)})$.

Then we need to count the contribution of the degree of the projection (π_s, p) restrict to each one of these components. Choose a generic parameter u and neighborhoods $U_2 \subset \mathbb{C}^s \times \mathbb{C}^n, U_1 \subset \mathbb{C}^s$ in such a way that for any point in U_1 there are $e_J(f)$ pre-images in $V \cap U_2$ counting multiplicities.

Hence

$$e_J(f) = \sum_{x\in S} \dim_{\mathbb{C}} \frac{\mathcal{O}_{s+n,x}}{(m_s, I^2_1(F|\Sigma(F)), J[(\pi_s,p)\circ f, I^2_1(F|\Sigma(F))])}$$
$$= \sum_{x\in S} \dim_{\mathbb{C}} \frac{\mathcal{O}_{n,x}}{(I^2_1(f_u|\Sigma(f_u)), J[p\circ f_u, I^2_1(f_u|\Sigma(f_u))])}$$

where $S = {\pi_s}^{-1}(0) \cap V$. From the genericity of the parameter u, we consider that f_u is stable. First we remark that $F^{-1}(P_1(F(D^2_1(F|\Sigma(F))),\pi_s))$ contributes $2m_1(f(D^2_1(f|\Sigma(f))))$ with this degree, we need now to count the contribution of the 0-stable types. To count this contribution, we use the fact that the stable germs in $\mathcal{O}(n,3)$, with $n > 3$ are the suspension of an A_k singularity which appears in $\mathcal{O}(3,3)$, and in this case, the double points set of this singularity in $\mathcal{O}(n,3)$ coincides with the double points set of the singularity in $\mathcal{O}(n,3)$, see [[10], p. 378] for more details.

Therefore we use the calculation done by Jorge Pérez in [[17], p. 912] to get that the contribution of the type A_3 is 3, the contribution of the type $A_{(1,2)}$ is also 3 and of the type $A_{(1,1,1)}$ is 6. Then, since

$$e_J(f) = \sum_{x\in S} \dim_{\mathbb{C}} \frac{\mathcal{O}_{n,x}}{(I^2_1(f_u|\Sigma(f_u)), J[p\circ f_u, I^2_1(f_u|\Sigma(f_u))])} = \deg(((\pi_s,p)\circ F)|V)$$

and $F|V$ is finite and bimeromorphic, we get

$$\deg(((\pi_s,p)\circ F)|V) = \deg((\pi_s,p)|P_1(F(D^2_1(F|\Sigma(F))),\pi_s)$$
$$= 2m_1(f(D^2_1(f|\Sigma(f)))) + 3\sharp A_3 + 3\sharp A_{(1,2)} + 6\sharp A_{(1,1,1)}.$$

On the other side,

$$e_J(f) = \dim_{\mathbb{C}} \frac{\mathcal{O}_n}{(I^2_1(f|\Sigma(f)), J[p\circ f, I^2_1(f|\Sigma(f))])}.$$

Then

$$\dim_{\mathbb{C}} \frac{\mathcal{O}_n}{(I^2_1(f|\Sigma(f)), J[p\circ f, I^2_1(f|\Sigma(f))])}$$
$$= 2m_1(f(D^2_1(f|\Sigma(f)))) + 3\sharp A_3 + 3\sharp A_{(1,2)} + 6\sharp A_{(1,1,1)}. \tag{4}$$

From these, we get the equality (IV)

$$2m_0(f(D^2_1(f|\Sigma(f)))) - 2m_1(f(D^2_1(f|\Sigma(f)))) + \mu(D^2_1(f|\Sigma(f)))$$
$$= 3\sharp A_{(1,2)} + 3\sharp A_3 + 6\sharp A_{(1,1,1)} + 1.$$ □

5. Relations among the invariants of the stable types in $\Sigma(f)$

To obtain the relations among the invariants in $\Sigma(f)$ we use the result of Teissier, given in [20] p. 481, where it is shown how the absolute polar multiplicities of a d-dimensional hypersurface X with isolated singularity are related to the Milnor numbers $\mu^{(k)}(H) = \mu(X \cap H^k)$, with H^k being a generic hyperplane of dimension k, we have

$$m_k(X) = \mu^{(k+1)}(X) + \mu^{(k)}(X), 0 \leq k \leq d-1. \qquad (*)$$

We remark that this result is also valid for ICIS, as we see in [5], p. 210.

To apply these results we remember that the strata of $\Sigma(f)$ are the regular parts of $\Sigma(f)$, $\Sigma^{(n-2,1)}(f)$ and $D_1^2(f|\Sigma(f))$, $\Sigma(f)$ is a two dimensional ICIS and for corank one map germs, the set $\Sigma^{(n-2,1)}(f)$ is also an ICIS.

Gaffney in [5] defines the d-th polar multiplicity for spaces that are ICIS as follows. Let $(X^d, 0)$ be an ICIS of dimension d, then the dth polar multiplicity of $(X^d, 0)$, is $m_d(X^d) = \dim_{\mathbb{C}} \frac{\mathcal{O}_X}{J(p_1,f)}$, where $f : (\mathbb{C}^n, 0) \to (\mathbb{C}^{n-d}, 0)$, $f^{-1}(0) = X^d$ and $p_1 : \mathbb{C}^n \to \mathbb{C}$ is a generic linear projection. We note that as $V(p_1, f)$ is ICIS, then by Lê-Greuel theorem, we have

$$m_d(X^d) = \mu(X^d) + \mu(X^d \cap p_1^{-1}(0)). \qquad (**)$$

We remark that this equality was first noticed by Gaffney in [5], p. 211.

Theorem 5.1. *Let $f \in \mathcal{O}(n, 3)$ with $n > 3$ be a finitely determined map germ, then:*

$$m_2(\Sigma(f)) - m_1(\Sigma(f)) + m_0(\Sigma(f)) = \mu(\Sigma(f)) + 1. \qquad \text{(I)}$$

Moreover, if f is of corank one,

$$m_1(\Sigma^{n-2,1}(f)) - m_0(\Sigma^{n-2,1}(f)) = \mu(\Sigma^{n-2,1}(f)) - 1, \qquad \text{(II)}$$

$$m_1(\Sigma(f)) = m_0(\Sigma^{n-2,1}(f)), \qquad \text{(III)}$$

$$m_1(D_1^2(f|\Sigma(f))) - m_0(D_1^2(f|\Sigma(f))) = \mu(D_1^2(f|\Sigma(f))) - 1. \qquad \text{(IV)}$$

Proof. To show the first equality we use the fact that $\Sigma(f)$ is an ICIS with dimension 2. From $(*)$

$$m_0(\Sigma(f)) = \mu^{(1)}(\Sigma(f)) + \mu^{(0)}(\Sigma(f)) \qquad (1)$$

$$m_1(\Sigma(f)) = \mu^{(2)}(\Sigma(f)) + \mu^{(1)}(\Sigma(f)). \qquad (2)$$

From (1) we obtain

$$\mu^{(1)}(\Sigma(f)) = m_0(\Sigma(f)) - 1 \qquad (3)$$

and from (2):

$$\mu^{(2)}(\Sigma(f)) = m_1(\Sigma(f)) - \mu^{(1)}(\Sigma(f)). \qquad (4)$$

Moreover, from $(**)$ we get

$$m_2(\Sigma(f)) = \mu(\Sigma(f) \cap p_2^{-1}(0)) + \mu(\Sigma(f)). \qquad (5)$$

Then we obtain

$$m_2(\Sigma(f)) - \mu(\Sigma(f) \cap p_2^{-1}(0)) = \mu(\Sigma(f)), \qquad (6)$$

But

$$\mu(\Sigma(f) \cap p_2^{-1}(0)) = \mu^{(2)}(\Sigma(f)). \tag{7}$$

From (7) and (6) we get

$$m_2(\Sigma(f)) - \mu^{(2)}(\Sigma(f)) = \mu(\Sigma(f)). \tag{8}$$

From (4) and (8), we obtain

$$m_2(\Sigma(f)) - m_1(\Sigma(f)) + \mu^{(1)}(\Sigma(f)) = \mu(\Sigma(f)). \tag{9}$$

From (3) and (9):

$$m_2(\Sigma(f)) - m_1(\Sigma(f)) + m_0(\Sigma(f)) - 1 = \mu(\Sigma(f))$$

and the equality (I) follows.

If we suppose that f is of corank 1, $\Sigma^{n-2,1}(f)$ is a 1-dimensional ICIS, then $(*)$ and $(**)$ hold, or

$$m_0(\Sigma^{n-2,1}(f)) = \mu^{(1)}(\Sigma^{n-2,1}(f)) + \mu^{(0)}(\Sigma^{n-2,1}(f)) \tag{10}$$

and this is equivalent to

$$\mu^{(1)}(\Sigma^{n-2,1}(f)) = m_0(\Sigma^{n-2,1}(f)) - 1 \tag{11}$$

and

$$m_1(\Sigma^{n-2,1}(f)) = \mu(\Sigma^{n-2,1}(f) \cap p_1^{-1}(0)) + \mu(\Sigma^{n-2,1}(f)). \tag{12}$$

But

$$\mu(\Sigma^{n-2,1}(f) \cap p_1^{-1}(0)) = \mu^{(1)}(\Sigma^{n-2,1}(f)) \tag{13}$$

from (13) and (12) it follows that

$$m_1(\Sigma^{n-2,1}(f)) = \mu^{(1)}(\Sigma^{n-2,1}(f)) + \mu(\Sigma^{n-2,1}(f)). \tag{14}$$

Then, from (11) and (14) we obtain

$$m_1(\Sigma^{n-2,1}(f)) - m_0(\Sigma^{n-2,1}(f)) = \mu(\Sigma^{n-2,1}(f)) - 1.$$

Therefore the equality (III) follows from the definition of the polar multiplicities and using the genericity of the projections, as we can see in the proof of 4.4.

The proof of the equality (IV) is analogous to the proof of the equality (II), since $D_1^2(f|\Sigma(f))$ is 1-dimensional and is also an ICIS. □

We remark that, as the set $f^{-1}(0)$ is also an ICIS of dimension $n-3$ we obtain, in a similar way to the proof of the above theorem, the following equation

$$\Sigma_{i=0}^{n-3}(-1)^i m_i(f^{-1}(0)) - 1 = (-1)^{n-3}\mu(\Sigma(f))$$

and from this equation we reduce the number of polar invariants needed for $f^{-1}(0)$. However, as this set is a subset of $X(f)$, we shall show in the next section that the associated Lê numbers of $X(f)$ control all these polar multiplicities.

6. The invariants in the stable types of $X(f_t)$

We consider now sets $X(f_t) = f_t^{-1}(\Delta(f_t))$ and $X(F) = F^{-1}(\Delta(F))$. Call $h_t : (\mathbb{C}^n, 0) \to (\mathbb{C}, 0)$ the function that defines $X(f_t) = f^{-1}(\Delta(f_t))$ in $\mathbb{C}^n$ and $H : \mathbb{C}^n \times \mathbb{C} \to \mathbb{C}$ the function that defines $X(F) = F^{-1}(\Delta(F))$.

As we are considering $n > 3$ the set $X(f_t)$ is an hypersurface, possibly with non-isolated singularities, then we cannot apply Theorem 2.4 to the hypersurface $X(f_t)$ in $\{t\} \times \mathbb{C}^n$.

On the other side, we apply the results of Lê and Teissier given in [12], and of Gaffney and Gassler in [6] to prove the Whitney equisingularity of the regular set of $X(F)$ along the parameter space $T = \mathbb{C} \times \{0\}$.

The main invariants used to obtain the Whitney equisingularity of the family $X(f_t)$ are the Lê numbers, which are the generalization of the Milnor number for handling non-isolated hypersurface singularities. See [15] for the definition of these numbers.

Proposition 6.1. *The pair $(X(F) - \Sigma(H), T)$ is Whitney equisingular if, and only if, the Lê numbers $\lambda^i(h_t/H^k)$ and $\lambda^i(h_t)$ are constants on T for all $i = 1, \ldots, k-1$, $k = 1, \ldots, n-2$, where H^k is a generic k-dimensional linear subspace in $\mathbb{C}^n$.*

Proof. To proof this result we follow the proof done in [18] for the case $n = p$. From Gaffney and Gassler in [6], p. 726 we see that

$$\chi^{(k)} = m_k(h_t) + \sum_{i=0}^{k} (-1)^{k-i} \lambda^i(h_t), \tag{6.1}$$

here $\chi^{(k)}$ denotes the reduced Euler characteristic of the Milnor fibre of h_t restricted to a generic k-dimensional linear subspace in $\mathbb{C}^n$.

We denote by χ^* the sequence $\chi^* = \left(\chi^{(n)}, \ldots, \chi^{(2)}\right)$. From Theorem (5.3.1) p. 95 of [21], we see that $X(F) - \Sigma(H)$ is Whitney equisingular along the parameter space if, and only if, χ^* is constant.

If the Lê numbers of $X(f_t)$ and the Lê numbers of all generic planar sections of $X(f_t)$ are constant on T, we apply the equality (6.1) for $\chi^{(k)}$ and the following equalities given by Gaffney and Gassler in [6], p. 710:

$$\begin{aligned} \lambda^0(h_t/H^j) &= \lambda^{n-j}(h_t) + m_{n-j}(h_t), \text{ for } j = 2, \ldots, n-1, \\ \lambda^{k-i}(h_t/H^k) &= \lambda^{n-i}(h_t) \text{ for } i = 1, \ldots, k-1, \end{aligned} \tag{6.2}$$

to obtain that χ^* is constant, hence the pair $(X(F) - \Sigma(H)), T)$ is Whitney equisingular.

We remark that here $m_k(h_t)$ denotes the *relative polar multiplicity*, defined by Teissier in [20].

On the other side, if the pair $(X(F) - \Sigma(H), T)$ is Whitney equisingular, then χ^* is constant. Moreover we obtain that all relative polar multiplicities are also constant, see [20], Chapter V, Theorem 2.1. From the equality (6.1) we obtain that λ^j are constant for all j. From the equality (6.2) we obtain that $\lambda^j(h_t/H^k)$ are constants for each generic k-plane H^k. □

Remark 6.2. From this proposition and also from Theorem (5.3.1) of [21] we obtain that the total space $X(F)$ has a stratification by the stable types and the parameter space T, such that the condition W_f, therefore the Whitney equisingularity, holds for every pair of strata, except possibly over T.

If the singular set of $X(F)$ is Cohen Macaulay (for the structure given by $J(H)$), however, we can use the Lê Numbers associated to the sets $X(f_t)$ to obtain the Whitney equisingularity of the full set $X(F)$ along the parameter space.

Let $F : (\mathbb{C} \times \mathbb{C}^n, (0,0)) \to (\mathbb{C} \times \mathbb{C}^3, (0,0))$ with $n > 3$ be a 1-parameter unfolding of a finitely determined map germ $f \in \mathcal{O}(n,3)$.

Theorem 6.3. *Suppose that the stratification by the stable types of F has only the parameter space $T = \mathbb{C} \times \{0\}$ in $\mathbb{C} \times \mathbb{C}^3$ as a locus of instability and the singular set of $X(F)$ is Cohen Macaulay.*

Then the pair $(X(F), T)$ is Whitney equisingular if, and only if, the sequence $(m_1(h_t), \ldots, m_{n-1}(h_t), \chi^*)$ is independent of t.

Proof. First we remark that if the closure of a stratum S of $V(H) = X(F)$ is the image of a component of the exceptional divisors of the Blow up $B = Bl_{J(H)}\mathbb{C}^{n+1}$, from the fact that the condition A_F of Thom is generic and also from dimensional reazons, this component is the conormal space of $\bar{S}$. Then, applying Theorem 6.3 of [7], the Blow up of this component by the pullback of m_T has a exceptional divisor which is equidimensional over T. From Theorem V.1.2 of [20]), the pair (S, T) satisfies the condition W_F, hence the Whitney equisingularity.

Therefore it is enough to show that each stable type is the image of a component of the exceptional divisor of the Blow up $B = Bl_{J(H)}\mathbb{C}^{n+1}$.

First we show this for the mono germs. Here it is sufficient to show that if f_Q represents a minimal stable type Q which appears with positive dimension in F, then the relative polar curve of $X(f_Q)$ at the origin is not empty, but this curve is empty if and only if the intersection $H_1 \cap \cdots \cap H_{n-1} \cap Bl_{J(h)}\mathbb{C}^n$ is empty if, and only if, the fiber over the origin is not a component.

Therefore we conclude that the origin is the image of a component of the exceptional divisor. Since $X(f_Q)$ is a cartesian product along the strata of $X(F)$ which are different from T, the stratum which represents the stable type $X(f_Q)$ is also the image of a component of the exceptional divisor.

We remember that if the multiplicity of the relative polar curve is not zero, then the polar curve is not empty. Massey showed in ([16], p. 365) that this multiplicity is the number of spheres in the homotopy type of the link of the singularity, independently of the dimension of the singular locus. Here we only need to show now that this number is greater than zero.

From the hypothesis we see that each set $X(f_Q)$ is Cohen Macaulay, then we apply Theorem 2.5 of Damon in [2], to get that $X(f_Q)$ is a free divisor, and from Theorem 3.3 of [2], we obtain that this number is the $\mathcal{A}_e$ codimension of $X(f_Q)$, which is greater than zero.

Therefore the multiplicity of the relative polar curve is not zero, and we finish the proof for mono germs.

Now we consider the stable types of the multi germs. Here we have that if two smooth sheets of $X(F)$ intersect, then this has codimension 0 in $\Sigma(H)$, so this stratum must be the image of a component of the exceptional divisor. Otherwise, we can assume that the set $X(f_t)$ is locally the union of two hypersurfaces X_1 and X_2 embedded in $\mathbb{C}^k \times \mathbb{C}^{n-k}$, with equations $g_1(z_1, \dots, z_k) = 0$ and $g_2(z_{k+1}, \dots, z_n) = 0$. There are two cases: one with $k = n-1$ and $g_2 = z_n$ and the other is with both g_1, g_2 defining singular hypersurfaces.

Suppose first that $h_2 = z_n$. Then we can assume that ϕ_1 parametrizes a branch of the polar curve of h_1 and $h_{1z_1} \circ \phi_1$ is zero for $1 \leq i \leq n-2$. As h_1 is in the integral closure of $J(h_1)$ we obtain that $(h_1/h_{1z_{n-1}}) \circ \phi_1$ is an analytic germ ψ^n. Call $\varphi = (\phi_1, \psi_n)$ then if $f = z_n h_1 f_{z_i} \circ \varphi = 0$ for $1 \leq i \leq n-2$ and $(f_{z_n} - f_{z_{n-1}}) \circ \varphi = (h_1 - z_n h_{1z_{n-1}}) \circ \varphi = 0$. Hence φ parametrizes the branch of the polar curve of f.

Now suppose that $f = h_1 h_2$ and assume that ϕ_1 and ϕ_2 are parametrizations of the branches of the polar curves h_1 and h_2, respectively. Then $\varphi = (\phi_1, \phi_2)$ parametrizes a polar surface of f and we obtain $f_{z_1} \circ \varphi = 0$ for $1 \leq i \leq k$ and $k+1 \leq i \leq n$. Therefore, for an appropriate choice of A and B, $(Af_{z_k} + Bf_{z_n}) \circ \varphi$ defines a curve of singularities which branch is parametrized by ψ, then $\varphi \circ \psi$ parametrizes a branch of the polar curve of f and we are done. □

From the equivalence between the sequences

$$(m_1(h_t), \dots, m_{n-1}(h_t), \chi^*)$$

and

$$(m_1(h_t), \dots, m_{n-1}(h_t), \lambda_2(h_t), \dots, \lambda_n(h_t))$$

and the equations 6.2, we obtain the following:

Corollary 6.4. *The pair par $(X(f_t), T)$ is Whitney equisingular if and only if the Lê numbers $\lambda^{n-j}(h_t)$, and $\lambda^0(h_t|H^j)$, $2 \leq j \leq n-1$ are constant in the origin for any t.*

7. The main results

From the main theorem of Gaffney, we need the constancy of $(4n+10)$ invariants to guarantee the Whitney equisingularity, for example, in the case $n = 4$ we need 26 invariants.

To minimize this number we apply all results shown here to obtain first a result for the case of map germs with any corank.

Theorem 7.1. *Let $F : (\mathbb{C} \times \mathbb{C}^n, (0,0)) \to (\mathbb{C} \times \mathbb{C}^3, (0,0))$ with $n > 3$ be a 1-parameter unfolding of a finitely determined map germ $f \in \mathcal{O}(n,3)$. Suppose that the stratification by the stable types of F has only the parameter space $T = \mathbb{C} \times \{0\}$ in $\mathbb{C} \times \mathbb{C}^3$ as a locus of instability and the singular set of $X(F)$ is Cohen Macaulay.*

Then the family is Whitney equisingular if and only if the following numbers are constant at the origin for all f_t*:*

1. *In the singular set:* $\mu(\Sigma(f_t)), m_1(\Sigma(f_t)), m_0(\Sigma^{(n-2,1)}(f_t)), m_1(\Sigma^{(n-2,1)}(f_t))$, $m_0(D^2_1(f_t|\Sigma(f_t))), m_1(D^2_1(f_t|\Sigma(f_t)))$.
2. *In the discriminant:* $m_1(\Delta(f_t)), m_0(f_t(\Sigma^{(n-2,1)}(f_t)))$, $m_1(f(\Sigma^{(n-2,1)}(f_t)))$, $m_0(f_t(D^2_1(f_t|\Sigma(f_t)))), m_1(f_t(D^2_1(f_t|\Sigma(f_t))))$.
3. *In the set* $X(f_t)$*: The Lê numbers* $\lambda^i(h_t)$*, for* $1 \leq i \leq n$ *and* $\lambda^k(h_t|H^{n-k})$*, for* $1 \leq k \leq n-1$.

In the case of corank one map germs, we minimize the number of invariants to obtain the following:

Theorem 7.2. *Let* $F(t,x) = (t, f_t(x))$ *be an unfolding of a finitely determined map germ and of corank one* $f \in \mathcal{O}(n,3), n > 3$*. Suppose that the stratification by the stable types of* F *has only the parameter space* $T = \mathbb{C} \times \{0\}$ *in* $\mathbb{C} \times \mathbb{C}^3$ *as a locus of instability and the singular set of* $X(F)$ *is Cohen Macaulay.*

Then the family is Whitney equisingular if and only if the following numbers are constant at the origin for all f_t*:*

1. *In the singular set:* $\mu(\Sigma(f_t))$, $\mu(\Sigma^{(n-2,1)}(f_t))$, $m_1(\Sigma(f_t))$, $m_0(D^2_1(f_t|\Sigma(f_t)))$, $m_1(D^2_1(f_t|\Sigma(f_t)))$,
2. *In the discriminant:* $m_1(\Delta(f_t))$, $m_0(f_t(D^2_1(f_t|\Sigma(f_t))))$,$m_1(f_t(D^2_1(f_t|\Sigma(f_t))))$.
3. *In the set* $X(f_t)$*: The Lê numbers* $\lambda^{n-j}(h_t)$*, and* $\lambda^0(h_t|H^j)$*,* $2 \leq j \leq n-1$.

Here we reduce the number of invariants from $4n+10$ to $2n+2$. For instance, if $n=4$ we reduce this number from 26 to 10, if $n=5$, we reduce the number from 30 to 17.

8. The local Euler obstruction of the stable types

The local Euler obstruction for nonsingular varieties, introduced in [14] by R. MacPherson in a purely obstructional way, is an invariant that is also associated to the polar invariants.

Lê and Teissier in [12], with the aid of Gonzales-Sprinberg's purely algebraic interpretation of the local Euler obstruction, showed that the local Euler obstruction is an alternate sum of the multiplicity of the local polar varieties.

Here we apply these results to obtain explicit and algebraic formulae for the Euler obstruction of the stable types which appear in stable mappings from $\mathbb{C}^n$ to $\mathbb{C}^3$.

Suppose that $X \subset \mathbb{C}^n$ is an analytic space of dimension d, ν the transformation of Nash of X. Let $p \in X$ and $z = (z_1, \dots, z_n)$ be local coordinates in $\mathbb{C}^n$ such that $z_i(p) = 0$.

Let $\| z \|^2 = \Sigma z_i \overline{z_i}$. Since $\| z \|^2$ is a real-valued function, $d\| z \|^2$ may be considered as a section of $(T\mathbb{C}^n)^*$ where $*$ denotes the real dual bundle retaining only its orientation from the complex structure. We can also consider $d\| z \|^2$

as a restriction to a section r of $(TX)^*$. In [1] it is showed that for small ϵ, the section r is non zero over ν^{-1} where $0 \leq \| z \| \leq \epsilon$. Therefore let $B_\epsilon = \{z/\| z \| \leq \epsilon\}$ and $S_\epsilon = \{z/\| z \| = \epsilon\}$. The obstruction to extending r as a non zero section of TX^* from $\nu^{-1}(S_\epsilon)$ to $\nu^{-1}(B_\epsilon)$, which we denote by $Eu(TX^*, r)$, lies in $H^d(\nu^{-1}(B_\epsilon), \nu^{-1}(S_\epsilon); \mathbb{Z})$. If $O_{(\nu^{-1}(B_\epsilon),\nu^{-1}(S_\epsilon))}$ denotes the orientation class in $H_d(\nu^{-1}(B_\epsilon), \nu^{-1}(S_\epsilon); \mathbb{Z})$, then we define the local Euler obstruction of X at p to be $Eu(TX^*, r)$ evaluated on $O_{(\nu^{-1}(B_\epsilon),\nu^{-1}(S_\epsilon))}$ or symbolically

$$Eu_p(X) = \langle Eu(TX^*, r), O_{(\nu^{-1}(B_\epsilon),\nu^{-1}(S_\epsilon))} \rangle$$

to $\nu^{-1}(B_\epsilon)$ (see [14] or [9] for the definition and more details).

The following result shows how the local Euler obstruction and the polar multiplicities are related.

Theorem 8.1. (Lê Dũng Trang et Teissier, [12]) *Let X be a reduced analytic space at $0 \in \mathbb{C}^{n+1}$ of dimension d. Then*

$$Eu_0(X) = \Sigma_{i=0}^{d-1}(-1)^{d-i-1} m_i(X)$$

where $m_i(X)$ denotes the absolute polar multiplicity of the polar variety $P_i(X)$.

From this theorem, we can now deduce the formulae for the Euler obstructions of the stable types at $\Delta(f)$ from Theorems 4.1, 4.3, 4.6 and 5.1.

Corollary 8.2. *Let $f \in \mathcal{O}(n,3), n > 3$ be a finitely determined map germ. Then:*

$$\begin{aligned}
Eu_0(\Delta(f)) &= m_2(f(\Sigma(f))) - \mu(\Sigma(f)) - 1,\\
Eu_0(f(\Sigma^{n-2,1}(f))) &= m_1(f(\Sigma^{n-2,1}(f))) - \mu(\Sigma^{n-2,1}(f)) + 1 + \sharp A_3,\\
2Eu_0(f(D_1^2(f|\Sigma(f)))) &= 2m_1(f(D_1^2(f|\Sigma(f)))) - \mu(D_1^2(f|\Sigma(f)))\\
&\quad + 3\sharp A_{(1,2)} + 3\sharp A_3 + 6\sharp A_{(1,1,1)} + 1,\\
Eu_0(\Sigma(f)) &= m_2(\Sigma(f)) - \mu(\Sigma(f)) - 1,\\
Eu_0(\Sigma^{n-2,1}(f)) &= m_1(\Sigma^{n-2,1}(f)) - \mu(\Sigma^{n-2,1}(f)) + 1,\\
Eu_0(D_1^2(f|\Sigma(f))) &= m_1(D_1^2(f|\Sigma(f))) - \mu(D_1^2(f|\Sigma(f))) + 1.
\end{aligned}$$

Acknowledgements

This work is part of the Ph.D. Thesis of Eliris Cristina Rizziolli, supported by CAPES, under supervision of the others authors. The authors thank USP and CAPES for this support. The authors also thank the referee for his several suggestions which helped to clarify this article.

References

[1] J.P. Brasselet and M.H. Schwartz, *Sur les classes de Chern d'un ensemble analytique complexe*, Asterisque **82-83**, 93–146, 1981.

[2] J. Damon, *Nonlinear sections of nonisolated complete intersections*, Kluwer Academic Publishers, 2001.

[3] T. Gaffney, *Polar Multiplicities and Equisingularity of Map Germs*, Topology, vol.32, No.1, 185–223, 1993.

[4] T. Gaffney, *Properties of Finitely Determined Germs*, Ph.D. Thesis, Brandeis University, 1975.

[5] T. Gaffney, *Multiplicities and Equisingularities of I.C.I.S Germs*, Invent. Math. 123, 200–220, 1996.

[6] T. Gaffney and R. Gassler, *Segre Numbers and Hypersurfaces Singularities*, Jour. Algebraic Geometry 08, 695–736, 1999.

[7] T. Gaffney and D. Massey, *Trends in Equisingularity Theory in Singularity Theory*, edited by Bruce, J.W. and Mond. D., Cambridge University Press, 1999.

[8] T. Gaffney and R. Vohra, *A numerical characterization of equisingularity for map germs from n-space, $(n \geq 3)$, to the plane.* J. Dyn. Syst. Geom. Theor. **2**, 43–55, 2004.

[9] G. Gonzalez-Sprinberg, *L'obstruction locale d'Euler et le théorème de Mac-Pherson*, Astérisque **82 et 83**, 7–33, ed. J.L. Verdier, Soc. Math. France, Paris, 1981.

[10] V.V. Goryunov, *Semi-simplicial resolutions and homology of images and discriminants of mappings*, Proc. London Math. Soc. **3**, 70, 363–385, 1995.

[11] Lê D.T., *Calculation of Milnor number of isolated singularity of complete intersection*, Funktsional'ny: Analizi Ego Prilozheniya, vol. 8, no. 2, 45–49, 1974.

[12] Lê D.T. and B. Teissier, *Variétés polaires locales et classes de Chern de variétés singulières.* Annals of Math., vol. 114, 457–491, 1981.

[13] E.J.N. Looijenga, *Isolated singular points on complete intersections*, London Mathematical Soc., Lecture Note series, 77, 1984.

[14] R.D. MacPherson, *Chern class for singular algebraic varieties*, Ann. of Math., **100**, 423–432, 1974.

[15] D. Massey, *Le Cycles and Hypersurface Singularities*, Springer Lecture Notes in Mathematics, No. 1615, 1995.

[16] D. Massey, *Numerical invariants of perverse sheaves*, Duke Math. Journal, 73, ♯2 307–369, 1994.

[17] V.H. Jorge Pérez, *Polar multiplicities and equisingularity of map germs from $\mathbb{C}^3$ to $\mathbb{C}^3$*, Houston Journal of Mathematics. **29** n: 04, 901–923, 2003.

[18] V.H. Jorge Pérez, D. Levcovitz and M.J. Saia, *Invariants, equisingularity and Euler obstruction of map germs from $\mathbb{C}^n$ to $\mathbb{C}^n$*, Journal fur die Reine und angewandte Mathematik, Crelle's Journal, **587** 2005, 145–167.

[19] V.H.J. Jorge Pérez and M.J. Saia. *Euler Obstruction, polar multiplicities and equisingularity of map germs in $O(n,p)$, $n < p$.* Int. Jour. Maths. **17**, n. 8, 887–903, 2006.

[20] B. Tessier, *Variétés Polaires 2: Multiplicités Polaires, Sections Planes, et Conditions de Whitney*, Actes de la conference de géometrie algébrique à la Rábida, Springer Lecture Notes, 961, 314–491, 1981.

[21] B. Tessier, *Cycles Evanescents, Sections Planes et Conditions de Whitney II.* Proc. Symposia Pure Math., vol. 40, part 2, Amer. Math. Soc., 65–104, 1983.

Victor H. Jorge Pérez
Instituto de Ciências Matemáticas e de Computação
Universidade de São Paulo,
Caixa Postal 668,
CEP 13560-970,
São Carlos (SP), Brazil
e-mail: vhjperez@icmc.usp.br

Eliris C. Rizziolli
Instituto de Geociências e Ciências Exatas
Universidade Estadual Paulista "Júlio Mesquita Filho"
Campus de Rio Claro,
Caixa Postal 178,
CEP 13506-700
Rio Claro, (SP), Brazil
e-mail: eliris@rc.unesp.br

Marcelo J. Saia
Instituto de Ciências Matemáticas e de Computação
Universidade de São Paulo,
Caixa Postal 668,
CEP 13560-970,
São Carlos (SP), Brazil
e-mail: mjsaia@icmc.usp.br.

Real and Complex Singularities
Trends in Mathematics, 289–298

Versality Properties of Projective Hypersurfaces

Andrew A. du Plessis

Abstract. Let X be a hypersurface of degree d in $P^n(\mathbb{C})$ with isolated singularities, and let $f : \mathbb{C}^{n+1} \to \mathbb{C}$ be a homogeneous equation for X.

The singularities of X can be simultaneously versally deformed by deforming the equation f, in an affine chart containing all of the singularities, by the addition of all monomials of degree at most r, for sufficiently large r; it is known (see, e.g., §1) that $r \geq n(d-2)$ suffices. Conversely, if the addition in the affine chart of all monomials of degree at most $n(d-2) - 1 - a$, $a \geq 0$, fails to simultaneously versally deform the singularities of X, then we will say that X is *a-non-versal*.

The first main result of this paper shows that X is a-non-versal if, and only if, there exists a homogeneous polynomial vector field with coefficients of degree a, which annihilates f but is not Hamiltonian for f.

Our second main result is a sufficient condition for an a-non-versal hypersurface to be *topologically* a-versal.

Let X be a hypersurface of degree d in $P^n(\mathbb{C})$ with isolated singularities, and let $f : \mathbb{C}^{n+1} \to \mathbb{C}$ be a homogeneous equation for X.

The singularities of X can be simultaneously versally deformed by deforming the equation f, in an affine chart containing all of the singularities, by the addition of all monomials of degree at most r, for sufficiently large r; it is known (see, e.g., §1) that $r \geq n(d-2)$ suffices. Conversely, if the addition in the affine chart of all monomials of degree at most $n(d-2) - 1 - a$, $a \geq 0$, fails to simultaneously versally deform the singularities of X, then we will say that X is *a-non-versal*.

The first main result of this paper shows that X is a-non-versal if, and only if, there exists a homogeneous polynomial vector field with coefficients of degree a, which annihilates f but is not Hamiltonian for f.

There has been much research both on versality properties of hypersurfaces and on vector fields preserving hypersurfaces; the connection between them is somewhat surprising, and appears to have many applications.

A case of particular interest is when $a = n(d-2) - 1 - d$; in this case a-non-versal hypersurfaces are those which fail to have their singularities versally unfolded by the family of all hypersurfaces of degree d.

Necessary conditions for this non-versality in terms of the sum of the Tjurina numbers of the singularities of the hypersurface have been studied by Shustin [11], Shustin and Tyomkin [12], and du Plessis and Wall [5], [9]. In [9], the result of this paper are used to conclude that the best possible conditions are obtained. The result is also essential in [3], where it is used to provide many examples of exotic geometry in families of hypersurfaces of as low codimension as possible.

We note also the case $a = n(d-2)-d$; in this case a-non-versal hypersurfaces X are those which fail to have their singularities versally unfolded by the family of all hypersurfaces of degree d which agree with X on a fixed hyperplane transverse to X. Best possible necessary conditions for this non-versality in terms of the sum of the Tjurina numbers of the singularities are also obtained in [9], again with help from the results of this paper.

Detecting non-versality in this situation gives information about the stratification of [4] for the stratum containing the restriction of f to a hyperplane transverse to X. The conclusion for the case $n = 2$, $d = 5$, $a = 1$ supplies the proof for the argument alluded to (but not used) in [13].

As regards more degenerate singularities, it follows easily from the stated result that X is 0-non-versal if and only if it is a cone, and that X is 1-non-versal if and only if it admits an algebraic one-parameter subgroup of $PGL_n(\mathbb{C})$ as symmetry group. It is not difficult, at least in principle, to enumerate symmetric hypersurfaces with isolated singularities; and they have many interesting properties; see [7] and [8].

Our second main result is a sufficient condition for an a-non-versal hypersurface to be *topologically* a-versal.

This result is applied in [7], §3, in discussing the topological 1-versality of curves.

1. Vector fields and non-versality

Let X be a hypersurface of degree d in $P^n(\mathbb{C})$ with isolated singularities, and let $f \in \mathbb{C}[x_0, \dots, x_n]$ be a homogeneous equation for X. The first aim of this section is to prove:

Theorem 1.1. *Let $a \geq 0$. Then X is a-non-versal if, and only if, there exists a homogeneous vector field on $\mathbb{C}^{n+1}$ with coefficients of degree a, which annihilates f but is not Hamiltonian for f.*

We will make use of some ideas from [2] in proving this.

We sketch the theory. Let $F_1, \dots, F_n \in \mathbb{C}[x_0, \dots, x_n]$ be homogeneous polynomials of degrees $d_1, \dots, d_n$, whose zero-set in $P^n(\mathbb{C})$ is zero-dimensional. Write

K for the ideal generated by these polynomials. Let $J \supset K$ be another ideal defined by homogeneous polynomials. Write $\tilde{J}$ for the *saturation* of J, that is, the ideal generated by homogeneous polynomials $\phi \in \mathbb{C}[x_0, \ldots, x_n]$ such that, for some $k \geq 0$, $(x_0, \ldots, x_n)^k \phi \in J$. Set $S = \mathbb{C}[x_0, \ldots, x_n]/\tilde{J}$, and write H_S for its Hilbert function. Finally, let $I = Ann_K(J)$.

We have:

Proposition 1.2. *Let ℓ be a linear form whose only common zero with K is $\underline{0} \in \mathbb{C}^{n+1}$, and let $\bar{S} = \mathbb{C}[x_0, \ldots, x_n]/(\tilde{J}, \ell)$. Write $m = \sum_{i=1}^{n}(d_i - 1)$. Then*

$$\dim_{\mathbb{C}} \bar{S} - H_S(m - a - 1) = \dim_{\mathbb{C}}(I/K)_a,$$

where $(I/K)_a$ is the ath graded piece of I/K.

This statement is in fact equivalent to the 'Cayley-Bacharach' theorem [2, CB7]. It looks very different from the statement of that theorem, however, so we prove it directly.

We begin with some information on saturated ideals.

Lemma 1.3. *Let $Q \subset \mathbb{C}[x_0, \ldots, x_n]$ be a homogeneous ideal whose zero-set $Z(Q) \subset P^n(\mathbb{C})$ is of dimension 0, $\ell \in \mathbb{C}[x_0, \ldots, x_n]$ a linear form such that $Z(\ell) \cap Z(Q) = \emptyset$.*

Then the following are equivalent:

1. *Q is a saturated ideal,*
2. *ℓ is not a zero divisor modulo Q,*
3. *$\mathbb{C}[x_0, \ldots, x_n]/Q$ is a free $\mathbb{C}[\ell]$-module.*

Proof. We prove (1)⇒(2), (2)⇒(3), and (3)⇒(1):

(1)⇒(2): Q and ℓ vanish together only at $\underline{0} \in \mathbb{C}^{n+1}$, so by the Nullstellensatz there exists $k > 0$ such that $(x_0, \ldots, x_n)^k \subset (Q, \ell)$. Thus if $\ell\phi \in Q$, then

$$(x_0, \ldots, x_n)^k \phi \subset (Q\phi, \ell\phi) \subset Q;$$

whence, since Q is saturated, $\phi \in Q$. So ℓ is not a zero divisor modulo Q.

(2)⇒(3): Let the homogeneous polynomials $\phi_1, \ldots, \phi_r \in \mathbb{C}[x_0, \ldots, x_n]$ project to a $\mathbb{C}$-basis for $\mathbb{C}[x_0, \ldots, x_n]/(Q, \ell)$.

Then any homogeneous polynomial $\psi \in \mathbb{C}[x_0, \ldots, x_n]$ can be written in the form $\sum_{i=0}^{r} \alpha_i \phi_i + \ell\psi'$ modulo Q, with the $\alpha_i \in \mathbb{C}$, and ψ' homogeneous and of lower degree than ψ. Thus the obvious induction shows that the ϕ_i $\mathbb{C}[\ell]$-span $\mathbb{C}[x_0, \ldots, x_n]$ modulo Q.

Now suppose $\sum_{i=1}^{r} \alpha(\ell)\phi_i \in Q$, with the $\alpha_i \in \mathbb{C}[\ell]$ and not all zero. Dividing through by a suitable power of ℓ gives, because ℓ is not a zero divisor modulo Q, a relation of the same form where at least one of the α_i has non-zero constant term. Reducing modulo ℓ then gives a non-trivial $\mathbb{C}$-linear combination of the ϕ_i in (Q, ℓ), which is impossible. So the ϕ_i give a free $\mathbb{C}[\ell]$-basis for $\mathbb{C}[x_0, \ldots, x_n]$ modulo Q.

(3)⇒(1): Let $\psi \in \mathbb{C}[x_0, \ldots, x_n] \setminus Q$. Let $\phi_1, \ldots, \phi_r \in \mathbb{C}[x_0, \ldots, x_n]$ project to a free basis for $\mathbb{C}[x_0, \ldots, x_n]/Q$ as $\mathbb{C}[\ell]$-module. We can write $\psi = \sum_{i=1}^{r} \alpha_i(\ell)\phi_i$

modulo Q, with at least one of the polynomials $\alpha_i(\ell)$ non-zero. Multiplying by ℓ^k, $\ell^k\psi = \sum_{i=1}^r \ell^k\alpha_i(\ell)\phi_i$ modulo Q. Since at least one of the $\ell^k\alpha(\ell)$ is non-zero, $\ell^k\psi \notin Q$. Thus $(x_0,\dots,x_n)^k\psi \notin Q$ for any $k \geq 0$. So Q is saturated. □

Proof of 1.2. We claim that K is saturated. Suppose $\ell \in \mathrm{Ann}_K(\phi)$ with $\phi \notin K$ Then ℓ is contained in an associated prime P of K. Since it also contains K, P's only zero in $\mathbb{C}^{n+1}$ is $\underline{0}$, so $\operatorname{codim} P = n+1$. But this contradicts the unmixedness theorem ([1], 18.14), which implies that all associated primes of K have the same codimension as K, i.e., n. So ℓ is not a zero divisor modulo K, and K is saturated.

It follows that I is also saturated. For if $(x_0,\dots,x_n)^k\phi \in I$ for some $k \geq 0$, then $(x_0,\dots,x_n)^k\phi J \subset K$, so that $\phi J \subset K$.

We may observe, also, that $I = \mathrm{Ann}_K(\tilde{J})$. For if $\phi \in \tilde{J}$, then for some $k \geq 0$ $(x_0,\dots,x_n)^k\phi \subset J$, whence $(x_0,\dots,x_n)^k\phi I \subset K$. Since K is saturated, $\phi I \subset K$.

With these details taken care of, the rest of the argument proceeds as in [2, pp. 315–6].

Let $A = \mathbb{C}[x_0,\dots,x_n]/(K,\ell)$, and define the ideals $\bar{I} = (I,\ell)/(K,\ell)$ and $\bar{J} = (\tilde{J},\ell)/(K,\ell)$ in A.

Because ℓ is not a divisor of zero modulo $\tilde{J}$, the Hilbert function $H_{A/\bar{J}}$ for $A/\bar{J}$ is the difference function of the Hilbert function H_S for S, so that

$$H_S(m) = \sum_{i=0}^{m} H_{A/\bar{J}}(i).$$

We can calculate the Hilbert functions of $\mathbb{C}[x_0,\dots,x_n]/K$, $\mathbb{C}[x_0,\dots,x_n]/I$ similarly, because ℓ is not a divisor of zero modulo I or K. Subtracting, we find

$$\dim(I/K)_a = \sum_{i=0}^{a} \bar{I}_i.$$

Now A is a graded Gorenstein ring, with (1-dimensional) socle in degree m: for a proof, see [2, CB8, CB9]. This means ([2, pp. 313, 316]) that if σ is any linear map $A \to \mathbb{C}$ vanishing on all the graded pieces of A except the socle, then composing σ with multiplication on A yields a perfect pairing Q of A with itself, such that A_j is paired with A_{m-j} for $0 \leq j \leq m$.

Now $\bar{I}$, $\bar{J}$ are orthogonal complements with respect to the pairing Q (they are clearly orthogonal; and the graded nature of the pairing shows that the sum of their dimensions is $\dim A$), so $\dim \bar{I}_j = \operatorname{codim}_{A_{m-j}} \bar{J}_{m-j} = H_{A/\bar{J}}(m-j)$.

Thus

$$H_S(m-1-a) + \dim(I/K)_a = \sum_{i=0}^{m-1-a} H_{A/\bar{J}}(i) + \sum_{j=0}^{a} \dim \bar{I}_j$$

$$= \sum_{i=0}^{m-1-a} H_{A/\bar{J}}(i) + \sum_{j=0}^{a} H_{A/\bar{J}}(m-j) = \sum_{i=0}^{m} H_{A/\bar{J}}(i) = \dim A/\bar{J}.$$

Since $A/\bar{J}$ is isomorphic to $\bar{S}$, the argument is complete. □

We return to the hypersurface $X \subset P^n(\mathbb{C})$ with isolated singularities, and equation f.

Let H be a hyperplane transverse to X; changing coordinates, we may suppose $H = \{x_0 = 0\}$. X has no singularities on H, so to study its singularities it suffices to work in the affine coordinate chart $x_0 = 1$; and define the polynomial function $g : \mathbb{C}^n \to \mathbb{C}$ by $g(x_1, \ldots, x_n) = f(1, x_1, \ldots, x_n)$.

The singularities of X are represented by the singularities $\Sigma(g) \cap g^{-1}(0)$ of $g^{-1}(0)$ as a set, and by the ring $R = \mathbb{C}[x_1, \ldots, x_n]/(g, J(g))$ as a variety, where $J(g) = (\partial g/\partial x_1, \ldots, \partial g/\partial x_n)$ is the Jacobian ideal of g.

There is a natural isomorphism of $\mathbb{C}$-vector spaces

$$R \to \bigoplus_{p \in \Sigma(g) \cap g^{-1}(0)} R_{m_p},$$

where m_p is the maximal ideal of polynomials vanishing at p. Since the germs of g at the points $p \in \Sigma(g) \cap g^{-1}(0)$ are finitely $\mathcal{R}$-determined, R_{m_p} is naturally isomorphic to $\mathcal{O}_p/(g, J(g))\mathcal{O}_p$.

We now contemplate deforming f in the affine chart $x_0 = 1$ by unfolding g by all monomials in $x_1, \ldots, x_n$ of degree $\leq k$. The Kodaira-Spencer map τ_k of the germ of this unfolding at $\Sigma(g) \cap g^{-1}(0)$ maps these monomials to $\oplus_{p \in \Sigma(g) \cap g^{-1}(0)} R_{m_p} = R$ by projection. Thus τ_k is surjective, which is equivalent to the unfolding being versal, if and only if the projections of the monomials span R.

Returning to homogeneous coordinates, let $\tilde{J}(f)$ be the saturation of $J(f) = (\partial f/\partial x_0, \ldots, \partial f/\partial x_n)$, and let $K(f) = (\partial f/\partial x_1, \ldots, \partial f/\partial x_n)$.

Proposition 1.4. *Set $s = n(d-2)$. Then*

$$\dim_{\mathbb{C}} \operatorname{coker} \tau_k = \dim_{\mathbb{C}} (\operatorname{Ann}_{K(f)}(J(f))/K(f))_{s-1-k}.$$

Proof. The Euler relation $df = \Sigma_{i=0}^n x_i \partial f/\partial x_i$ shows that $\tilde{J}(f)$ is also the saturation of $(f, \partial f/x_1, \ldots, \partial f/\partial x_n)$. By 1.3, x_0 is a not a zero divisor modulo $\tilde{J}(f)$, so setting $x_0 = 1$ sends $\tilde{J}(f)$ to $(g, J(g))$, and hence induces a surjection from the ring $S(f) = \mathbb{C}[x_0, \ldots, x_n]/\tilde{J}(f)$ to R. This surjection maps the kth graded piece $S(f)_k$ of $S(f)$ bijectively to the image of τ_k, so that

$$\dim_{\mathbb{C}} \operatorname{coker} \tau_k = \dim_{\mathbb{C}} R - H_{S(f)}(k).$$

Let $\bar{S}(f)_t = \mathbb{C}[x_0, \ldots, x_n]/(\tilde{J}(f), \ell - t)$ $t \in \mathbb{C}$. By 1.3 $S(f)$ is a free $\mathbb{C}[\ell]$-module, so $\dim_{\mathbb{C}} \bar{S}_t$ is constant for all $t \in \mathbb{C}$. In particular, since $R \cong \bar{S}(f)_1$, $\dim_{\mathbb{C}} R = \dim_{\mathbb{C}} \bar{S}(f)_0$; and we have

$$\dim_{\mathbb{C}} \operatorname{coker} \tau_k = \dim_{\mathbb{C}} \bar{S}(f)_0 - H_{S(f)}(k).$$

Applying 1.2 with $K = K(f)$, $J = J(f)$ and $a = s-1-k$ shows that the left-hand side of this equation is equal to $\dim_{\mathbb{C}}(\operatorname{Ann}_{K(f)}(J(f))/K(f))_{s-1-k}$, completing the proof. □

The $\mathbb{C}[x_0, \ldots, x_n]$-module $H(f)$ of polynomial vector fields *Hamiltonian* for f is generated by the vector fields $H_{i,j} = \partial f/\partial x_i \partial/\partial x_j - \partial f/\partial x_j \partial/\partial x_i$ for $0 \leq i < j \leq n$. These annihilate f, so $H(f)$ is a submodule of $\Delta(f)$, the $\mathbb{C}[x_0, \ldots, x_n]$-module of all polynomial vector fields annihilating f.

We have:

Lemma 1.5. *Projection on the coefficient of $\partial/\partial x_0$ induces a graded isomorphism of $\mathbb{C}[x_0, \ldots, x_n]$-modules*

$$\Delta(f)/H(f) \to \mathrm{Ann}_{K(f)}(J(f))/K(f).$$

Proof. It is clear that $\mathrm{Ann}_{K(f)}(J(f)) = \mathrm{Ann}_{K(f)}(\partial f/\partial x_0)$, and equally clear that projection to the coefficient of $\partial/\partial x_0$ maps $\Delta(f)$ onto $\mathrm{Ann}_{K(f)}(\partial f/\partial x_0)$. Furthermore, the coefficient of $\partial/\partial x_0$ of any vector field Hamiltonian for f is contained in $K(f)$. It follows that the projection induces a graded surjection $\Delta(f)/H(f) \to \mathrm{Ann}_{K(f)}(J(f))/K(f)$. To see that this is injective, we need to see that any vector field ξ annihilating f whose $\partial/\partial x_0$-coefficient ξ_0 is in $K(f)$, is Hamiltonian. So suppose that we can write ξ_0 in the form $\sum_{i=1}^{n} \alpha_i \partial f/\partial x_i$. Define $\eta = \xi - \sum_{i=1}^{n} \alpha_i H_{0,i}$. Then η annihilates f, and has zero $\partial/\partial x_0$-coefficient. Thus its other coefficients give a relation amongst the $\partial f/\partial x_i$, $i = 1, \ldots, n$. Since these form a regular sequence, there are only trivial relations amongst them, so that η has the form $\sum_{1 \leq i < j \leq n} \beta_{i,j} H_{i,j}$; in particular it is Hamiltonian for f. So ξ is Hamiltonian for f, and the proof is complete. □

Proof of 1.1. Combining 1.4 (for $k = s - 1 - a$) with 1.5 shows that

$$\dim_{\mathbb{C}}(\Delta(f)/H(f))_a = \dim_{\mathbb{C}} \operatorname{coker} \tau_{s-1-a}. \tag{1}$$

Since X is a-non-versal if and only if τ_{s-1-a} is not surjective, 1.1 follows. □

It is also possible to derive 1.1 from the existence of symmetric discriminant matrices for versal unfoldings of weighted homogeneous functions; indeed the derivation from [5, 1.4] was the original proof, and is rather shorter than the arguments above. However, this derivation would not yield the application of 1.2 which will be required in the proof of 2.1.

Consider first the case $a = 0$. Then 1.1 shows that X being 0-non-versal is equivalent to f being annihilated by a non-zero constant vector field, which we may take as $\partial/\partial x_0$. But this in turn is equivalent to f being independent of x_0, so to X being a cone singularity.

Now consider the case $a = 1$.

Corollary 1.6. *X is 1-non-versal if and only if it is invariant under a 1-parameter algebraic subgroup of $PGL_{n+1}(\mathbb{C})$.*

Proof. If X is invariant under a 1-parameter subgroup of $PGL_{n+1}(\mathbb{C})$, then the infinitesimal generator of the group is a non-zero linear vector field annihilating f, and 1.1 gives the failure of versality claimed.

Conversely, if X is 1-non-versal, then the theorem yields a non-zero linear vector field ξ which annihilates f. Since the linear vector fields form the tangent space to $GL_{n+1}(\mathbb{C})$, exponentiating ξ yields a 1-parameter subgroup of $GL_{n+1}(\mathbb{C})$ which preserves f, and hence X.

If this subgroup is not algebraic, f must also be invariant under its Zariski closure, which has positive dimension, and so contains algebraic 1-parameter subgroups. □

2. Topological versality

A projective hypersurface with isolated singularities is *topologically a-versal* if, in some affine chart containing its singularities, deforming its equation by adding all monomials of degree at most $n(d-2)-a-1$ induces a simultaneous topologically versal deformation of the singularities.

In this section we will prove the following sufficient condition for an a-non-versal hypersurface to be topologically a-versal.

Theorem 2.1. *Let $X \in P^n(\mathbb{C})$ be a hypersurface with isolated singularities, and let $f = 0$ be an equation for X.*

Suppose that $\xi \in \Delta(f)_a \backslash H(f)$ generates $(\Delta(f)/H(f))_a$. If there exists a non-simple singular point P of X at which ξ does not vanish, then X is topologically a-versal.

We can suppose, as in §1, that $x_0 = 0$ is transverse to X; and we will use the notation developed in §1 for this situation. We will write p for the representation of P in the affine chart $x_0 = 1$. We write ξ in coordinates as $\sum_{i=0}^n \xi_i \partial/\partial x_i$; and write $\bar{\xi}_i$ for the polynomial obtained from ξ_i by setting $x_0 = 1$.

We will prove 2.1 via four lemmas.

Lemma 2.2. $\bar{\xi}_0(p) \neq 0$.

Proof. Since $df = \sum_{i=0}^n x_i \partial f/\partial x_i$, we have

$$0 = x_0 \xi \cdot f = \xi_0 (df - \sum_{i=1}^n \partial f/\partial x_i) + x_0 \sum_{i=1} \xi_i \partial f/\partial x_i.$$

Setting $x_0 = 1$ gives

$$\bar{\xi}_0 dg = \sum_{i=1}^n (\bar{\xi}_0 x_i - \bar{\xi}_i) \partial g/\partial x_i;$$

so that $\sum_{i=1}^n (\bar{\xi}_0 x_i - \bar{\xi}_i) \partial/\partial x_i$ is a vector field tangent to $g^{-1}(0)$. It must thus vanish at the singular points of $g^{-1}(0)$, so in particular at p.

Suppose $\bar{\xi}_0(p) = 0$. Then $\bar{\xi}_i(p) = 0$ for $i = 1, \ldots, n$ also; so ξ vanishes at P, a contradiction. So $\bar{\xi}_0(p) \neq 0$. □

Lemma 2.3. *The germ g_p of g at p is quasi-homogeneous.*

Proof. Since $\bar{\xi}_0(p) \neq 0$, $\bar{\xi}$ is invertible in $R_{m_p} \cong \mathcal{O}_p/(g, J(g))\mathcal{O}_p$. Now $\xi_0 \in \mathrm{Ann}_{K(f)}(f)$, so $\bar{\xi}_0 \in \mathrm{Ann}_{J(g)}(g)$. Thus, taking germs at p, $g_p \in J(g_p)$. Thus by Saito's theorem [10], g_p is quasi-homogeneous. □

Lemma 2.4. *Let $h \in \mathbb{C}[x_1, \dots, x_n]$ represent the Hessian of g_p in R_{m_p}, and represent 0 in R_{m_q} for $q \in \Sigma(g) \cap g^{-1}(0) \setminus \{p\}$. Then h projects to a generator for Coker τ_{s-a-1}.*

Proof. Since the germ of g at p is quasi-homogeneous, its Hessian generates the socle of R_{m_p}. It follows that $\dim_{\mathbb{C}}((h, g, J(g))/(g, J(g)) = 1$.

Write $\bar{R}$ for the ring $\mathbb{C}[x_1, \dots, x_n]/(h, g, J(g))$. Showing that h projects to a $\mathbb{C}$-generator for Coker τ_{s-1-a} is equivalent to showing that the projection $\pi : R \to \bar{R}$ maps Im τ_{s-1-a} onto $\bar{R}$.

Let H be the homogenisation of h with respect to x_0, L the saturation of $(H, J(f))$, and $T = \mathbb{C}[x_0, \dots, x_n]/L$. Then setting $x_0 = 1$ maps L onto $(h, g, J(g))$, and so induces a surjection $T \to \bar{R}$, which maps the $(s-1-a)$th graded piece of T injectively onto the $\mathbb{C}$-subspace of $\bar{R}$ spanned by the images of all monomials in $x_1, \dots, x_n$ of degree $\leq s-1-a$; that is, onto $\pi(\mathrm{Im}\,\tau_{s-1-a})$. It follows that showing that $\pi(\mathrm{Im}\,\tau_{s-1-a}) = \bar{R}$ is equivalent to showing that $H_T(s-1-a) = \dim_{\mathbb{C}} \bar{R}$.

Let $\bar{T} = \mathbb{C}[x_0, \dots, x_n]/(L, x_0)$. According to 1.3, $\dim_{\mathbb{C}} \bar{T} = \dim_{\mathbb{C}} \bar{R}$, whilst according to 1.2,

$$\dim_{\mathbb{C}} \bar{T} - H_T(s-a-1) = \dim_{\mathbb{C}}(\mathrm{Ann}_{K(f)}(H, J(f))/K(f))_a,$$

so that showing $H_T(s-1-a) = \dim_{\mathbb{C}} \bar{R}$ is equivalent to showing

$$(\mathrm{Ann}_{K(f)}(H, J(f))/K(f))_a = \{0\}.$$

Now $\bar{\xi}_0 h$ is non-zero in R, so $\bar{\xi}_0 \notin \mathrm{Ann}_{(g,J(g))}(h)$, and so, a fortiori, $\bar{\xi}_0 \notin \mathrm{Ann}_{J(g)}(h, g, J(g))$. Homogenising this gives $\xi_0 \notin \mathrm{Ann}_{K(f)}(H, f, K(f))$. Now $\mathrm{Ann}_{K(f)}(H, f, K(f)) = \mathrm{Ann}_{K(f)}(H, J(f))$, because $(H, f, K(f))$ and $(H, J(f))$ have the same saturation. Thus $(\mathrm{Ann}_{K(f)}(H, J(f))/K(f))_a$ is a proper subspace of the one-dimensional space $(\mathrm{Ann}_{K(f)}(J(f))/K(f))_a$, so is $\{0\}$, completing the proof. □

Write $\{G : N \to P; i, j\}$ for the unfolding of g by all monomials of degree at most $s-a-1$ in $x_1, \dots, x_n$. Here $N = \mathbb{C}^n \times \mathbb{C}^m$, $P = \mathbb{C} \times \mathbb{C}^m$, m being the number of monomials, and i, j are the natural embeddings of $\mathbb{C}^n$, $\mathbb{C}$ in N, P, respectively, as $\mathbb{C}^n \times 0_m$, $\mathbb{C} \times 0_m$. Set $S = \Sigma(g) \cap g^{-1}(0)$.

Lemma 2.5. *The germ of G at $i(S)$ has jet-extension multi-transverse to the $\mathcal{K}$-invariant submanifold generated by $\{g_p + t \cdot H(g) : t \in \mathbb{C}\}$ and the $\mathcal{K}$-classes of g_q, $q \in S \setminus \{p\}$.*

Proof. We will make the usual translation of multi-transversality in jet-space of sufficiently high order to transversality in the target of a versal unfolding. Let $\{\tilde{G}, I, J\}$ be a versal unfolding of the germ of G at $i(S)$. Then the presentations by $\tilde{G}$ of the $\mathcal{K}$-invariant submanifolds named intersect transversely in the target of $\tilde{G}$, and the result claimed is equivalent to showing that J is transverse to their intersection P.

Since h projects to a generator for the Kodaira-Spencer map for the germ of G at $i(S)$, such an unfolding $(\tilde{G} : (N \times \mathbb{C}, i(S) \times 0) \to (P \times \mathbb{C}, j(0) \times 0); I, J)$ is obtained by unfolding the germ of G at $i(p)$ with $H(g_p)$, and the germs of G at the other points of $i(\Sigma(g) \cap g^{-1}(0))$ trivially; here we take I, J as the appropriate germs of the natural embeddings of N, P in $N \times \mathbb{C}$, $P \times \mathbb{C}$, respectively, as $N \times 0$, $P \times 0$. With this choice of $\tilde{G}$, P contains the unfolding axis $(j(0) \times \mathbb{C}, j(0) \times 0)$. It follows that J is transverse to P, as required. □

Proof of 2.1. We make use of the complex avatars of a definition and result of [4]. We define civilization for a $\mathcal{K}$-invariant complex submanifold of complex jet-space as in [4, p.347, ll.1-7], except that $\mathbb{R}$ is replaced by $\mathbb{C}$ whenever it occurs.

The proof of [4, 9.1.3] (in [4, 9.4]) can then be followed more or less word for word to conclude that, if a complex multi-germ G has jet extension multi-transverse to a collection of civilized submanifolds containing the $\mathcal{K}$-classes of the germs at its base-points, then there exists a V-tame retraction of any versal unfolding of G to G; it follows immediately that G is topologically versal.

It is clear that $\mathcal{K}$-classes are civilized; thus to prove 2.1, it remains to show that the $\mathcal{K}$-invariant submanifold S generated by $\{g_p + t \cdot H(g) \,:\, t \in \mathbb{C}\}$ is civilized.

Since g_p is quasi-homogeneous but not simple, Wirthmüller's theorem [14, 3.6] gives the required model retraction. For the retraction in the target is obtained by integrating a pair of continuous vector fields (the real and imaginary parts of a complex vector field), analytic off the presentation of S, which are weighted homogeneous in appropriate coordinates. It follows that these vector fields can be chosen to preserve level sets of the associated weighted distance from this presentation. Thus the retraction is smooth, so tame, on these level sets, and the conclusion follows from [4, 9.6.4]. The proof is complete. □

References

[1] Eisenbud, D., *Commutative algebra with a view towards algebraic geometry*, Graduate Texts in Mathematics 150, Springer-Verlag. 1995.

[2] Eisenbud, D., M. Green and J. Harris, Cayley-Bacharach theorems and conjectures, Bull. Amer. Math. Soc **33** (1997) 295–324.

[3] du Plessis, A.A., Minimal intransigent hypersurfaces, *this volume.*

[4] du Plessis, A.A. and C.T.C. Wall, *The geometry of topological stability*, Oxford University Press. 1995.

[5] du Plessis, A.A. and C.T.C. Wall, Versal deformations in a space of polynomials of fixed weight, Compositio Math. **114** (1998), 113–124.

[6] du Plessis, A.A. and C.T.C. Wall, Application of the theory of the discriminant to highly singular plane curves, Math. Proc. Camb. Phil. Soc. **126** (1999), 256–266.

[7] du Plessis, A.A. and C.T.C. Wall, Curves in $P^2(\mathbb{C})$ with 1-dimensional symmetry, Revista Math. Complutense **12** (1999), 117–132.

[8] du Plessis, A.A. and C.T.C. Wall, Hypersurfaces in $P^n(\mathbb{C})$ with one-parameter symmetry groups, Proc. R. Soc. Lond. A **456** (2000), 2515–2541.

[9] du Plessis, A.A. and C.T.C. Wall, Singular hypersurfaces, versality, and Gorenstein algebras, J. Algebraic Geometry **9** (2000), 309–322.

[10] Saito, K., Quasihomogene isolierte Singularitäten von Hyperflächen, Invent. Math. **14** (1971), 123–142.

[11] Shustin, E., Versal deformations in the space of planar curves of fixed degree, Functional analysis and its applications **21** (1987) 82–84.

[12] Shustin, E., and I. Tyomkin, Versal deformation of algebraic hypersurfaces with isolated singularities, Math. Ann. **313** (1999), 297–314.

[13] Wall, C.T.C., Highly singular quintic curves, Math. Proc. Camb. Phil. Soc., **119** (1996) 257–277.

[14] Wirthmüller, K., *Universell topologisch triviale Deformationen*, doctoral thesis, Universität Regensburg, 1979.

Andrew A. du Plessis
Department of Mathematical Sciences
University of Aarhus
Building 530, Ny Munkegade
DK-8000 Aarhus C, Denmark
e-mail: `matadp@imf.au.dk`

Real and Complex Singularities
Trends in Mathematics, 299–310

Minimal Intransigent Hypersurfaces

Andrew A. du Plessis

Abstract. We give examples of hypersurfaces of degree d in $P^n(\mathbb{C})$, whose singularities are *not* versally deformed by the family $H_d(n)$ of all hypersurfaces of degree d in $P^n(\mathbb{C})$, and which are of minimal codimension with this property.

In the three cases $(n, d) = (2, 6), (3, 4)$ and $(5, 3)$, such hypersurfaces necessarily have one-parameter symmetry. We list the possibilities. The singularities of these hypersurfaces are not all simple, and they are simultaneously *topologically* versally deformed by $H_d(n)$.

In less degenerate cases the examples we give are hypersurfaces with only simple singularities. The failure of versality can be expected to show itself in the geometry of $H_d(n)$, either because the μ-constant stratum S containing the hypersurface is of codimension less than μ in $H_d(n)$, or because S is not smooth. We will see elsewhere that this is the case for the examples we consider here. In particular, the singularities of these hypersurfaces are *not* topologically versally deformed by $H_d(n)$.

1. Introduction

Let X be a hypersurface of degree d in $P^n(\mathbb{C})$ with only isolated singularities, and let $f \in \mathbb{C}[x_0, \ldots, x_n]$ be a homogeneous equation for X. We will say that X is *intransigent* if the family $H_d(n)$ of all hypersurfaces of degree d does not simultaneously versally deform the singularities of X, and *transigent* otherwise. (The use of "intransigent" here is patterned on the use of "rigid" for varieties with no deformations within the class under consideration). We recall some earlier results on such hypersurfaces.

The $\mathbb{C}[x_0, \ldots, x_n]$-module $H(f)$ of polynomial vector fields *Hamiltonian* for f is generated by the vector fields

$$H_{i,j} = \partial f/\partial x_i \partial/\partial x_j - \partial f/\partial x_j \partial/\partial x_i \quad \text{for} \quad 0 \leq i < j \leq n.$$

These annihilate f, so $H(f)$ is a submodule of $\Delta(f)$, the $\mathbb{C}[x_0, \ldots, x_n]$-module of all polynomial vector fields annihilating f. The grading of $\mathbb{C}[x_0, \ldots, x_n]$ by degree induces a grading on these modules in the obvious way.

Proposition 1.1. [6, 1.1] *X is intransigent if and only if there exists a homogeneous vector field which annihilates f, is not Hamiltonian for f, and is of degree $(n-1)(d-2)-3$.*

We will denote by $\tau(X)$ the sum of the Tjurina numbers of the singularities of X.

Proposition 1.2. [9, 1.1] *If X is intransigent then $\tau(X) \geq \delta(d)$, where*

$$\delta(3) = 16, \delta(4) = 18, \quad \text{and} \quad \delta(d) = 4(d-1) \quad \text{for } d \geq 5.$$

An intransigent hypersurface X for which $\tau(X) = \delta(d)$ will be called *minimal.*

In cases where $(n-1)(d-2) \leq 2$, so when $d = 2$ or $d = 3$, $n = 2$, no intransigent hypersurfaces can exist. In all other cases, minimal intransigent hypersurfaces do exist: algebraically simple, but for the most part geometrically rather uninteresting, examples are given in [9, §3]. The aim of this paper is to present rather more revealing examples, which will illustrate some unusual geometry in the families $H_d(n)$.

If $(n-1)(d-2) = 3$ – there are just two cases, $(n, d) = (2, 5), (4, 3)$ – the annihilating vector field must be of degree 0, so a constant vector field. Changing coordinates, this may be taken as $\partial/\partial x_0$, showing that f is independent of x_0, so X is a cone. Conversely, since any cone is annihilated by a constant vector field, it is intransigent when $(n-1)(d-2) \geq 3$ – but not minimal when $(n-1)(d-2) > 3$. It follows from a theorem of Wirthmüller [17] (see also [7, 9.1.4]) that the singularities of the cones in these two cases are *topologically* versally deformed by the family $H_d(n)$ – we say that these hypersurfaces are *topologically* transigent.

If $(n-1)(d-2) = 4$ – there are just three cases, $(n, d) = (2, 6), (3, 4)$ and $(5, 3)$ – the annihilating vector field must be linear. It was shown in [6, 1.6] that the existence of such a vector field is equivalent to the hypersurface admitting a one-dimensional subgroup of $PGL_n(\mathbb{C})$ as symmetries. Thus when $(n-1)(d-2) \geq 4$, any symmetric hypersurface is intransigent, but is not minimal when $(n-1)(d-2) > 4$. All the symmetric hypersurfaces in the three cases above (and many others) have been enumerated (see [8], [10]). The following can be extracted from the enumeration.

Proposition 1.3. *In the three cases $(n, d) = (2, 6), (3, 4), (5, 3)$, a minimal intransigent hypersurface has a semi-simple symmetry group, of form $[w_0, 0, \ldots, 0, w_n]$.*

The notation for the symmetry groups used above reflects the fact that if a one-dimensional algebraic subgroup of $PGL_n(\mathbb{C})$ is semi-simple, then its infinitesimal generator has, in appropriate coordinates, diagonal form $\sum_{r=0}^{n} w_r x_r \partial/\partial x_r$, where the *weights* w_r are integers. We denote such a subgroup by $[w_0, w_1, \ldots, w_n]$.

We list below the symmetry groups, their invariant monomials, and the singularities possible for minimal intransigent hypersurfaces.

Group	Monomials	Singularities
$[-2,0,3]$	$x_1^6,\ x_0^3x_1x_2^2$	$Z_{13}+D_7$
$[-1,0,5]$	$x_1^6,\ x_0^5x_2$	NF_{20}
$[-1,0,4]$	$x_1^6,\ x_0^4x_1x_2$	$NC_{19}+A_1$
$[-1,0,2]$	$\{x_1^3,\ x_0^2x_2\}^2$	$W_{1,0}+A_5$
$[-1,0,1]$	$\{x_1^2,\ x_0x_2\}^3$	$2T_{2,3,6}$
$[-1,0,0,3]$	$x_0^3x_3,\ \{x_1,\ x_2\}^4$	$V'(1,1,1,1)$
$[-1,0,0,2]$	$x_0^2x_3\{x_1,\ x_2\},\{x_1,\ x_2\}^4$	$VA_{0,0}+A_3$
$[-1,0,0,1]$	$x_0^2x_3^2,\ x_0x_3\{x_1,\ x_2\}^2,\ \{x_1,\ x_2\}^4$	$2T_{2,4,4}$
$[-1,0,0,0,0,2]$	$x_0^2x_5,\ \{x_1,\ x_2,\ x_3,\ x_4\}^3$	O_{16}
$[-1,0,0,0,0,1]$	$x_0x_5\{x_1,\ x_2,\ x_3,\ x_4\},\ \{x_1,\ x_2,\ x_3,\ x_4\}^3$	$2T_{3,3,3}$

In the lists of invariant monomials, the expression $\{X_1,\dots,X_k\}^r$ (the superscript is omitted when $r=1$) denotes the set of homogeneous monomials of degree r in $X_1,\dots,X_k$; and, if m is a monomial, then $m\{X_1,\dots,X_k\}^r$ denotes the set of products of m with elements of $\{X_1,\dots,X_k\}^r$. A generic linear combination of the monomials listed gives a minimal intransigent hypersurface with exactly the singularities shown. The notations for these singularities, where they are not standard, are those of [16].

We observe that in all these cases a non-simple singularity is present. It follows from [6, 2.1], [10, 2.8] that the singularities of these hypersurfaces are topologically transigent. Indeed, it seems likely that all intransigent hypersurfaces in these cases, including the cones, are topologically transigent; for more information on this see [8], [10].

The cases where $(n-1)(d-2)>4$ are the subject of the remainder of the paper. In §2 we will exhibit, in each case, minimal intransigent hypersurfaces with isolated singularities, all of which are simple.We will see elsewhere how to describe the intransigence of these hypersurfaces geometrically: we will see that *either* the μ-constant stratum S containing the hypersurface is of codimension strictly less than μ in $H_d(n)$, *or* S is not smooth near the hypersurface. In particular, the hypersurfaces we exhibit are *not* topologically transigent. Such μ-constant strata have previously been observed. For the first type, an early example in $P^2(\mathbb{C})$ is due to Segre [11], whilst in higher dimensions the number of nodes on some hypersurfaces of degree d discovered recently (see, e.g., [1, pp. 419-20], [2], [14]) exceeds the dimension of $H_d(n)$. For the second type there are examples (though only of curves) due to Luengo [5] and Greuel, Lossen and Shustin [4], [13]. However, these examples are of higher codimension than those we exhibit, where, of course, the codimension is the least possible.

The possibilities of the constructions described in §2 are far from exhausted by the examples we describe; and comparison with exceptional deformations in other situations (see, e.g., [7, pp. 510-4]) suggests than many other mechanisms will also give rise to intransigence. It seems that intransigence is very widespread.

It is a pleasure to acknowledge T. Wall's help with the preparation of this article.

2. Construction of examples

The following simple lemma is the key to our constructions.

Lemma 2.1. *Let $F : \mathbb{C}^m \to \mathbb{C}$ be an analytic function, annihilated by the vector fields ξ_i, for $i = 1, \ldots, r$, let $g : \mathbb{C}^{n+1} \to \mathbb{C}^m$ be an analytic map, and let $f = F \circ g$.*

Suppose $m = n + r$. Then the determinant of the $(m+1) \times (m+1)$ matrix

$$M = \begin{pmatrix} \frac{\partial f}{\partial x_0} & \cdots & \frac{\partial f}{\partial x_n} & 0 & \cdots & 0 \\ \frac{\partial g_1}{\partial x_0} & \cdots & \frac{\partial g_1}{\partial x_n} & \xi_{11} \circ g & \cdots & \xi_{r1} \circ g \\ \vdots & & \vdots & \vdots & & \vdots \\ \frac{\partial g_m}{\partial x_0} & \cdots & \frac{\partial g_m}{\partial x_n} & \xi_{1m} \circ g & \cdots & \xi_{rm} \circ g \end{pmatrix}$$

vanishes identically, so that the determinants of the $m \times m$ minors adjoint to the first $n+1$ entries in the first row of M give the coefficients of a vector field θ annihilating f. Moreover, θ is non-zero at $p \in \mathbb{C}^{n+1}$ if, and only if, the vectors $\xi_1(g(p)), \ldots, \xi_r(g(p))$ are linearly independent and $\operatorname{Im} dg_p + \mathbb{C} \cdot \{\xi_1(g(p)), \ldots, \xi_r(g(p))\} = \mathbb{C}^m$.

Proof. Let $p \in \mathbb{C}^{n+1}$. By the chain rule, $\frac{\partial f}{\partial x_j}(p) = \sum_{l=1}^m \frac{\partial F}{\partial y_l}(g(p)) \cdot \frac{\partial g_l}{\partial x_j}(p)$. Also, since ξ_k annihilates F, we have $0 = \sum_{l=1}^m \frac{\partial F}{\partial y_l} \cdot \xi_{kl}$; so $0 = \sum_{l=1}^m \frac{\partial F}{\partial y_l}(g(p)) \cdot \xi_{kl}(g(p))$. Hence the rows $M_1, \ldots, M_{m+1}$ of M satisfy $M_1 = \sum_{l=}^m \frac{\partial F}{\partial y_l}(g(p)) \cdot M_{l+1}$ at p. Thus M has rank less than $m+1$ at p, so its determinant vanishes there. Since $p \in \mathbb{C}^{n+1}$ was arbitrarily chosen, $\det M$ vanishes identically, as claimed.

Write M' for the matrix obtained by removing the first row of M. Then, at $p \in \mathbb{C}^{n+1}$, at least one of the minors adjoint to the first $n+1$ entries in the first row of M has rank m if, and only if, M' has rank m and the final r columns of M' are linearly independent. Thus $\theta(p) \neq 0$ exactly when the stated conditions hold. □

Specializing to a (weighted) homogeneous situation yields a somewhat sharper statement. In what follows, $G_{n+1}(w_1, \ldots, w_m)$ will denote the vector space of maps $\mathbb{C}^{n+1} \to \mathbb{C}^m$ whose coordinate functions are homogeneous polynomials of positive degrees $w_1, \ldots, w_m$.

Proposition 2.2. *Let $F : \mathbb{C}^m \to \mathbb{C}$ be a polynomial function, weighted-homogeneous with respect to weights $w_1, \ldots w_m/d$, and annihilated by r linearly independent weighted-homogeneous vector fields $\xi_1, \ldots, \xi_r$, of weights $c_1, \ldots, c_r$ respectively. Write $n = m - r$, $s = \sum_{i=1}^m w_i - n + \sum_{j=1}^r c_j$.*

Let $g \in G_{n+1}(w_1, \ldots, w_m)$, and set $f = F \circ g$, so f is a homogeneous polynomial of degree d. Then $\Delta(f)_s \neq 0$ for generic choice of g from $G_{n+1}(w_1, \ldots, w_m)$. Moreover, if either $s < d - 1$ or *there exists a point $q \in \Sigma(F)$ such that $\xi_1(q)$, …, $\xi_r(q)$ are linearly independent, then $\Delta(f)_s/H(f)_s \neq 0$ for any $g \in G_{n+1}(w_1, \ldots, w_m)$ such that the projective hypersurface $\{f = 0\}$ in $P^n(\mathbb{C})$ has only isolated singularities.*

Proof. Let $M(g)$ be the matrix of 2.1 for $F, \xi_1, \dots, \xi_r$, and g. For $1 \le i \le m$, $0 \le j \le n$, $\frac{\partial g_i}{\partial x_j}$ has degree $w_i - 1$, while for $1 \le i \le m$, $1 \le k \le r$, ξ_{ki} has degree $w_i + c_k$. It is immediate that the determinants of the $m \times m$ minors of $M(g)$ adjoint to the first n entries in the first row of $M(g)$ are homogeneous of degree s; so that the vector field $\theta(g)$ annihilating $f = F \circ g$ implied by $\det M(g) = 0$ is of degree s.

To simplify notation, we will write G for $G_{n+1}(w_1, \dots, w_m)$ for the rest of the proof. Let $C \subset G$ consist of of those $g \in G$ for which $\theta(g) \neq 0$, and, for any $q \in \mathbb{C}^m$, let $C_q \subset C$ consist of of those $g \in G$ for which there exists $p \in \mathbb{C}^{n+1}$ with $g(p) = q$ and $\theta(g)(p) \neq 0$. Each C_q is clearly a Zariski-open subset of G, so $C = \cup_{q \in \mathbb{C}^m} C_q$ is too. Thus to see that $\theta(g)$ is non-zero for generic choice of $g \in G$, we only need to see that C_q is non-empty for some $q \in \mathbb{C}^m$. We show that C_q is non-empty for any q such that the vectors $\xi_1(q), \dots, \xi_r(q)$ are linearly independent. For this, write such q as $(q_1, \dots, q_m)$ and let $L = (L_1, \dots, L_m)$ be an injective linear map $\mathbb{C}^n \to \mathbb{C}^m$ such that $\operatorname{Im} L + \mathbb{C} \cdot \{\xi_1(q), \dots, \xi_r(q)\} = \mathbb{C}^m$. Define $g = (g_1, \dots, g_m) : \mathbb{C}^{n+1} \to \mathbb{C}^m$ by setting $g_i(x_0, x_2, \dots, x_n) = x_0^{w_i} q_i + x_0^{w_i - 1} L_i(x_1, \dots, x_n)$ for $1 \le i \le m$, and set $p = (1, 0, \dots, 0)$. Then $q = g(p)$, and $dg_p|_{\mathbb{C}^n} = L$, so that $\operatorname{Im} dg_p \supset \operatorname{Im} L$. Thus, by 2.1, $\theta(g)(p) \neq 0$, and $g \in C_q$, as required.

For the final statement, let $D \subset G$ consist of those $g \in G$ such that the zeroes of $f = F \circ g$ define a projective hypersurface with isolated singularities. Clearly D is Zariski-open in G. We suppose in what follows that D is not empty. Let E be the subset of D such that if $g \in X$, then $f_g = F \circ g$ is not annihilated by a vector field of degree s not Hamiltonian for f_g. Then E too is Zariski-open in G. For suppose $g \in E$. Choose coordinates $x_0, \dots, x_n$ in $\mathbb{C}^{n+1}$ so that $x_0 = 0$ is transverse to $\{f_g = 0\}$, define the function k_g by setting $x_0 = 1$ in f_g, and let K_g be the deformation of k_g obtained by adding arbitrary monomials in $x_1, \dots, x_n$ of degree at most $n(d-2) - 1 - s$. Then, according to 1.1, K_g induces a simultaneous versal deformation of the singularities of k_g. By openness of versality ([15, p. 640]), the corresponding construction induces a simultaneous versal deformation $K_{g'}$ of $k_{g'}$, for g' in some Zariski-open neighborhood E_g of g in G. Thus, by 1.1 again, $E_g \subset E$.

If $s < d-1$ let $D' = C \cap D$. Then D' is Zariski-dense in G, and if $g \in D'$, then $\theta(g)$ is non-zero. It is thus non-Hamiltonian for f, because the module of vector fields Hamiltonian for f is generated by vector fields of degree $d-1$. Thus in this case E must be empty.

If there exists a point $q \in \Sigma(F)$ such that $\xi_1(q), \dots, \xi_r(q)$ are linearly independent, then let $D' = D \cap C_q$. Thus D' is Zariski-dense in G. If $g \in C_q$, let $p \in \mathbb{C}^n$ be the corresponding point with $g(p) = q$. Then $\theta(g)$ is non-zero at p, which is a critical point of $f = F \circ g$. Thus $\theta(g)$ is not Hamiltonian for f, since vector fields Hamiltonian for f have coefficients in $J(f)$, so vanish at $\Sigma(f)$. Thus E is empty in this case too, and the proof is complete. □

All our examples of intransigent hypersurfaces are derived from a special case of 2.2:

Corollary 2.3. *Suppose $m \geq 2r$. Let $F : \mathbb{C}^m \to \mathbb{C}$ be a function of form*

$$F = F_0 + \sum_{i=1}^{r} y_{2i-1}^{a_i} y_{2i}^{b_i} F_i,$$

where $a_i \geq 1$, $b_i \geq 2$ for $1 \leq i \leq r$, and $F_0, \dots, F_r$ are polynomials in $y_{2r+1}, \dots, y_m$ alone. Suppose also that F is weighted-homogeneous with respect to weights $w_1, \dots, w_m/d$, where $w_i < d$ for $i \leq i \leq m$. Write $n = m - r$, $s = \sum_{i=1}^{m} w_i - n$.

Let $g \in G_{n+1}(w_1, \dots, w_m)$, and set $f = F \circ g$, so f is a homogeneous polynomial of degree d. Then $\Delta(f)_s / H(f)_s \neq 0$ for any g such that the projective hypersurface $\{f = 0\}$ has only isolated singularities.

Proof. Let $p \in \mathbb{C}^m$ have y_{2i-1}-coordinate 1 for $1 \leq i \leq r$, and its other coordinates 0. It is clear that $p \in \Sigma(F)$.

Now observe that F is annihilated by the vector fields $\xi_r = b_i y_{2i-1} \frac{\partial}{\partial y_{2i-1}} - a_i y_{2i} \frac{\partial}{\partial y_{2i}}$, for $1 \leq i \leq r$; and that the vectors $\xi_1(p), \dots, \xi_r(p)$ are linearly independent.

Since the vector fields $\xi_1, \dots, \xi_r$ are of weight zero, 2.2 now gives the result. □

In the examples based on 2.3 which follow, we exhibit just one $g_0 \in G_{n+1}(w_1, \dots, w_m)$ such that $\{F \circ g_0 = 0\}$ is a minimal intransigent hypersurface. This will be enough to ensure that $\{F \circ g = 0\}$ is a minimal intransigent hypersurface for generic choice of g:

Lemma 2.4. *Let m, r and F satisfy the hypotheses of 2.3; and define also s, n as in those hypotheses. Suppose that $s = (n-1)(d-2) - 3$, and that $g_0 \in G_{n+1}(w_1, \dots, w_m)$ is such that $\{F \circ g_0 = 0\}$ is a minimal intransigent hypersurface. Then $\{F \circ g = 0\}$ is a minimal intransigent hypersurface in $P^n(\mathbb{C})$ for a generic choice of g from $G_{n+1}(w_1, \dots, w_m)$. Moreover, if the singularities of $\{F \circ g_0 = 0\}$ are simple, then the singularities of $\{F \circ g = 0\}$ are of the same analytic types as those of $\{F \circ g_0 = 0\}$.*

Proof. Write G for $G_{n+1}(w_1, \dots, w_m)$, and $D \subset G$ for the Zariski-open subset consisting of those $g \in G$ for which the projective hypersurface $\{F \circ g = 0\}$ in $P^n(\mathbb{C})$ has only isolated singularities. According to 1.1 and 2.3, $\{F \circ g = 0\}$ is intransigent if $g \in D$. Thus, according to 1.2, $\tau(\{F \circ g = 0\}) \geq \delta(d)$. Now $g \to \tau(\{F \circ g = 0\})$ is an upper semi-continuous function on D, so $D_0 = \{g \in D \,|\, \tau(\{F \circ g = 0\}) = \delta(d)\}$ is a Zariski-open subset of D, so of G. Since $g_0 \in D_0$, D_0 is non-empty, and hence dense in G. But D_0 consists exactly of those $g \in G$ for which $\{F \circ g = 0\}$ is a minimal intransigent hypersurface, so the first statement is proved. For the second, we note that D_0 is connected, and that a path γ in D_0 containing g_0 yields, taking germs at the singularities of $\{F \circ \gamma(t) = 0\}$, a τ-constant family of multi-germs. Since this family contains a $\mathcal{K}$-simple multi-germ, all its elements lie in the same multi-$\mathcal{K}$-class, as claimed. □

As a final preparation for the examples, we describe a kind of suspension. Let m, r and F satisfy the hypotheses of 2.3; and define also n, s as in 2.3. Define $\tilde{F} : \mathbb{C}^{m+2} \to \mathbb{C}$ by

$$\tilde{F}(y_1, \ldots, y_{m+2}) = F(y_1, \ldots, y_m) + y_{m+1}^2 y_{m+2};$$

so $\tilde{F}$ is weighted-homogeneous with respect to $w_1, \ldots, w_m$, 1, $d - 2/d$. Then $m + 2$, $r + 1$ and $\tilde{F}$ also satisfy the hypotheses of 2.3; and hence its conclusion, with n and s replaced by $n + 1$ and $s + d - 2$.

Lemma 2.5. *Let F, $\tilde{F}$ be as above, and suppose that $g \in G_{n+1}(w_1, \ldots, w_m)$ is such that $X = \{F \circ g = 0\}$ is a projective hypersurface in $P^n(\mathbb{C})$ with isolated singularities. Let ℓ be a linear function on $\mathbb{C}^n$ such that $\{\ell = 0\}$ is transverse to X. Define $\tilde{g} : \mathbb{C}^{n+2} \to \mathbb{C}^{m+2}$ by*

$$\tilde{g}_i = \begin{cases} g_i & \text{if } 1 \leq i \leq m, \\ x_{n+1} & \text{if } i = m+1, \\ c\,(x_{n+1}^{d-2} - \ell^{d-2}) & \text{if } i = m+2, \end{cases}$$

where $c \in \mathbb{C} \setminus \{0\}$.

Then, for all but finitely many values of c, $\tilde{X} = \{\tilde{F} \circ \tilde{g} = 0\}$ is a projective hypersurface in $P^{n+1}(\mathbb{C})$ whose isolated singularities are analytically isomorphic to the Thom suspensions of the singularities of X; so $\tau(\tilde{X}) = \tau(X)$. In particular, if $s = (n-1)(d-2)-3$ and X is a minimal intransigent hypersurface, then so is $\tilde{X}$.

Proof. Changing coordinates in $\mathbb{C}^{n+1}$, we can suppose that $\ell = x_0$. Write $f = F \circ g$, $\tilde{f} = \tilde{F} \circ \tilde{g}$; we have

$$\tilde{f}(x_0, \ldots, x_{n+1}) = f(x_0, \ldots, x_n) + c\,(x_{n+1}^d - x_0^{d-2} x_{n+1}^2).$$

Thus at a singularity of $\tilde{X}$, $\partial f/\partial x_1, \ldots, \partial/\partial x_n$ must vanish, together with *either* x_{n+1} and $\partial f/\partial x_0$ *or* $dx_{n+1}^{d-2} - 2x_0^{d-2}$ and $\partial f/\partial x_0 - c(d-2)x_{n+1}^2 x_0^{d-3}$.

By the transversality assumption, $\partial f/\partial x_1, \ldots, \partial/\partial x_n$ have only finitely many common zeroes, $[z_1], \ldots, [z_k]$ say, in $P^n(\mathbb{C})$; and none of these are also zeroes of both x_0 and $\partial f/\partial x_0$. Hence the second possibility occurs only for finitely many values of c; these are given by $c = (\partial f/\partial x_0(z_i))/((d-2)\alpha^2 x_0^{d-3})$, for $i = 1, \ldots, k$ and α such that $\alpha^{d-2} = 2/d$.

Thus, ruling out this finite collection of values for c, the first possibility for singularities of $\tilde{X}$ is the only one, and $[p]$ is a singular point of X if and only if $[(p, 0)]$ is a singular point of $\tilde{X}$. Also, because $x_0 \neq 0$ at p, the singularity of $\tilde{X}$ at $[(p, 0)]$ is analytically equivalent to the Thom suspension of the singularity of X at $[p]$. Since Thom suspension does not affect Tjurina number, the first statement is proved; the second statement follows from 1.1, 1.2 and 2.3. □

Examples 2.6. *We suppose* $k \geq 0$.

(i) (a) *Let* $F : \mathbb{C}^{2k+2} \to \mathbb{C}$ *be given by*

$$F(y_1, \ldots, y_{2k+2}) = y_1^8 + y_2^2 + \sum_{j=1}^{k} y_{2j+1}^2 y_{2j+2}.$$

For generic $g \in G_{k+3}(1, 4, \ldots, 1, 6)$, $\{F \circ g = 0\}$ *is a minimal intransigent hypersurface of degree* 8 *in* $P^{k+2}(\mathbb{C})$, *with simple singularities* $4A_7$.

(b) *Let* $F : \mathbb{C}^{2k+3} \to \mathbb{C}$ *be given by*

$$F(y_1, \ldots, y_{2k+3}) = y_1^d + y_2^2 y_3 + \sum_{j=1}^{k} y_{2j+2}^2 y_{2j+3},$$

where $d > 8$. *For generic* $g \in G_{k+3}(1, 4, d-8, \ldots, 1, d-2)$, $\{F \circ g = 0\}$ *is a minimal intransigent hypersurface of degree* d *in* $P^{k+2}(\mathbb{C})$, *with simple singularities* $4A_{d-1}$.

(ii) (a) *Let* $F : \mathbb{C}^{2k+2} \to \mathbb{C}$ *be given by*

$$F(y_1, \ldots, y_{2k+2}) = y_1^7 + y_1 y_2^2 + \sum_{j=1}^{k} y_{2j+1}^2 y_{2j+2}.$$

For generic $g \in G_{k+3}(1, 3, \ldots, 1, 5)$, $\{F \circ g = 0\}$ *is a minimal intransigent hypersurface of degree* 7 *in* $P^{k+2}(\mathbb{C})$, *with simple singularities* $3D_8$.

(b) *Let* $F : \mathbb{C}^{2k+3} \to \mathbb{C}$ *be given by*

$$F(y_1, \ldots, y_{2k+3}) = y_1^d + y_1 y_2^2 y_3 + \sum_{j=1}^{k} y_{2j+2}^2 y_{2j+3},$$

where $d > 7$. *For generic* $g \in G_{k+3}(1, 3, d-7, \ldots, 1, d-2)$, $\{F \circ g = 0\}$ *is a minimal intransigent hypersurface of degree* d *in* $P^{k+2}(\mathbb{C})$, *with simple singularities* $3D_{d+1} + (d-7)A_1$.

Proof. By 2.5, it suffices to treat the cases $k = 0$. Setting $k = 0$, we apply 2.3, and find that in each case n is 2 and s is $d - 5$, d being the degree of $F \circ g$. Since $d - 5 = (n-1)(d-2) - 3$, it follows from 1.2 and 2.4 that it suffices to exhibit just one example for each case with exactly the singularities claimed. These examples follow:

(i) (a) Define $g_0(x_0, x_1, x_2) = (x_1, x_2^4 - x_0^4)$, so that

$$F \circ g_0(x_0, x_1, x_2) = x_1^8 + (x_2^4 - x_0^4)^2.$$

Then $C = \{F \circ g_0 = 0\}$ has just four singularities, at the points with coordinates satisfying $x_1 = 0$, $x_2^4 = x_0^4$, and these singularities are of type A_7.

(b) Define $g_0(x_0, x_1, x_2) = (x_1, x_2^4 - x_0^4, 2x_2^{d-8} - x_0^{d-8})$, so that

$$F \circ g_0(x_0, x_1, x_2) = x_1^d + (x_2^4 - x_0^4)^2(2x_2^{d-8} - x_0^{d-8}).$$

Then $C = \{F \circ g_0 = 0\}$ has just four singularities, at the points with coordinates satisfying $x_1 = 0$, $x_2^4 = x_0^4$, and these singularities are of type A_{d-1}.

(ii) (a) Define $g_0(x_0, x_1, x_2) = (x_1, x_2^3 - x_0^3)$, so that

$$F \circ g_0(x_0, x_1, x_2) = x_1^7 + x_1(x_2^3 - x_0^3)^2.$$

Then $C = \{F \circ g_0 = 0\}$ has just three singularities, of $\mathcal{K}$-type D_8, at the points with coordinates $x_1 = 0$, $x_2^3 = x_0^3$.

(b) Define $g_0(x_0, x_1, x_2) = (x_1, x_2^3 - x_0^3, 2x_2^{d-7} - x_0^{d-7})$, so that

$$F \circ g_0(x_0, x_1, x_2) = x_1^d + x_1(x_2^3 - x_0^3)^2(2x_2^{d-7} - x_0^{d-7}).$$

Then $C = \{F \circ g_0 = 0\}$ has just $d - 4$ singularities; three, of $\mathcal{K}$-type D_{d+1}, at the points with coordinates $x_1 = 0$, $x_2^3 = x_0^3$, and $d - 7$, all nodes, at the points with coordinates $x_1 = 0$, $2x_2^{d-7} = x_0^7$. □

Interesting intransigent (but not minimal) hypersurfaces are also revealed by non-generic choice of g in the above.

For example, take $k = 0$ in (i), and define

$$g(x_0, x_1, x_2) = \begin{cases} (x_1, x_1x_0^3 + x_2^4) & \text{in case } (a), \text{ when } d = 8, \\ (x_1, x_1x_0^3 + x_2^4, 2x_2^{d-8} - x_0^{d-8}) & \text{in case } (b), \text{ when } d > 8. \end{cases}$$

Then the curve $\{F \circ g = 0\}$ has a single A_{4d-1}-singularity; the case $d = 9$ is one of the examples of Luengo [5]. Here the line $\{g_1 = 0\}$ intersects the quartic $\{g_2 = 0\}$ at $[(1, 0, 0)]$ with multiplicity four. Less degenerate intersections yield less degenerate singularities; replacing g_2 by $x_1x_0^3 + x_2^3(x_2 - x_0)$, $x_1x_0^3 + (x_2^2 - x_0^2)^2$, and $x_1x_0^3 + x_2^2(x_2^2 - x_0^2)$ yields singularities $A_{3d-1} + A_{d-1}$, $2A_{2d-1}$, and $A_{2d-1} + 2A_{d-1}$, respectively.

Further examples will be presented in more compact form. The equation f_0 of a minimal intransigent hypersurface of degree d in $P^n(\mathbb{C})$ with the constellation Δ of simple singularities will be given, together with a function $F : \mathbb{C}^m \to \mathbb{C}$ satisfying the hypotheses of 2.3. Comparison of the forms of f_0 and F will make it clear for which $w_1, \ldots, w_m$ and $g_0 \in G_n(w_1, \ldots, w_m)$, the identity $f_0 = F \circ g_0$ holds. It will be left to the reader to apply 1.2 and 2.3 to see that $\{f_0 = 0\}$ is indeed minimal and intransigent, and to apply 2.4 and 2.5 to obtain the full range of associated examples, also in higher dimensions. Extended in this way, the examples 2.7 below give, together with 2.6, minimal intransigent hypersurfaces with simple singularities in all cases $(n - 1)(d - 2) > 4$, as promised in the introduction.

Examples 2.7.

	Δ	f_0	F
$n = 3$, $d = 5$	$4A_4$	$x_1^5 + x_2(x_2^4 + x_3^4 - x_0^4)$	$y_1^5 + y_2y_3$
$d > 5$	$(d-1)A_4$	$x_1^5(x_1^{d-5} + x_0^{d-5}) + x_2(x_2^{d-1} + x_3^{d-1} - x_0^{d-1})$	$y_1^5y_2 + y_3y_4$
$n = 4$, $d = 4$	$6A_3$	$x_1^4 + x_2(x_2^3 + x_3^3 - x_0^3) + (x_4^2 - x_0^2)^2$	$y_1^4 + y_2y_3 + y_4^2$
	$3E_6$	$x_1^4 + x_2(x_2^3 + x_3^3 - x_0^3) + x_0x_4^3$	$y_1^4 + y_2y_3 + y_4y_5^3$
$n = 6$, $d = 3$	$4D_4$	$x_1^3 + x_2^3 + x_3(x_3^2 + x_4^2 - x_0^2)$ $+x_5(x_5^2 + x_6^2 - 2x_0^2)$	$y_1^3 + y_2^3 + y_3y_4 + y_5y_6$

A natural question arises: which other constellations of simple singularities are possible on minimal intransigent hypersurfaces? For example, can all the singularities be nodes? No intransigent *curve* with all singularities nodes can exist, by a classical result of Severi [12]; the examples below show, however, that minimal intransigent hypersufaces of this type exist when $n = 3$ and $d \geq 8$, $n = 4$ and $d \geq 5$, $n \geq 5$ and $d = 4$, and $n \geq 8$ and $d = 3$.

Examples 2.8. *In each case,* $\Delta = \delta(d)A_1$.

	f_0	F
$n=3,\ d=8$	$x_1(x_1^7+x_2^7-x_0^7)+(x_3^4-x_0^4)^2$	$y_1y_2+y_3^2$
$d>8$	$x_1(x_1^{d-1}+x_2^{d-1}-x_0^{d-1})+c_1(2x_3^{d-8}-x_0^{d-8})(x_3^4-x_0^4)^2$	$y_1y_2+y_3y_4^2$
$n=4,$	$x_1(x_1^{d-1}+x_2^{d-1}-x_0^{d-1})+c_2(2x_3^{d-4}-x_0^{d-4})(x_3^2-x_0^2)^2$	
$d\geq 5$	$+c_3(2x_4^{d-4}-x_0^{d-4})(x_4^2-x_0^2)^2$	$y_1y_2+y_3y_4^2+y_5y_6^2$
$n=5,\ d=4$	$x_1(x_1^3+x_2^3-x_0^3)+x_3(x_3^3+x_4^3-x_0^3)+(x_5^2-x_0^2)^2$	$y_1y_2+y_3y_4+y_5^2$
$n=8,\ d=3$	$\sum_{i=1}^4 x_{2i-1}(x_{2i-1}^2+x_{2i}^2-ix_0^2)$	$\sum_{i=1}^4 y_{2i-1}y_{2i}$

To ensure that the examples in which they appear do not have more singularities than claimed, the constants c_1, c_2, $c_3 \in \mathbb{C}$ must be chosen so that they, and the quotient c_2/c_3, avoid, for each relevant d, a finite set of algebraic numbers. In particular, then, it suffices to choose c_1 to be transcendental, and c_2, c_3 to be algebraically independent and transcendental.

There are similar examples with all singularities cusps:

Examples 2.9. *In each case,* $\Delta = \{\frac{1}{2}\delta(d)\}A_2$.

	f_0	F
$n=3,\ d=6$	$x_1(x_1^5+x_2^5-x_0^5)+(x_3^2-x_0^2)^3$	$y_1y_2+y_3^3$
$d>6$	$x_1(x_1^{d-1}+x_2^{d-1}-x_0^{d-1})+c_1(2x_3^{d-6}-x_0^{d-6})(x_3^2-x_0^2)^3$	$y_1y_2+y_3y_4^3$
$n=4,$	$x_1(x_1^{d-1}+x_2^{d-1}-x_0^{d-1})+c_2(2x_3^{d-3}-x_0^{d-3})x_3^3$	
$d\geq 5$	$+c_3(2x_4^{d-4}-x_0^{d-4})(x_4^2-x_0^2)^2$	$y_1y_2+y_3y_4^3+y_5y_6^2$
$n=5,\ d=4$	$x_1(x_1^3+x_2^3-x_0^3)+x_3(x_3^3+x_4^3-x_0^3)+x_0x_5^3$	$y_1y_2+y_3y_4+y_5y_6^3$
$n=7,\ d=3$	$\sum_{i=1}^3 x_{2i-1}(x_{2i-1}^2+x_{2i}^2-ix_0^2)+x_7^3$	$\sum_{i=1}^3 y_{2i-1}y_{2i}+y_7^3$

As in the remark following 2.8, c_1, c_2, $c_3 \in \mathbb{C}$ must be chosen so that they, and the quotient c_2/c_3, avoid, for each relevant d, a finite set of algebraic numbers; again it suffices to choose c_1 to be transcendental, and c_2, c_3 to be algebraically independent and transcendental.

There are also intransigent *curves* with just cusp singularities: Segre gave, in [11], the examples $f = g_1^2 + g_2^3$, where g_1, g_2 are generic homogeneous polynomials in three variables, of degrees $2k$, $3k$, respectively. Then $\{f = 0\}$ has just $6k^2$ cusp singularities; applying 2.3 with $F(y_1, y_2) = y_1^2 + y_2^3$, we see that this curve is intransigent when $k \geq 3$. It is not, however, minimal; indeed, no minimally intransigent curve with just cusp singularities exists. For by 1.2 such a curve would have

$2(d-1)$ cusps, whilst a result of Greuel and Karras [3, 6.3, 6.4 (1)], generalising the result of Segre cited above and results of Zariski [18], shows that any curve with p nodes and q cusps as its singularities is transigent whenever $q < 3d$.

References

[1] Arnold, V.I., S.M. Gusein-Zade and A.N. Varchenko, *Singularities of differentiable maps* **II**, Monographs in Mathematics **83**, Birkhäuser, 1985.

[2] Goryunov, V.V., Symmetric quartics with many nodes, pp. 147–161 in *Singularities and curves*, ed. V.I. Arnol'd, Advances in Soviet Mathematics **21**, AMS, Providence, RI, 1994.

[3] Greuel, G.-M., and U. Karras, Families of varieties with prescribed singularities, Compositio Math. **69** (1989), 83–110.

[4] Greuel, G.-M., C. Lossen and E. Shustin, New asymptotics in the geometry of equisingular families of curves, International Mathematics Research Notices **13** (1997), 595–611.

[5] Luengo, I., On the existence of complete families of projective plane curves, which are obstructed, J. London Math. Soc. (2) **36** (1987), 33–43.

[6] du Plessis, A. A., Versality properties of projective hypersurfaces, *this volume.*

[7] du Plessis, A.A., and C.T.C. Wall, *The geometry of topological stability*, London Math. Soc. monographs new series **9**, Oxford University Press, 1995.

[8] du Plessis, A.A. and C.T.C. Wall, Curves in $P^2(\mathbb{C})$ with one-dimensional symmetry, Revista Mat. Complutense **12** (1999), 117–132.

[9] du Plessis, A.A. and C.T.C. Wall, Singular hypersurfaces, versality, and Gorenstein algebras, J. Alg. Geom. **9** (2000), 309–322.

[10] du Plessis, A.A. and C.T.C. Wall, Hypersurfaces in $P^n(\mathbb{C})$ with 1-parameter symmetry groups, Proc. Roy. Soc. Lond. A **456** (2000), 2515–2541.

[11] Segre, B., Esistenza e dimensione di sistemi continui di curve piane algebriche con dati caraterri, Atti. Acad. naz. Lincei Rend. (ser. 6) **10** (1929), 31–38.

[12] Severi, F., *Vorlesung über algebraische Geometrie*, Teubner, 1921, and Johnson, 1968.

[13] Shustin, E., Smoothness of equisingular families of plane algebraic curves, International Math. Research Notices **2** (1997), 67–82.

[14] van Straten, D., A quintic hypersurface in P^4 with 130 nodes, Topology **32** (1993), 857–864.

[15] Teissier, B. The hunting of invariants in the geometry of discriminants, pp. 565–678 in *Real and complex singularities*, ed. P. Holm, Sijthoff & Noordhoff International Publishers, 1977.

[16] Wall, C.T.C., Notes on trimodal singularities, Manuscripta Math. **100** (1999), 131–157.

[17] Wirthmüller, K., *Universell topologisch triviale Deformationen*, doctoral thesis, Universität Regensburg, 1979.

[18] Zariski, O., *Algebraic Surfaces* (2nd ed.), Springer Verlag, 1971.

Andrew A. du Plessis
Department of Mathematical Sciences
University of Aarhus
Building 530, Ny Munkegade
DK-8000 Aarhus C, Denmark
e-mail: matadp@imf.au.dk

Real and Complex Singularities
Trends in Mathematics, 311–325

On the Link Space of a $\mathbb{Q}$-Gorenstein Quasi-Homogeneous Surface Singularity

Anna Pratoussevitch

Abstract. In this paper we prove the following theorem: Let M be the link space of a quasi-homogeneous hyperbolic $\mathbb{Q}$-Gorenstein surface singularity. Then M is diffeomorphic to a coset space $\tilde{\Gamma}_1\backslash\tilde{G}/\tilde{\Gamma}_2$, where $\tilde{G}$ is the 3-dimensional Lie group $\widetilde{\mathrm{PSL}}(2,\mathbb{R})$, while $\tilde{\Gamma}_1$ and $\tilde{\Gamma}_2$ are discrete subgroups of $\tilde{G}$, the subgroup $\tilde{\Gamma}_1$ is co-compact and $\tilde{\Gamma}_2$ is cyclic. Conversely, if M is diffeomorphic to a coset space as above, then M is diffeomorphic to the link space of a quasi-homogeneous hyperbolic $\mathbb{Q}$-Gorenstein singularity. We also prove the following characterisation of quasi-homogeneous $\mathbb{Q}$-Gorenstein surface singularities: A quasi-homogeneous surface singularity is $\mathbb{Q}$-Gorenstein of index r if and only if for the corresponding automorphy factor (U, Γ, L) some tensor power of the complex line bundle L is Γ-equivariantly isomorphic to the rth tensor power of the tangent bundle of the Riemannian surface U.

Mathematics Subject Classification (2000). Primary 32S25; Secondary 14J17, 14J60.

Keywords. $\mathbb{Q}$-Gorenstein singularity, quasi-homogeneous singularity.

1. Introduction

Graded affine coordinate rings of quasi-homogeneous surface singularities can be identified with graded rings of generalised automorphic forms. The description in terms of automorphy factors was found in 1975–77 by Dolgachev, Milnor, Neumann and Pinkham [Dol75, Dol77, Mil75, Neu77, Pin77].

For some special classes of quasi-homogeneous surface singularities as Gorenstein and $\mathbb{Q}$-Gorenstein singularities one can obtain more precise descriptions of the corresponding automorphy factors.

In Theorem 3 we obtain a characterisation of hyperbolic and spherical $\mathbb{Q}$-Gorenstein quasi-homogeneous surface singularities in terms of their automorphy

Research partially supported by SFB 611 of the DFG.

factors. This characterisation leads to a description of their links as biquotients of certain 3-dimensional Lie groups by discrete subgroups. More precisely, we prove the following statement:

Theorem 1. *The link space of a hyperbolic $\mathbb{Q}$-Gorenstein quasi-homogeneous surface singularity of index r is diffeomorphic to a biquotient*

$$\tilde{\Gamma}_1\backslash\tilde{G}/\tilde{\Gamma}_2,$$

where $\tilde{G}$ is the universal cover $\widetilde{\mathrm{PSL}}(2,\mathbb{R})$ of the 3-dimensional Lie group $\mathrm{PSL}(2,\mathbb{R})$, while $\tilde{\Gamma}_1$ and $\tilde{\Gamma}_2$ are discrete subgroups of the same level in $\tilde{G}$, $\tilde{\Gamma}_1$ is co-compact, and the image of $\tilde{\Gamma}_2$ in $\mathrm{PSL}(2,\mathbb{R})$ is a cyclic subgroup of order r. Hereby the level of a discrete subgroup $\tilde{\Gamma}\subset\tilde{G}$ is the index of $\tilde{\Gamma}\cap Z(\tilde{G})$ as a subgroup of $Z(\tilde{G})$.

Conversely, any biquotient as above is diffeomorphic to the link space of a hyperbolic $\mathbb{Q}$-Gorenstein quasi-homogeneous surface singularity.

These statements are generalisations of the results of Dolgachev [Dol83] on Gorenstein quasi-homogeneous surface singularities. The Gorenstein quasi-homogeneous surface singularities correspond to the case of the trivial group $\tilde{\Gamma}_2$.

Similar statements are also true in the case of Euclidean automorphy factors and corresponding singularities. This case was already discussed by Dolgachev in [Dol83].

The description of the link space of a hyperbolic Gorenstein quasi-homogeneous surface singularity as a quotient of the Lie group $\widetilde{\mathrm{PSL}}(2,\mathbb{R})$ by the action of a discrete subgroup was the motivation for the study in [Pra], [BPR03] of a certain construction of fundamental domains for such actions. This construction leads to interesting results on the combinatorial geometry of the link spaces of Gorenstein quasi-homogeneous surface singularities.

We expect that our construction of fundamental domains can be generalised in order to study the combinatorial geometry of the link spaces of $\mathbb{Q}$-Gorenstein quasi-homogeneous surface singularities. We shall discuss the combinatorial geometry of the link spaces in the $\mathbb{Q}$-Gorenstein case in an ongoing paper.

The paper is organised as follows: In Section 2 we explain the description of quasi-homogeneous surface singularities via automorphy factors. In Section 3 we define $\mathbb{Q}$-Gorenstein quasi-homogeneous surface singularities and introduce our characterisation of the corresponding automorphy factors (Theorem 3). Then in Section 4 we prove some technical results needed to prove this characterisation. After that we prove Theorem 3 in Section 5. Finally we prove Theorem 1 in Section 6.

Notation: In this paper we use $\mathbb{R}_+$ for $\{x\in\mathbb{R}\mid x>0\}$. We denote by L^* the associated $\mathbb{C}^*$-bundle of a complex line bundle L, while $L^\vee$ is the dual bundle of L.

2. Automorphy factors

In this section we recall the results of Dolgachev, Milnor, Neumann and Pinkham [Dol75, Dol77, Mil75, Neu77, Pin77] on the graded affine coordinate rings, which correspond to quasi-homogeneous surface singularities.

Definition. *A (negative unramified) automorphy factor* (U, Γ, L) is a complex line bundle L over a simply connected Riemann surface U together with a discrete co-compact subgroup $\Gamma \subset \mathrm{Aut}(U)$ acting compatibly on U and on the line bundle L, such that the following two conditions are satisfied:

1) The action of Γ is free on L^*, the complement of the zero section in L.
2) Let $\Gamma' \triangleleft \Gamma$ be a normal subgroup of finite index, which acts freely on U, and let $\bar{L} \to C$ be the complex line bundle $\bar{L} = L/\Gamma'$ over the compact Riemann surface $C = U/\Gamma'$. Then $\bar{L}$ is a negative line bundle.

A simply connected Riemann surface U can be $\mathbb{C}\mathrm{P}^1$, $\mathbb{C}$, or H, the real hyperbolic plane. We call the corresponding automorphy factor and the corresponding singularity *spherical*, *Euclidean*, resp. *hyperbolic*.

Remark. There always exists a normal freely acting subgroup of Γ of finite index. In the hyperbolic case the existence follows from the theorem of Fox-Bundgaard-Nielsen. If the second assumption in the last definition holds for some normal freely acting subgroup of finite index, then it holds for any such subgroup.

The simplest examples of such a complex line bundle with group action are the cotangent bundle of the complex projective line $U = \mathbb{C}\mathrm{P}^1$ and the tangent bundle of the hyperbolic plane $U = H$ equipped with the canonical action of a subgroup $\Gamma \subset \mathrm{Aut}(U)$.

Let (U, Γ, L) be a negative unramified automorphy factor. Since the bundle $\bar{L} = L/\Gamma'$ is negative, one can contract the zero section of $\bar{L}$ to get a complex surface with one isolated singularity corresponding to the zero section. There is a canonical action of the group Γ/Γ' on this surface. The quotient is a complex surface $X(U, \Gamma, L)$ with an isolated singular point o, which depends only on the automorphy factor (U, Γ, L).

The following theorem summarises the results of Dolgachev, Milnor, Neumann, and Pinkham:

Theorem 2. *The surface $X(U, \Gamma, L)$ associated to a negative unramified automorphy factor (U, Γ, L) is a quasi-homogeneous affine algebraic surface with a normal isolated singularity. Its affine coordinate ring is the graded $\mathbb{C}$-algebra of generalised Γ-invariant automorphic forms*

$$A = \bigoplus_{m \geqslant 0} H^0(U, L^{-m})^{\Gamma}.$$

All normal quasi-homogeneous surface singularities (X, x) are obtained in this way, and the automorphy factors with $(X(U, \Gamma, L), o)$ isomorphic to (X, x) are uniquely determined by (X, x) up to isomorphism.

3. Q-Gorenstein quasi-homogeneous surface singularities

In this section we recall the definition of $\mathbb{Q}$-Gorenstein singularities and the characterisation of the corresponding automorphy factors.

A normal isolated singularity of dimension n is Gorenstein if and only if there is a nowhere vanishing n-form on a punctured neighborhood of the singular point. For example all isolated singularities of complete intersections are Gorenstein.

A natural generalisation of Gorenstein singularities are the $\mathbb{Q}$-Gorenstein singularities (compare [Rei87, Ish87, Ish00]). A normal isolated singularity of dimension at least 2 is $\mathbb{Q}$-*Gorenstein* if there is a natural number r such that the divisor $r \cdot \mathcal{K}_X$ is defined on a punctured neighborhood of the singular point by a function. Here $\mathcal{K}_X$ is the canonical divisor of X. The smallest such number r is called the *index* of the singularity. A normal isolated surface singularity is Gorenstein if and only if it is $\mathbb{Q}$-Gorenstein of index 1.

In Section 5 we prove the following characterisation of hyperbolic and spherical $\mathbb{Q}$-Gorenstein quasi-homogeneous surface singularities in terms of the corresponding automorphy factors:

Theorem 3. *A hyperbolic resp. spherical quasi-homogeneous surface singularity is $\mathbb{Q}$-Gorenstein of index r if and only if for the corresponding automorphy factor (U, Γ, L) there is an integer m (called the level or the exponent of the automorphy factor) without common divisors with r and a Γ-invariant isomorphism $L^m \cong T_U^r$.*

We call an automorphy factor with properties as in Theorem 3 a $\mathbb{Q}$-*Gorenstein* automorphy factor *of level m and index r.*

4. The associated bundle of the quotient bundle

Let (U, Γ, L) be a spherical or hyperbolic negative unramified automorphy factor. As in the definition let $\Gamma' \triangleleft \Gamma$ be a normal subgroup of Γ acting freely on U, and let $p : \bar{L} \to C$ be the complex line bundle with total space $\bar{L} = L/\Gamma'$ and base $C = U/\Gamma'$. In this section we consider the associated $\mathbb{C}^*$-bundle of the bundle p, i.e., $p|_{\bar{L}^*} : \bar{L}^* \to C$. For ease of notation we set $W := \bar{L}^*$ and $q := p|_W$. We first present some technical lemmas, which will be used later to determine

$$\Omega^{2,r}(W) := (\Omega^2(W))^{\otimes r}.$$

Lemma 4. *The following $\mathcal{O}_C$-algebras are isomorphic*

$$q_*(\mathcal{O}_W) \cong \bigoplus_{m \in \mathbb{Z}} \mathcal{O}_C(\bar{L}^m).$$

Lemma 5. *We have $\Omega^2(W) \cong q^*(\Omega^1_C)$.*

Lemma 6. *If the bundle $\bar{L}$ is non-trivial and the sheaf $\Omega^{2,r}(W)$ is trivial, then there exists up to complex multiples only one nowhere vanishing section in $\Omega^{2,r}(W)$.*

We postpone the proofs of these lemmas until the end of this section and discuss first the main result of the section, the description of (Γ/Γ')-invariant sections in $\Omega^{2,r}(W)$.

Proposition 7. *The sheaf $\Omega^{2,r}(W)$ is trivial if and only if there exists an integer m and an isomorphism $\bar{L}^m \cong T_C^r$.*

Assume that $\Omega^{2,r}(W)$ is trivial and let m be the integer such that $\bar{L}^m \cong T_C^r$. Then the global nowhere vanishing sections in $\Omega^{2,r}(W)$ are (Γ/Γ')-invariant if and only if the isomorphism $\bar{L}^m \cong T_C^r$ is (Γ/Γ')-equivariant.

Proof. We first prove that $\Omega^{2,r}(W) \cong \mathcal{O}_W$ implies $\bar{L}^m \cong T_C^r$ for some $m \in \mathbb{Z}$. Then we prove that $\bar{L}^m \cong T_C^r$ for some $m \in \mathbb{Z}$ implies $\Omega^{2,r}(W) \cong \mathcal{O}_W$. To this end we first consider the cases $m = 1$ and $m = -1$, and then we look at the case $m \neq 0$.

1) Let us prove that $\Omega^{2,r}(W) \cong \mathcal{O}_W$ implies $\bar{L}^m \cong T_C^r$ for some $m \in \mathbb{Z}$. Assume that $\Omega^{2,r}(W) \cong \mathcal{O}_W$. On the one hand, using Lemma 4, we have

$$q_*(\Omega^{2,r}(W)) \cong q_*(\mathcal{O}_W) \cong \bigoplus_{i\in\mathbb{Z}} \mathcal{O}_C(\bar{L}^i).$$

On the other hand, using Lemma 5, we have

$$q_*(\Omega^{2,r}(W)) \cong q_*((q^*(\Omega_C^1))^{\otimes r}).$$

Now we obtain for $\Omega_C^{1,r} := (\Omega_C^1)^{\otimes r}$

$$q_*((q^*(\Omega_C^1))^{\otimes r}) \cong q_*(q^*(\Omega_C^{1,r})),$$

because q^* is compatible with tensor products. The projection formula implies

$$q_*(q^*(\Omega_C^{1,r})) \cong q_*(\mathcal{O}_W \otimes_{\mathcal{O}_W} q^*(\Omega_C^{1,r})) \cong q_*(\mathcal{O}_W) \otimes_{\mathcal{O}_C} \Omega_C^{1,r}.$$

Using Lemma 4 again, we obtain finally

$$\begin{aligned} q_*(\Omega^{2,r}(W)) &\cong q_*(\mathcal{O}_W) \otimes_{\mathcal{O}_C} \Omega_C^{1,r} \\ &\cong \Big(\bigoplus_{m\in\mathbb{Z}} \mathcal{O}_C(\bar{L}^m)\Big) \otimes_{\mathcal{O}_C} \Omega_C^{1,r} \\ &\cong \bigoplus_{m\in\mathbb{Z}} \mathcal{O}_C(\bar{L}^m \otimes T_C^{-r}). \end{aligned}$$

Comparing both equations for $q_*(\Omega^{2,r}(W))$ we obtain

$$\bigoplus_{i\in\mathbb{Z}} \mathcal{O}_C(\bar{L}^i) \cong \bigoplus_{m\in\mathbb{Z}} \mathcal{O}_C(\bar{L}^m \otimes T_C^{-r}),$$

hence

$$\mathcal{O}_C(\bar{L}^m \otimes T_C^{-r}) \cong \mathcal{O}_C(\bar{L}^0) \cong \mathcal{O}_C$$

for some $m \in \mathbb{Z}$. This implies $\bar{L}^m \cong T_C^r$.

2) We now assume that $\bar{L} \cong T_C^r$, i.e., $m = 1$. The Riemann surface U is then the real hyperbolic plane H. We study the tangent bundle T_C of the hyperbolic surface $C = H/\Gamma'$. We define a 2-form on $(T_C)^*$ in local coordinates by

$$\eta = \frac{1}{t^2} \cdot (dz \wedge dt).$$

Using the fact that any change of coordinates is of the form

$$(z, t) \mapsto (\varphi(z), \varphi'(z) \cdot t)$$

we can verify that this local definition gives rise to a global nowhere vanishing 2-form on $(T_C)^*$, and that this 2-form is invariant under an action of $g \in \Gamma$ if and only if the action is given by

$$(z, t) \mapsto (g(z), g'(z) \cdot t),$$

i.e., the action coincides with the canonical action of g on T_C. The 2-form on $(T_C)^*$ induces a nowhere vanishing section η in $\Omega^{2,r}((T_C^r)^*)$, which is invariant under an action of $g \in \Gamma$ if and only if the action is given in local coordinates by

$$(z, t) \mapsto (g(z), (g'(z))^r \cdot t),$$

i.e., the action coincides with the canonical action of g on T_C^r. Hence if the isomorphism $\bar{L} \cong T_C^r$ is (Γ/Γ')-equivariant, there exists a (Γ/Γ')-invariant nowhere vanishing section in $\Omega^{2,r}(W)$.

3) We now assume that $\bar{L} \cong T_C^{-r} \cong (T_C^\vee)^r$, i.e., $m = -1$. The Riemann surface U is then the complex projective line $\mathbb{C}\mathrm{P}^1$. We study the cotangent bundle $T_C^\vee$ of the surface $C = \mathbb{C}\mathrm{P}^1/\Gamma'$. We define a 2-form on $(T_C^\vee)^*$ in local coordinates by

$$\eta = dz \wedge dt.$$

Using the fact that any change of coordinates is of the form

$$(z, t) \mapsto (\varphi(z), \frac{1}{\varphi'(z)} \cdot t)$$

we can verify that this local definition gives rise to a global nowhere vanishing 2-form on $(T_C^\vee)^*$. We continue in the proof as for $m = 1$ and obtain a nowhere vanishing section η in $\Omega^{2,r}((T_C^{-r})^*)$, which is invariant under an action of $g \in \Gamma$ if and only if the action coincides with the canonical action of g on $(T_C^\vee)^r$. Hence if the isomorphism $\bar{L} \cong T_C^{-r}$ is (Γ/Γ')-equivariant, there exists a (Γ/Γ')-invariant nowhere vanishing section in $\Omega^{2,r}(W)$.

4) As the next step of the proof we consider the case $\bar{L}^m \cong T_C^r$ with $m \neq 0$. Let η be the nowhere vanishing section in $\Omega^{2,r}((T_C^{\pm r})^*)$, i.e., in $\Omega^{2,r}((\bar{L}^{|m|})^*)$, constructed in Subsections 2 and 3. We consider the covering $\tau : \bar{L} \to \bar{L}^m$. The pull-back $\tau^*(\eta)$ of the section η under the covering τ is a nowhere vanishing section in $\Omega^{2,r}(\bar{L}^*) = \Omega^{2,r}(W)$. If the isomorphism $\bar{L}^m \cong T_C^r$ is (Γ/Γ')-equivariant, the induced section in $\Omega^{2,r}(W)$ is (Γ/Γ')-invariant.

5) We now assume that $\Omega^{2,r}(W)$ is trivial and that there exists a nowhere vanishing section ω in $\Omega^{2,r}(W)$, which is (Γ/Γ')-invariant. Then in particular there exists an integer m such that $\bar{L}^m \cong T_C^r$. Let η be the nowhere vanishing section in $\Omega^{2,r}((T_C^{\pm r})^*)$ constructed before. By Lemma 6 the sections ω and η are complex multiples of each other, hence the isomorphism $\bar{L}^m \cong T_C^r$ is (Γ/Γ')-equivariant. □

Now it remains to show the technical lemmas, which we have used in the proof of Lemma 7. We first prove Lemma 4:

Proof. We have to prove that the following $\mathcal{O}_C$-algebras are isomorphic

$$q_*(\mathcal{O}_W) \cong \bigoplus_{m\in\mathbb{Z}} \mathcal{O}_C(\bar{L}^m).$$

Let us consider a local trivialisation $\varphi : \bar{L}|_V \to V \times \mathbb{C}$ of the complex line bundle $\bar{L}$ over an open affine subset $V \subset C$. This trivialisation induces trivialisations

$$\varphi : W|_V \to V \times \mathbb{C}^*$$

of the $\mathbb{C}^*$-bundle $W \to C$ and

$$\varphi^{\otimes m} : \bar{L}^m|_V \to V \times \mathbb{C}^{\otimes m}$$

of the complex line bundle $\bar{L}^m = \bar{L}^{\otimes m} \to C$. Then we obtain

$$\begin{aligned} q_*(\mathcal{O}_W)(V) &= (\mathrm{pr}\circ\varphi)_*(\mathcal{O}_W)(V) = \varphi_*\mathcal{O}_W(V \times \mathbb{C}^*) \\ &\cong \mathcal{O}_{V\times\mathbb{C}^*}(V \times \mathbb{C}^*) \cong \mathcal{O}_C(V) \otimes_{\mathbb{C}} \mathcal{O}_{\mathbb{C}^*}(\mathbb{C}^*) \\ &\cong \mathcal{O}_C(V) \otimes_{\mathbb{C}} \mathbb{C}[t, t^{-1}] \cong \bigoplus_{m\in\mathbb{Z}} \mathcal{O}_C(V) \cdot t^{-m}. \end{aligned}$$

This implies that any section of $q_*(\mathcal{O}_W)$ can be locally uniquely represented as a finite sum of the form $\sum f_m \cdot t^{-m}$ with $f_m \in \mathcal{O}_C(V)$. Using the induced local trivialisations of W and $\bar{L}^{\otimes m}$ over V together with the identifications $\mathbb{C}^{\otimes m} \cong \mathbb{C}$ and $\mathbb{C}^{\otimes(-1)} \cong \mathrm{Hom}(\mathbb{C}, \mathbb{C}) \cong \mathbb{C}$ we can construct a bijection between sections in $\bar{L}^m$ over V and functions in $\mathcal{O}_C(V)$. We obtain a family of isomorphisms

$$(q_*(\mathcal{O}_W)(V) \to \bigoplus_{m\in\mathbb{Z}} \mathcal{O}_C(V) \cdot t^{-m})_V,$$

which does not depend on the chosen trivialisations, is compatible with the restriction maps and hence induces an isomorphism of $\mathcal{O}_C$-algebras

$$q_*(\mathcal{O}_W) \cong \bigoplus_{m\in\mathbb{Z}} \mathcal{O}_C(\bar{L}^m).$$ □

Now we prove Lemma 5:

Proof. We have to prove that $\Omega^2(W) \cong q^*(\Omega^1_C)$. To this end we consider the sheaf of relative forms $\Omega^1_{W|C}$. This sheaf is trivial and generated by a relative form given in local coordinates by $\frac{dt}{t}$. The following short exact sequence of locally free sheaves of ranks 1, 2, and 1

$$0 \to q^*(\Omega^1_C) \to \Omega^1_W \to \Omega^1_{W|C} \to 0$$

implies

$$\Lambda^2(\Omega^1_W) \cong \Lambda^1(q^*(\Omega^1_C)) \otimes \Lambda^1(\Omega^1_{W|C}) \cong q^*(\Omega^1_C). \qquad \square$$

Finally we prove Lemma 6:

Proof. We have to prove that if the bundle $\bar{L}$ is non-trivial and the sheaf $\Omega^{2,r}(W)$ is trivial, then the nowhere vanishing section in $\Omega^{2,r}(W)$ is unique up to complex multiples. Consider two nowhere vanishing sections in $\Omega^{2,r}(W)$. There quotient is a nowhere vanishing regular function. It remains to prove that all nowhere vanishing regular functions on W are constant. Using Lemma 4 we obtain

$$\mathcal{O}_W(W) \cong q_*(\mathcal{O}_W)(C) \cong \bigoplus_{m\in\mathbb{Z}} \mathcal{O}_C(\bar{L}^m)(C).$$

Nowhere vanishing functions on W correspond to nowhere vanishing sections in $\bar{L}^m$. A non-homogeneous section in $\bar{L}^m$ can not be nowhere vanishing. A homogeneous nowhere vanishing section in $\bar{L}^m$ exists if and only if $\bar{L}^m$ is trivial. But the bundle $\bar{L}$ is negative, hence $\bar{L}^m$ is trivial if and only if $m = 0$. $\square$

5. Automorphy factors of Q-Gorenstein quasi-homogeneous surface singularities

In this section we use the results of Section 4 to prove Theorem 3:

Theorem. *A hyperbolic resp. spherical quasi-homogeneous surface singularity is $\mathbb{Q}$-Gorenstein of index r if and only if for the corresponding automorphy factor (U, Γ, L) there is an integer m (called the level or the exponent of the automorphy factor) without common divisors with r and a Γ-equivariant isomorphism*

$$L^m \cong T^r_U.$$

Proof. We first assume that for some positive integer r and integer m there is a Γ-equivariant isomorphism $L^m \cong T^r_U$. This isomorphism induces a (Γ/Γ')-equivariant isomorphism $\bar{L}^m \cong T^r_C$. Then according to proposition 7 there exist global nowhere vanishing (Γ/Γ')-invariant sections in $\Omega^{2,r}(W)$. Such a section induces a nowhere vanishing section in $\Omega^{2,r}(W/(\Gamma/\Gamma')) = \Omega^{2,r}(X^*)$, hence the corresponding singularity $(X(U, \Gamma, L), o)$ is $\mathbb{Q}$-Gorenstein.

Now let us assume that singularity (X, x) with automorphy factor (U, Γ, L) is $\mathbb{Q}$-Gorenstein of index r, i.e., there exist nowhere vanishing sections in

$$\Omega^{2,r}(X^*) \cong \Omega^{2,r}(W/(\Gamma/\Gamma')).$$

We consider the singularity $(\bar{X}, \bar{x})$, which corresponds to the automorphy factor (U, Γ', L). For this singularity we have $X \cong \bar{X}/(\Gamma/\Gamma')$. The pull-back of a nowhere vanishing section in $\Omega^{2,r}(X^*)$ along the unramified covering $\bar{X}^* \to X^*$ is a nowhere vanishing (Γ/Γ')-invariant section in $\Omega^{2,r}(\bar{X}^*)$, hence the singularity $(\bar{X}, \bar{x})$ is also $\mathbb{Q}$-Gorenstein of index r.

A nowhere vanishing (Γ/Γ')-invariant section in $\Omega^{2,r}(\bar{X}^*)$ induces a nowhere vanishing (Γ/Γ')-invariant section in $\Omega^{2,r}(W) = \Omega^{2,r}(\bar{L}^*)$. Then proposition 7 implies the existence of an (Γ/Γ')-equivariant isomorphism $\bar{L}^m \cong T_C^r$ for some integer m, i.e., the induced action of (Γ/Γ') on $\bar{L}^m \cong T_C^r$ coincides with the canonical action of (Γ/Γ') on T_C^r. Hence the action of Γ on $L^m \cong T_U^r$ also coincides with the canonical action of Γ on T_U^r, i.e., there exists a Γ-equivariant isomorphism $L^m \cong T_U^r$. □

Remark. Theorem 3 also follows from the following result of K. Watanabe [Wat81], appearing in the context of the theory of commutative rings: Let $R = R(X, D)$ be a normal graded ring, which is presented by the Pinkham-Demazure method. Then the canonical module K_R of R is Q-Cartier of index r, if and only if, there exists a rational function ϕ on X such that $r(K + D') - mD = \mathrm{div}_X(\phi)$ for some integer m and r is the minimum of such r. In our case, $X = C = U/\Gamma'$ and $\pi : U \to C$ is the Galois cover associated to the automorphy factor, and the result is translated to the relation on U as equivariant isomorphism $T_U^{-r} \cong L^m$. Our proofs of Proposition 7 and Theorem 3 give a more direct description of the automorphy factors in question.

6. From hyperbolic automorphy factors to biquotients

In this section we prove Theorem 1:

Theorem. *The link space of a hyperbolic $\mathbb{Q}$-Gorenstein quasi-homogeneous surface singularity of index r is diffeomorphic to a biquotient*

$$\tilde{\Gamma}_1 \backslash \tilde{G} / \tilde{\Gamma}_2,$$

where $\tilde{G}$ is the universal cover $\widetilde{\mathrm{PSL}}(2,\mathbb{R})$ of the 3-dimensional Lie group $\mathrm{PSL}(2,\mathbb{R})$, while $\tilde{\Gamma}_1$ and $\tilde{\Gamma}_2$ are discrete subgroups of the same level in $\tilde{G}$, $\tilde{\Gamma}_1$ is co-compact, and the image of $\tilde{\Gamma}_2$ in $\mathrm{PSL}(2,\mathbb{R})$ is a cyclic subgroup of order r. Conversely, any biquotient as above is diffeomorphic to the link space of a quasi-homogeneous hyperbolic $\mathbb{Q}$-Gorenstein singularity.

Before we explain the proof of this theorem, we give a description of the Lie group $G = \mathrm{PSL}(2,\mathbb{R})$ and its coverings.

As topological space $\mathrm{PSL}(2,\mathbb{R})$ is homeomorphic to the solid torus $\mathbb{S}^1 \times \mathbb{C}$. The fundamental group of the solid torus G is infinite cyclic. Therefore, for each natural number m there is a unique connected m-fold covering

$$G_m = \tilde{G}/(m \cdot Z(\tilde{G}))$$

of G, where $Z(\tilde{G})$ is the central subgroup of $\tilde{G}$. For $m = 2$ this is the group $G_2 = \mathrm{SL}(2,\mathbb{R})$.

We identify the group $G = \mathrm{PSL}(2,\mathbb{R})$ with the group $\mathrm{Aut}(H)$ of automorphisms of the hyperbolic plane. We think of an element $g \in G$ as a map $g : H \to H$.

We use the following description of the covering groups G_m of $G = \mathrm{PSL}(2,\mathbb{R})$, which fixes a group structure. Let $\mathrm{Hol}(H,\mathbb{C}^*)$ be the set of all holomorphic functions $H \to \mathbb{C}^*$.

Proposition 8. *The m-fold covering group G_m of G can be described as*

$$\{(g,\delta) \in G \times \mathrm{Hol}(H,\mathbb{C}^*) \mid \delta^m(z) = g'(z) \text{ for all } z \in H\}$$

with multiplication

$$(g_2,\delta_2)\cdot(g_1,\delta_1) = (g_2 \cdot g_1, (\delta_2 \circ g_1)\cdot \delta_1).$$

Remark. This description of G_m and the description of $\tilde{G}$ that we give later are inspired by the notion of automorphic differential forms of fractional degree, introduced by J. Milnor in [Mil75]. For a more detailed discussion of this fact see [LV80], Section 1.8.

We now explain the connection between automorphy factors in question and lifts of Fuchsian groups into the finite coverings of $\tilde{G}$.

Definition. A *lift* of the Fuchsian group Γ into G_m is a subgroup Γ^* of G_m such that the restriction of the covering map $G_m \to G$ to Γ^* is an isomorphism between Γ^* and Γ.

Proposition 9. *There is a* 1-1*-correspondence between hyperbolic* $\mathbb{Q}$*-Gorenstein automorphy factors* $(H,\Gamma,H\times\mathbb{C})$ *of level* m *and index* r *and the lifts of* Γ *into* G_m.

Proof. Using the description of the covering G_m from Proposition 8, we see, on the one hand, that there is a 1-1-correspondence between lifts of Γ into G_m and families $\{\delta_g\}_{g\in\Gamma}$ of holomorphic functions $\delta_g : H \to \mathbb{C}^*$ such that for any $g \in \Gamma$

$$\delta_g^m = g'$$

and for any $g_1, g_2 \in \Gamma$

$$\delta_{g_2\cdot g_1} = (\delta_{g_2}\circ g_1)\cdot\delta_{g_1}.$$

Let $\mathcal{D}$ be the set of all such families (δ_g).

On the other hand there is a 1-1-correspondence between hyperbolic $\mathbb{Q}$-Gorenstein automorphy factors $(H,\Gamma,H\times\mathbb{C})$ of level m and index r and families $\{e_g\}_{g\in\Gamma}$ of holomorphic functions $e_g : H \to \mathbb{C}^*$ such that for any $g \in \Gamma$

$$e_g^m = (g')^r$$

and for any $g_1, g_2 \in \Gamma$

$$e_{g_2\cdot g_1} = (e_{g_2}\circ g_1)\cdot e_{g_1}.$$

Let $\mathcal{E}$ be the set of all such families $(e_g)_{g\in\Gamma}$.

It remains to establish a 1-1-correspondence between the sets $\mathcal{D}$ and $\mathcal{E}$. This correspondence is defined as follows: Let us assign to a family $(\delta_g)\in\mathcal{D}$ the family (e_g) given by $e_g := \delta_g^r$. One checks easily that $(e_g)\in\mathcal{E}$.

If the images $(e_g), (\tilde{e}_g) \in \mathcal{E}$ of $(\delta_g), (\tilde{\delta}_g) \in \mathcal{D}$ coincide then on the one hand

$$\left(\frac{\tilde{\delta}_g}{\delta_g}\right)^r = \frac{\tilde{\delta}_g^r}{\delta_g^r} = \frac{\tilde{e}_g}{e_g} = 1,$$

on the other hand

$$\left(\frac{\tilde{\delta}_g}{\delta_g}\right)^m = \frac{\tilde{\delta}_g^m}{\delta_g^m} = \frac{g'}{g'} = 1.$$

But the integers m and r are relatively prime, hence there exists only one complex number ξ with the property $\xi^m = \xi^r = 1$, namely $\xi = 1$. Hence for any $g \in \Gamma$

$$\frac{\tilde{\delta}_g}{\delta_g} \equiv 1,$$

i.e., the families (δ_g) and $(\tilde{\delta}_g)$ coincide. So we have shown that the mapping $\mathcal{D} \to \mathcal{E}$ is injective.

Now let us consider a family $(e_g) \in \mathcal{E}$. It holds $g'(z) \notin \mathbb{R}_+ \cup \{0\}$ for all $z \in H$, hence there exist functions $\rho_g : H \to \mathbb{R}_*$ and $\varphi_g : H \to (0, 1)$ such that

$$g' = \rho_g \cdot \exp(2\pi i \varphi_g).$$

The chain rule implies

$$\rho_{g_2 \cdot g_1} = (\rho_{g_2} \circ g_1) \cdot \rho_{g_1}$$

and

$$\varphi_{g_2 \cdot g_1} - \varphi_{g_2} \circ g_1 - \varphi_{g_2} \in \mathbb{Z}.$$

The function e_g is then of the form

$$e_g = \rho_g^{\frac{r}{m}} \cdot \exp\left(2\pi i \cdot \frac{r \cdot \varphi_g + k_g}{m}\right)$$

for some function $k_g : H \to \mathbb{Z}$. The function k_g is continuous and hence constant. The integers m and r are relatively prime, hence there is an integer n_g such that

$$r \cdot n_g \equiv k_g \mod m.$$

Let us define a family (δ_g) by setting

$$\delta_g = \rho_g^{\frac{1}{m}} \cdot \exp\left(2\pi i \cdot \frac{\varphi_g + n_g}{m}\right).$$

We now prove that the family (δ_g) is in $\mathcal{D}$. The first property

$$\delta_g^m = \rho_g \cdot \exp\left(2\pi i \cdot (\varphi_g + n_g)\right) = \rho_g \cdot \exp\left(2\pi i \cdot \varphi_g\right) = g'$$

is satisfied. The second property

$$\delta_{g_2 \cdot g_1} = (\delta_{g_2} \circ g_1) \cdot \delta_{g_1}$$

is equivalent to

$$\rho_{g_2 \cdot g_1}^{\frac{1}{m}} = (\rho_{g_2} \circ g_1)^{\frac{1}{m}} \cdot \rho_{g_1}^{\frac{1}{m}}$$

and

$$(\varphi_{g_2\cdot g_1} - \varphi_{g_2}\circ g_1 - \varphi_{g_1}) + (n_{g_2\cdot g_1} - n_{g_2}\circ g_1 - n_{g_1}) \equiv 0 \mod m.$$

The first of these equations follows from

$$\rho_{g_2\cdot g_1} = (\rho_{g_2}\circ g_1)\cdot\rho_{g_1}.$$

To prove the second equations we observe that

$$e_{g_2\cdot g_1} = (e_{g_2}\circ g_1)\cdot e_{g_1}$$

implies that m is a divisor of the integer

$$r\cdot(\varphi_{g_2\cdot g_1} - \varphi_{g_2}\circ g_1 - \varphi_{g_1}) + (k_{g_2\cdot g_1} - k_{g_2}\circ g_1 - k_{g_1}).$$

Because of $r\cdot n_g \equiv k_g \mod m$ also the integer

$$r\cdot\Big((\varphi_{g_2\cdot g_1} - \varphi_{g_2}\circ g_1 - \varphi_{g_1}) + (n_{g_2\cdot g_1} - n_{g_2}\circ g_1 - n_{g_1})\Big)$$

is divisible by m. Since m and r are relatively prime, the number m must be a divisor of the integer

$$(\varphi_{g_2\cdot g_1} - \varphi_{g_2}\circ g_1 - \varphi_{g_1}) + (n_{g_2\cdot g_1} - n_{g_2}\circ g_1 - n_{g_1}).$$

So the family (δ_g) is in $\mathcal{D}$. The image of the family (δ_g) under the map $\mathcal{D}\to\mathcal{E}$ is

$$\delta_g^r = \rho_g^{\frac{r}{m}}\cdot\exp\left(2\pi i\cdot\frac{r\cdot\varphi_g + r\cdot n_g}{m}\right) = \rho_g^{\frac{r}{m}}\cdot\exp\left(2\pi i\cdot\frac{r\cdot\varphi_g + k_g}{m}\right) = e_g.$$

So we have proved that the mapping $\mathcal{D}\to\mathcal{E}$ is surjective. □

Now we explain the connection between lifts of Fuchsian groups into the finite coverings of $\tilde{G}$ and discrete subgroup of finite index in $\tilde{G}$.

We use the following description of the covering groups $\tilde{G}$ of $G = \mathrm{PSL}(2,\mathbb{R})$, which fixes a group structure. Let $\mathrm{Hol}(H,\mathbb{C})$ be the set of all holomorphic functions $H\to\mathbb{C}$.

Proposition 10. *The universal covering group $\tilde{G}$ of G can be described as*

$$\{(g,\delta)\in G\times\mathrm{Hol}(H,\mathbb{C}) \mid e\circ\delta = g'\},$$

where $e(w) = \exp(2\pi i w)$. The multiplication is given by

$$(g_2,\delta_2)\cdot(g_1,\delta_1) = (g_2\cdot g_1, \delta_2\circ g_1 + \delta_1).$$

The covering map $\tilde{G}\to G_m$ is given by

$$(g,\delta)\mapsto(g, e(\delta/m)).$$

Remark. The center of the group $\tilde{G}$ is infinite cyclic and is equal to the preimage of the unit element in G:

$$Z(\tilde{G}) = \{(g,\delta)\in\tilde{G} \mid g = \mathrm{Id},\quad \delta \text{ is an integer constant}\}.$$

Definition. The *level* of a discrete subgroup $\tilde{\Gamma}\subset\tilde{G}$ is the index of $\tilde{\Gamma}\cap Z(\tilde{G})$ as a subgroup of $Z(\tilde{G})$.

The following fact is well known (see for example Section 4 in [KR85]):

Proposition 11. *There is a one-to-one correspondence between discrete co-compact subgroups of level m in $\tilde{G}$ and liftings of discrete co-compact subgroups in* $\mathrm{PSL}(2,\mathbb{R})$ *into the m-fold covering of* $\mathrm{PSL}(2,\mathbb{R})$. *The correspondence is given by mapping a subgroup in $\tilde{G}$ into its image under the covering map $\tilde{G} \to G_m$.*

We now prove Theorem 1.

Proof. Let (X,x) be a hyperbolic $\mathbb{Q}$-Gorenstein quasi-homogeneous surface singularity. Let (H,Γ_1,L) be the corresponding $\mathbb{Q}$-Gorenstein automorphy factor of level m and index r. Let us consider a trivialisation $L \simeq H \times \mathbb{C}$ of the bundle L. Combining the results of Propositions 9 and 11 we see that there is a discrete co-compact subgroup $\tilde{\Gamma}_1$ of level m in $\tilde{G}$ such that the action of the group Γ_1 can be described as

$$g \cdot (z,t) = (g(z), e(\delta(z)r/m)\cdot t),$$

where $\delta : H \to \mathbb{C}$ is a holomorphic function such that (g,δ) is an element of $\tilde{\Gamma}_1$. This action of $\tilde{\Gamma}_1$ can be obtained as a restriction of the action of the group $\tilde{G}$ on L via

$$(g,\delta)\cdot(z,t) = (g(z), e(\delta(z)r/m)\cdot t).$$

It is easy to check, that this is an action of $\tilde{G}$. The unit subbundle of L can be identified with the subbundle

$$S = \{(z,t) \in H \times \mathbb{C} \mid |t|^m = (\mathrm{Im}(z))^r\}.$$

The bundle S is invariant under $\tilde{G}$: For

$$(z',t') = (g,\delta)\cdot(z,t) = (g(z), e(\delta(z)r/m)\cdot t)$$

we have

$$\frac{|t'|^m}{|t|^m} = \left|e\left(\delta(z)\cdot\frac{r}{m}\right)\right|^m = |e(\delta(z))|^r = |g'(z)|^r = \left(\frac{\mathrm{Im}\, g(z)}{\mathrm{Im}\, z}\right)^r = \frac{(\mathrm{Im}(z'))^r}{(\mathrm{Im}(z))^r}.$$

The stabiliser of a point $(z_0,t_0) \in S$ is

$$\tilde{\Gamma}_2 := \mathrm{Stab}_{\tilde{G}}((z_0,t_0)) = \left\{(g,\delta) \in \tilde{G} \mid g(z_0) = z_0, \quad \delta(z)\cdot\frac{r}{m} \in \mathbb{Z}\right\}.$$

We now determine the level of the subgroup $\tilde{\Gamma}_2$:

$$\begin{aligned}\tilde{\Gamma}_2 \cap Z(\tilde{G}) &= \{(g,\delta) \in Z(\tilde{G}) \mid \delta \text{ is an integer constant divisible by } m\} \\ &= m \cdot Z(\tilde{G}).\end{aligned}$$

The map $(g,\delta) \mapsto (g,\delta)\cdot(i,1)$ defines a $\tilde{\Gamma}_1$-equivariant diffeomorphism $\tilde{G}/\tilde{\Gamma}_2 \to S$. Here $\tilde{\Gamma}_1$ acts on $\tilde{G}$ by left multiplication. We obtain the following commutative diagram

$$\begin{array}{ccc}\tilde{G}/\tilde{\Gamma}_2 & \longrightarrow & L^*/\mathbb{R}_+ \cong S \\ \downarrow & & \downarrow \\ \tilde{\Gamma}_1\backslash\tilde{G}/\tilde{\Gamma}_2 & \longrightarrow & X^*/\mathbb{R}_+ \cong M.\end{array}$$

Hence we have

$$M \cong \tilde{\Gamma}_1 \backslash \tilde{G} / \tilde{\Gamma}_2.$$

Conversely, let $\tilde{\Gamma}_1$ and $\tilde{\Gamma}_2$ be discrete subgroups of level m in $\tilde{G}$, let $\tilde{\Gamma}_1$ be co-compact, and let the image of $\tilde{\Gamma}_2$ in $\mathrm{PSL}(2,\mathbb{R})$ be a cyclic subgroup of order r. Then $\Gamma_1 = \tilde{\Gamma}_1/(\tilde{\Gamma}_1 \cap Z(\tilde{G}))$ is a discrete co-compact subgroup of $\mathrm{PSL}(2,\mathbb{R})$. We can define an automorphy factor $(H, \Gamma_1, H \times \mathbb{C})$ by setting

$$g \cdot (z,t) = (g(z), e(\delta(z)r/m) \cdot t),$$

where $\delta : H \to \mathbb{C}$ is a holomorphic function such that (g,δ) is an element of $\tilde{\Gamma}_1$. ¿From the first part of the proof we know that the link of the corresponding quasi-homogeneous $\mathbb{Q}$-Gorenstein surface singularity is diffeomorphic to $\tilde{\Gamma}_1 \backslash \tilde{G} / \tilde{\Gamma}_2$. □

Acknowledgment

I am grateful to Egbert Brieskorn, Vicente Cortes, Claus Hertling, Shihoko Ishii, Walter Neumann, and Jacob Stix for helpful discussions related to this work. I would like to thank the referee for useful remarks.

References

[BPR03] Egbert Brieskorn, Anna Pratoussevitch, and Frank Rothenhäusler, *The Combinatorial Geometry of Singularities and Arnold's Series E, Z, Q*, Moscow Mathematical Journal **3** (2003), no. 2, 273–333, the special issue dedicated to Vladimir I. Arnold on the occasion of his 65th birthday.

[Dol75] Igor V. Dolgachev, *Automorphic forms and quasihomogeneous singularities*, Funct. Anal. Appl. **9** (1975), 149–151.

[Dol77] ———, *Automorphic forms and weighted homogeneous equations*, provisional version, unpublished typed manuscript, 1977.

[Dol83] ———, *On the Link Space of a Gorenstein Quasihomogeneous Surface Singularity*, Math. Ann. **265** (1983), 529–540.

[Ish87] Shihoko Ishii, *Isolated Q-Gorenstein singularities of dimension three*, Complex analytic singularities, Adv. Stud. Pure Math., vol. 8, North-Holland, Amsterdam, 1987, pp. 165–198.

[Ish00] Shihoko Ishii, *The quotients of log-canonical singularities by finite groups*, Singularities — Sapporo 1998, Adv. Stud. Pure Math., vol. 29, Kinokuniya, Tokyo, 2000, pp. 135–161.

[KR85] Ravi S. Kulkarni and Frank Raymond, *3-dimensional Lorentz space forms and Seifert fiber spaces*, J. Diff. Geometry **21** (1985), 231–268.

[LV80] Gérard Lion and Michèle Vergne, *The Weil representation, Maslov index and theta series*, Progress in Mathematics, vol. 6, Birkhäuser Boston, Mass., 1980.

[Mil75] John Milnor, *On the 3-dimensional Brieskorn manifolds $M(p,q,r)$*, Knots, groups and 3-manifolds (L. P. Neuwirth, ed.), Annals of Math. Studies, vol. 84, Princeton University Press, Princeton, 1975, pp. 175–225.

[Neu77] Walter D. Neumann, *Brieskorn complete intersections and automorphic forms*, Invent. Math. **42** (1977), 285–293.

[Pin77] Henry Pinkham, *Normal surface singularities with $\mathbb{C}^*$-action*, Math. Ann. **227** (1977), 183–193.

[Pra] Anna Pratoussevitch, *Fundamental Domains in Lorentzian Geometry*, to appear in Geom. Dedicata, `math.DG/0308279`.

[Rei87] Miles Reid, *Young person's guide to canonical singularities*, Algebraic geometry, Bowdoin, 1985 (Brunswick, Maine, 1985) (S. Bloch, ed.), Proceedings of Symposia in Pure Mathematics, vol. 46, Amer. Math. Soc., Providence, RI, 1987, pp. 345–414.

[Wat81] Kei-ichi Watanabe, *Some remarks concerning Demazure's construction of normal graded rings*, Nagoya Math. J. **83** (1981), 203–211.

Anna Pratoussevitch
Mathematisches Institut
Universität Bonn
Beringstraße 1
D-53115 Bonn, Germany
e-mail: `anna@math.uni-bonn.de`

Real and Complex Singularities
Trends in Mathematics, 327–342

Singularity Exchange at the Frontier of the Space

Dirk Siersma and Mihai Tibăr

Abstract. In deformations of polynomial functions one may encounter "singularity exchange at infinity" when singular points disappear from the space and produce "virtual" singularities which have an influence on the topology of the limit polynomial. We find several rules of this exchange phenomenon, in which the total quantity of singularity turns out to be not conserved in general.

Mathematics Subject Classification (2000). 32S30, 14B07, 58K60, 70G55, 70G40.

Keywords. Deformation of polynomials, singularities at infinity, non-conservation of number of vanishing cycles.

1. Introduction

The study of the singular fibration produced by a polynomial in the affine space was pioneered by Broughton [6] more than 20 years ago, and took a certain ampleness ever since (over 40 papers). More recently, several papers go beyond this and consider the more complex situation of *families of polynomial functions* [8, 17, 18, 11, 4, 15, 5, 7]. In this topic, one may pose many natural questions in analogy to the ones which have been treated in the local case, that of families of local holomorphic functions. Moreover, the first classification results in the global affine case and the first lists of families of polynomials [13, 3] showed new phenomena and gave rise to more open questions.

We study here families of polynomial functions by focussing on the transformation of singularities in the neighborhood of infinity, a phenomenon which we have already remarked in [15]. This is a natural and challenging topic inside mathematics since the atypical fibres of a polynomial turn out to be not only due to the "visible" singularities, but also to the "bad" asymptotic behavior at the infinite frontier of the space. We deal here with the evolution and interaction of singularities in deformations at the infinite frontier of the space, in what concerns the phenomenon of conservation or non-conservation of certain numbers attached to singularities (that we recall below).

Let $\{f_s\}_s$ be a holomorphic family of complex polynomial functions $f_s : \mathbb{C}^n \to \mathbb{C}$, for s in a small neighborhood of $0 \in \mathbb{C}$. For a fixed polynomial function f_s there is a well defined general fibre G_s, since the set of atypical values $\Lambda(f_s)$ is a finite set. When specializing to f_0, the number of atypical values may vary (decrease, increase or be constant) and the topology of the general fibre may change. We consider constant degree families within certain classes of polynomials (F-class $\subset$ B-class $\subset$ W-class, cf. Definition 3.1) which have the property that the *vanishing cycles of* f_s (i.e., the generators of the reduced homology of G_s) are concentrated in dimension $n-1$ and are localisable at finitely many points, in the affine space or in the part at infinity of the projective compactification of some fibre of f_s. In the affine space $\mathbb{C}^n$, such a point is a singular point of f_s. The sum of all affine Milnor numbers is the total Milnor number $\mu(s)$, which has an algebraic interpretation as the dimension of the quotient algebra $\mathbb{C}[x_1, \ldots, x_n]/(\frac{\partial f}{\partial x_1}, \ldots, \frac{\partial f}{\partial x_n})$. Singularities at infinity are equipped with so-called Milnor-Lê numbers (cf. [14]) and their sum is denoted by $\lambda(s)$. Then the Euler characteristic of the generic fiber G_s is $1 + (-1)^{n-1}(\mu(s) + \lambda(s))$.

A natural problem which arises is to understand the behavior, when $s \to 0$, of the μ and λ-singularities, which support the vanishing cycles of f_s. It is well known and easy to see that, for singularities which tend to a μ-singular point, the total number of local vanishing cycles is constant, in other words the local balance law is *conservative.* However some μ-singularities may tend to infinity and change into λ-singularities; this is the phenomenon we address here. First, in full generality, for any deformation, we get the:

- *global lower semi-continuity* of the highest Betti number:
  $b_{n-1}(G_0) \leq b_{n-1}(G_s)$ (Proposition 2.1).

Next, focussing on constant degree deformations inside the B-class, we prove several facts on the singularity exchange at infinity:

- the number of local vanishing cycles of μ and λ-singularities tending to a λ-singular point is *lower semi-continuous*, but it is not conserved in general (Theorem 4.2). The proof consists in counting vanishing cycles after surrounding the problem of the non-isolated singularities which might appear.
- in $(\mu + \lambda)$-constant deformations, the local balance law at any λ-singularity of f_0 is *conservative* and *atypical values cannot escape to infinity* (Corollary 6.1).
- in $(\mu + \lambda)$-constant deformations, the monodromy fibrations over any admissible loop (in particular, the monodromy fibrations at infinity) are isotopic in the family, whenever $n \neq 3$ (Theorem 6.5).
- in deformations with constant generic singularity type at infinity, λ-singularities of f_0 are *locally persistent* in f_s but cannot split such that more than one λ-singularity occurs in the same fibre (Theorem 5.2).

– in deformations inside the F-class, a λ-singularity cannot be deformed into only μ-singularities (Corollary 6.4).

The semi-continuity results (first two of the above list) are certainly related to the semi-continuity of the *spectrum*, a result proved by Némethi and Sabbah [11] for the class of "weakly tame" polynomials. Their class excludes by definition the λ-singularities, but on the other hand the spectrum (defined with Hodge theoretical ingredients) gives more refined information than the total Milnor number. It is also interesting to remark that the lower semi-continuity in all these results is opposite to the *upper semi-continuity* in case of deformations of holomorphic function germs.

We end by supplying with several examples which illustrate the above described aspects of the exchange phenomenon.

2. Deformations in general

It is well known that the $(n-1)$th Betti number of the Milnor fibre of a holomorphic function germ is upper semi-continuous, i.e., it does not decrease under specialization. In case of a polynomial $f_s : \mathbb{C}^n \to \mathbb{C}$, the role of the Milnor fibre is played by the general fibre G_s of f_s. This is a Stein manifold of dimension $n-1$ and therefore it has the homotopy type of a CW complex of dimension $\leq n-1$, which is also finite, since G_s is algebraic. Moreover, the $(n-1)$th homology group with integer coefficients is free. We prove the following general specialization result.

Proposition 2.1. *Let* $P : \mathbb{C}^n \times \mathbb{C}^k \to \mathbb{C}$ *be any holomorphic deformation of a polynomial* $f_0 := P(\cdot, 0) : \mathbb{C}^n \to \mathbb{C}$. *Then the general fibre* G_0 *of* f_0 *can be naturally embedded into the general fibre* G_s *of* f_s, *for* $s \neq 0$ *close enough to* 0. *The embedding* $G_0 \subset G_s$ *induces an inclusion* $H_{n-1}(G_0) \hookrightarrow H_{n-1}(G_s)$ *which is compatible with the intersection form.*

Proof. It is enough to consider a 1-parameter family of hypersurfaces $\{f_s^{-1}(t)\}_{s\in L} \subset \mathbb{C}^n$, for fixed t, where L denotes some parametrised complex curve through 0. We denote by X_t the total space over a small neighborhood L_ε of 0 in L. By choosing t generic enough, we may assume that $f_s^{-1}(t)$ is a generic fibre of f_s, for s in a small enough neighborhood of 0. Let $\sigma : X_t \to L_\varepsilon$ denote the projection. Now X_t is the total space of a family of non-singular hypersurfaces. Since $\sigma^{-1}(0)$ is an affine hypersurface, by taking a large enough radius R, we get $\partial \bar{B}_{R'} \pitchfork \sigma^{-1}(0)$, for all $R' \geq R$. Moreover, the sphere $\partial \bar{B}_R$ is transversal to all nearby fibres $\sigma^{-1}(s)$, for small enough s. It follows that the projection σ from the pair of spaces $(X_t \cap (B_R \times \mathbb{C}), X_t \cap (\partial \bar{B}_R \times \mathbb{C}))$ to L_ε is a proper submersion and hence, by Ehresmann's theorem, it is a trivial fibration. By the above transversality argument, we have $B_R \cap \sigma^{-1}(0) \stackrel{\text{diff}}{\simeq} B_R \cap \sigma^{-1}(s)$. This shows the first claim.

The affine hypersurfaces $\sigma^{-1}(s)$ are finite cell complexes of dimension $\leq n-1$. By the classical Andreotti-Frankel [2] argument for the distance function, the hypersurface $\sigma^{-1}(s)$ is obtained from $B_R \cap \sigma^{-1}(s)$ by adding cells of index at most $n-1$. This shows that $H_n(G_s, G_0) = 0$, so the second claim. The compatibility with the intersection form is standard. □

Under certain conditions we can also compare the "monodromy fibrations at infinity" in the family, see §6.2. Proposition 2.1 will actually be exploited through the *semi-continuity of the highest Betti number*, as a consequence of the inclusion of homology groups:

$$b_{n-1}(G_s) \geq b_{n-1}(G_0), \text{ for } s \text{ close enough to } 0. \tag{2.1}$$

3. Compactification of families of polynomials

We shall now focus on polynomials for which the singularities at infinity are isolated, in a sense that we make precise here.

Let P be a deformation of f_0, i.e., $P : \mathbb{C}^n \times \mathbb{C}^k \to \mathbb{C}$ is a family of polynomial functions $P(x, s) = f_s(x)$ such that $f_0 = f$. We assume in the following that our deformation depends holomorphically on the parameter $s \in \mathbb{C}^k$. We also assume that $\deg f_s$ is independent on s, for s in some neighborhood of 0, and we denote it by d. We attach to P the following hypersurface:

$$\mathbb{Y} = \{([x : x_0], s, t) \in \mathbb{P}^n \times \mathbb{C}^k \times \mathbb{C} \mid \tilde{P}(x, x_0, s) - t x_0^d = 0\},$$

where $\tilde{P}$ denotes the homogenized of P by the variable x_0, considering s as parameter varying in a small neighborhood of $0 \in \mathbb{C}^k$. Let $\tau : \mathbb{Y} \to \mathbb{C}$ be the projection to the t-coordinate. This extends the map P to a proper one in the sense that $\mathbb{C}^n \times \mathbb{C}^k$ is embedded in $\mathbb{Y}$ (via the graph of P) and $\tau_{|\mathbb{C}^n \times \mathbb{C}^k} = P$. Let $\sigma : \mathbb{Y} \to \mathbb{C}^k$ denote the projection to the s-coordinates.

Notations. $\mathbb{Y}_{s,*} := \mathbb{Y} \cap \sigma^{-1}(s)$, $\mathbb{Y}_{*,t} := \mathbb{Y} \cap \tau^{-1}(t)$ and $\mathbb{Y}_{s,t} := \mathbb{Y}_{s,*} \cap \tau^{-1}(t) = \mathbb{Y}_{*,t} \cap \sigma^{-1}(s)$. Note that $\mathbb{Y}_{s,t}$ is the closure in $\mathbb{P}^n$ of the affine hypersurface $f_s^{-1}(t) \subset \mathbb{C}^n$.

Let $\mathbb{Y}^\infty := \mathbb{Y} \cap \{x_0 = 0\} = \{P_d(x, s) = 0\} \times \mathbb{C}$ be the hyperplane at infinity of $\mathbb{Y}$, where P_d is the degree d homogeneous part of P in variables $x \in \mathbb{C}^n$. Remark that for any fixed s, $\mathbb{Y}^\infty_{s,t} := \mathbb{Y}_{s,t} \cap \mathbb{Y}^\infty$ does not depend on t.

Definition 3.1. We consider the following classes of polynomials:

(i) f is a *F-type* polynomial if its compactified fibres and their restrictions to the hyperplane at infinity have at most isolated singularities.

(ii) f is a *B-type* polynomial if its compactified fibres have at most isolated singularities.

It follows that F-class $\subset$ B-class. They are both contained into the W-class, which consists polynomials for which the proper extension $\tau : \mathbb{X} \to \mathbb{C}$ has only isolated singularities with respect to some Whitney stratification of $\mathbb{X}$ such that $\mathbb{X}^\infty$ is a union of strata, see [14]. The notation $\mathbb{X}$ stands for $\mathbb{Y}$ when a single polynomial is considered (i.e., there is no parameter s).

In two variables, if f has isolated singularities in $\mathbb{C}^2$, then it is automatically of F-type. Deformations inside the F-class were introduced in [15] under the name $\mathcal{F}$ISI deformations. Broughton [6] considered for the first time B-type polynomials and studied the topology of their general fibers. The W-class of polynomials appears in [14]. In deformations of a polynomial f_0 we usually require to stay inside the same class but we may also deform into a "less singular" class (like B-type into F-type, Example 8.6).

The singular locus of $\mathbb{Y}$, $\operatorname{Sing}\mathbb{Y} := \{x_0 = 0, \frac{\partial P_d}{\partial x}(x,s) = 0, P_{d-1}(x,s) = 0, \frac{\partial P_d}{\partial s}(x,s) = 0\} \times \mathbb{C}$ is included in $\mathbb{Y}^\infty$ and is a product-space by the t-coordinate. It depends only on the degrees d and $d-1$ parts of P with respect to the variables x.

Let $\Sigma := \{x_0 = 0, \frac{\partial P_d}{\partial x}(x,s) = 0, P_{d-1}(x,s) = 0\} \subset \mathbb{P}^{n-1} \times \mathbb{C}^k$. If we fix s, the singular locus of $\mathbb{Y}_{s,*}$ is the analytic set $\Sigma_s \times \mathbb{C}$, where $\Sigma_s := \Sigma \cap \{\sigma = s\}$, and it is the union of the singularities at the hyperplane at infinity of the hypersurfaces $\mathbb{Y}_{s,t}$, for $t \in \mathbb{C}$.

We denote by $W_s := \{[x] \in \mathbb{P}^{n-1} \mid \frac{\partial P_d}{\partial x}(x,s) = 0\}$ the set of points at infinity where $\mathbb{Y}^\infty_{s,t}$ is singular, in other words where $\mathbb{Y}_{s,t}$ is either singular or tangent to $\{x_0 = 0\}$. It does not depend on t and we have $\Sigma_s \subset W_s$.

REMARK 3.2. From the above definition and the expressions of the singular loci we have the following characterisation:

(i) f_0 is a B-type polynomial $\Leftrightarrow \dim \operatorname{Sing} f_0 \leq 0$ and $\dim \Sigma_0 \leq 0$,
(ii) f_0 is a F-type polynomial $\Leftrightarrow \dim \operatorname{Sing} f_0 \leq 0$ and $\dim W_0 \leq 0$.

Let us also remark that $\dim \Sigma_0 \leq 0$ (respectively $\dim W_0 \leq 0$) implies that $\dim \Sigma_s \leq 0$ (respectively $\dim W_s \leq 0$), whereas $\dim \operatorname{Sing} f_0 \leq 0$ does not imply automatically $\dim \operatorname{Sing} f_s \leq 0$ for $s \neq 0$.

4. Semi-continuity at infinity

Let P be a deformation of f_0 such that f_s is of W-type, for all s close enough to 0. It is shown in [12, 14] that the vanishing cycles of f_s (for fixed s) are concentrated in dimension $n-1$ and are localized at well-defined points, either in the affine space or at infinity. We shall call them *μ-singularities* and *λ-singularities* respectively. To such a singular point $p \in \mathbb{Y}_{s,*}$ one associates its local Milnor number denoted $\mu_p(s)$ or its Milnor-Lê number $\lambda_p(s)$. Let $\mu(s)$ be the *total Milnor number*, respectively $\lambda(s)$ be the *total Milnor-Lê number at infinity*, where $b_{n-1}(G_s) = \mu(s) + \lambda(s)$.

By [14], the atypical fibers of a W-type polynomial f_s are exactly those fibers which contain μ or λ-singularities; equivalently, those of which the Euler characteristic is different from $\chi(G_s)$. We denote by $\Lambda(f_s)$ the set of atypical values of f_s.

The above-cited facts together with our semi-continuity result (2.1) show that, for s close to 0 we have:

$$\mu(s) + \lambda(s) \geq \mu(0) + \lambda(0).$$

REMARK 4.1. The total Milnor number $\mu(s)$ is lower semi-continuous under specialization $s \to 0$. In case $\mu(s)$ decreases, we say that there is *loss of μ at infinity*, since this may only happen when one of the two following phenomena occur:

(a) the modulus of some critical point tends to infinity and the corresponding critical value is bounded ([15, Example 8.1]);
(b) the modulus of some critical value tends to infinity ([15, Examples (8.2) and (8.3)]).

In contrast to $\mu(s)$, it turns out that $\lambda(s)$ is not semi-continuous; under specialization, it can increase or decrease (Example 8.1, 8.3). Moreover, the λ-values may behave like the critical values in case (b) above, see Example 8.6.

To understand the behavior of $\lambda(s)$ in more detail, we focus on the B-class. The following result extends our [15, Theorem 5.4] and needs a more involved proof, which will be given in §7.

Theorem 4.2. (Lower semi-continuity at λ-singularities)
Let P be a constant degree one-parameter deformation inside the B-class. Then, locally at any λ-singularity $p \in \mathbb{Y}_{0,t}$ of f_0, we have:

$$\lambda_p(0) \le \sum_i \lambda_{p_i}(s) + \sum_j \mu_{p_j}(s),$$

where p_i are the λ-singularities and p_j are the μ-singularities of f_s which tend to the point p as $s \to 0$.

5. Persistence of λ-singularities

In order to get further information on the $\mu \mapsto \lambda$ exchanges we focus on two subclasses of the B-class. In this section we define cgst-type deformations and in the next section we study deformation with constant $\mu + \lambda$.

Let us first remark that for a deformation $\{f_s\}_s$ inside the B-class the compactified fibres of f_s have only isolated singularities. The positions of these singularities depend only on s (and not on t). When $s \to 0$ these singularities can split or disappear.

Let us take some $x(0) \in \Sigma_0$. Take $t \notin \Lambda(f_0)$ and assume without dropping generality that $t \notin \Lambda(f_s)$ for all small enough s. Lazzeri's non-splitting argument, see [9] and also [1, 10], tells us that the Milnor number of $\mathbb{Y}_{0,t}$ at $(x(0), t)$ is strictly larger than the sum of the Milnor numbers of $\mathbb{Y}_{s,t}$ at all points $(x_i(s), t) \in \Sigma_s \times \{t\}$ such that $x_i(s) \to x(0)$, if these points are more than one. In other words, if we have $x(s) \to x(0)$ and $t \notin \Lambda(f_0)$ such that the Milnor number of $\mathbb{Y}_{s,t}$ at $(x(s), t)$ is constant, then there can be only one point $x(s) \in \Sigma_s$ which tends to $x(0) \in \Sigma_0$ as $s \to 0$.

Definition 5.1. We say that a constant degree deformation inside the B-class has *constant generic singularity type at infinity at some point* $x(0) \in \Sigma_0$ if we have the constancy of the Milnor number of $\mathbb{Y}_{s,t}$ at $(x(s), t)$ for s varying in some small neighbourhood of 0, where $x(s) \to x(0)$ and $t \notin \Lambda(f_0)$ is fixed. We also say in this case that the *cgst assumption* holds at $x(0)$.

If the cgst assumption holds at all points in Σ_0, then we say that the germ at $s = 0$ of the deformation P has (or is) *cgst at infinity*.

We shall see that the cgst assumption does not imply that $b_{n-1}(G_s)$ is constant (see Example 8.6). We send to Remark 6.2 for further comments on cgst.

We want to see what happens with a λ-singularity of f_0 in a deformation f_s. We say that a λ-singularity is *persistent* if it splits into one or more singularities, out of which at least a λ-singularity. For such a splitting we refer to §8. We have the following result:

Theorem 5.2. *Let P be a constant degree deformation, inside the B-class, with constant generic singularity type at infinity. Then:*

(a) *the λ-singularities of f_0 are persistent in f_s.*

(b) *a λ-singularity of f_0 cannot split such that two or more λ-singularities belong to the same fiber.*

REMARK 5.3. The case which is not covered by part (b) can indeed occur, i.e., some λ-singularity may split into λ-singularities in several fibres, see Example 8.2.

Proof. **(a)** Let $(z, t_0) \in \Sigma_0 \times \mathbb{C}$ be a λ-singularity of f_0. Let us denote by $G(y, s, t)$ the localization of the map $\tilde{P}(x, x_0, s) - tx_0^d$ at the point $(z, 0, t_0) \in \mathbb{Y}$. Let $y_0 = 0$ be the local equation of the hyperplane at infinity of $\mathbb{P}^n$. The idea is to consider the 2-parameter family of functions $G_{s,t} \colon \mathbb{C}^n \to \mathbb{C}$, where $G_{s,t}(y) = G(y, s, t)$. Then $G(y, s, t)$ is the germ of a deformation of the function $G_{0,t_0}(y)$.

We consider the germ at $(z, 0, t_0)$ of the singular locus Γ of the map $(G, \sigma, \tau) : \mathbb{C}^n \to \mathbb{C}^3$. This is the union of the singular loci of the functions $G_{s,t}$, for varying s and t. We claim that Γ is a surface, more precisely, that every irreducible component Γ_i of Γ is a surface. We secondly claim that the projection $D \subset \mathbb{C}^3$ of Γ by the map (y_0, σ, τ) is a surface, in the sense that all its irreducible components are surfaces. Moreover, the projections $\Gamma \stackrel{(y_0,\sigma,\tau)}{\longrightarrow} D$ and $D \stackrel{(s,t)}{\longrightarrow} \mathbb{C}^2$ are finite (ramified) coverings.

All our claims follow from the following fact: the local Milnor number conserves in deformations of functions. The function germ G_{0,t_0} with Milnor number, say μ_0, deforms into a function $G_{s,t}$ with finitely many isolated singularities, and the total Milnor number is conserved, for any couple (s, t) close to $(0, t_0)$.

Let us now remark that the germ at $(z, 0, t_0)$ of $\Sigma \times \mathbb{C}$ is a union of components of Γ and projects by (y_0, s, t) to the plane $D_0 := \{y_0 = 0\}$ of $\mathbb{C}^3$. However, the inclusion $D_0 \subset D$ cannot be an equality, by the above argument on the total "quantity of singularities" and since we have a jump $\lambda > 0$ at the point of origin $(z, 0, t_0)$. So there must exist some other components of D. Every such component being a surface in $\mathbb{C}^3$, has to intersect the plane $D_0 \subset \mathbb{C}^3$ along a curve. Therefore, for every point (s', t') of such a curve, the sum of Milnor numbers of the function G on the hypersurface $\{y_0 = 0, \sigma = s', \tau = t'\}$ (where the sum is taken over the singular points that tend to the original point $(z, 0, t_0)$ when $s' \to 0$) is therefore strictly higher than the one computed for a generic point of the plane D_0. Therefore our claim (a) will be proved if we prove two things:

(i) the singularities of G on the hypersurface $\{y_0 = 0, \sigma = s', \tau = t'\}$ that tend to the original point $(z, 0, t_0)$ when $s' \to 0$ are included into $G = 0$, and

(ii) there exists a component $D_i \subset D$ such that $D_i \cap D_0 \neq D \cap \{s = 0\}$.

To show (i), let $g_k(y, s)$ denote the degree k part of P after localizing it at p and note that $G(y, s, t) = g_d(y, s) + y_0(g_{d-1}(y, s) + \cdots) - ty_0^d$. Then observe that the set:

$$\Gamma \cap \{y_0 = 0\} = \{\frac{\partial g_d}{\partial y} = 0, g_{d-1} = 0\} \tag{5.1}$$

does not depend on the variable t and its slice by $\{\sigma = s, \tau = t\}$ consists of finitely many points. These points may fall into two types: (I). points on $\{g_d = 0\}$, and therefore on $\{G = 0\}$, and (II). points not on $\{g_d = 0\}$. We show that type II points do not actually occur. This is a consequence of our hypothesis on the constancy of generic singularity type at infinity, as follows. By choosing a generic $\hat{t}$ such that $\hat{t} \notin \Lambda(s)$ for all s, and by using the independence on t of the set (5.1), this condition implies that type II points cannot collide with type I points along the slice $\{y_0 = 0, \sigma = s, \tau = \hat{t}\}$ as $s \to 0$. By absurd, if there were collision, then there would exist a singularity in the slice $\{G = 0, y_0 = 0, \sigma = 0, \tau = \hat{t}\}$ with Milnor number higher than the generic singularity type at infinity. It then follows that:

$$\Gamma \cap \{y_0 = 0\} = \Gamma \cap \{G = y_0 = 0\} \tag{5.2}$$

which proves (i). Now observe that the equality (5.2) also proves (ii), by a similar reason: if there were a component D_i such that $D_i \cap D_0 = D \cap \{s = 0\}$ then there would exist a singularity in the slice $\{G = 0, y_0 = 0, \sigma = 0, \tau = \hat{t}\}$ with Milnor number higher than the generic singularity type at infinity. Notice that we have in fact proved more, namely:

(ii′) there is no component $D_i \neq D_0$ such that $D_i \cap D_0 = D \cap \{s = 0\}$.

This ends the proof of (a).

(b) Suppose that there were collision of some singularities out of which two or more λ-singularities are in the same fibre. Then there are at least two different points $z_i \neq z_j$ of Σ_s which collide as $s \to 0$. This situation is excluded by the cgst assumption (Definition 5.1). □

6. Specificity of $\mu + \lambda$ constant deformations

6.1. Local conservation and behavior of λ

In §8 we comment a couple of examples where the of Theorem 4.2 is strict. Let us first show that, when imposing the constancy of $\mu + \lambda$, this inequality turns into an equality.

Corollary 6.1. *Let P be a constant degree deformation inside the B-class such that $\mu(s) + \lambda(s)$ is constant. Then:*

(a) *As $s \to 0$, there cannot be loss of μ or of λ with corresponding atypical values tending to infinity.*

(b) λ *is upper semi-continuous, i.e.,* $\lambda(s) \leq \lambda(0)$.
(c) *there is local conservation of* $\mu + \lambda$ *at any* λ*-singularity of* f_0.

Proof. (a) If there is loss of μ or of λ, then this must necessarily be compensated by increase of λ at some singularity at infinity of f_0. But Theorem 4.2 shows that the local $\mu + \lambda$ cannot increase in the limit.
(b) is clear since $\mu(s) + \lambda(s)$ is constant and $\mu(s)$ can only decrease when $s \to 0$.
(c) Global conservation of $\mu + \lambda$ together with local semi-continuity (by Theorem 4.2) imply local conservation. □

REMARK 6.2. It is interesting to point out that within the class of B-type polynomials there is no inclusion relation between the properties "constant generic singularity type" and "$\mu(s) + \lambda(s)$ constant", see Examples 8.4, 8.6. We shall see in the following that in the F-class the two conditions are equivalent because of the relation (6.2).

For B-type polynomials, we have the formula:

$$b_{n-1}(G_s) = \mu(s) + \lambda(s) = (-1)^{n-1}(\chi^{n,d} - 1) - \sum_{x \in \Sigma_s} \mu_{x,\text{gen}}(s) - (-1)^{n-1}\chi^{\infty}(s), \tag{6.1}$$

where $\chi^{n,d} = \chi(V_{\text{gen}}^{n,d}) = n+1-\frac{1}{d}\{1+(-1)^n(d-1)^{n+1}\}$ is the Euler characteristic of the smooth hypersurface $V_{\text{gen}}^{n,d}$ of degree d in $\mathbb{P}^n$ and $\chi^{\infty}(s) := \chi(\{f_d(x,s) = 0\})$. We denote by $\mu_{x,\text{gen}}(s)$ the Milnor number of the singularity of $\mathbb{Y}_{s,t}$ at the point $(x,t) \in \Sigma_s \times \mathbb{C}$, for a generic value of t. The change in $b_{n-1}(G_s)$ can be described in terms of change in $\mu_{x,\text{gen}}(s)$ and $\chi^{\infty}(s)$. Since the latter is not necessarily semi-continuous (cf. Examples 8.4–8.6), we may expect interesting exchange of data between the two types of contributions.

Proposition 6.3. *Let* $\Delta\chi^{\infty}$ *denote* $(-1)^n(\chi^{\infty}(s) - \chi^{\infty}(0))$.
(a) *If* $\Delta\chi^{\infty} < 0$ *then the deformation is not cgst.*
(b) *If* $\Delta\chi^{\infty} = 0$ *and the deformation has constant* $\mu + \lambda$ *then, for all* $x \in \Sigma_s$, $\mu_{x,\text{gen}}(s)$ *is constant.*
(c) *If* $\Delta\chi^{\infty} > 0$ *then the deformation cannot have constant* $\mu + \lambda$. □

For F-type polynomials, formula (6.1) takes the following form, see also [15, (2.1) and (2.4)]:

$$\mu(s) + \lambda(s) = (d-1)^n - \sum_{x \in \Sigma_s} \mu_{x,\text{gen}}(s) - \sum_{x \in W_s} \mu_x^{\infty}(s), \tag{6.2}$$

where $\mu_x^{\infty}(s)$ denotes the Milnor number of the singularity of $\mathbb{Y}_{s,t} \cap H^{\infty}$ at the point $(x,t) \in W_s \times \mathbb{C}$, which is actually independent on the value of t. Note that in the F-class we have $\Delta\chi^{\infty} \geq 0$.

The relation 6.2 shows that the change in the Betti number $b_{n-1}(G_s)$ can be described in terms of change in the $\mu_{x,\text{gen}}(s)$ and change in $\mu_x^{\infty}(s)$. Both are semi-continuous, so they are forced to be constant in $\mu + \lambda$ constant families.

Consequently, the class of F-type polynomials such that $\mu+\lambda$ =const. verifies the hypotheses of Theorem 5.2. It has been noticed by the first named author that in the deformations with constant $\mu+\lambda$ which occur in Siersma-Smeltink's lists [13] the value of λ cannot be dropped to 0. Since these deformations are in the F-class and in view of the above observation, this behavior is now completely explained by Theorem 5.2(a). More precisely, we have proved:

Corollary 6.4. *Inside the F-class, a λ-singularity cannot be deformed into only μ-singularities by a constant degree deformation with $\mu + \lambda =$ const.* □

6.2. Monodromy in families with constant $\mu + \lambda$

For some polynomial f_0, one calls *monodromy at infinity* the monodromy around a large enough disc D containing all the atypical values of f_0. The locally trivial fibration above the boundary $\partial \bar{D}$ of the disc is called *monodromy fibration at infinity.*

The global Lê-Ramanujam problem consists in showing the constancy of the monodromy fibration at infinity in a family with constant $\mu + \lambda$. Actually one can state the same problem for any *admissible loop* γ in $\mathbb{C}$, i.e., a simple loop (homeomorphic to a circle) such that it does not contain any atypical value of f_s, for all s close enough to 0.

The second named author proved a Lê-Ramanujam type result for a large class of polynomials, including the B-class (cf. [17, 18]), with the supplementary condition that there is no loss of μ at infinity of type 4.1(b). This hypothesis can now be removed, due to our Corollary 6.1(a). Moreover, the same result clearly holds over any admissible loop. Therefore, by revisiting the statement [18, Theorem 5.2], we get the following more general one:

Theorem 6.5. *Let P be a constant degree deformation inside the B-class. If $\mu + \lambda$ is constant and $n \neq 3$ then:*

(a) *the monodromy fibrations over any admissible loop are isotopic in the family.*
(b) *the monodromy fibrations at infinity are isotopic in the family.* □

7. Proof of Theorem 4.2

For the proof, we need to define a certain critical locus. First endow $\mathbb{Y}$ with the coarsest Whitney stratification $\mathcal{W}$. Note that (unlike the case of a single polynomial and its attached space $\mathbb{X}$ treated in [14]) *we do not require here* that $\mathbb{Y}^\infty$ is a union of strata. Let $\Psi := (\sigma, \tau) : \mathbb{Y} \to \mathbb{C} \times \mathbb{C}$ be the projection. The *critical locus* Crit Ψ is the locus of points where the restriction of Ψ to some stratum of $\mathcal{W}$ is not a submersion. When writing Crit Ψ we usually understand a small representative of the germ of Crit Ψ at $\mathbb{Y}_{0,*}$. It follows that Crit Ψ is a closed analytic set and that its affine part Crit $\Psi \cap (\mathbb{C}^n \times \mathbb{C} \times \mathbb{C})$ is the union, over $s \in \mathbb{C}$, of the affine critical loci of the polynomials f_s. Notice that both Crit Ψ and its affine part Crit $\Psi \cap (\mathbb{C}^n \times \mathbb{C} \times \mathbb{C})$ are in general not product spaces by the t-variable. In case of a constant degree one-parameter deformation in the B-class, the stratification

$\mathcal{W}$ has a maximal stratum which contains the complement of the 2-surface $\Sigma \times \mathbb{C}$. At any point of this complement, all the spaces $\mathbb{Y}$, $\mathbb{Y}_{s,*}$ and $\mathbb{Y}_{s,t}$ are nonsingular in the neighborhood of infinity. Therefore $\mathrm{Crit}\,\Psi \cap (\mathbb{Y}^\infty \setminus \Sigma \times \mathbb{C}) = \emptyset$. Since our deformation is in the B-class, it follows that the affine part $\mathrm{Crit}\,\Psi \cap (\mathbb{C}^n \times \mathbb{C} \times \mathbb{C})$ is of dimension at most 1. Next, the map is Ψ is submersive over a Zariski-open subset of any 2-dimensional stratum included in $\Sigma \times \mathbb{C}$. It follows that the part at infinity of $\mathrm{Crit}\,\Psi$ has dimension < 2. We altogether conclude that $\dim \mathrm{Crit}\,\Psi \leq 1$.

Nevertheless, this fact does not insure that the functions σ and τ have isolated singularity with respect to our stratification $\mathcal{W}$. (It is precisely not the case in "almost all" examples.) Nevertheless, in the pencil $\sigma + \varepsilon\tau$, $\varepsilon \in \mathbb{C}$, all the functions except finitely many of them are functions with isolated singularity at p with respect to the stratification $\mathcal{W}$. Let us fix some ε close to zero and consider locally, in some good neighborhood $\mathcal{B}$ of $(p, 0) \in \mathbb{Y}$, the couple of functions $\Psi_\varepsilon = (\sigma + \varepsilon\tau, \tau)\colon \mathcal{B} \to \mathbb{C}^2$.

The function $\tau_|\colon (\sigma + \varepsilon\tau)^{-1}(0) \to \mathbb{C}$ defines an isolated singularity at p and $(\sigma + \varepsilon\tau)^{-1}(0)$ is a germ of a complete intersection at p. By applying the stratified Bouquet Theorem of [16] we get that the Milnor-Lê fibre of $\tau_|$ is homotopy equivalent to a bouquet of spheres $\bigvee S^{n-1}$. It follows that the general fiber of Ψ_ε – that is $\mathcal{B} \cap \Psi_\varepsilon^{-1}(s, t)$, for some $(s, t) \notin \mathrm{Disc}\,\Psi$ – is homotopy equivalent to the same bouquet $\bigvee S^{n-1}$; let ρ denote the number of S^{n-1} spheres in this bouquet.

On the other hand, the Milnor fiber at p of the function $\sigma + \varepsilon\tau$ is homotopy equivalent to a bouquet $\bigvee S^n$, by the same result *loc. cit.*; let ν denote the number of S^n spheres.

In the remainder, we count the vanishing cycles (I): along $(\sigma + \varepsilon\tau)^{-1}(0)$, respectively (II): along $(\sigma + \varepsilon\tau)^{-1}(u)$, for $u \neq 0$ close enough to 0, and we compare the results. The vanishing cycles are all in dimension $n - 2$. One may use Figure 1 in order to follow the computations; in this picture, the germ of the discriminant locus $\mathrm{Disc}\,\Psi$ at $\Psi(p)$ is the union of the τ-axis, σ-axis and some other curves.

(I) We start with the fiber $\mathcal{B} \cap \Psi_\varepsilon^{-1}(0, \delta)$, where δ is close enough to 0. To obtain $\mathcal{B} \cap (\sigma + \varepsilon\tau)^{-1}(0)$, which is contractible, one attaches to $\mathcal{B} \cap \Psi_\varepsilon^{-1}(0, \delta)$ a certain number of $(n-1)$ cells corresponding to the vanishing cycles at infinity, as $t \to 0$, in the family of fibers $\Psi_\varepsilon^{-1}(0, t)$. This is exactly the number ρ defined above and it is here the sum of two numbers, corresponding to the attaching in two steps, as we detail in the following. One is the number of cycles in $\mathcal{B} \cap \Psi_\varepsilon^{-1}(s, \delta)$, vanishing, as $s \to 0$, at points that tend to p when δ tends to 0; we denote this number by ξ. The other number is the number of cycles in $\mathcal{B} \cap \Psi^{-1}(0, t)$, vanishing as $t \to 0$; this number is $\lambda_p(0)$, by definition. From this one may draw the inequality: $\lambda_p(0) \leq \rho$.

(II) Here we start with the fiber $\mathcal{B} \cap \Psi_\varepsilon^{-1}(u, \delta)$, which is homeomorphic to $\mathcal{B} \cap \Psi_\varepsilon^{-1}(0, \delta)$ and to $\mathcal{B} \cap \Psi^{-1}(u, \delta)$. The Milnor fiber $\mathcal{B} \cap \{\sigma + \varepsilon\tau = u\}$ cuts the critical locus $\mathrm{Crit}\Psi$ at certain points p_k. The number of points, counted with multiplicities, is equal to the local intersection number $\mathrm{int}_p(\{\sigma + \varepsilon\tau = 0\}, \mathrm{Crit}\Psi)$. When walking along $\mathcal{B} \cap \{\sigma + \varepsilon\tau = u\}$, one has to add to the fiber $\mathcal{B} \cap \Psi_\varepsilon^{-1}(u, \delta)$ a number of cells corresponding to the vanishing cycles at points $\{\sigma + \varepsilon\tau = u\} \cap \{\sigma = 0\}$, which is

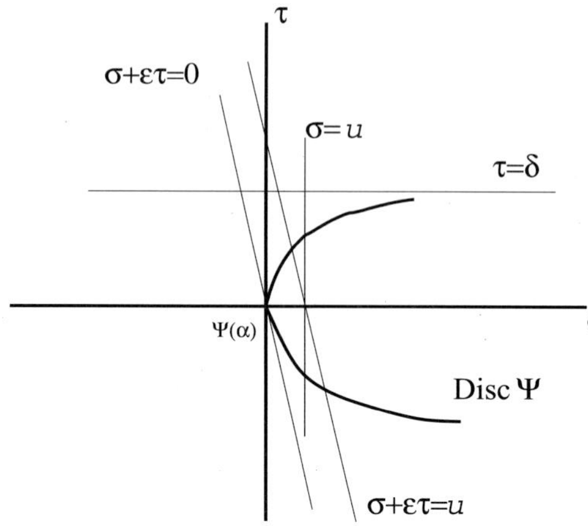

FIGURE 1. *Counting vanishing cycles.*

just the number ξ defined above, and to the vanishing cycles at points $\{\sigma + \varepsilon\tau = u\} \cap \overline{\mathrm{Crit}\Psi \setminus \{\sigma = 0\}}$. The intersection number $\mathrm{int}_p(\{\sigma + \varepsilon\tau = 0\}, \overline{\mathrm{Crit}\Psi \setminus \{\sigma = 0\}})$ is less or equal to the intersection number $\mathrm{int}_p(\{\sigma = 0\}, \overline{\mathrm{Crit}\Psi \setminus \{\sigma = 0\}})$. Now, when walking along $\mathcal{B} \cap \{\sigma = u\}$, one has to add to $\mathcal{B} \cap \Psi^{-1}(u, \delta)$ a number of cells corresponding to the vanishing cycles at points p_i and p_j, which number is, by definition, $\sum_i \lambda_{p_i}(u) + \sum_j \mu_{p_j}(u)$. We get the inequality: $\xi + \sum_i \lambda_{p_i}(u) + \sum_j \mu_{p_j}(u) \geq \rho + \nu$.

Finally, by collecting the inequalities obtained at steps (I) and (II), we obtain:

$$\lambda_p(0) = \rho - \xi \leq \rho + \nu - \xi \leq \sum_i \lambda_{p_i}(u) + \sum_j \mu_{p_j}(u), \tag{7.1}$$

which proves our claim. □

8. Examples

8.1. F-class examples; behavior of λ

EXAMPLE 8.1. $f_s = (xy)^3 + sxy + x$, see Figure 2(a).
This is a deformation inside the F-class, with constant $\mu + \lambda$, where λ increases. For $s \neq 0$: $\lambda = 1 + 1$ and $\mu = 1$. For $s = 0$: $\lambda = 3$ and $\mu = 0$.

EXAMPLE 8.2. $f_s = (xy)^4 + s(xy)^2 + x$, see Figure 2(b).
This deformation has constant $\mu = 0$, $\lambda(0) = 2$ at one point and $\lambda(s) = 1 + 1$ at two points at infinity which differ by the value of t only, namely $([0 : 1], s, 0)$ and $([0 : 1], s, -s^2/4)$.

EXAMPLE 8.3. $f_s = xy^4 + s(xy)^2 + y$, see Figure 2(c).
Here λ decreases. For $s \neq 0$: $\lambda = 2$ and $\mu = 5$. For $s = 0$: $\lambda = 1$ and $\mu = 0$.

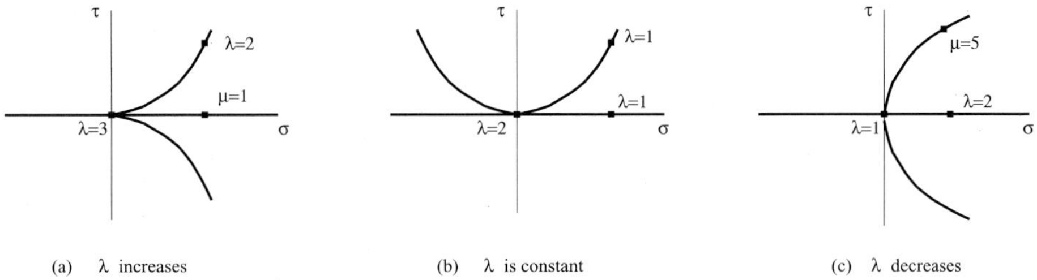

FIGURE 2. *Mixed splitting in* (a) *and* (c); *pure* λ*-splitting in* (b).

8.2. B-class examples

We use in this section formula (6.1). We pay special attention to the sign of $\Delta\chi^\infty$ and illustrate the difference between cgst-type deformations and $(\mu+\lambda)$-constant deformations.

EXAMPLE 8.4. $f_s = x^4 + sz^4 + z^3 + y$.
This is a deformation inside the B-class with constant $\mu+\lambda$, which is not cgst at infinity (Definition 5.1). We have $\lambda = \mu = 0$ for all s. Next, $\mathbb{Y}_{s,t}$ is singular only at $p := [0:1:0]$ and the singularities of $\mathbb{Y}^\infty_{0,t}$ change from a single smooth line $\{x^4 = 0\}$ with a special point p on it into the isolated point p which is a $\tilde{E}_7$ singularity of $\mathbb{Y}^\infty_{s,t}$. We use the notation $\oplus$ for the Thom-Sebastiani sum of two types of singularities in separate variables.

We have:

$s = 0$: the generic type is $A_3 \oplus E_7$ with $\mu = 21$ and $\chi(\mathbb{Y}^\infty_{0,t}) = 2$.
$s \neq 0$: the generic type is $A_3 \oplus E_6$ with $\mu = 18$ and $\chi(\mathbb{Y}^\infty_{s,t}) = 5$.

The jumps of $+3$ and -3 compensate each other.

EXAMPLE 8.5. $f_s = x^4 + sz^4 + z^2y + z$.
This is a $\mu+\lambda$ constant B-type family, with two different singular points of $\mathbb{Y}_{0,t}$ at infinity, and where the change in one point interacts with the other. It is locally cgst in one point, but not in the other. We have that $\lambda = 3$ and $\mu = 0$ for all s, $\mathbb{Y}_{s,t}$ is singular at $p := [0:1:0] \in H^\infty$ for all s (see types below) and at $q := [1:0:0] \in H^\infty$ with type A_3. The singularities of $\mathbb{Y}^\infty_{s,t}$ change from a single smooth line $\{x^4 = 0\}$ into the isolated point p with $\tilde{E}_7$ singularity.

For the point p we have for all s the generic type $A_3 \oplus D_5$ if $t \neq 0$, which jumps to $A_3 \oplus D_6$ if $t = 0$. This causes $\lambda = 3$.

At q, the A_3-singularity for $s = 0$ gets smoothed (independently of t) and here the deformation is not locally cgst. The change on the level of $\chi(\mathbb{Y}^\infty_{s,t})$ is from 2 to 5, so $\Delta\chi^\infty = -3$, which compensates the disappearance of the A_3-singularity from $\mathbb{Y}_{0,t}$ to $\mathbb{Y}_{s,t}$.

EXAMPLE 8.6. $f_s = x^2y + x + z^2 + sz^3$.
This is a cgst B-type family, where $\mu + \lambda$ is not constant. Notice that f_s is F-type for all $s \neq 0$, whereas f_0 is not F-type (but still B-type). The generic type at infinity is D_4 for all s and there is a jump $D_4 \to D_5$ for $t = 0$ and all s. For $s \neq 0$ a second jump $D_4 \to D_5$ occurs for $t = c/s^2$, for some constant c.

There are no affine critical points, i.e., $\mu(s) = 0$ for all s, but $\lambda(s) = 2$ if $s \neq 0$ and $\lambda(0) = 1$. We have that $\Lambda(f_s) = \{0, c/s^2\}$ for all $s \neq 0$, and that χ^∞ changes from 3 if $s = 0$ to 2 if $s \neq 0$, so $\Delta\chi^\infty = +1$.

There is a persistent λ-singularity in the fibre over $t = 0$ and there is a branch of the critical locus $\mathrm{Crit}\Psi$ which is asymptotic to $t = \infty$.

8.3. Cases of lower semi-continuity at λ-singularities

In Theorem 4.2 we have an inequality which we may write in short-hand as follows, by referring to its proof (formula 7.1):

$$\lambda = I_{\text{gen}} - \nu \leq I_{\text{gen}} \leq I_{s=0} \tag{8.1}$$

This inequality can have two different sources:

- the nongeneric intersection number $I_{s=0}$ and its difference to the generic one I_{gen},
- the number ν, which is related to the equisingularity properties of $\mathbb{Y}$.

So the excess in the formula is $\nu + (I_{s=0} - I_{\text{gen}})$. The following examples illustrate the different types of excess: $\nu \neq 0$, respectively $\nu = 0$ and $I_{s=0} - I_{\text{gen}} > 0$. In the latter case, the space $\mathbb{Y}$ is singular.

EXAMPLE 8.7. We start with a F-type polynomial f_0 and consider a Yomdin deformation $f_0 - sx_1^d$ for sufficient general x_1. In this case the space $\mathbb{Y}$ is non-singular and the function $\sigma + \varepsilon\tau$ behaves locally as a linear function. It follows that $\nu = 0$. Moreover in this case $I_{s=0} - I_{\text{gen}}$ turns out to be positive because of the tangency of some components of the discriminant set to the s-axis. Compare to [15, Theorem 5.4], where the local lower semi-continuity was proved in the case of Yomdin deformations.

EXAMPLE 8.8. $f_s = x^2y^b + x + sxy^k$.
In the range $\frac{b}{2} < k \leq b$, this has the following data:

$s = 0$: $\lambda = b$, $\mu = 0$, $\lambda + \mu = b$;
$s \neq 0$: $\lambda = 0$, $\mu = 2k$, $\lambda + \mu = 2k$.

Both intersection numbers I_{gen} and $I_{s=0}$ are the same and equal to $2k$. We read the inequality (8.1) as: $b = 2k - \nu \leq 2k \leq 2k$. So $\nu = 2k - b$ and this is positive in case $\frac{b}{2} < k \leq b$.

For the complementary range $1 < k < \frac{b}{2}$ we have a family with an extra λ-discriminant branch at $t = 0$.

There is the following data here:

$s = 0$: $\lambda = b$, $\mu = 0$, $\lambda + \mu = b$;

$s \neq 0$: $\lambda = b - 2k$, $\mu = 2k$, $\lambda + \mu = b$.

In this range one has $\nu = 0$, $\lambda = b = I_{s=0}$, which gives equality in Theorem 4.2. This local conservation is characteristic to families with constant global $\mu + \lambda$, see Corollary 6.1(c).

References

[1] N. A'Campo, *Le nombre de Lefschetz d'une monodromie*, Indag. Math. 35 (1973), 113–118.

[2] A. Andreotti, T. Frankel, *The Lefschetz theorem on hyperplane sections*, Ann. of Math. (2) 69 (1959), 713–717.

[3] V.I. Arnol'd, *Singularities of fractions and the behavior of polynomials at infinity*, Tr. Mat. Inst. Steklova 221 (1998), 48–68; translation in Proc. Steklov Inst. Math. 1998, no. 2 (221), 40–59.

[4] A. Bodin, *Invariance of Milnor numbers and topology of complex polynomials*, Comment. Math. Helv. 78 (2003), no. 1, 134–152.

[5] A. Bodin, M. Tibăr, *Topological triviality of families of complex polynomials*, Adv. Math. 199 (2006), no. 1, 136–150.

[6] S.A. Broughton, *On the topology of polynomial hypersurfaces*, Proceedings A.M.S. Symp. in Pure. Math., vol. 40, I (1983), 165–178.

[7] S.M. Gusein-Zade, D. Siersma, *Deformations of polynomials and their zeta functions*, math.AG/0503450.

[8] Hà H.V., A. Zaharia, *Families of polynomials with total Milnor number constant*, Math. Ann. 304 (1996), no. 3, 481–488.

[9] F. Lazzeri, *A theorem on the monodromy of isolated singularities*, in: Singularités à Cargèse, 1972, pp. 269–275. Astérisque, Nos. 7 et 8, Soc. Math. France, Paris, 1973.

[10] Lê D.T., *Une application d'un théorème d'A'Campo à l'équisingularité*, Nederl. Akad. Wetensch. Proc. (Indag. Math.) 35 (1973), 403–409.

[11] A. Némethi, C. Sabbah, *Semicontinuity of the spectrum at infinity*, Abh. Math. Sem. Univ. Hamburg 69 (1999), 25–35.

[12] A. Parusiński, *On the bifurcation set of complex polynomial with isolated singularities at infinity*, Compositio Math. 97 (1995), no. 3, 369–384.

[13] D. Siersma, J. Smeltink, *Classification of singularities at infinity of polynomials of degree* 4 *in two variabales*, Georgian Math. J. 7 (1) (2000), 179–190.

[14] D. Siersma, M. Tibăr, *Singularities at infinity and their vanishing cycles*, Duke Math. Journal 80 (3) (1995), 771-783.

[15] D. Siersma, M. Tibăr, *Deformations of polynomials, boundary singularities and monodromy*, Mosc. Math. J. 3 (2) (2003), 661–679.

[16] M. Tibăr, *Bouquet decomposition of the Milnor fibre*, Topology 35, 1 (1996), 227–241.

[17] M. Tibăr, *On the monodromy fibration of polynomial functions with singularities at infinity*, C.R. Acad. Sci. Paris, 324 (1997), 1031–1035.

[18] M. Tibăr, *Regularity at infinity of real and complex polynomial maps*, in: Singularity Theory, The C.T.C Wall Anniversary Volume, LMS Lecture Notes Series 263 (1999), 249–264. Cambridge University Press.

Dirk Siersma
Mathematisch Instituut
Universiteit Utrecht
PO Box 80010
NL-3508 TA Utrecht, The Netherlands
e-mail: `siersma@math.uu.nl`

Mihai Tibăr
Mathématiques
Université des Sciences
et Technologies de Lille
F-59655 Villeneuve d'Ascq, France
e-mail: `tibar@math.univ-lille1.fr`

Real and Complex Singularities
Trends in Mathematics, 343–359

Mackey Functors on Provarieties

Shoji Yokura

Abstract. MacPherson's Chern class transformation on complex algebraic varieties is a certain unique natural transformation from the constructible function covariant functor to the integral homology covariant functor, and it can be extended to a category of provarieties. In this paper, as further extensions of this we consider natural transformations among Mackey functors on provarieties and also on "indvarieties" and discuss some notions and examples related to these extensions.

Mathematics Subject Classification (2000). Primary 14C17, 18F30; Secondary 14F99, 18F30, 55N99.

Keywords. provariety, constructible function, Chern class, Mackey functor, Grothendieck ring of varieties, motivic measure, bivariant theory, ind-variety.

1. Introduction

In [Y2] we considered a Chern class of proalgebraic varieties and in particular we have related its construction to motivic measures ([Cr], [DL1, DL2], [Lo], [Ve], etc.). To define such a Chern class, we first need to consider a "proalgebraic" version of constructible function, which is called *proconstructible function* in [Y2]. One key for defining such a Chern class is that the correspondence F assigning to a variety X the abelian group $F(X)$ of constructible functions satisfies "base change isomorphism" or what is called "Mackey property" of the Mackey functor, which was for the first time introduced by Andreas W. M. Dress ([Dr1], [Dr2]) in the theory of representations of finite groups. This is a bifunctor from the category of G-sets (where G is a finite group) to the abelian category. For Mackey functors, also see [Bo], [Lin], [TW], [Yo], etc.

In this paper, we shall discuss extensions of Mackey functor to proalgebraic varieties and discuss some notions and examples related to these extensions.

Partially supported by Grant-in-Aid for Scientific Research (No.17540088), the Ministry of Education, Culture, Sports, Science and Technology (MEXT), Japan.

2. Constructible functions and MacPherson's Chern class transformations

In the following we work in the context of complex algebraic varieties.

Let $F(X)$ be the abelian group of constructible functions on a variety X. Then the assignment $F : \mathcal{V} \to \mathcal{A}$ is a contravariant functor (from the category of varieties to the category of abelian groups) by the usual functional pullback: for a morphism $f : X \to Y$

$$f^* : F(Y) \to F(X) \quad \text{defined by} \quad f^*(\alpha) := \alpha \circ f.$$

For a constructible set $A \subset X$, we define

$$\chi(A; \alpha) := \sum_{n \in \mathbf{Z}} n\chi(A \cap \alpha^{-1}(n)).$$

Here the Euler–Poincaré characteristic of a constructible set A is defined by $\chi(A) := \sum_i \chi(A_i)$, if A is the disjoint union of finitely many locally closed algebraic subsets A_i, for which we use the Euler–Poincaré characteristic of the cohomology with compact support. This is well defined by the additivity of the Euler–Poincaré characteristic (with compact support). In addition, it has the following "generalized multiplicativity property": for an algebraic morphism $f : A \to B$ such that the Euler–Poincaré characteristics of all the fibers $\chi(f^{-1}(y))$ are equal, $\chi(A) = \chi(B) \times \chi(f^{-1}(y))$.

It turns out that the assignment $F : \mathcal{V} \to \mathcal{A}$ also becomes a covariant functor by the following pushforward:

$$f_* : F(X) \to F(Y) \quad \text{defined by} \quad f_*(\alpha)(y) := \chi(f^{-1}(y); \alpha).$$

For more details on constructible functions and, in particular, for comparison with standard Grothendieck operations on constructible sheaves, see [Sch2, Chapter 2] (and also [Dim], [KS], [Scha]).

The main importance of this functor F is that we need it to define the so-called MacPherson's Chern class transformation. P. Deligne and A. Grothendieck conjectured and R. MacPherson [Mac] solved the following:

Theorem 2.1. *Let H_* be the Borel-Moore homology theory and the morphisms we consider are proper morphisms. Then there exists a unique natural transformation*

$$c_* : F \to H_*$$

from the constructible function covariant functor F to the homology covariant functor H_* satisfying the "normalization" that the value of the characteristic function $\mathbb{1}_X$ of a smooth complex algebraic variety X is the Poincaré dual of the total Chern cohomology class:

$$c_*(\mathbb{1}_X) = c(T_X) \cap [X].$$

The above natural transformation $c_* : F \to H_*$ is called *MacPherson's Chern class transformation.*

J.-P. Brasselet and M.-H. Schwartz [BrSc] showed that the distinguished value $c_*(1\!\!1_X)$ of the characteristic function of a variety under this transformation is isomorphic to the Schwartz class [Sc1, Sc2] via the Alexander duality, if X is embedded into a complex manifold. Thus, for a complex algebraic variety X, singular or nonsingular, the homology class $c_*(1\!\!1_X)$ is called the total *Chern–Schwartz–MacPherson class of* X and denoted simply by $c_*(X)$.

3. "Indconstructible" functions and motivic measures

Let I be a directed set and let $\mathcal{C}$ be a given category. Then a projective system is, by definition, a system $\{X_i, \pi_{ii'} : X_{i'} \to X_i(i < i'), I\}$ consisting of objects $X_i \in \mathrm{Obj}(\mathcal{C})$, morphisms $\pi_{ii'} : X_{i'} \to X_i \in \mathrm{Mor}(\mathcal{C})$ for each $i < i'$ and the index set I. The object X_i is called a *term* and the morphism $\pi_{ii'} : X_{i'} \to X_i$ a *bonding morphism* or *structure morphism* ([MS]). The projective system $\{X_i, \pi_{ii'} : X_{i'} \to X_i(i < i'), I\}$ is sometimes simply denoted by $\{X_i\}_{i\in I}$.

Given a category $\mathcal{C}$, Pro-$\mathcal{C}$ is the category whose objects are projective systems $X = \{X_i\}_{i\in I}$ in $\mathcal{C}$ and whose set of morphisms from $X = \{X_i\}_{i\in I}$ to $Y = \{Y_j\}_{j\in J}$ is

$$\text{Pro-}\mathcal{C}(X, Y) := \varprojlim_J (\varinjlim_I \mathcal{C}(X_i, Y_j)).$$

Given a projective system $X = \{X_i\}_{i\in I} \in \text{Pro-}\mathcal{C}$, the projective limit $X_\infty := \varprojlim X_i$ may not belong to the source category $\mathcal{C}$. For a certain sufficient condition for the existence of the projective limit in the category $\mathcal{C}$, see [MS] for example. For a study of pro-objects, also see [AM] and [Grot].

An object in Pro-$\mathcal{C}$ is called a *pro-object*. A projective system of algebraic varieties is called a *pro-variety* and its projective limit is called a *provariety*, which may not be an algebraic variety but simply a topological space.

Let $T : \mathcal{C} \to \mathcal{D}$ be a covariant (contravariant, resp.) functor between two categories $\mathcal{C}, \mathcal{D}$. Obviously the covariant (contravariant, resp.) functor T extends to a covariant (contravariant, resp.) pro-functor

$$\text{Pro-}T : \text{Pro-}\mathcal{C} \to \text{Pro-}\mathcal{D}$$

defined by $\text{Pro-}T(\{X_i\}_{i\in I}) := \{T(X_i)\}_{i\in I}$. Let $T_1, T_2 : \mathcal{C} \to \mathcal{D}$ be two covariant (contravariant, resp.) functors and $N : T_1 \to T_2$ be a natural transformation between the two functors T_1 and T_2. Then the natural transformation $N : T_1 \to T_2$ extends to a natural pro-transformation

$$\text{Pro-}N : \text{Pro-}T_1 \to \text{Pro-}T_2.$$

From here on, for the sake of simplicity, we only deal with the case when the directed set Λ is the natural numbers $\mathbf{N}$ and a pro-morphism $\{f_n\}$ of two provarieties $\{X_n\}$ and $\{Y_n\}$ is such that for each n the following diagram commutes:

$$\begin{array}{ccc} X_{n+1} & \xrightarrow{f_{n+1}} & Y_{n+1} \\ {\scriptstyle \rho_{n(n+1)}}\downarrow & & \downarrow{\scriptstyle \pi_{n(n+1)}} \\ X_n & \xrightarrow[f_n]{} & Y_n. \end{array}$$

Definition 3.1. For a proalgebraic variety $X_\infty = \varprojlim_n X_n$, the inductive limit of the inductive system $\{F(X_n), \pi^*_{nm} : F(X_n) \to F(X_m)(n < m)\}$ is denoted by $F^{\mathrm{ind}}(X_\infty)$;

$$F^{\mathrm{ind}}(X_\infty) := \varinjlim_n F(X_n) = \bigcup_n \pi^n\left(F(X_n)\right)$$

where $\pi^n : F(X_n) \to \varinjlim_n F(X_n)$ is the homomorphism sending α_n to its equivalence class $[\alpha_n]_\infty$ of α_n. An element of the group $F^{\mathrm{ind}}(X_\infty)$ is called *an "indconstructible" function* on the proalgebraic variety X_∞. As a function on X_∞, the value of $[\alpha_n]_\infty$ at a point $(x_m) \in X_\infty$ is defined by

$$[\alpha_n]_\infty\big((x_m)\big) := \alpha_n(x_n),$$

which is well defined.

Remark 3.2. The above "indconstructible" was called "proconstructible" in [Y2], but in this paper we name it so since it is defined via the "inductive limit" and also in order to avoid some possible confusions; later we also discuss similar ones defined via the projective limits.

Remark 3.3. $F^{\mathrm{ind}}(X_\infty)$ certainly depends on the given projective system $\mathcal{S} = \{X_n, \pi_{nm} : X_m \to X_n(n < m)\}$, so in this sense it should be denoted by something like $F^{\mathrm{ind}}_{\mathcal{S}}(X_\infty)$ with reference to the projective system $\mathcal{S}$, but for the sake of simplicity the subscript $\mathcal{S}$ is dropped.

Lemma 3.4. *For each positive integer n, let $G_n = \mathbf{Z}$ be the integers and $h_{n(n+1)} : G_n \to G_{n+1}$ be the homomorphism defined by multiplication by a non-zero integer p_n, i.e., $h_{n(n+1)}(m) = mp_n$. Then there exists a unique (injective) homomorphism*

$$\Psi : \varinjlim_n G_n \to \mathbf{Q}$$

such that

$$\Psi\left(h^n(r_n)\right) = \frac{r_n}{p_0 p_1 \cdots p_{n-1}}.$$

Here $h^n : G_n \to \varinjlim_n G_n$ is the homomorphism sending r_n to its equivalence class $g^n(r_n)$ of r_n and we set $p_0 := 1$.

Using this lemma, together with the "generalized multiplicativity property" of the Euler–Poincaré characteristic, we can show the following theorem:

Theorem 3.5. *Let $X_\infty = \varprojlim_{n\in\mathbf{N}} X_n$ be a provariety such that for each n the structure morphism $\pi_{n(n+1)} : X_{n+1} \to X_n$ satisfies the condition that the Euler–Poincaré characteristics of the fibers of $\pi_{n(n+1)}$ are non-zero (which implies the surjectivity of the morphism $\pi_{n(n+1)}$) and the same; for example, $\pi_{n(n+1)} : X_{n+1} \to X_n$ is a locally trivial fiber bundle with fiber variety being F_n and $\chi(F_n) \neq 0$. Let us denote the constant Euler–Poincaré characteristic of the fibers of the morphism $\pi_{n(n+1)} : X_{n+1} \to X_n$ by χ_n and we set $\chi_0 := 1$.*

(i) *The canonical Euler–Poincaré (ind)characteristic homomorphism, i.e., a "canonical realization" of the inductive limit of the Euler–Poincaré characteristic homomorphisms $\{\chi : F(X_n) \to \mathbf{Z}\}_{n\in\mathbf{N}}$, is described as the homomorphism*
$$\chi^{\mathrm{ind}} : F^{\mathrm{ind}}(X_\infty) \to \mathbf{Q}$$
defined by
$$\chi^{\mathrm{ind}}([\alpha_n]_\infty) = \frac{\chi(\alpha_n)}{\chi_0 \cdot \chi_1 \cdot \chi_2 \cdots \chi_{n-1}}.$$
(Here "canonical realization" means "through the injective homomorphism in the above lemma".)

(ii) *In particular, if the Euler–Poincaré characteristics χ_n are all the same, say $\chi_n = \chi$ for any n, then the canonical Euler–Poincaré (ind)characteristic homomorphism $\chi^{\mathrm{ind}} : F^{\mathrm{ind}}(X_\infty) \to \mathbf{Q}$ is described by*
$$\chi^{\mathrm{ind}}([\alpha_n]_\infty) = \frac{\chi(\alpha_n)}{\chi^{n-1}}.$$
In this special case, the target ring $\mathbf{Q}$ can be replaced by the ring $\mathbf{Z}\left[\frac{1}{\chi}\right]$.

In a more special case, the target ring $\mathbf{Q}$ in the above theorem can be replaced by the Grothendieck ring of varieties.

Let $K_0(\mathcal{V}_{\mathbf{C}})$ be the Grothendieck ring of algebraic varieties, i.e., the free abelian group generated by the isomorphism classes of varieties modulo the subgroup generated by elements of the form $[V] - [V'] - [V \setminus V']$ for a closed subset $V' \subset V$ with the ring structure $[V] \cdot [W] := [V \times W]$. There are distinguished elements in $K_0(\mathcal{V}_{\mathbf{C}})$: $\mathbb{1}$ is the class [p] of a point p and $\mathbf{L}$ is the Tate class $[\mathbf{C}]$ of the affine line $\mathbf{C}$. From this definition, we can see that any constructible set of a variety determines an element in the Grothendieck ring $K_0(\mathcal{V}_{\mathbf{C}})$. Provisionally the element $[V]$ in the Grothendieck ring $K_0(\mathcal{V}_{\mathbf{C}})$ is called the *Grothendieck "motivic" class of V* and let us denote it by $\Gamma(V)$. Hence we get the following homomorphism, called the *Grothendieck "motivic" class homomorphism*: for any variety X
$$\Gamma : F(X) \to K_0(\mathcal{V}_{\mathbf{C}}), \quad \text{which is defined by} \quad \Gamma(\alpha) = \sum_{n\in\mathbf{Z}} n\left[\alpha^{-1}(n)\right].$$
Or $\Gamma\left(\sum a_V \mathbb{1}_V\right) := \sum a_V [V]$ where V is a constructible set in X and $a_V \in \mathbf{Z}$. From now on, we sometimes write $[\alpha]$ for $\Gamma(\alpha)$ for a constructible function α.

This Grothendieck "motivic" class homomorphism is *tautological* and a more "geometric" one is the Euler–Poincaré characteristic homomorphism $\chi : F(X) \to \mathbf{Z}$. Note that the homomorphism $\chi : F(X) \to \mathbf{Z}$ factors into the homomorphism $\Gamma : F(X) \to K_0(\mathcal{V}_{\mathbf{C}})$ followed by the (universal Euler–Poincaré characteristic) homomorphism $\chi^{\mathrm{Gr}} : K_0(\mathcal{V}_{\mathbf{C}}) \to \mathbf{Z}$ defined by $\chi^{\mathrm{Gr}}([V]) := \chi(V)$, which is well defined; namely, we have $\chi = \chi^{\mathrm{Gr}} \circ \Gamma$.

The above theorem is about extending the Euler–Poincaré characteristic homomorphism $\chi : F(X) \to \mathbf{Z}$ to the category of proalgebraic varieties. Thus a very natural problem is to generalize the Grothendieck "motivic" class homomorphism $\Gamma : F(X) \to K_0(\mathcal{V}_{\mathbf{C}})$ to the category of proalgebraic varieties.

Theorem 3.6. *Let $X_\infty = \varprojlim_{n \in \mathbf{N}} X_n$ be a proalgebraic variety such that each structure morphism $\pi_{n,n+1} : X_{n+1} \to X_n$ satisfies the condition that for each n there exists a $\gamma_n \in K_0(\mathcal{V}_{\mathbf{C}})$ such that $[\pi_{n(n+1)}{}^{-1}(S_n)] = \gamma_n \cdot [S_n]$ for any constructible set $S_n \subset X_n$, for example, $\pi_{n(n+1)} : X_{n+1} \to X_n$ is a Zariski locally trivial fiber bundle with fiber variety being F_n (in which case $\gamma_n = [F_n] \in K_0(\mathcal{V}_{\mathbf{C}})$).*

(i) *The canonical proalgebraic Grothendieck class homomorphism,*

$$\Gamma^{\mathrm{ind}} : F^{\mathrm{ind}}(X_\infty) \to K_0(\mathcal{V})_{\mathcal{G}}$$

is described by

$$\Gamma^{\mathrm{ind}}\left([\alpha_n]_\infty\right) = \frac{[\alpha_n]}{\gamma_0 \cdot \gamma_1 \cdot \gamma_2 \cdots \gamma_{n-1}}.$$

Here $\gamma_0 := \mathbb{1}$ and $K_0(\mathcal{V})_{\mathcal{G}}$ is the ring of fractions of $K_0(\mathcal{V})$ with respect to the multiplicatively closed set consisting of finite products of powers of γ_m $(m = 1, 2, 3, \ldots)$, i.e.,

$$\mathcal{G} := \left\{\gamma_{j_1}^{m_1}\gamma_{j_2}^{m_2}\cdots\gamma_{j_s}^{m_s} | j_i \in \mathbf{N}, m_i \in \mathbf{N}\right\}.$$

(ii) *In particular, if γ_n are all the same, say $\gamma_n = \gamma$ for any n, then the canonical proalgebraic Grothendieck class homomorphism*

$$\Gamma^{\mathrm{ind}} : F^{\mathrm{ind}}(X_\infty) \to K_0(\mathcal{V})_{\mathcal{G}}$$

is described by

$$\Gamma^{\mathrm{ind}}\left([\alpha_n]_\infty\right) = \frac{[\alpha_n]}{\gamma^{n-1}}.$$

In this special case the quotient ring $K_0(\mathcal{V})_{\mathcal{G}}$ shall be simply denoted by $K_0(\mathcal{V})_\gamma$.

Remark 3.7. In the above theorem one should be a bit careful: the target ring $K_0(\mathcal{V}_{\mathbf{C}})_{\mathcal{G}}$ cannot be replaced by the "total quotient ring" of the Grothendieck ring $K_0(\mathcal{V}_{\mathbf{C}})$ unlike in the previous theorem, because the Grothendieck ring $K_0(\mathcal{V}_{\mathbf{C}})$ is not a domain unlike the ring $\mathbf{Z}$ of integers as shown recently by B. Poonen [Po, Theorem 1] and thus one cannot define the total quotient ring of $K_0(\mathcal{V}_{\mathbf{C}})$.

For more generalized versions of the above Theorem 3.5 and Theorem 3.6 , see [Y2] and also [Y3].

4. Mackey functors and bivariant theories

In [Y2] we extended the above Euler–Poincaré characteristic $\chi^{\mathrm{ind}} : F^{\mathrm{ind}}(X_\infty) \to \mathbf{Q}$ to "classes", in other words a MacPherson's Chern class transformation on proalgebraic varieties, in the study of which we needed two keys:

(†) The first one is the following simple fact about constructible functions: for any fiber square

$$\begin{array}{ccc} X' & \xrightarrow{g'} & X \\ {\scriptstyle f'}\downarrow & & \downarrow{\scriptstyle f} \\ Y' & \xrightarrow{g} & Y, \end{array}$$

the following diagram commutes ([Er, Proposition 3.5], [FM, Axiom (A_{23})])

$$\begin{array}{ccc} F(X) & \xrightarrow{g'^*} & F(X') \\ {\scriptstyle f_*}\downarrow & & \downarrow{\scriptstyle f'_*} \\ F(Y) & \xrightarrow{g^*} & F(Y'). \end{array}$$

(‡) The second one is the so-called Verdier–Riemann–Roch formula for Chern class (e.g., [FM], [Sch1] and [Y1]).

In this section, instead of reviewing the construction of MacPherson's Chern class transformation on proalgebraic varieties, we consider the abstract situation for which these two properties are satisfied.

Let $\mathcal{C}_1$ and $\mathcal{C}_2$ be two categories. A *bifunctor* $M : \mathcal{C}_1 \to \mathcal{C}_2$ is defined to be a pair $M = (M_*, M^*)$ of a covariant functor $M_* : \mathcal{C}_1 \to \mathcal{C}_2$ and a contravariant functor $M^* : \mathcal{C}_1 \to \mathcal{C}_2$ such that they agree on objects of $\mathcal{C}_1$; thus for any object $X \in obj(\mathcal{C}_1)$ we have one object $M_*(X) = M^*(X) =: M(X) \in obj(\mathcal{C}_2)$ and for any morphism $f : Y \to X$ in $\mathrm{mor}\,(\mathcal{C}_1)$ we have two morphisms

$$f_* := M_*(f) : M(Y) \to M(X) \quad \text{and} \quad f^* := M^*(f) : M(X) \to M(Y).$$

A natural transformation $\tau : M \to N$ of bifunctors M and N is a family of morphisms $\tau_X : M(X) \to N(X)$ such that τ is both a natural transformation from M_* to N_* and from M^* to N^* ([Dr2]). For bifunctors, e.g., see [Be].

In this paper, we restrict ourselves to $\mathcal{C}_1 =$ the category $\mathcal{V}$ of complex algebraic varieties and $\mathcal{C}_2 =$ the category $\mathcal{A}$ of abelian groups. In this sense, our bifunctor is a more geometrical one.

Definition 4.1 (Mackey functors). If a bifunctor $M = (M_*, M^*) : \mathcal{V} \to \mathcal{A}$ satisfies the following two conditions (M-I) and (M-II), then the bifunctor M is called *a Mackey functor*:

(M-I) for any fiber square in the category $\mathcal{V}$

$$\begin{array}{ccc} X' & \xrightarrow{g'} & X \\ {\scriptstyle f'}\downarrow & & \downarrow{\scriptstyle f} \\ Y' & \xrightarrow[g]{} & Y \end{array}$$

it holds that ${f'}_* {g'}^* = g^* f_*$. (This property is called the *Mackey property.*)

(M-II) $M : \mathcal{V} \to \mathcal{A}$ preserves finite coproducts, i.e.,

$$M(X \sqcup Y) = M(X) \oplus M(Y).$$

Remark 4.2. If M satisfies only (M-I), then M is called *a pre-Mackey functor.*

A typical model of a Mackey functor for us in this paper is the Mackey functor F of constructible functions.

There is another notion of "bifunctor" (not in the true sense) dealing with both a covariant theory and a contravariant theory at the same time. That is the Bivariant Theory which was introduced by W. Fulton and R. MacPherson [FM].

Let $\mathcal{C}$ be a category with fiber products, a final object pt and a class of "proper" maps, which is closed under composition and base change, and contains all identity maps. A bivariant theory $\mathbf{B}$ on such a category $\mathcal{C}$ with values in the category of abelian groups is an assignment to each morphism $X \xrightarrow{f} Y$ in the category $\mathcal{C}$ an abelian group $\mathbf{B}(X \xrightarrow{f} Y)$, which is equipped with the following three basic operations:

Product operations: For morphisms $f : X \to Y$ and $g : Y \to Z$, the product operation

$$\bullet : \mathbf{B}(X \xrightarrow{f} Y) \otimes \mathbf{B}(Y \xrightarrow{g} Z) \to \mathbf{B}(X \xrightarrow{gf} Z)$$

is defined.

Pushforward operations: For morphisms $f : X \to Y$ and $g : Y \to Z$ with f proper, the pushforward operation

$$f_\star : \mathbf{B}(X \xrightarrow{gf} Z) \to \mathbf{B}(Y \xrightarrow{g} Z)$$

is defined.

Pullback operations: For a fiber square

$$\begin{array}{ccc} X' & \xrightarrow{g'} & X \\ {\scriptstyle f'}\downarrow & & \downarrow{\scriptstyle f} \\ Y' & \xrightarrow[g]{} & Y, \end{array}$$

the pullback operation

$$g^\star : \mathbf{B}(X \xrightarrow{f} Y) \to \mathbf{B}(X' \xrightarrow{f'} Y')$$

is defined. Here, for simplicity, we take all fiber squares as "independent squares" in the sense of [FM].

And these three operations are required to satisfy the seven compatibility axioms (see [FM, Part I, §2.2] for details).

Let $\mathbf{B}, \mathbf{B}'$ be two bivariant theories on a category $\mathcal{C}$. Then a *Grothendieck transformation* from $\mathbf{B}$ to $\mathbf{B}'$

$$\gamma : \mathbf{B} \to \mathbf{B}'$$

is a collection of homomorphisms

$$\mathbf{B}(X \to Y) \to \mathbf{B}'(X \to Y)$$

for a morphism $X \to Y$ in the category $\mathcal{C}$, which preserves the above three basic operations:

$$\begin{aligned} &\text{(i)} && \gamma(\alpha \bullet_{\mathbf{B}} \beta) = \gamma(\alpha) \bullet_{\mathbf{B}'} \gamma(\beta), \\ &\text{(ii)} && \gamma(f_\star \alpha) = f_\star \gamma(\alpha), \qquad \text{and} \\ &\text{(iii)} && \gamma(g^\star \alpha) = g^\star \gamma(\alpha). \end{aligned}$$

$\mathbf{B}^*(X) := \mathbf{B}(X \xrightarrow{id} X)$ becomes a contravariant functor, whereas $\mathbf{B}_*(X) := \mathbf{B}(X \to pt)$ is covariantly functorial for proper maps. And a Grothendieck transformation $\gamma : \mathbf{B} \to \mathbf{B}'$ induces natural transformations $\gamma_* : \mathbf{B}_* \to \mathbf{B}'_*$ and $\gamma^* : \mathbf{B}^* \to \mathbf{B}'^*$. If we have a Grothendieck transformation $\gamma : \mathbf{B} \to \mathbf{B}'$, then via a bivariant class $b \in \mathbf{B}(X \xrightarrow{f} Y)$ we get the commutative diagram

$$\begin{array}{ccc} \mathbf{B}_*(Y) & \xrightarrow{\gamma_*} & \mathbf{B}'_*(Y) \\ {\scriptstyle b\bullet}\downarrow & & \downarrow{\scriptstyle \gamma(b)\bullet} \\ \mathbf{B}_*(X) & \xrightarrow[\gamma_*]{} & \mathbf{B}'_*(X). \end{array}$$

This is called *the Verdier-type Riemann–Roch formula associated to the bivariant class b.*

Definition 4.3. Let $\mathcal{S}$ be a class of maps in $\mathcal{C}$, closed under compositions and containing all identity maps. If there exists an assignment θ assigning to each map $f : X \to Y$ an element $\theta(f) \in \mathbf{B}(X \to Y)$ such that

(i) $\theta(g \circ f) = \theta(f) \bullet \theta(g)$ for $f : X \to Y$ and $g : Y \to Z$ in $\mathcal{S}$ and
(ii) $\theta(id_X) = 1_X$ for all X

where 1_X is the unit of $\mathbf{B}(X \xrightarrow{id_X} X)$ which satisfies that $\alpha \bullet 1_X = \alpha$ for all maps $g : W \to X$ and for any $\alpha \in \mathbf{B}(W \xrightarrow{g} X)$, then the assignment θ is called *a canonical orientation for $\mathcal{S}$.*

Given a fiber square

$$\begin{array}{ccc} X' & \xrightarrow{g'} & X \\ {\scriptstyle f'}\downarrow & & \downarrow{\scriptstyle f} \\ Y' & \xrightarrow[g]{} & Y \end{array}$$

with $g, g' \in \mathcal{S}$, it is not necessarily true that $f^{\star}\theta(g) = \theta(g')$ (see [FM, Remark., page 28]). If such a relation holds, the canonical orientation θ shall be called *a nice canonical orientation for* $\mathcal{S}$.

Example. Let $\mathcal{C}$ be the category of complex algebraic varieties with "proper" meaning the usual one and let $\mathbf{B} = F$ be the abelian group of constructible function of the source variety of a morphism, i.e., $\mathbf{B}(X \to Y) := F(X)$. We associate to a morphism $f : X \to Y$ the orientation $1_f := \mathbb{1}_X$. Then for any such a class $\mathcal{S}$ of maps the association $\theta(X \xrightarrow{f} Y) := 1_f$ is a nice canonical orientation.

Example. Let $\mathcal{C}$ be as in the above example and let $\mathbf{B}$ be the Fulton–MacPherson's bivariant homology theory [FM] or the operational bivariant theory of Chow groups [FM] (also see [Fu]). If we consider the class $\mathcal{S}$ of "smooth" morphisms, then taking the corresponding relative orientation or fundamental class $[f]$ of a smooth morphism gives rise to a "nice canonical orientation" for $\mathcal{S}$.

Let $\mathbf{B}$ be a bivariant theory on $\mathcal{C}$ and θ a nice canonical orientation for $\mathcal{S}$. Such data give rise to a *"conditional"* pre-Mackey functor in the following sense:

(i) For a map $g : Y' \to Y$ in $\mathcal{S}$, the "pullback" homomorphism $\theta(g)\bullet : \mathbf{B}_*(Y) \to \mathbf{B}_*(Y')$ is contravariant (because of the requirement (i) of the above definition), and

(ii) For a fiber square

$$\begin{array}{ccc} X' & \xrightarrow{g'} & X \\ {\scriptstyle f'}\downarrow & & \downarrow{\scriptstyle f} \\ Y' & \xrightarrow[g]{} & Y \end{array}$$

with $g, g' \in \mathcal{S}$, and f, f' proper, we get the following commutative diagram:

$$\begin{array}{ccc} \mathbf{B}_*(X') & \xleftarrow{\theta(g')\bullet} & \mathbf{B}_*(X) \\ {\scriptstyle f'_\star}\downarrow & & \downarrow{\scriptstyle f_\star} \\ \mathbf{B}_*(Y') & \xleftarrow[\theta(g)\bullet]{} & \mathbf{B}_*(Y) \end{array}$$

which is a "conditional" Mackey property because g and g' have to belong to $\mathcal{S}$. Here we note that this conditional Mackey property, i.e., $f'_\star(\theta(g')\bullet ?) = \theta(g) \bullet f_\star ?$, follows from the fact that θ is a nice canonical orientation for $\mathcal{S}$, i.e., $f^\star\theta(g) = \theta(g')$, and the projection formula [FM, Axiom (A_{123})] of the bivariant theory $\mathbf{B}$. Such a conditional pre-Mackey functor shall be denoted by $\mathbf{B}_*^\theta$; namely $\mathbf{B}_*^\theta(X) := \mathbf{B}_*(X)$ and the pushforward $f_* : \mathbf{B}_*(X) \to \mathbf{B}_*(Y)$ is considered for any proper morphism $f : X \to Y$ and the pullback homomorphism $g^* := \theta(g)\bullet : \mathbf{B}_*(Y) \to \mathbf{B}_*(Y')$ is considered only for a morphism $g : Y' \to Y$ belonging to the chosen class $\mathcal{S}$.

Let $(\mathbf{B}, \theta)$ and $(\mathbf{B}', \theta')$ be two bivariant theories on a category $\mathcal{C}$ with nice canonical orientations for a class $\mathcal{S}$. If a natural transformation $\tau : \mathbf{B}_* \to \mathbf{B}'_*$ between the associated covariant functors satisfies the condition that for any morphism $g : Y' \to Y \in \mathcal{S}$ the following diagram commutes

$$\begin{array}{ccc} \mathbf{B}_*(Y) & \xrightarrow{\tau} & \mathbf{B}'_*(Y) \\ {\scriptstyle \theta(g)\bullet}\downarrow & & \downarrow{\scriptstyle \theta'(g)\bullet} \\ \mathbf{B}_*(Y') & \xrightarrow[\tau]{} & \mathbf{B}'_*(Y'), \end{array}$$

which is called the *Verdier-type Riemann–Roch formula associated to the nice canonical orientations θ and θ' for $\mathcal{S}$*, then the transformation $\tau : \mathbf{B}_* \to \mathbf{B}'_*$ becomes a natural transformation between two conditional pre-Mackey functors $\mathbf{B}_*^\theta$ and ${\mathbf{B}'_*}^{\theta'}$.

Example. Let $\mathcal{C}$ be as in the above examples with $\mathcal{S}$ being the class of smooth morphisms. Let $\mathbf{B} = F$ and $\mathbf{B}'$ the Fulton–MacPherson's bivariant homology theory [FM] or the operational bivariant theory of Chow groups ([FM] and [Fu]) and let θ be the nice canonical orientation described in the above Example 4. Let $\theta'(g)$ be the nice canonical orientation $c(T_g) \bullet [g]$ for a smooth morphism g, where $c(T_g)$ is the total Chern class of the relative tangent bundle T_g of the smooth morphism g. Then MacPherson's Chern class transformation $c_* : F \to H_*$, where H_* is the homology theory or the Chow group (or the Chow homology group, i.e., the image of the cycle map from the Chow group to the homology group), becomes a natural transformation of the two conditional pre-Mackey functors $\mathbf{B}_*^\theta(= F)$ and ${\mathbf{B}'_*}^{\theta'}(= H_*)$. And the Verdier-type Riemann–Roch formula associated to the nice canonical orientations θ and θ' for $\mathcal{S}$ is the following commutative diagram (which is called the Verdier-type Riemann–Roch formula for Chern class for a smooth morphism):

$$\begin{array}{ccc} F(Y) & \xrightarrow{c_*} & H_*(Y) \\ {\scriptstyle f^*}\downarrow & & \downarrow{\scriptstyle c(T_f)\cap f^*} \\ F(Y') & \xrightarrow[c_*]{} & H_*(Y'). \end{array}$$

5. (Pre-)Mackey functors on provarieties

There are at least two possible extensions of (pre-)Mackey functors to provarieties.

First, using the covariance of a (pre-)Mackey functor M, we can define the following:

Definition 5.1. Let $M : \mathcal{V} \to \mathcal{A}$ be a (pre-)Mackey functor. For a provariety $X_\infty = \varprojlim_{n\in\mathbf{N}} X_n$,

$$M^{\mathrm{proj}}(X_\infty) := \varprojlim_{n\in\mathbf{N}} M(X_n)$$

and for a promorphism $f_\infty : X_\infty \to Y_\infty$

$$f_{\infty *} : M^{\mathrm{proj}}(X_\infty) \to M^{\mathrm{proj}}(Y_\infty)$$

is defined by the projective limit

$$f_{\infty *} := \varprojlim_{n\in\mathbf{N}} \{f_{n*} : M(X_n) \to M(Y_n)\}.$$

It is obvious that M^{proj} is a covariant functor. When it comes to contravariance of M^{proj}, we cannot expect it for an arbitrary promorphism $f_\infty : X_\infty \to Y_\infty$. We need a certain requirement for the promorphism:

Definition 5.2. A promorphism $f_\infty : X_\infty \to Y_\infty$ is called a *fiber-square promorphism* (abbr., f.s. promorphism) if for each $n \in \mathbf{N}$ the following commutative diagram is a fiber square

$$\begin{array}{ccc} X_{n+1} & \xrightarrow{f_{n+1}} & Y_{n+1} \\ {\scriptstyle \rho_{n(n+1)}}\downarrow & & \downarrow{\scriptstyle \pi_{n(n+1)}} \\ X_n & \xrightarrow[f_n]{} & Y_n. \end{array}$$

With this definition, the Mackey property of M implies the following

Lemma 5.3. *For a f.s. promorphism $f_\infty : X_\infty \to Y_\infty$, we define*

$${f_\infty}^* := \varprojlim_{n\in\mathbf{N}} \{{f_n}^* : M(Y_n) \to M(X_n)\}.$$

Then M^{proj} is a contravariant functor for f.s.promorphisms.

Hence we can have the following

Proposition 5.4. *Let $\mathcal{P}rov$ be the category of provarieties. Then the correspondence $M^{\mathrm{proj}} : \mathcal{P}rov \to \mathcal{A}$ is a bifunctor provided that the contravariance is considered only for f.s. promorphisms.*

Definition 5.5. The following commutative diagram of promorphisms

$$\begin{array}{ccc} X'_\infty & \xrightarrow{g'_\infty} & X_\infty \\ {\scriptstyle f'_\infty}\downarrow & & \downarrow{\scriptstyle f_\infty} \\ Y'_\infty & \xrightarrow[g_\infty]{} & Y_\infty \end{array}$$

with $g_\infty : Y'_\infty \to Y_\infty$ being a f.s. promorphism is called a fiber square if for each $n \in \mathbf{N}$ the following diagram is a fiber square

$$\begin{array}{ccc} X'_n & \xrightarrow{g'_n} & X_n \\ {\scriptstyle f'_n}\downarrow & & \downarrow{\scriptstyle f_n} \\ Y'_n & \xrightarrow[g_n]{} & Y_n . \end{array}$$

Remark 5.6. Being a f.s.promorphism is preserved by the base change, i.e., in the above commutative diagram $g_\infty : Y'_\infty \to Y_\infty$ being a f.s. promorphism implies that $g'_\infty : X'_\infty \to X_\infty$ becomes also a f.s. promorphism.

If we consider the Mackey property for the above fiber square, we can show the following

Theorem 5.7. *$M^{\mathrm{proj}} : \mathcal{P}rov \to \mathcal{A}$ is a (pre-)Mackey functor.*

On the other hand, using the contravariance of a (pre-)Mackey functor M, we can define the following:

Definition 5.8. Let $M : \mathcal{V} \to \mathcal{A}$ be a (pre-)Mackey functor. For a provariety $X_\infty = \varprojlim_{n\in\mathbf{N}} X_n$

$$M^{\mathrm{ind}}(X_\infty) := \varinjlim_{n\in\mathbf{N}} M(X_n)$$

and for a promorphism $f_\infty : X_\infty \to Y_\infty$

$$f_\infty{}^* : M^{\mathrm{ind}}(Y_\infty) \to M^{\mathrm{ind}}(X_\infty)$$

is defined by the inductive limit

$$f_\infty{}^* := \varinjlim_{n\in\mathbf{N}} \left\{ f_n{}^* : M(Y_n) \to M(X_n) \right\}.$$

It is obvious that M^{ind} is a contravariant functor.

Lemma 5.9. *For a f.s.promorphism $f_\infty : X_\infty \to Y_\infty$, we define*

$$f_{\infty *} := \varinjlim_{n\in\mathbf{N}} \left\{ f_{n*} : M(Y_n) \to M(X_n) \right\}.$$

Then M^{ind} is a covariant functor for f.s.promorphisms.

Hence we can have the following

Proposition 5.10. *The correspondence $M^{\mathrm{ind}} : \mathcal{P}rov \to \mathcal{A}$ is a bifunctor provided that the covariance is considered only for f.s.promorphisms.*

As in the case of M^{proj}, if we consider the Mackey property for the above fiber square, we can show the following

Theorem 5.11. *$M^{\mathrm{ind}} : \mathcal{P}rov \to \mathcal{A}$ is a (pre-)Mackey functor.*

Theorem 5.12. *Any natural transformation $\tau : M \to N$ of two (pre-)Mackey functors $M, N : \mathcal{V} \to \mathcal{A}$ can be extended to two natural transformations of (pre-)Mackey functors of provarieties*

$$(IND) \quad \tau^{\text{ind}} : M^{\text{ind}} \to N^{\text{ind}}, \quad (PROJ) \quad \tau^{\text{proj}} : M^{\text{proj}} \to N^{\text{proj}}$$

such that

(IND-i) (covariance) for a f.s.promorphism $f_\infty : X_\infty \to Y_\infty$ the following diagram commutes:

$$\begin{array}{ccc} M^{\text{ind}}(X_\infty) & \xrightarrow{\tau^{\text{ind}}} & N^{\text{ind}}(X_\infty) \\ {\scriptstyle f_{\infty *}}\downarrow & & \downarrow{\scriptstyle f_{\infty *}} \\ M^{\text{ind}}(Y_\infty) & \xrightarrow[\tau^{\text{ind}}]{} & N^{\text{ind}}(Y_\infty), \end{array}$$

(IND-ii) (contravariance) for any promorphism $f_\infty : X_\infty \to Y_\infty$ the following diagram commutes:

$$\begin{array}{ccc} M^{\text{ind}}(Y_\infty) & \xrightarrow{\tau^{\text{ind}}} & N^{\text{ind}}(Y_\infty) \\ {\scriptstyle f_\infty{}^*}\downarrow & & \downarrow{\scriptstyle f_\infty{}^*} \\ M^{\text{ind}}(X_\infty) & \xrightarrow[\tau^{\text{ind}}]{} & N^{\text{ind}}(X_\infty), \end{array}$$

(PROJ-i) (covariance) for any promorphism $f_\infty : X_\infty \to Y_\infty$ the following diagram commutes:

$$\begin{array}{ccc} M^{\text{proj}}(X_\infty) & \xrightarrow{\tau^{\text{proj}}} & N^{\text{proj}}(X_\infty) \\ {\scriptstyle f_{\infty *}}\downarrow & & \downarrow{\scriptstyle f_{\infty *}} \\ M^{\text{proj}}(Y_\infty) & \xrightarrow[\tau^{\text{proj}}]{} & N^{\text{proj}}(Y_\infty), \end{array}$$

(PROJ-ii) (contravariance) for a f.s. promorphism $f_\infty : X_\infty \to Y_\infty$ the following diagram commutes:

$$\begin{array}{ccc} M^{\text{proj}}(Y_\infty) & \xrightarrow{\tau^{\text{proj}}} & N^{\text{proj}}(Y_\infty) \\ {\scriptstyle f_\infty{}^*}\downarrow & & \downarrow{\scriptstyle f_\infty{}^*} \\ M^{\text{proj}}(X_\infty) & \xrightarrow[\tau^{\text{proj}}]{} & N^{\text{proj}}(X_\infty). \end{array}$$

The homology theory or the Chow group (or the Chow homology group) H_* is not a pre-Mackey functor. However, as observed in the previous section, it becomes a conditional pre-Mackey functor with $\mathcal{S}$ being the class of smooth morphisms and MacPherson's Chern class transformation $c_* : F \to H_*$ becomes a natural transformation of the two (pre-)Mackey functors. Hence we get the following theorem:

Theorem 5.13. *Let* $X_\infty = \varprojlim_{n\in\mathbf{N}} X_n$ *and* $Y_\infty = \varprojlim_{n\in\mathbf{N}} Y_n$ *be provarieties of smooth proper morphisms* $\pi_{n(n+1)} : X_{n+1} \to X_n$ *and* $\rho_{n(n+1)} : Y_{n+1} \to Y_n$.

(i) (*covariance*) *for a "proper" f.s.promorphism* $f_\infty : X_\infty \to Y_\infty$ (*i.e., all* f_n *are proper*) *the following diagram commutes:*

$$\begin{array}{ccc} F^{\mathrm{ind}}(X_\infty) & \xrightarrow{c_*^{\mathrm{ind}}} & H_{**}^{\mathrm{ind}}(X_\infty) \\ {\scriptstyle f_{\infty *}}\downarrow & & \downarrow{\scriptstyle f_{\infty *}} \\ F^{\mathrm{ind}}(Y_\infty) & \xrightarrow[c_*^{\mathrm{ind}}]{} & H_{**}^{\mathrm{ind}}(Y_\infty), \end{array}$$

(ii) (*contravariance*) *for any "smooth" promorphism* $f_\infty : X_\infty \to Y_\infty$ (*i.e., all* f_n *are smooth*) *the following diagram commutes:*

$$\begin{array}{ccc} F^{\mathrm{ind}}(Y_\infty) & \xrightarrow{c_*^{\mathrm{ind}}} & H_{**}^{\mathrm{ind}}(Y_\infty) \\ {\scriptstyle f_\infty{}^*}\downarrow & & \downarrow{\scriptstyle f_\infty{}^*} \\ F^{\mathrm{ind}}(X_\infty) & \xrightarrow[c_*^{\mathrm{ind}}]{} & H_{**}^{\mathrm{ind}}(X_\infty). \end{array}$$

Here $H_{**}^{\mathrm{ind}}(X_\infty) := \varinjlim_n \{c(T_{\pi_{n(n+1)}}) \cap \pi_{n(n+1)}^* : H_*(X_n) \to H_*(X_{n+1})\}$.

Remark 5.14. Instead of projective systems, we can consider the dual notion, i.e., inductive systems. The inductive limit of an inductive system $\{\pi_{nm} : X_n \to X_m (n < m)\}$ shall be denoted by X_∞^{ind}. Such a variety shall be called *an indvariety*, and our provariety shall be denote by X_∞^{proj} to avoid confusion. We can consider the same things as above for indvarieties: for a (pre-)Mackey functor M, we can consider at least two things

$$M^{\mathrm{proj}}(X_\infty^{\mathrm{ind}}) \quad \text{and} \quad M^{\mathrm{ind}}(X_\infty^{\mathrm{ind}}).$$

The details are left for the reader.

In fact, a special indvariety has been already studied (e.g., see [DP], [DPW], [Kum1], [Kum2], [Sha], etc.). Let k be an algebraic closed filed. A set X is called an *ind-variety* over k if there exists a filtration $X_0 \subset X_1 \subset X_2 \subset \cdots$ such that (i) $X = \bigcup_{n\geq 0} X_n$ and (ii) each X_n is a finite-dimensional variety over k and the inclusion $X_n \hookrightarrow X_{n+1}$ is a closed embedding. And the ring of regular functions $k[X]$ is defined by

$$k[X] := \varprojlim_{n\in\mathbf{N}} k[X_n].$$

Hence, in our terminology, the above ind-variety is an indvariety and $k[X]$ corresponds to $M^{\mathrm{proj}}(X_\infty^{\mathrm{ind}})$.

Further investigations on general objects

$$M^{\mathrm{proj}}(X_\infty^{\mathrm{proj}}), M^{\mathrm{ind}}(X_\infty^{\mathrm{proj}}), M^{\mathrm{proj}}(X_\infty^{\mathrm{ind}}), M^{\mathrm{ind}}(X_\infty^{\mathrm{ind}})$$

and their applications will be done in a different paper.

Acknowledgement

The author would like to thank Jörg Schürmann and the referee for their comments and suggestions.

References

[AM] M. Artin and B. Mazur, *Etale Homotopy*, Springer Lecture Notes in Math. **100**, Springer-Verlag, Berlin, 1969.

[Be] J. Benabou, *Introduction to bicategories*, Springer Lecture Notes in Math. **47** (1967), 1–77.

[Bo] S. Bouc, *Green functors and G-sets*, Springer Lecture Notes in Math. **1671**, 1997.

[BrSc] J.-P. Brasselet and M.-H. Schwartz, *Sur les classes de Chern d'une ensemble analytique complexe,* Astérisque **82–83** (1981), 93–148.

[Cr] A. Craw, *An introduction to motivic integration*, in *Strings and Geometry,* Clay Math. Proc., **3** (2004), Amer. Math. Soc., 203–225.

[DL1] J. Denef and F. Loeser, *Germs of arcs on singular algebraic varieties and motivic integration*, Invent. Math. **135** (1999), 201–232.

[DL2] J. Denef and F. Loeser, *Geometry on arc spaces of algebraic varieties*, European Congress of Mathematicians (Barcelona, 2000), **1** (2001) Birkhäuser, 327–348.

[Dim] A. Dimca, *Sheaves in Topology*, Springer-Verlag, 2004.

[DP] I. Dimitrov and I. Penkov, *Ind-varieties of generalized flags as homogeneous spaces for classical ind-groups*, IMRN **55** (2004), 2935–2953.

[DPW] I. Dimitrov, I. Penkov and J.A. Wolf, *A Bott-Borel-Weil theory for direct limits of algebraic groups*, Amer. J. Math. **124** (2002), 955–998.

[Dr1] A.W.M. Dress, *Notes on the theory of representations of finite groups, Part I*, Mimeogrphed Lecture Notes. Univ. of Bielefeld, 1971.

[Dr2] A.W.M. Dress, *Contributions to the theory of induced representations*, in *Algebraic K-Theory II* (*H. Bass, ed.*), Springer Lecture Notes in Math. **342**, Springer-Verlag, 1973, 183–240.

[Er] L. Ernström, *Topological Radon transforms and the local Euler obstruction,* Duke Math. J. **76** (1994), 1–21.

[Fu] W. Fulton, *Intersection Theory*, Springer-Verlag, 1981.

[FM] W. Fulton and R. MacPherson, *Categorical frameworks for the study of singular spaces*, Memoirs of Amer. Math. Soc. **243**, 1981.

[Grot] A. Grothendieck, *Technique de descente et théorèmes d'existence en géométrie algébrique, II*, Séminaire Bourbaki, 12ème année **exposé** 190-195, 1959–60.

[KS] M. Kashiwara and P. Schapira, *Sheaves on Manifolds*, Springer-Verlag, Berlin, Heidelberg, 1990.

[Kum1] S. Kumar, *Infinite Grassmannians and moduli spaces of G-bundles*, in *Vector Bundles on Curves – New Directions* (*M.S. Narasimhan, ed.*), Springer Lecture Notes in Math. **1647** (1997), 1–49.

[Kum2] S. Kumar, *Kac-Moody Groups, Their Flag Varieties and Representation Theory,* Progress in Mathematics **204** (Birkhäuser), 2003.

[Lin] H. Lindner, *A remark on Mackey functors*, Manuscripta Math. **18** (1976), 273–278.

[Lo] E. Looijenga, *Motivic measures*, Séminaire Bourbaki, Astérisque **276** (2002), 267–297.

[Mac] R. MacPherson, *Chern classes for singular algebraic varieties*, Ann. of Math. **100** (1974), 423–432.

[MS] S. Mardesić and J. Segal, *Shape Theory*, North-Holland, 1982.

[Po] B. Poonen, *The Grothendieck ring of varieties is not a domain*, Math. Res. Letters **9** (2002), 493–498.

[Scha] P. Schapira, *Operations on constructible functions*, J. Pure Appl. Algebra, **72** (1991), 83–93.

[Sch1] J. Schürmann, *A generalized Verdier-type Riemann–Roch theorem for Chern–Schwartz–MacPherson classes*, math.AG/0202175.

[Sch2] J. Schürmann, *Topology of singular spaces and constructible sheaves*, Monografie Matematyczne **63** (New Series), Birkhäuser, Basel, 2003.

[Sc1] M.-H. Schwartz, *Classes caractéristiques définies par une stratification d'une variété analytique complex*, C. R. Acad. Sci. Paris **t. 260** (1965), 3262–3264, 3535–3537.

[Sc2] M.-H. Schwartz, *Classes et caractères de Chern des espaces linéaires*, C. R. Acad. Sci. Paris Sér. I. Math. **295**(1982), 399–402.

[Sha] I.R. Shafarevich, *On some infinite-dimensional groups, II*, Izv. Akad. Nauk USSR Ser. Mat. **45** (1981), 214–226.

[Ve] W. Veys, *Arc spaces, motivic integration and stringy invariants*, , Proceedings of 12th MSJ-IRI symposium "Singularity Theory and Its Applications" (Sapporo, Japan, 16-25 September, 2003), Advanced Studies in Pure Mathematics (to appear), (math.AG/0401374).

[TW] J. Thevenaz and P. Webb, *The structure of Mackey functors*, Transactions Amer. Math. Soc. **347(6)** (1995), 1865–1961.

[Y1] S. Yokura, *On a Verdier-type Riemann–Roch for Chern–Schwartz–MacPherson class*, Topology and Its Applications **94** (1999), 315–327.

[Y2] S. Yokura, *Chern classes of proalgebraic varieties and motivic measures*, math.AG/0407237.

[Y3] S. Yokura, *Characteristic classes of proalgebraic varieties and motivic measures*, (preprint 2005).

[Yo] T. Yoshida, *Idempotents and Transfer Theorems of Burnside Rings, Character Rings and Span Rings*, Algebraic and Topological Theories (to the memory of Dr.Takehiko MIYATA), (1985), 589–615.

Shoji Yokura
Department of Mathematics and Computer Science
Faculty of Science, University of Kagoshima
21-35 Korimoto 1-chome
Kagoshima 890-0065, Japan
e-mail: `yokura@sci.kagoshima-u.ac.jp`

Addresses

Eric Dago AKEKE,
Université de Provence
akeke@cmi.univ-mrs.fr

Begoña ALARCON COTILLAS,
ICMC-USP
begonha@icmc.usp.br

Paolo ALUFFI,
Florida State University
aluffi@math.fsu.edu

Sergei ANISOV
Utrecht University
anisov@math.uu.nl

Fuensanta AROCA,
IMATE-UNAM
fuen@matcuer.unam.mx

Daniel BARLET,
Université de Nancy
barlet@iecn.u-nancy.fr

Valentina BARUCCI,
Università di Roma
barucci@mat.uniroma1.it

Jose BASTO-GONÇALVES,
Universidade do Porto
jbgoncal@fc.up.pt

Karim BEKKA,
Univ. de Rennes1
karim.bekka@univ-rennes1.fr

Maria Alice BERTOLIM,
Universidade de Campinas
bertolim@ime.unicamp.br

Lev BIRBRAIR,
Univ. Federal do Ceará
lev@agt.uva.es

Carles BIVIÀ-AUSINA,
Universitat Politècnica de València
carbivia@mat.upv.es

Arnaud BODIN,
Université de Lille I
Arnaud.Bodin@math.univ-lille1.fr

Jean-Paul BRASSELET,
IML - CNRS
jpb@iml.univ-mrs.fr

Thomas BRÉLIVET,
Universidad de Valladolid
brelivet@agt.uva.es

Ragnar-Olaf BUCHWEITZ,
University of Toronto
at Scarborough (UTSC)
ragnar@math.toronto.edu

Pierrette CASSOU-NOGUES,
Université de Bordeaux I
cassou@math.u-bordeaux1.fr

Sofia CASTRO,
Universidade do Porto,
sdcastro@fep.up.pt

Helena COBO PABLOS,
Universidad de Madrid
hcobopab@mat.ucm.es

João Carlos Ferreira COSTA,
ICMC-USP
joaocar@icmc.usp.br

Nivaldo de GÓES GRULHA JR.,
ICMC-USP
njunior@icmc.usp.br

James DAMON,
University of North Carolina
jndamon@math.unc.edu

André DIATTA,
University of Liverpool
adiatta@liv.ac.uk

Peter DONELAN,
Victoria University of Wellington
peter.donelan@mcs.vuw.ac.nz

Daniel DREIBELBIS,
University of North Florida
ddreibel@unf.edu

Andrew DU PLESSIS,
Aarhus Universitet
matadp@imf.au.dk

Wolfgang EBELING,
University of Hannover
ebeling@math.uni-hannover.de

Javier FERNANDEZ DE BOBADILLA,
Utrecht University
bobadilla@math.uu.nl

Massimo FERRAROTTI,
Politecnico di Torino
ferrarotti@polito.it

Jacques FURTER,
Brunel University
mastjef@brunel.ac.uk

Terence GAFFNEY,
Northeastern University
gaff@neu.edu

Peter GIBLIN,
University of Liverpool
pjgiblin@liv.ac.uk

Victor GORYUNOV,
University of Liverpool
goryunov@liv.ac.uk

Vincent GRANDJEAN,
Univ. de Bath
cssvg@bath.ac.uk

Gert-Martin GREUEL,
University of Kaiserslautern
greuel@mathematik.uni-kl.de

Marcelo Escudeiro HERNANDES,
Universidade Estadual de Maringá
mehernandes@uem.br

Maria Elenice Rodrigues HERNANDES,
ICMC - USP
elenice@icmc.usp.br

Claus HERTLING,
Université, Nancy 1
hertling@iecn.u-nancy.fr

Kevin HOUSTON,
University of Leeds
k.houston@leeds.ac.uk

Alain JACQUEMARD,
Université de Bourgogne
jacmar@u-bourgogne.fr

Stanislaw JANECZKO,
Polish Academy of Sciences
janeczko@impan.gov.pl

Victor Hugo JORGE PEREZ,
ICMC- USP
vhjperez@icmc.usp.br

Maxim KAZARIAN,
University of Moscow
kazarian@mccme.ru

León KUSHNER,
Facultad de Ciencias
UNAM
kushner@servidor.unam.mx

Isabel LABOURIAU,
Universidade do Porto
islabour@fc.up.pt

Dung Trang LE,
ICTP, Trieste, Italy
ledt@ictp.trieste.it

Daniel LEHMANN,
Université de Montpellier II
lehmann@math.univ-montp2.fr

Daniel LEVCOVITZ,
ICMC - USP
lev@icmc.usp.br

Anatoly LIBGOBER,
University of Illinois at Chicago.
libgober@math.uic.edu

Vanda Maria LUCHESI,
ICMC - USP
vmluc@icmc.usp.br

Ignacio LUENGO,
Universidad de Madrid
iluengo@mat.ucm.es

Matilde MARCOLLI,
Max Planck Institute,
marcolli@mpim-bonn.mpg.de

Bernd MARTIN,
University Cottbus
martin@math.tu-cottbus.de

Rodrigo MARTINS,
ICMC-USP
rmartins@icmc.usp.br

David MASSEY,
Northeastern University
dmassey@neu.edu

Alejandro MELLE HERNANDEZ,
Universidad de Madrid
amelle@mat.ucm.es

David MOND,
University of Warwick
mond@maths.warwick.ac.uk

Piotr MORMUL,
Warsaw University
mormul@mimuw.edu.pl

Hossein MOVASATI,
Universität Göttingen
movasati@uni-math.gwdg.de

Claudio MUROLO,
Université Aix-Marseille I
murolo@cmi.univ-mrs.fr

Ana Claudia NABARRO,
ICMC-USP
anaclana@icmc.usp.br

Andras NEMETHI,
Alfred Renyi Inst. of Math.
nemethi@renyi.hu

Donald O'SHEA,
Mount Holyoke College
doshea@mtholyoke.edu

Regilene D.S. OLIVEIRA
ICMC-USP
regilene@icmc.usp.br

Adriana ORTIZ RODRIGUEZ,
UNAM
aortiz@matem.unam.mx

Anne PICHON,
Université de la Méditerranée
pichon@iml.univ-mrs.fr

Eliana Manuel PINHO,
Universidade do Porto
eopinho@fc.up.pt

Anna PRATOUSSEVITCH,
University of Bonn
anna@math.uni-bonn.de

Oswald RIEMENSCHNEIDER,
University of Hamburg
riemenschneider@math.uni-hamburg.de

Marcelo Jose SAIA,
ICMC - USP
mjsaia@icmc.usp.br

Lizandro SANCHEZ CHALLAPA,
ICMC - USP
challapa@icmc.usp.br

Ioana Cristina SERBAN,
Università di Roma
serban@mat.uniroma1.it

Eugenii SHUSTIN,
Tel Aviv University
shustin@post.tau.ac.il

Dirk SIERSMA,
Universiteit Utrecht
siersma@math.uu.nl

Angela Maria SITTA
Departamento de Matematica
IBILCE-UNESP
angela@ibilce.unesp.br

Jawad SNOUSSI,
IMATE-UNAM
jsnoussi@matcuer.unam.mx

Marcio SOARES,
UFMG
msoares@mat.ufmg.br

Maria Aparecida SOARES RUAS,
ICMC-USP
maasruas@icmc.usp.br

Andras SZUCS,
Univ. of Budapest
szucs@cs.elte.hu

Farid TARI,
University of Durham
farid.tari@durham.ac.uk

Bernard TEISSIER,
Université Paris 7
teissier@math.jussieu.fr

João Nivaldo TOMAZELLA,
UFSCar
tomazella@dm.ufscar.br

Tristan TORRELLI,
Universidad de Valladolid
torrelli@math.unice.fr

David TROTMAN,
Université de Provence
trotman@cmi.univ-mrs.fr

Gioia M. VAGO,
Université de Bourgogne
vago@u-bourgogne.fr

Adrian VARLEY,
Universidad de Valladolid
varley@agt.uva.es

Willem VEYS,
K.U.Leuven
wim.veys@wis.kuleuven.be

Renato VIDAL MARTINS,
UFMG
renato@mat.ufmg.br

Terry WALL,
University of Liverpool
ctcw@liv.ac.uk

Roberta WIK ATIQUE,
ICMC-USPrwik@icmc.usp.br

Minoru YAMAMOTO,
Kurume National College of Technology
minomoto@kurume-nct.ac.jp

Michail ZHITOMIRSKII,
Technion, Department of Mathematics,
mzhi@techunix.technion.ac.il